A HISTORY OF AMERICA IN 100 MAPS

A HISTORY OF AMERICA IN 100 MAPS

Susan Schulten

THE UNIVERSITY OF CHICAGO PRESS

The University of Chicago Press, Chicago 60637

For more information, contact the University of Chicago Press, 1427 E. 60th St., Chicago, IL 60637.
Published 2018
Printed in Italy by Printer Trento

27 26 25 24 23 22 21 20 19   3 4 5 6 7 8 9

ISBN-13: 978-0-226-45861-8 (cloth)
ISBN-13: 978-0-226-45875-5 (e-book)
DOI: https://doi.org/10.7208/chicago/9780226458755.001.0001

Published outside North and South America by the British Library, 2018.

Pages 2–3: *President Franklin Roosevelt consulting the massive globe presented to him in 1942 by the US Army. The Army designed the globe to rotate freely—without an axis—in order to facilitate strategic thinking in an age of aviation. World War II prompted new and radically different understandings of world geography, as shown in chapter 8.*

Library of Congress Cataloging-in-Publication Data

Names: Schulten, Susan, author.
Title: A history of America in 100 maps / Susan Schulten.
Description: Chicago : The University of Chicago Press, 2018. | Includes bibliographical references and index.
Identifiers: LCCN 2018011133 | ISBN 9780226458618 (cloth) | ISBN 9780226458755 (e-book)
Subjects: LCSH: United States—Historical geography—Maps. | America—Historical geography—Maps. |America—Discovery and exploration—Maps.
Classification: LCC E179.5 .S36 2018 | DDC 911/.73—dc23
LC record available at https://lccn.loc.gov/2018011133

CONTENTS

Edward Savage's portrait of President George Washington and the first family gathered around a map of the proposed national capital on the Potomac River. Maps of Washington, D.C. circulated widely in the 1790s, and became instantly recognizable and iconic symbols of national promise in the new United States. *For more on this map see page 106.*

INTRODUCTION: SEEING THE PAST THROUGH MAPS

In November 1864 General William Sherman began the most audacious campaign of the Civil War. After subduing the Army of the Tennessee in Atlanta, his men cut loose from supply lines to inflict as much damage as possible on the Confederacy. Sherman commissioned several maps to prepare for the operation, one of which was particularly innovative. This map of Georgia made at the general's request by the Census Office identified not only rivers and roads but also *resources*: each county was annotated with data regarding the white and slave population, agriculture, and livestock (as shown at right).

Sherman himself later testified to the importance of this map. It showed him where to look for food to feed his men and starve the enemy. It detailed the presence of slaves, the strongest of whom proved an asset to the Union Army as it moved through Georgia. But even more essential was the way the map helped Sherman envision this ambitious operation in the first place, a campaign that struck some of his fellow generals as downright absurd. The census map of Georgia was a groundbreaking attempt to harness data for strategy. Yet it also shaped that strategy by enabling the general to think differently about warfare. Simply put, the map *mattered.*

Sherman's testimony points to the rich yet often overlooked role of maps in history. Whether made for military strategy or urban reform, to encourage settlement or to investigate disease, maps both reflect and mediate change. They record efforts to make sense of the world in physical terms. They capture what people knew, what they thought they knew, what they hoped for, and what they feared. They invest information with meaning by translating it into visual form, and in so doing reveal decisions about how the world ought to be seen. Above all, they demonstrate that the past was not just a chronological story but a spatial one as well.

What follows is a visual tour of American history through maps, one that searches the main roads as well as the back-alleys of the past. It is not a comprehensive survey of American history or mapmaking, nor does it replicate the many excellent histories of exploration. Instead, it is an eclectic and selective discussion of the many ways in which maps have been used in the past: to master and claim territory, defeat an enemy, advance a cause, investigate a problem, learn geography, advertise a destination, entertain an audience, or navigate terrain. It features both official and ephemeral material, including maps of reconnaissance, political conflict, and territorial control as well as of education, science, and tourism.

A word on nomenclature: this "American" history focuses on the region that became the United States, from the voyages of discovery down to our own day. But it also includes maps that highlight the permeability of borders and the place of the nation in the wider world. Many of the maps also restore a degree of contingency that is often obscured by our modern vantage point. In the early 1500s, mapmakers in Europe tried to reconcile new geographical discoveries with Christopher Columbus' claim to have reached the East Indies. Maps of this era illustrate that confusion, showing that the western hemisphere came into view only slowly. More generally, the first three chapters do less to document the inevitable rise of Anglo-America than to remind us of the ongoing contest between Spanish, French, Dutch, and English powers in North America, and the indigenous presence that preceded them all.

Many of the maps reproduced in this volume have been deemed important for their role in statecraft and diplomacy. But readers will also find lesser-known artifacts made by soldiers on the front, Native American tribal leaders, and the first generation of girls to be publicly educated. For instance, John Mitchell's 1755 map of the colonies has long been regarded as a crucial document, and the copy featured on page 94 was used to negotiate the boundaries of the new United States at the Treaty of Paris in 1783. As such, Mitchell's map changed the course of history by establishing the nation's borders. Equally captivating is a map drawn by a Cherokee leader in the 1720s to negotiate the increasingly competitive deerskin trade in the Carolinas (page 72). The map is initially disorienting, for it represents space in terms of relationships rather than physical distance. But, once deciphered, it reveals as much about colonial America as the celebrated Mitchell map.

By exploring iconic as well as unfamiliar treasures we can also gain fresh perspective on the past. For instance, July 1776 is primarily remembered by Americans for the signing of the Declaration of Independence in Philadelphia. Yet, at that very moment, a group of Spanish missionaries set off from Santa Fe to assert control over the region that would become Arizona, Utah, and Colorado. The Dominguez and Escalante expedition sought an overland route to California, but with limited geographical knowledge it soon became lost. Were it not for Ute guides, the expedition would not have survived the trek, much less produced one of the most influential maps of the Southern Rockies (page 88). Bernardo de Miera y Pacheco drew the map to claim the Southwest for the Spanish Crown, just one of the many instances where territory was taken

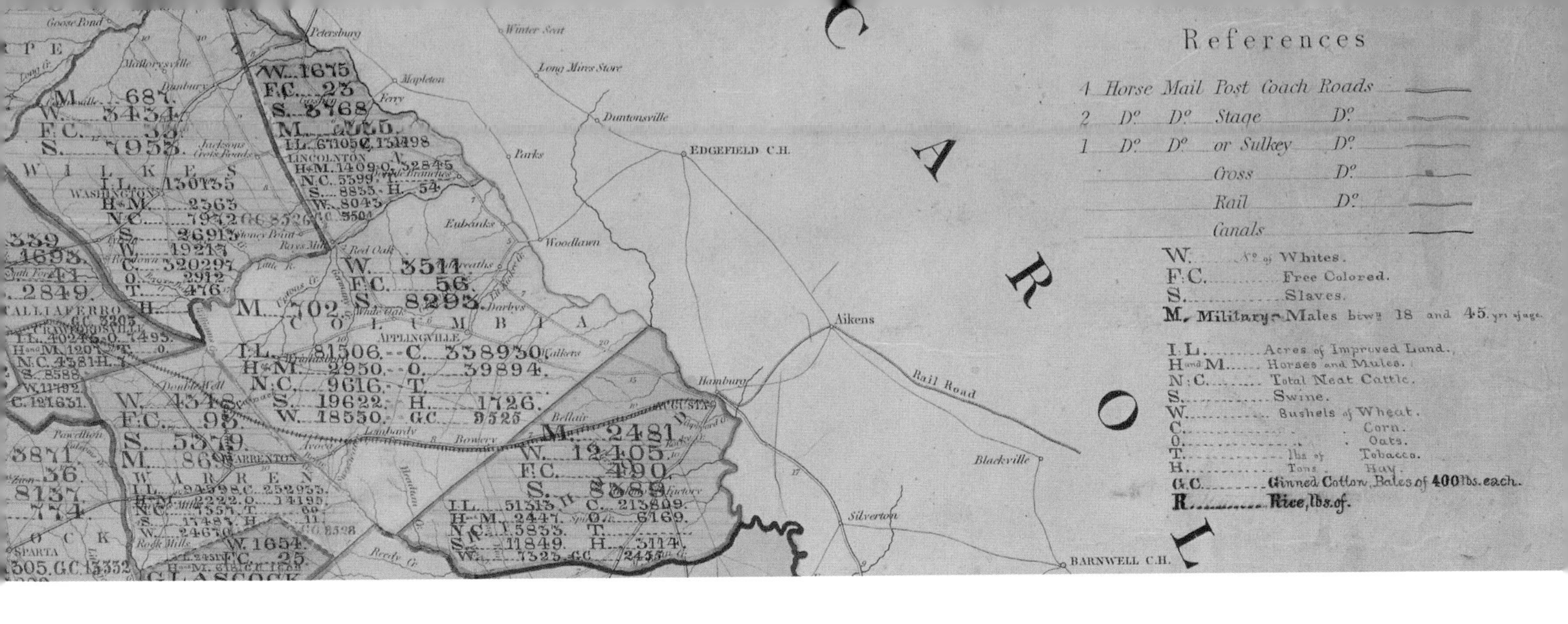

During the Civil War, the Census Office adapted several existing maps to aid Union military strategy. This detail is taken from a large map of Georgia, which was annotated to include census data on population and resources for each county. For the full map and its relevance during the war, see page 147.

on paper long before any real control was exercised on the ground. The map is also evidence that while the Revolutionary War consumed the east, the Southwest was engulfed in a set of very different imperial and geopolitical struggles.

The historical role of maps extends far beyond exploration and diplomacy. The explosion of cities in the late nineteenth century, for example, sparked the widespread use of maps as a tool of reform. W. E. B. DuBois was among the first sociologists to examine the dynamics of segregation in his landmark study, *The Philadelphia Negro* (1899) (see page 168). DuBois sought to explain why even elite and established African Americans in Philadelphia exercised so little social mobility relative to immigrants. By attempting to map the problem, DuBois joined a much larger conversation about urban space: Charles Booth mapped London's poor, Florence Kelley mapped Chicago's immigrants, New York reformers mapped the density of Manhattan's tenement districts, and—most notoriously—San Francisco's Board of Supervisors mapped Chinatown in a vicious effort to control the Chinese population. This frenzy of urban mapping reveals not just the complex hierarchy of race and ethnicity in the Gilded Age, but also the more general emergence of social science.

The maps in this volume were made in vastly different contexts. Yet when considered together, they underscore the persuasive power of cartography. Some do this explicitly, such as Malachy Postlethwayt's attempt to advance British control over the slave trade in the mid-eighteenth century (page 74). Similarly, the Federalists pointedly satirized Republican redistricting efforts in the Gerry-mander map of 1813, just as the Wilson administration mocked German peace efforts in 1917 (pages 116 and 180). But these are just the obvious cases of an observation that applies to every map in this book: each was made in a particular moment and for a specific end. However authoritative their claims or scientific their appearance, maps are the agents of their authors.

The significance of these maps is often embedded within their design. Catharine Cook's charming map of 1818 (page 118) speaks volumes about female education in the early republic. Likewise, John Wiltberger's imaginary map of sin and temptation on page 126 captures not just contemporary arguments for temperance but an antebellum understanding of Christianity and moral reform. Many of these images are compelling, even seductive, while others go about their business quietly. Who can resist the elaborate graphic style of Richard Edes Harrison, who masterfully conveyed the new realities of geography in the age of aviation? His maps on pages 206 and 208 integrated geography and design in service to a new posture of American internationalism during World War II. Similarly, Heinrich Berann's breathtaking painting of the ocean floor on page 242 helped an entire generation make sense of the emerging theory of plate tectonics. In both cases, the artistry and visual design were integral to the power of the map.

Each of the chapters that follow opens with a brief overview of the period before focusing on the maps themselves. Some stories are better served by maps than others: exploration and settlement, imperial rivalry, military conflict, infrastructure, and territorial expansion all figure prominently. But just as salient are themes of migration, slavery, politics, education, reform, and even leisure. Some of these maps have never been published, while others have long drawn attention. Each grew out of contemporary circumstances and concerns, and as such has the potential to both illuminate and complicate our understanding of history. Examined in context and with care, these artifacts offer unrivaled windows into the past.

1. 1490-1600

Contact and Discovery

Among the most enduring myths of American history is that in 1492 Christopher Columbus reached a land that was sparsely populated. Decades of research by scientists, anthropologists, and historians, however, have established that the Americas were home to between 50 and 70 million people organized into multiple and diverse societies by the time Columbus arrived. By comparison, Europe had a population of between 70 and 90 million. The most sophisticated of these societies were the Aztecs of Mexico, whose civilization included well-developed cities, complex social and political structures, and above all a dense population. Further north, the continental expanse that would become the United States was home to several native tribes, from the Iroquois Confederation in the Northeast to the tribes of the Mississippi Valley and the Pueblo and Hopi villages of the Southwest. The population of these tribes together numbered perhaps 7 million. The Pueblo villages had developed sophisticated agricultural systems that involved canals, dams, and terracing.

If we acknowledge these established societies throughout the Americas, then the traditional account of "discovery" becomes far more complicated. This chapter examines the first century of European exploration in the Americas by tracing the struggle to map this unknown territory. We open with the world as it was understood by Europeans in 1490, using a map that influenced Columbus as he planned his voyage to the Far East. Just a few years earlier Bartolomeu Dias had sailed down the western coast of Africa in an effort to reach the Indian Ocean. Dias demonstrated the possibility of reaching the East by sailing south, though the sheer length of the journey reinforced Columbus' belief that it would be quicker to reach the Indies by sailing west from Spain. In his mind, Japan was 2,400 miles from the Canary Islands, though his advisers believed it was at least four times further.

Our first map shows the logic behind Columbus' decision to sail west, for contemporary knowledge framed the world as consisting of Africa, Asia, and Europe. He was wrong, of course, for it was not Asia but the New World that lay west of the Atlantic Ocean. This error led to the European discovery of America. Columbus sought Asia, and only inadvertently discovered the Caribbean. Yet, because of his worldview, Columbus went to his death believing that he had in fact reached the East Indies rather than an entirely separate hemisphere. In his mind, the islands of the Caribbean were near the land of Japan that Marco Polo had described centuries earlier. This conviction that Asia lay west of the Atlantic Ocean shaped the maps drawn in the early 1500s, as geographers and mapmakers tried to assimilate the information brought back by the Atlantic voyages. For some, it was clear that this was an entirely separate land, but others tried to reconcile this new geography with the existing knowledge of Asia. This assumption produced a number of fascinating and profoundly confusing maps drawn in the early sixteenth century.

One of the most notable of these was issued in 1506 by Giovanni Contarini and Francesco Rosselli (page 14). Their map attempted to integrate knowledge brought back by the voyages of discovery within existing geographical frameworks. Europe and Africa appear in their familiar form, yet to the west we see an open sea that bears little resemblance to the western hemisphere. If we remember, however, that Contarini and Rosselli followed Columbus' belief that Asia lay west of Europe, then the picture begins to make sense. A few years later, the world map drawn by Vesconte Maggiolo on page 20 shows continued uncertainty. One the one hand, the coastlines of North and South America have begun to take shape. Yet at the same time Maggiolo seems unclear as to whether these landmasses are separate from—or connected to—Asia.

The naming of North and South America also came somewhat inadvertently, when Martin Waldseemüller sought to honor Amerigo Vespucci by attaching the name America to the southern continent in his ambitious map of 1507 (page 16). A comparison of these three maps made after 1500 highlights just how fluid geographical knowledge of the western hemisphere remained until the 1520s. While Waldseemüller pictured a narrow western hemisphere, Maggiolo and Contarini suggested a wide landmass that was perhaps connected to Asia. Throughout this period, we see the struggle to integrate new information with inherited worldviews. That conflict is symbolized at the top of Waldseemüller's map on page 18, with a portrait of the classical geographer Claudius Ptolemy alongside the living explorer Vespucci.

Within a few decades it became clear that what lay west of the Atlantic Ocean was a land heretofore unknown to Europeans, and subsequent maps show the effort to assess the western hemisphere. Sebastian Münster's map on page 24 was among the first to definitively separate North America from Asia, using knowledge gleaned from Ferdinand Magellan's circumnavigation of the globe. Once the Americas had been established, however, European explorers and their patrons continued to seek a portage or waterway that would enable them to pass *through* these continents to the Far East. For the next three centuries, exploration of North America was informed by a drive to find the fabled Northwest Passage.

Maps in the early chapters of this book reveal the simultaneous quest to exploit the riches of North America and also to reach *beyond* it. While the Spanish invested heavily in the extraction of resources from South and Central America in the sixteenth century, colonization efforts in North America by the French, Dutch, and English would not begin until the seventeenth century. As such, the maps in this chapter focus primarily on exploration rather than on settlement. They are wildly erroneous, but in those errors we find the motives behind the early voyages of discovery.

Historians use the phrase "Columbian exchange" to describe the complex interplay between the Americas and the Old World across the 1500s. That exchange brought new crops to the New World such as wheat, barley, and sugar, as well as horses, cattle, swine, and sheep. But the arrival of Europeans also devastated the Americas. Through forced labor and violent subjugation, but also through the transmission of smallpox, typhus, cholera, and measles, indigenous life was fundamentally altered. Not all Europeans accepted this arrangement. Bartolomé de las Casas devoted his life to exposing the iniquities of the Spanish colonial system. He rejected the belief that natives were savages, and instead portrayed them as victims of Spanish cruelty and theft.

When Columbus arrived on San Salvador in 1492, he regarded the Taíno people as little more than subjects, and failed to appreciate their religion, social structure, farming system, or navigation skills. Within a half century, the Taíno population had plummeted from thousands to fewer than 500. In 1519 Hernán Cortés brought 600 men from the Gulf Coast inland to the Aztec city of Tenochtitlan (page 22). The Spanish were greeted by the Aztec king Montezuma, but relations grew hostile once Cortés demanded gold. The Aztecs resisted Spanish control for nearly two years, but they were severely weakened by European diseases and their weapons were no match against European technology. In 1521 the Spanish conquered and destroyed Tenochtitlan, which became the site of Mexico City and the base of New Spain. The maps on pages 26 and 30 illustrate this long period of Spanish dominance in the sixteenth century. Historians and anthropologists have estimated that by 1600 the native population of the Americas had been reduced by as much as 80 percent.

We close this chapter with a map by John Dee, an advisor to Queen Elizabeth who strongly advocated an expanded role for the English in North America. The ink on Dee's chart of the northern hemisphere on page 32 has faded, yet his dogged and forceful endorsement of English power stimulated a wave of voyages at the end of the sixteenth century. Though Dee would not live to see the realization of English control in the New World, his map reminds us that throughout the sixteenth century, geographical knowledge was gained first and foremost through imperial rivalry.

THE WORLD THAT COLUMBUS KNEW

Henricus Martellus Germanus, Ptolemaic world map, 1489 or 1490

Generations of school children learned that "In fourteen hundred ninety-two Columbus sailed the ocean blue." But what compelled Christopher Columbus to set sail west from Spain across the Atlantic Ocean? Where did he think he was going? Columbus sought an ocean route to Cathay—or China—pursuing riches on behalf of the Spanish Crown. In 1492 he "discovered" the Caribbean island of Hispaniola, but lived the rest of his life believing that he had reached the East Indies. This conviction stemmed from the maps and globes of the era, all of which suggested that Asia lay west of Europe.

Among the most influential of these contemporary mapmakers was Henricus Martellus Germanus, whose long career in Florence lasted from 1459 to 1496. Martellus was especially prolific in those later years, and produced this world map as part of his manuscript atlas of the Mediterranean islands. Drawn on vellum in 1489 or 1490, the map shows the world as it had been depicted centuries earlier by Claudius Ptolemy, the classical geographer whose maps were rediscovered in the fifteenth century. Ptolemy's knowledge of the world spread throughout Europe before the discoveries of the western hemisphere rendered them irrelevant.

Though his world map was based on a classical model, Martellus took care to include the discoveries of his own time. His representation of Asia reflects knowledge brought back by Marco Polo. The extensive information along the coast of Africa was gained through the 1488 voyage of Bartolomeu Dias. In fact this was the first map to show the African continent as described by Dias after rounding the Cape of Good Hope. By including the entire African continent, Martellus implicitly suggested the possibility of an eastward route from Europe to Asia. The fabled source of the Nile is depicted as the "Montes Lune," or Mountains of the Moon. The dark ink used to mark the western coastline of Africa endures, though the extensive place names across the rest of the world have faded from the map.

This manuscript world map was printed and distributed by Francisco Rosselli, and became a standard view of the world in the 1490s. It incorporated the best and most recent knowledge of the day, which also explains its influence and circulation. A manuscript map circulating in the English court in 1502 directly reflects Martellus' view of the world, suggesting that it reached far beyond Florence. More importantly, Martin Behaim used a Martellus map to create his terrestrial globe of the world in 1492, which in turn guided Columbus as he sought a route to Asia.

Martellus positioned Japan just 3,500 miles west of Europe, and China 1,500 miles further. Behaim's globe replicated the geography of Martellus, and further stimulated interest in Asia by describing Japan as abundant in gold and spices. Behaim's globe led Columbus to believe that it would be easier to reach Asia by sailing west than by navigating around Africa. Moreover, Martellus' map reinforced what Columbus had learned from Paolo Toscanelli's world map of 1474, which also placed Asia directly west of Europe. All of this was incorrect, for none of these men were aware of the western hemisphere. But, ironically, the flaws in this geographical worldview led Columbus to believe he might reach Asia by sailing west.

Martellus' influence over geographical knowledge—and Columbus' worldview in particular—makes him crucial to American history. But recent research has amplified his importance even further. The map scholar Chet Van Duzer used new imaging technology to recover faded details from a large and more comprehensive Martellus world map housed at Yale University. These details demonstrate that Martellus' view of the world was the model for large parts of Martin Waldseemüller's 1507 map of the world (page 16). Van Duzer's research indicates that Waldseemüller drew extensively from Martellus' depiction of eastern Asia, southern Africa, and Japan, even though, by 1507, he recognized the limits of Martellus' map of the New World. Martellus made maps that mattered, and that encouraged navigators to explore the Atlantic Ocean. Those voyages would in turn render his worldview completely outdated.

A GENERATION OF CONFUSION

Giovanni Matteo Contarini and Francesco Rosselli, "Mundu Spericum," 1506

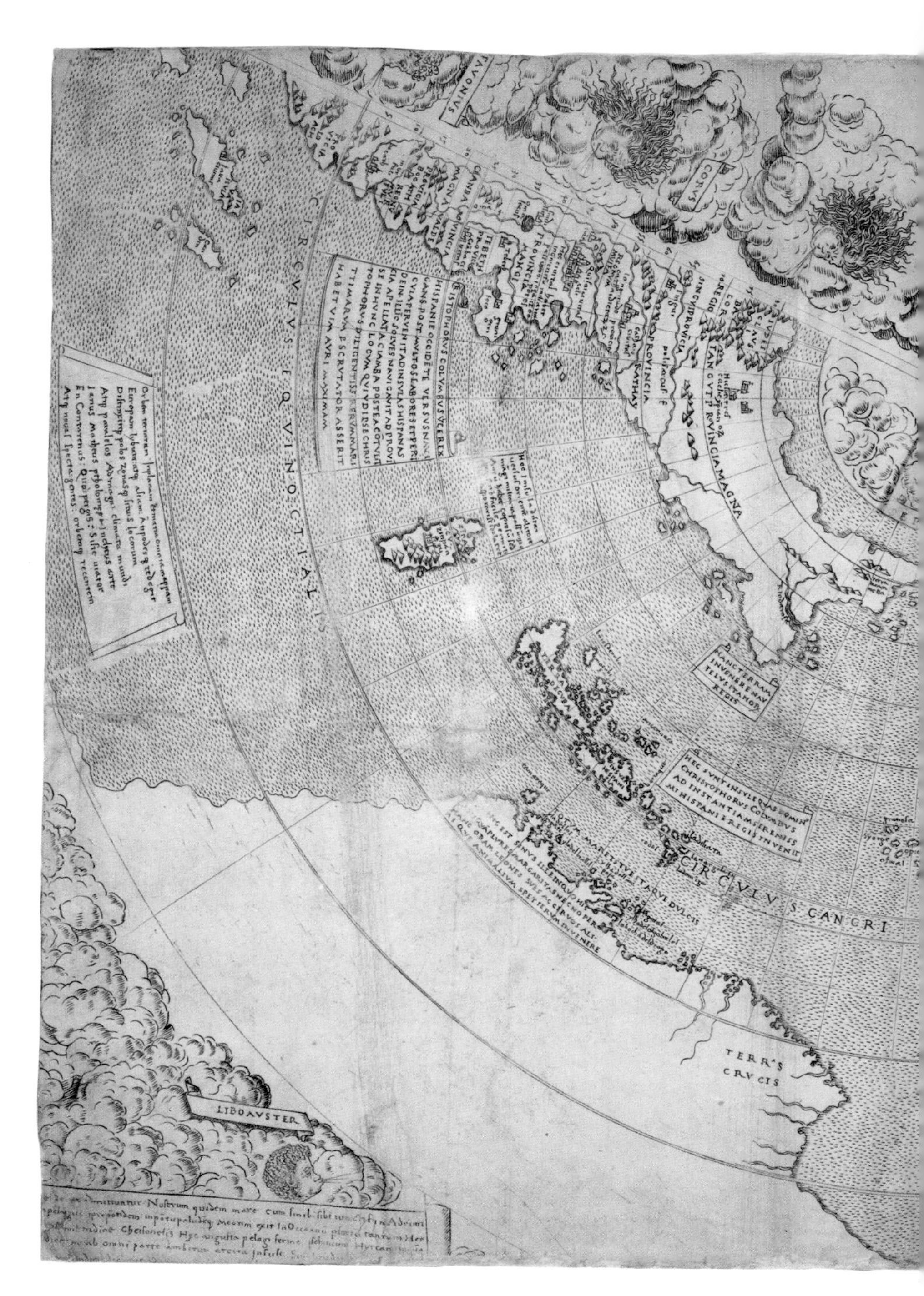

Imagine you were living in Europe around 1500. The voyages of discovery that began with Columbus brought back an avalanche of information, and eventually revealed an unknown world. But the new hemisphere came into view very slowly. Columbus went to his death insisting that he had reached the East Indies. Thereafter, John Cabot, Amerigo Vespucci, Pedro Álvares Cabral, and Vasco da Gama each brought back pieces of a geographical puzzle that still did not quite fit together. The problem was that new information was difficult to reconcile with existing geographical frameworks. Many Europeans continued to believe that Asia lay west of Europe, so when mapmakers initially tried to integrate new discoveries with older assumptions they generated more than a little confusion.

That confusion is on display in the first printed map to show any part of America, designed by Giovanni Matteo Contarini and engraved by Francesco Rosselli in 1506. Contarini designed the map in either Florence or Venice, most likely as the opening image of a new atlas that was never realized. He adopted a conic projection oriented around the North Pole, one that is simulated by placing a cone over the earth and then unwrapping it. Contarini may have used this as a way of working around the uncertainty that came with integrating the new discoveries and assertions of Columbus with the Far Eastern voyages of Vasco da Gama.

Encircling the North Pole at left is a large peninsula. Most likely this is Contarini's attempt to square Cabot's voyages to Greenland and Newfoundland in 1497 with existing knowledge of the Far East. In other words, the new discoveries were understood to be explorations of Asia rather than an entirely separate continent. These attempts to reconcile Asia and America demonstrate just how fluid geographical knowledge was during these voyages of discovery.

Toward the south at the bottom of the map we see an enormous landmass that reflects the voyage of Pedro Álvares Cabral to the eastern coast of Brazil in 1500. The most revealing aspect of the map is the depiction of islands just north of this landmass. Contarini drew the West Indies as a chain extending

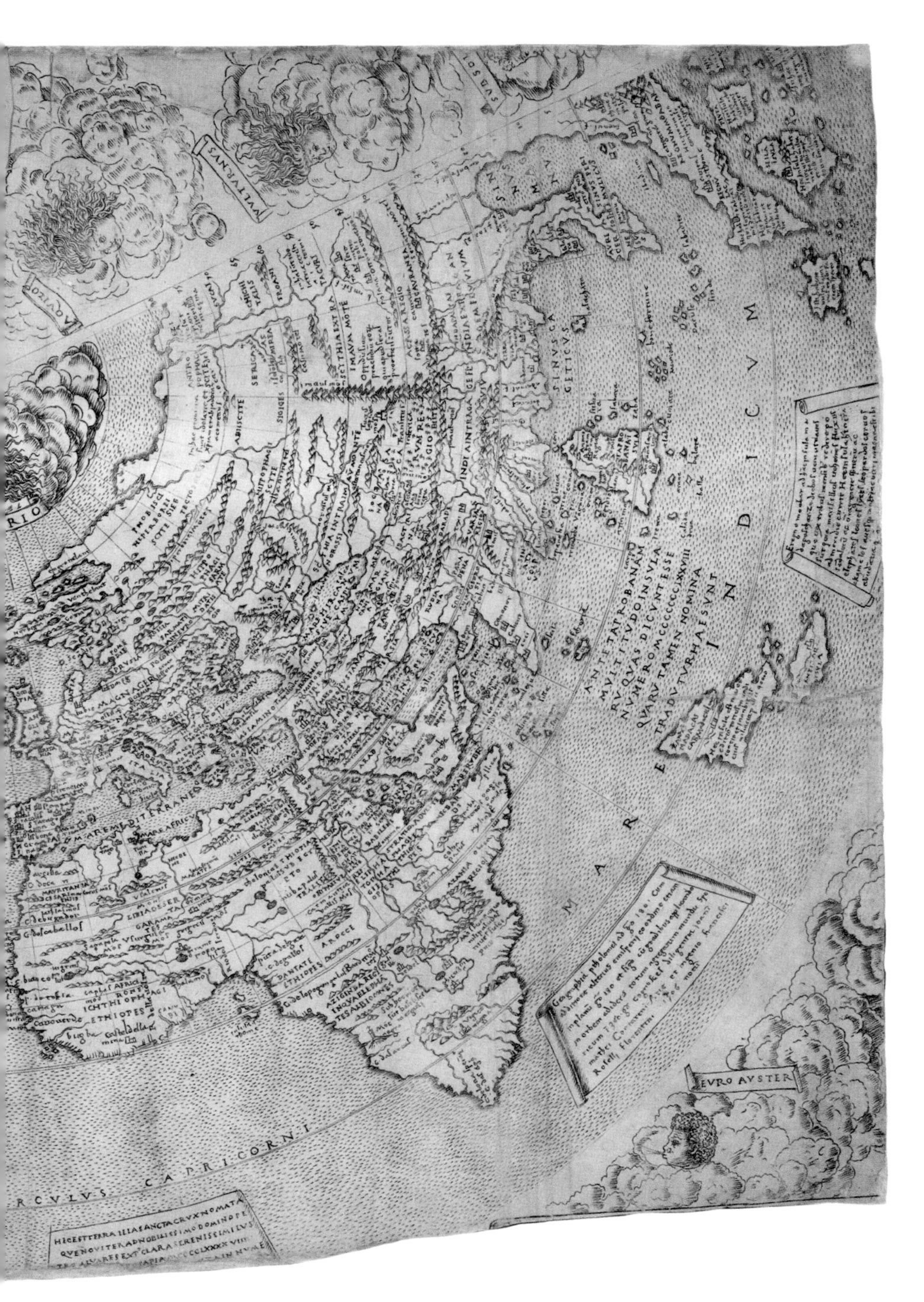

from east to west, and identified these as the "islands that Master Christopher Columbus discovered." And because he also believed that Columbus had reached the East Indies, Contarini placed the elongated island of "Zipangu" (Japan) just west of the West Indies. No doubt he struggled to resolve the existing geography of Cathay (China) with new reports from recent voyages.

The wide and open sea that surrounds these islands reveals the belief either that Columbus had reached Asia, or that a passage to Cathay existed to the west of these islands. The map mysteriously omits geographical knowledge of the North American coast that was attributed to Vespucci's second voyage. Perhaps this was left out because Contarini and Rosselli—like many—doubted that voyage had taken place. Or they may have rejected Vespucci's new information when it did not fit their existing picture of the world. Considered together, their dogged attempt to integrate disparate and partial information captures a moment before the eastern outline of North and South America was fully understood.

The Contarini–Rosselli map, which now exists only in the British Library, documents a fundamental paradox. The arrival of new knowledge forced a reconsideration of world geography, but that knowledge was understood within existing frameworks. These circumstances directly shaped the earliest attempts to map the voyages of discovery.

This presents a thorny but fascinating question: Is the Contarini–Rosselli map in fact the first map of America, when the mapmakers themselves believed they were representing Asia rather than a new and unknown land? Within just a few months, as we shall see in the next map, Martin Waldseemüller would delineate a more recognizable picture of the western hemisphere. But the Contarini-Rosselli map captures the intellectual challenge posed by the voyages of discovery. The map itself states this in a way that was truer than its makers knew: "behold new nations and a new-found world."

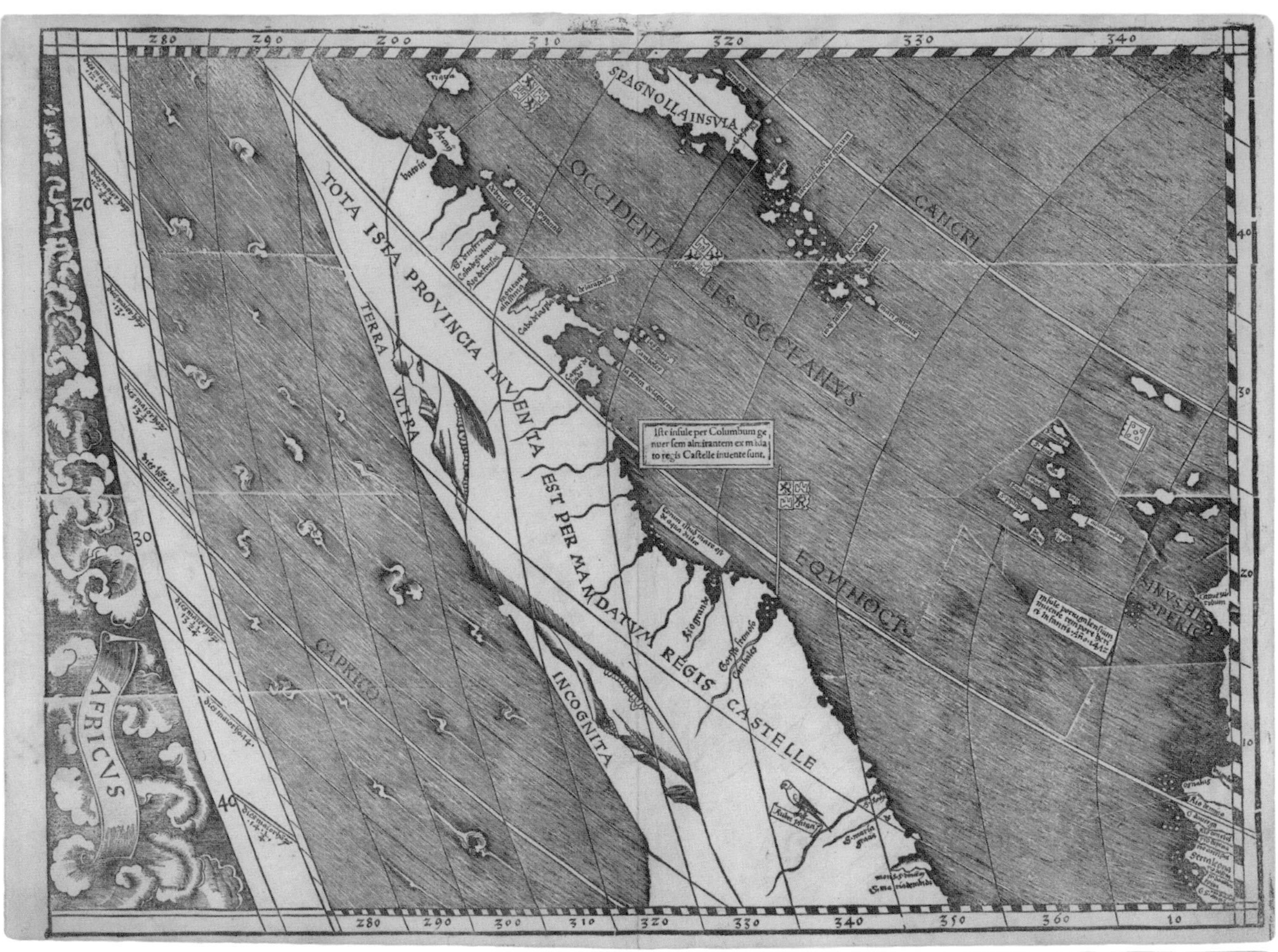
SPAGNOLLA INSVLA
OCCIDENTALES OCCEANVS
CANCRI
TOTA ISTA PROVINCIA INVENTA EST PER MANDATVM REGIS CASTELLE
TERRA VLTRA INCOGNITA
Iste insule per Columbum genuensem almirantem ex mandato regis Castelle inuente sunt.
EQVINOCT
CAPRICO
AFRICVS

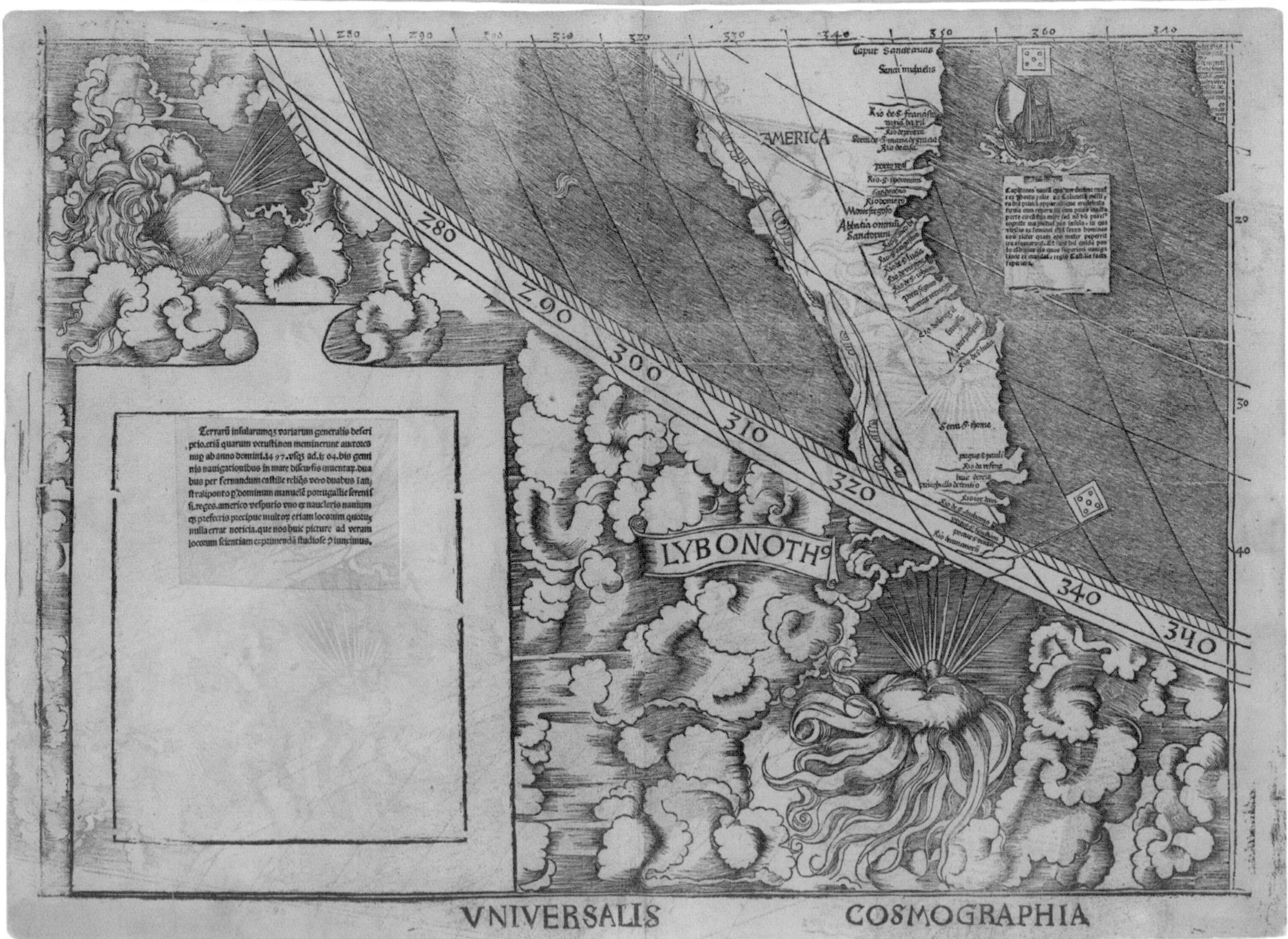
AMERICA
Terrarū insularumq3 variarum generalis descriptio.etiā quarum vetusti non meminerunt auctores nup ab anno domini.14 97.vsq3 ad.15 04.bis genis nauigationibus in mare discursis inuentar.duabus per fernandum castille reliq̄s vero duabus iam strali ponto p dominum manuelē portugallie serenissi.reges.americo vespurio vno ex naucleris nauium q3 prefectis precipue multor etiam locorum quorum nulla erat noticia.que nos huic picture ad veram locorum scientiam exprimendā studiose coniūcimus.
LYBONOTH9
VNIVERSALIS COSMOGRAPHIA

HOW AMERICA (INADVERTENTLY) GOT ITS NAME

Martin Waldseemüller, "Universalis Cosmographia," 1507

If the previous map leaves us searching for any sign of the western hemisphere, here and on the following pages a more familiar picture begins to emerge. We recognize a vaguely identifiable southern continent, and an outsized Caribbean off the coast of a second, northern landmass. Our own struggle to identify this geography mirrors that of Martin Waldseemüller, who published this massive New World map in 1507. Nearly five feet high and over seven feet wide, the map reflects Waldseemüller's effort to present the entire world based on inherited information as well as the latest news from Spain and Portugal.

Waldseemüller was part of a new school of cartography and cosmography formed in the Rhineland, where he and his colleagues were riveted by the Spanish and Portuguese voyages of discovery. They devoted themselves to integrating this new geographical information with the classical worldviews of Claudius Ptolemy, which framed the world as made up of Europe, Africa, and Asia. The overseas explorations of the 1490s and 1500s gradually disrupted this view, forcing Waldseemüller and others to reconcile what they thought they knew with the new information that was circulating through Europe.

The map is an ambitious attempt to synthesize this information on a single page. The use of latitude and longitude allowed Waldseemüller to depict the continents more precisely. In depicting India and southern Africa, he drew on the maps of Henricus Martellus, but it is his rendering of the New World that sets his map apart from its contemporaries.

Waldseemüller adopted the information brought back by Amerigo Vespucci, whose voyages convinced him that the continents of the western hemisphere were separate from Asia. On Giovanni Matteo Contarini's map issued the year before (page 14), for instance, the landmasses and islands to the west of Europe are understood as Asia and not as a "new world" at all. By contrast, at the left edge of Waldseemüller's map both the northern and southern continents have distinct western coasts, though confusion persisted regarding their relationship to Asia. And Waldseemüller revealed his own uncertainty regarding the relationship between North and South America: on the large map, he depicted a break between the two, stoking the hope of a navigable passage to the Far East. On the smaller inset map at the top, however, he joined the two continents firmly together.

Waldseemüller so strongly admired Vespucci that he attached the name "America" to the southern continent (see detail at lower left). To drive the point home, he depicted Vespucci alongside Ptolemy at the top of the map, the two figures presiding over the world. In that pairing, Waldseemüller symbolically connected the world of classical geography with the discoveries of his own day. In the narrative that accompanied the map, Waldseemüller further honored Vespucci's contribution to geographical knowledge.

Waldseemüller sought to disseminate his new picture of the world through central Europe. Within a few years, however, he experienced a change of heart, and in his 1516 world map he rejoined America to Asia. Moreover, his admiration for Vespucci had cooled, and he removed the name "America" from the map altogether. But by that time the name had not only caught on, but had spread to both continents of the western hemisphere. Gerard Mercator, for example, attached the names "N. America" and "S. America" to these respective continents in 1538, and Abraham Ortelius used the same in the many editions of his popular atlas *Theatrum Orbis Terrarum*.

Though scholars had long been aware that Waldseemüller made a large world map in 1507, the sole known surviving copy was not discovered until 1901. It had originally been acquired by a German globe maker, and later passed to the family of Prince Waldburg-Wolfegg. Inadvertently responsible for "naming" America, Waldseemüller's 1507 map commanded tremendous attention and value from the time it was rediscovered down to our own day. In 2003 the Library of Congress paid $10 million for this copy of the map, which John Hébert termed the nation's "birth certificate."

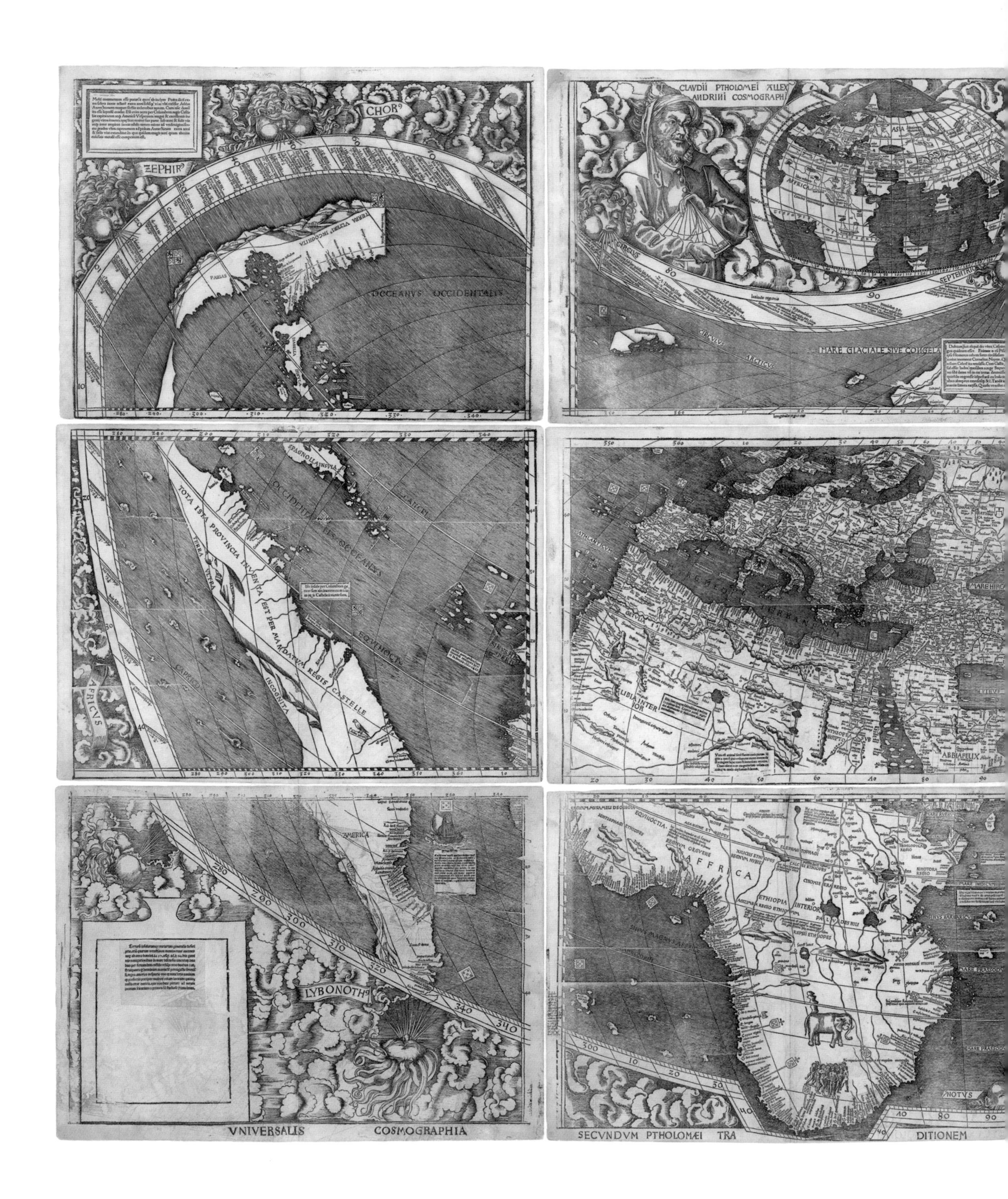
CHOR
ZEPHIR
OCCEANVS OCCIDENTALIS
PARIAS
CLAVDII PTHOLOMEI ALLEX
ANDRINI COSMOGRAPHI
ASIA
AFFRICA
CIRCVLVS ARTICVS
MARE GLACIALE SIVE CONGELA
TOTA ISTA PROVINCIA INVENTA EST PER MANDATVM REGIS CASTELLE
TERRA VLTRA INCOGNITA
SPAGNOLLA INSVLA
OCCIDENTALES OCCEANES
EQVINOCTIALIS
AFRICVS
LIBIA INTERIOR
MARE MEDITERRANEVM
ABBIAFELIX
AMERICA
LYBONOTH
AFFRICA
ETHIOPIA INTERIOR
NOTVS
VNIVERSALIS COSMOGRAPHIA
SECVNDVM PTHOLOMAEI TRA DITIONEM

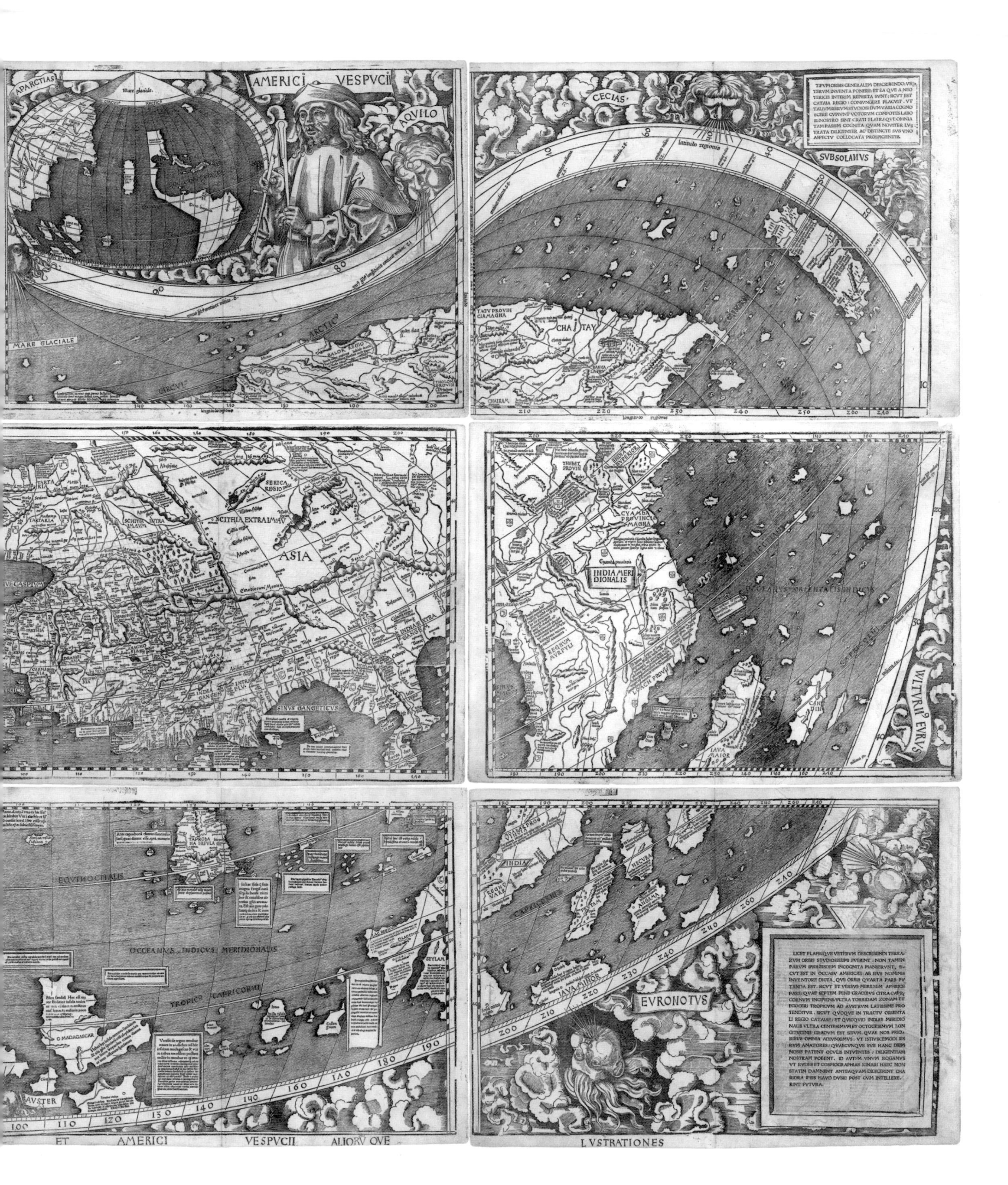
APARCTIAS
AMERICI VESPUCII
AQUILO
CECIAS
SUBSOLANUS
MARE GLACIALE
ASIA
INDIA MERIDIONALIS
SINUS GANGETICUS
EQUINOCTIALIS
OCCEANUS INDICUS MERIDIONALIS
TROPICUS CAPRICORNI
MADAGASCAR
AUSTER
EURONOTUS
VULTURUS EURUS
ET AMERICI VESPUCII ALIORUQUE LUSTRATIONES

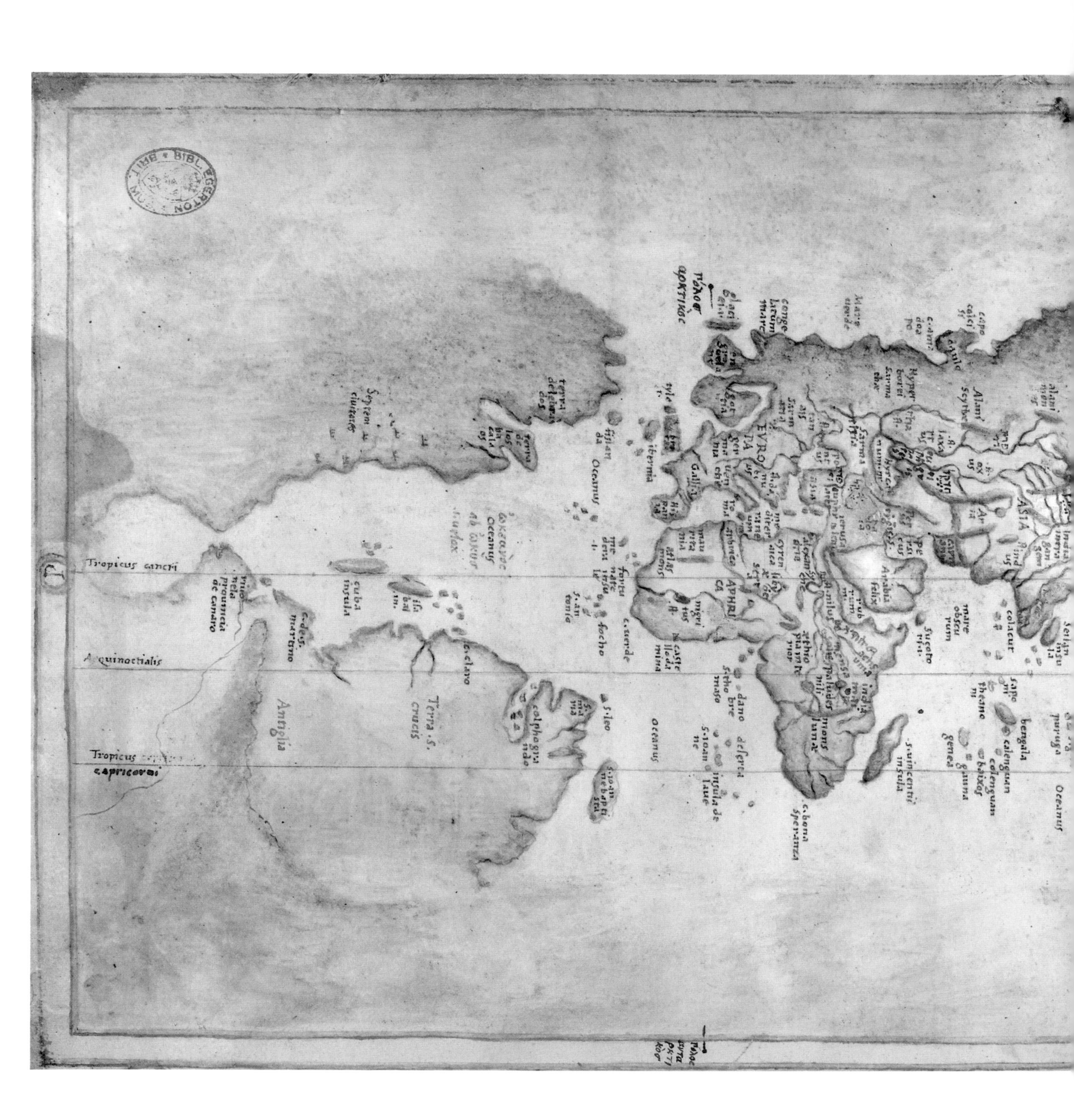
Tropicus cancri
Aequinoctialis
Tropicus
capricorni
Antiglia
cuba insula
Terra S. crucis
Oceanus
ASIA
Arabia felix
bengala
c. bona speranza

AMERICA LOOMS INTO VIEW

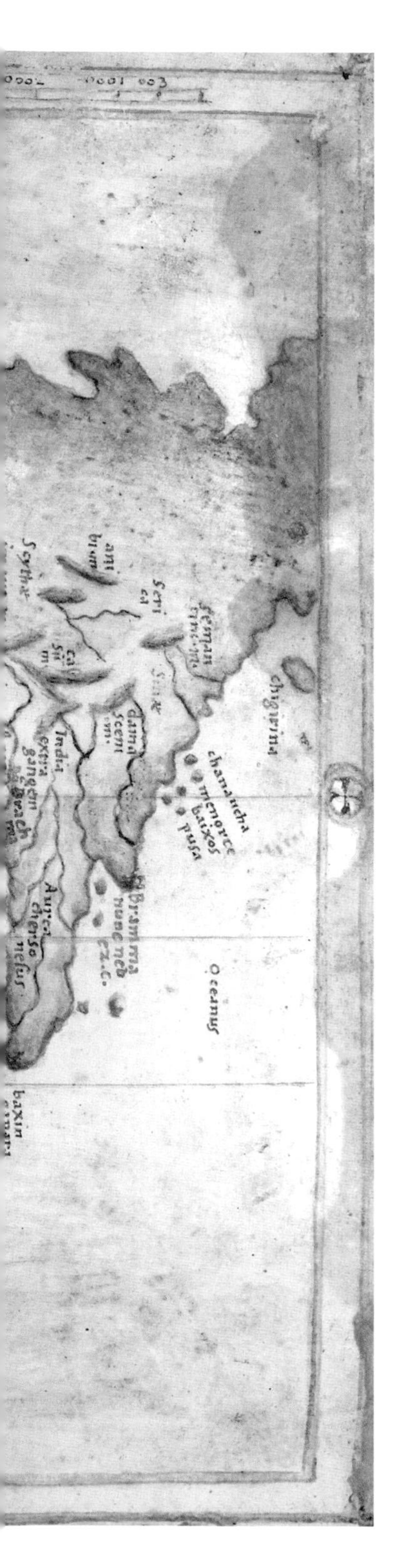

Vesconte Maggiolo, map of the world, from his atlas of portolan charts, circa 1508

At the end of the nineteenth century, the British Museum (whose library later became the nucleus of the British Library) acquired a mysterious portolan atlas measuring about eight inches by eleven inches. Portolan atlases were designed to aid navigators by charting coastlines through compass directions and distances. In the fifteenth and early sixteenth centuries, such charts were among the most jealously guarded of state secrets, for they recorded geographical intelligence at a time of constantly shifting—and potentially erroneous—knowledge about the New World. The quest to control this information was so intense that in 1504 the king of Portugal issued a death sentence for anyone who removed Portuguese maps of areas south of the equator from his kingdom.

The competition was fiercest, of course, between Spain and Portugal. That is precisely what makes this particular *portolano* so unique, for it is one of the few to integrate knowledge of the New World from both sources. The atlas is believed to be the work of the Genoese sailor and mapmaker Vesconte Maggiolo (1478–1530), or perhaps a copy of the same. Drawn between 1508 and 1510, the atlas opens with this captivating map of the world followed by thirteen portolan charts, including the first Italian one to identify any part of the North American coast.

At the far left Maggiolo used the letter P (Ponente) to indicate "west," while a Greek cross at the far right denotes the east. The right half of the map is an easily recognizable and delicately lettered eastern hemisphere. By contrast, west of the Atlantic Ocean we find a geography that is simultaneously strange and familiar. At upper left is a ghostly landmass that seems to emerge out of a fog. Along its eastern edge Maggiolo identified Labrador and Greenland—which were well known by then. Yet he and his contemporaries still struggled to understand the nature of this landmass. Was this newly discovered land the eastern part of Asia, as Columbus had claimed, or its own continent, as Amerigo Vespucci was to insist a few years later? Notice that at the opposite end of the page, at the Far East, Maggiolo depicted the Asian landmass as extending past the end of the map. This no doubt reveals Maggiolo's own confusion as to whether Asia was a separate continent or one connected to the regions described by Vespucci. Like many mapmakers encountering unknown terrain, he fudged a bit to cover his uncertainty.

Maggiolo named the northern land in the western hemisphere "Septem Civitates" (seven cities). This term was likely taken from Juan de la Cosa's portolan chart of 1500, the first to identify the New World. But, beyond that, Maggiolo's depiction of this geography remained vague. The sheer size of the landmass nicely—if unwittingly—captures the width of North America, whereas Martin Waldseemüller had depicted it as a small peninsula (page 18). Note that Maggiolo did not identify this land as "America," for Waldseemüller had attached that name only a year earlier and the practice had yet to spread.

Further south is an enclosed Gulf of Mexico, which was speculation on Maggiolo's part since this was not yet definitively established. Especially revealing is his depiction of a large bay in what is now northern Brazil, most likely the mouth of the Amazon River. Given that the Spanish had yet to explore the region, Maggiolo must have been relying upon Portuguese sources to guide his depiction of the area. This is important, because his general outline of the Americas was taken from Spanish sources such as Juan de la Cosa's portolan chart.

This integration of disparate and confidential sources is a kind of smoking gun, indicating that Maggiolo was working with maps influenced by Amerigo Vespucci. In 1508 Vespucci was hired as the Spanish pilot major, bringing his knowledge of earlier voyages undertaken on behalf of Portugal to his new role. One aspect of that role was his contribution to Spain's first "Padrón Real," a highly secret master map of the world created in 1507 or 1508. Vespucci was the only man in the world who had access to both Spanish and Portuguese intelligence, a "human bridge between Portuguese and Spanish" geographical knowledge. The Maggiolo manuscript map and atlas—held only at the British Library—captures that unique hybrid intelligence better than any world map of its time.

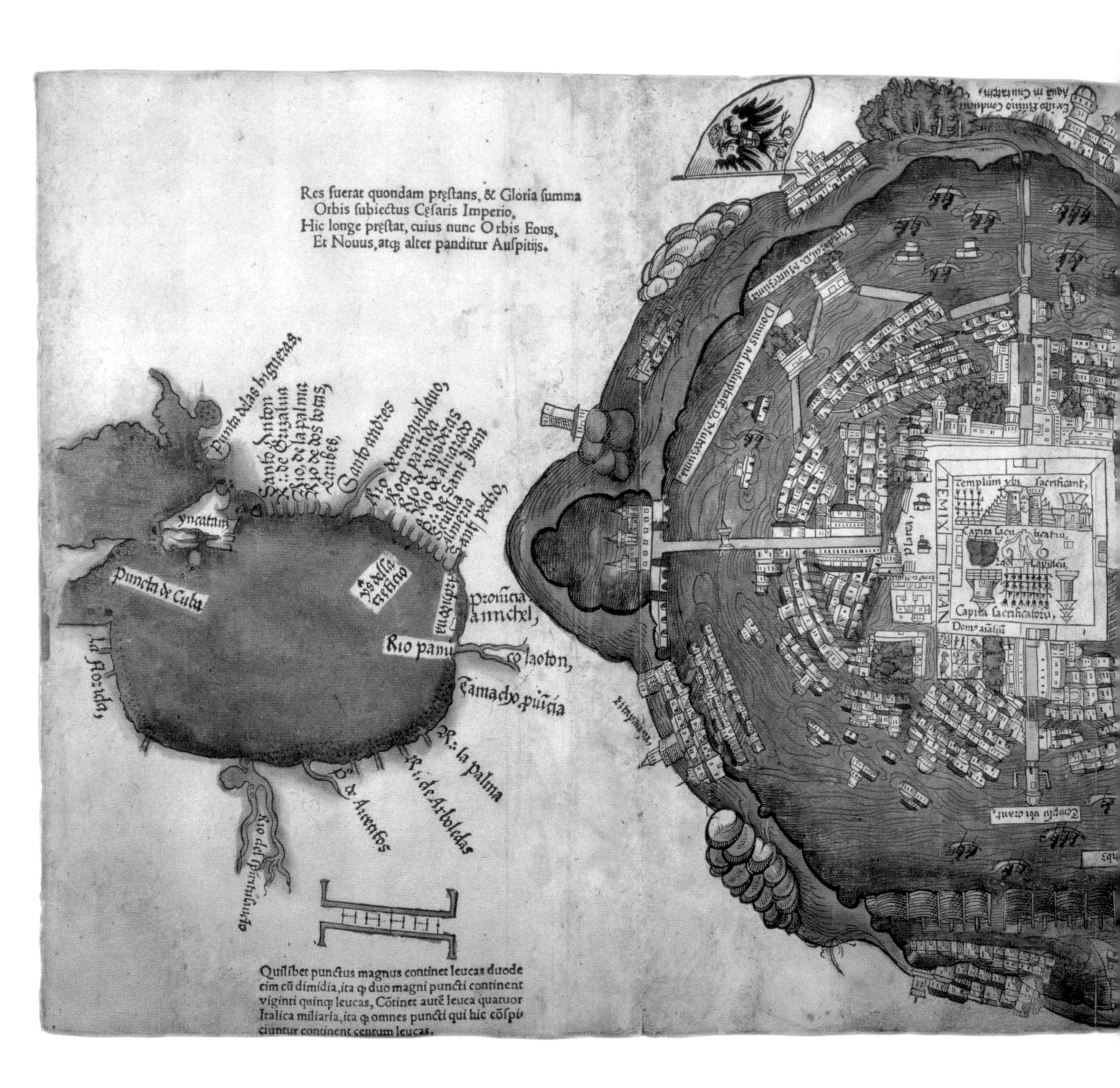

Res fuerat quondam prȩstans, & Gloria summa
Orbis subiectus Cȩsaris Imperio,
Hic longe prȩstat, cuius nunc Orbis Eous,
Et Nouus, atq; alter panditur Auspitijs.
Puncta de Cuba
yncatan
Santo andres
Rio panu
co laoton,
Tamacho puĩcia
Ri. la palma
Ri. de Arboledas
Rio del spiritusancto
La Florida,
Templum ubi sacrificant,
TEMIXTITAN
plaza,
Capita sacrificatoru,
Quilibet punctus magnus continet leucas duodecim cũ dimidia, ita q duo magni puncti continent viginti quinq; leucas, Cõtinet autẽ leuca quatuor Italica miliaria, ita q omnes puncti qui hic cõspiciuntur continent centum leucas.

THE INVASION AND DESTRUCTION OF MEXICO

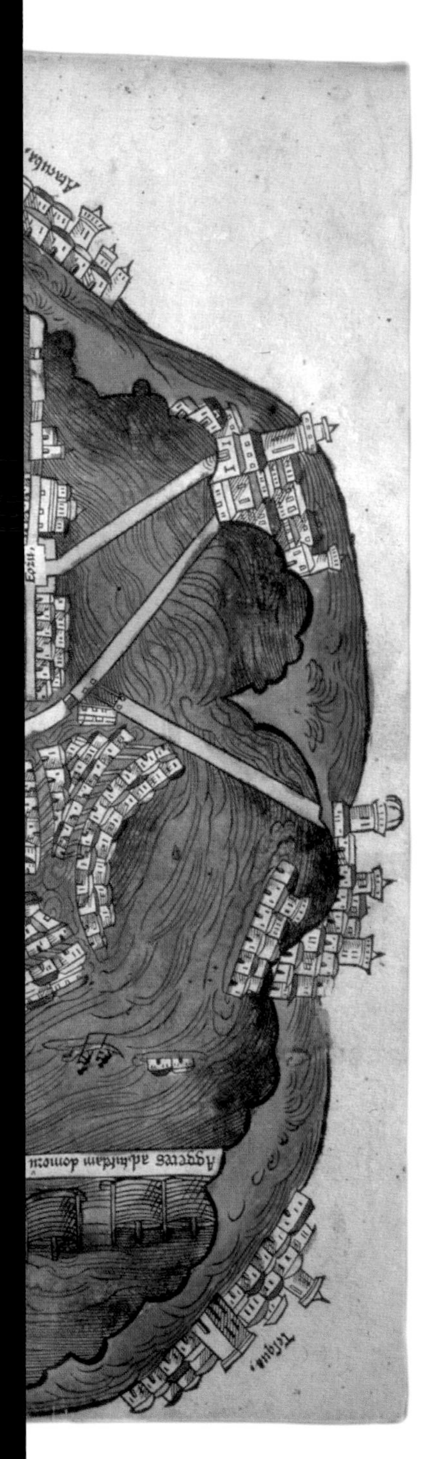

Hernán Cortés, map of the Gulf of Mexico and plan of Tenochtitlan, in *Praeclara Ferdinãdi Cortesii de Noua Maris Oceani Hyspania narratio sacratissimo*, 1524

The last three maps underscore the radical disruption that the early era of exploration brought to European worldviews. These two maps mark the next stage of that disruption, when Europeans began to penetrate the continental interior and to develop sustained encounters with the native population. In this case, that penetration is more accurately described as an invasion.

In June 1519 Hernán Cortés arrived on the eastern coast of Mexico with the aim of conquest, in defiance of the orders of the Spanish governor in Cuba. Cortés made his way inland, subjugating villages or forging alliances along the way. Six months after landing on the Mexican coast, he arrived at the Aztec capital of Tenochtitlan, the most important population center in North America at that time.

Cortés described this as an astonishing city of "unbelievable" complexity and sophistication. In letters to King Charles V he characterized Tenochtitlan as stronger than Granada, which the Spanish had recently reconquered from the Moors. In his view, it was as large and grand as Seville or Cordova, situated on a salt lake and accessed by four artificial causeways that were as "wide as two cavalry lances." This unique system of fortifications would challenge Cortés as he planned his attack. Within the city he found beautiful temples of worship built along straight, wide streets and waterways. With large and pleasing squares and a huge abundant daily market that attracted 30,000 people, this was a thriving society enmeshed in trade. It was that wealth—particularly gold—that drove the Spanish conquest.

In November Cortés met the Aztec leader Montezuma; their cordial relationship quickly grew hostile when the former demanded the gold that he believed was kept in the city. Cortés had Montezuma put under house arrest, which further inflamed the Aztec people. When the conquistador briefly left the city in April 1520, violence broke out between the Spanish and the Aztecs, and in June Montezuma was killed. The Aztecs drove the Spanish out, but a year later they again besieged the city and cut off its water supply. In July 1521 the Spanish burned Tenochtitlan to the ground and subsequently built Mexico City. Cortés overcame fierce Aztec resistance through more advanced weaponry and technology, but also because the Aztecs had been devastated by diseases he and his men had brought with them.

When the second of Cortés' letters to the king was published in Nuremberg in 1524, it included these two revealing maps. To the right we see a detailed plan of Tenochtitlan, the first published map of any urban center in the New World. Cortés depicted a lively city with a main temple, palaces, smaller dwellings, and grand squares, all surrounded by a great lake with several natives paddling canoes. For many years, scholars believed this map to be derived from Cortés' second letter to the king, for it resembled contemporary European iconography.

But Barbara Mundy has recently argued that the map contains elements not described in the letter, and thus was probably influenced by indigenous sources. This map gave Europeans the first glimpse of Tenochtitlan, though by the time it was drawn the city had been destroyed.

To the left was an equally important map, depicting the Gulf of Mexico. The scales of the two maps are completely different. The gulf is depicted with south at the top of the page, where "La Florida" appears at lower left. The map was sketched by Alonzo Álvarez Pineda, who was sent by the governor of Jamaica to find a westward ocean passage to the Pacific Ocean. Álvarez spent much of 1519 exploring the gulf, just as Cortés was making his way toward the Aztec capital. Though Álvarez found no water route to the west, he used his explorations to produce a qualitatively more accurate picture of the gulf as an enclosed body of water. This led him to conclude that North and South America were connected by land, settling a question that had bedeviled explorers and mapmakers for decades. He was also the first to depict Florida as a peninsula rather than an island, and the first European to come across the Mississippi River. On the map he recorded "Rio del spiritusancto," naming the river for the feast date on which he came upon its mouth.

While the gulf map documents a leap forward in geographical knowledge, the city plan profiles a civilization that was brutally destroyed through warfare and disease. That paradox aptly characterizes the Spanish legacy in North America throughout the sixteenth and seventeenth centuries.

THE HEMISPHERE TAKES SHAPE

Sebastian Münster, "Novae Insulae XVII Nova Tabula," 1540

By 1540 European geographers and mapmakers acknowledged that the landmasses of the western hemisphere were separate from Asia. Though debate persisted as to whether a narrow land bridge connected the two, this geographical consensus marked a shift toward a more accurate depiction of the Americas. This consensus is apparent in Sebastian Münster's "Novae Insulae," the first separate map of the western hemisphere. Münster was a German scholar at the University of Basel who closely followed the voyages of exploration. In 1540 he issued *Universal Geography*, with dozens of maps drawn from the Ptolemaic world as well as from modern geographical discoveries. With its wide circulation, the volume became one of the most accessible geographical pictures of the New World at the time.

As the map shows, contemporary understandings of North and South America remained highly fluid. Münster outlined a continuous coastline for the Americas, and gave them a somewhat recognizable shape. Throughout the map, older knowledge appears alongside newer discoveries. At the western edge of the map is a depiction of "Zipangri" (Japan) and the "7448" islands derived directly from the narratives of Marco Polo. Directly south, though, we see the oversized *Victoria* at sea, the sole surviving ship in Ferdinand Magellan's circumnavigation of the globe from 1519 to 1522. The Straits of Magellan are marked at the southern tip of the map, while further up the coast a fantastical depiction of cannibalism dominates what is now Brazil. To the northeast Münster identifies a fully enclosed Gulf of Mexico near Temistitan (Tenochtitlan), an explicit acknowledgment of the expeditions (and invasions) of Alonzo Álvarez Pineda and Hernán Cortés.

The discovery of an enclosed gulf ended the hope of a navigable passage through Mexico to the Pacific. Most of these hopes centered on Mexico, the Caribbean, and South America, which meant that areas further north were largely left unexplored. Once the Gulf of Mexico was understood to offer no passage, Europeans began to pin their fantasies of a passage to the Far East on North America. Münster's map reflects that shift of attention. The eastern coast of the North America bore some resemblance to its actual geographical contour, thanks to Giovanni da Verrazzano's explorations of the Atlantic coast on behalf of the French king in 1524. Like most explorers,

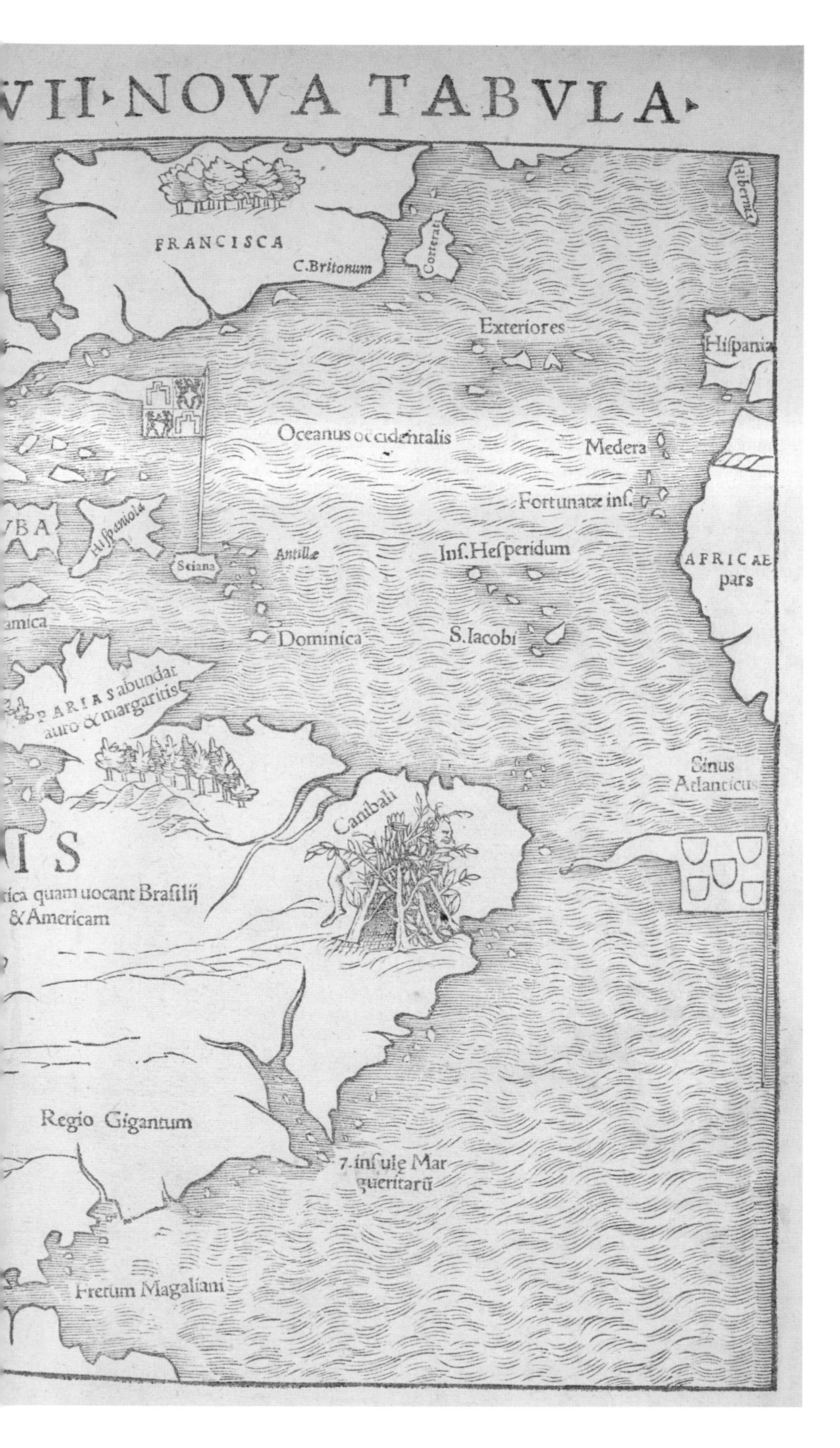

Verrazzano sought a route to the Far East. While exploring the coast of North Carolina, he mistook Pamlico and Albemarle Sounds for the Pacific Ocean. In the map of his expedition Verrazzano depicted a vast inland sea—the fabled Sea of Verrazzano—just west of the North Carolina coast.

Münster reproduced that geographical error on this map. A small spit of land suggests an easy passage between the Atlantic and the Pacific, with China and the Spice Islands just beyond. With this map, Münster effectively publicized French hopes of a Northwest Passage, and a view of the continent that was informed by these voyages. And, because Münster's maps circulated so widely, Verrazzano's vision of North America shaped subsequent voyages.

Münster's map was influential in other ways. In the second half of the sixteenth century the English joined the quest to find the Northwest Passage to the Far East. Münster's widely circulating picture of the world forcefully influenced Richard Hakluyt's argument for English colonization efforts in North America. In fact, Verrazzano's imagined passage to the Pacific, along the coast of North Carolina at Pamlico and Albemarle Sounds, was precisely the spot where the English eventually founded their earliest settlement, the failed colony of Roanoke.

THE SPANISH REACH NORTHWARD

Giacomo Gastaldi, with Paolo Forlani and Matteo Pagano, "Cosmographia Universalis et Exactissima ...," 1561

The Spanish were the first Europeans to explore any part of the North American interior. Among the most famous—or perhaps infamous—of these explorers were Hernando de Soto and Francisco Vázquez de Coronado, both of whom ventured through what is now the southern United States in search of wealth in the 1540s. Today Coronado is enshrined across the Southwest as a heroic explorer, though others consider him the archetypical Spanish invader. With Hernán Cortés as his model, Coronado set out for the Great Plains to find the fabled cities of gold that he believed were even richer than Mexico City. De Soto had similar visions of wealth, but also sought a passage to Asia. Though neither Coronado nor de Soto found riches, both of their expeditions treated the native populations brutally and left an enduring legacy of hostility.

News of Spanish expeditions found its way onto a large and gloriously detailed woodcut map designed by Giacomo Gastaldi in 1561. Gastaldi's map was one of the most complete pictures of the world at the time, though it was relatively unknown until the British Library purchased it at auction in 1978. Here we have reproduced the upper left quadrant of the map, which encompasses North America. A closer look here and on the next page reveals the ambitions of the Spanish as they ventured north from Mexico into what is now the southwestern United States.

The map was most likely authorized by the Venetian Senate; blank cartouches indicate that it was either unfinished or a printer's proof. Gastaldi was a well-regarded mapmaker in Venice, and this was among his last and most ambitious efforts. Like many maps of the sixteenth century, it includes elements of both confirmed geographical knowledge and persistent fantasy.

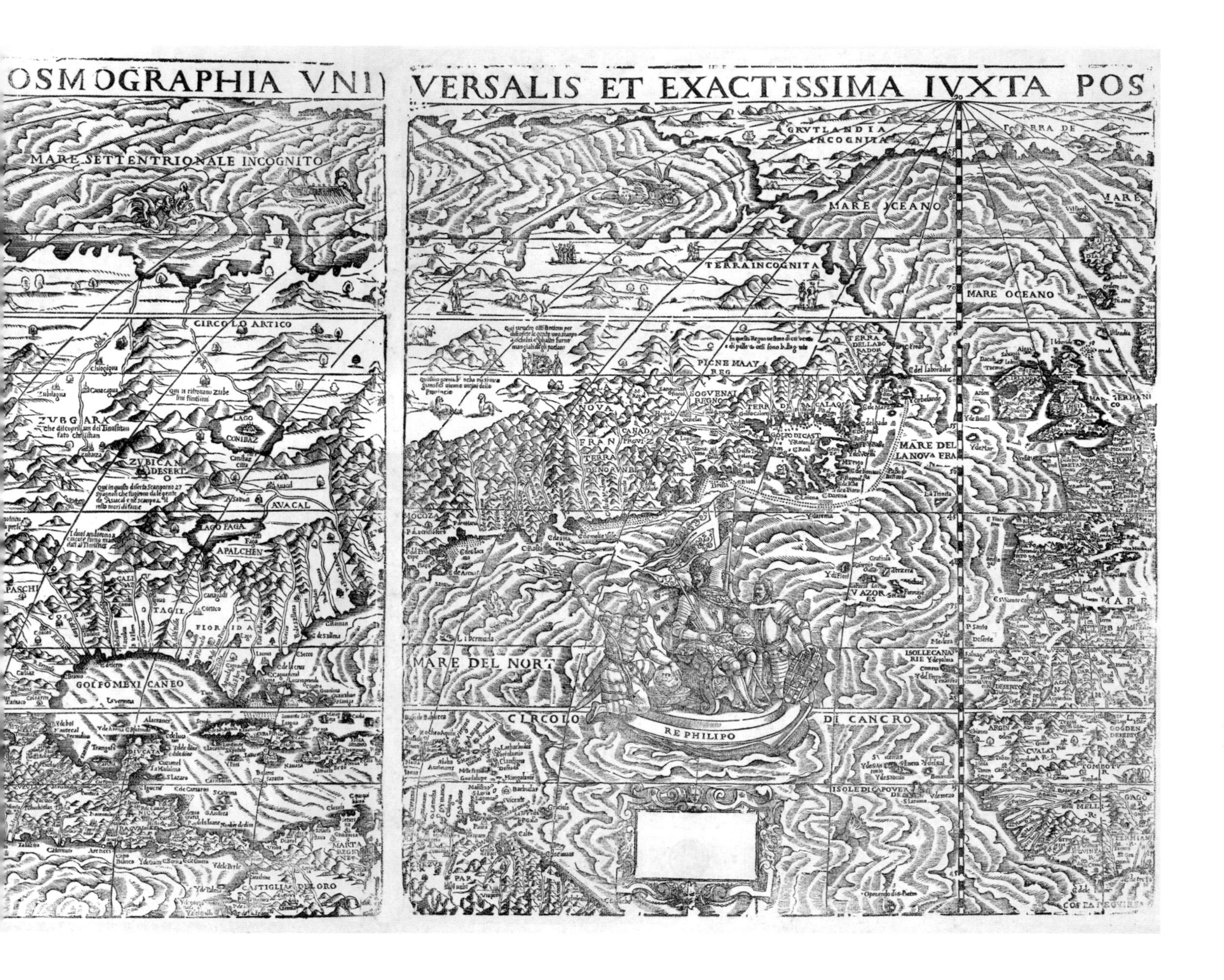
OSMOGRAPHIA VNI
VERSALIS ET EXACTISSIMA IVXTA POS
MARE SETTENTRIONALE INCOGNITO
GRVTLANDIA INCOGNITA
MARE OCEANO
TERRA INCOGNITA
MARE OCEANO
CIRCOLO ARTICO
ZVBGARA
LAGO CONIBAZ
ZVBICAN DESERT
AVACAL
LAGO PAGA
APALCHEN
FLORIDA
MARE DEL NORT
TERRA DEL LABORADOR
MARE DEL LA NOVA FRA
CIRCOLO
DI CANCRO
RE PHILIPO

Clima 17
Clima 16
MARE SETTENTRIONALE INCO
NITO
CIRCOLO ART
Pagul
GAMA
Selbot
Cubirag
Tolgaga
MAN
Chiogigua
Canacigua
Zubilagua
Qui si ritrouano Zube lini finissimi
ZUBGARA
che discoprì un del Tinistitan fato christian
LAGO
CONIBAZ
Conibaz citta
Zubarza
ZUBICAN DESERTO
Compagnie di gente che stano alla campagna a modo de Tartari detti ICLUBIRI
Qui in questo deserto Scamporno 27 spagnoli che fugirono da le gente de Auacal e ne scampo 2 il resto mori di fame
Qui scamporno la moltitudine quando spagnoli presero il paese ca... di notte recuperorno il detto paese
QUIRIR
LAGO FAGA
APALCHEN
Chichucho
Axa
Tiguas
Totonteac
Vachus
CAPASCHI
TERLCHICHI MECHI
Petatlan
Vacapa
AZATLA
Corazanes
FLORIDA
Coruco
Canagadi
TOPIRA
TOVA
Culiaz
GOLFO MEXI CANEO
VERMEO
Xalisco
XALISCO
C delle corrente
Ydel Xalisco
NOVA GALITIA
NOVA SPA
MECVACA
Alacranes
Trangoso
IVCATA
Cozumel
S Lazaro
C de Cantaron
Scatarina
S Andrea
Nonbre de dios
Racha partida
S Tomas
C dele corente
C Hermolo
Arcifices
Capo Bianco
Y de Gatti
C Borica
C de Guera
Y de le Perle
Y de Palma
CASTIG

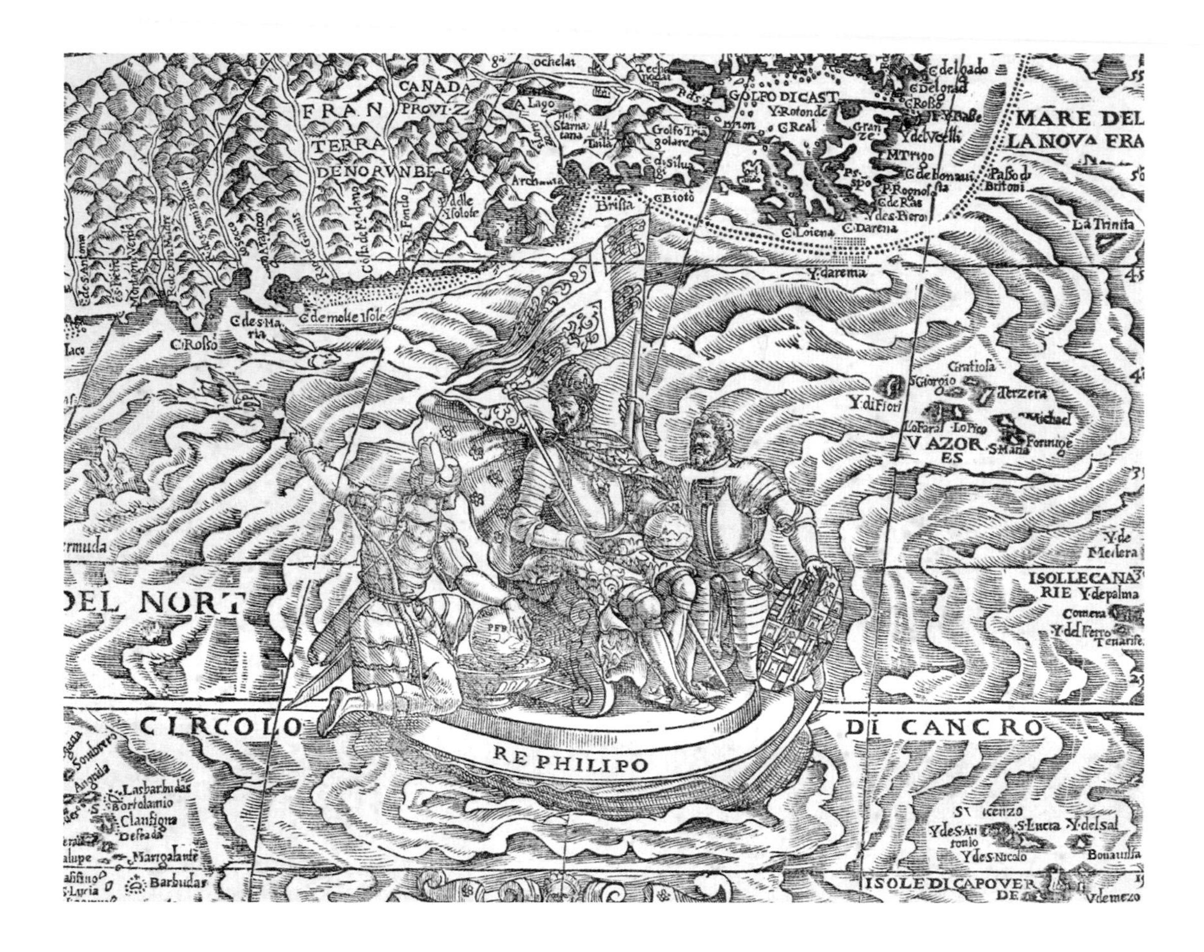

At the western edge of the map (on page 26) Gastaldi delineated the Strait of Anián. This was the first appearance of the passage between Asia and North America on any map. This fabled strait had been located throughout the North American west in the sixteenth century, and was not definitively identified as the Bering Strait until two centuries later. Gastaldi's depiction was especially important, for only a few years earlier on a different map he had connected Asia and North America as a single landmass, with no separation. Here, the Anián Strait connects the Pacific Ocean with a massive inland northern sea, which in turn allows passage to the Atlantic Ocean. These connected bodies of water fueled the hope of a Northwest Passage from Europe to the Far East.

The geography at the western edge of the map also suggested an easy voyage between North America and Asia. Japan is placed midway between China and North America, which transforms the massive Pacific Ocean into a smallish lake. Both the Atlantic and the Pacific oceans are spaces of imagination, with mermaids and mermen swimming alongside dragons and ocean vessels.

The details of Gastaldi's map shown at left and above are full of evidence of Spanish ambitions beyond Mexico. At the southern edge of the detail at left we find a densely settled Mexican landscape, reflecting the growing presence of Spain over the prior half century. Moving north, that density gives way to a wide open continent with limited human settlement. Bands of conquistadores march across the interior, interspersed with a few native villages. Here and there are strangely disfigured cattle, creatures that had been described by Coronado. No doubt these were some of the first European sightings of the bison that roamed the plains. The details along the left edge indicate that Gastaldi was aware of Coronado's brutal exploits in this region, for he identified the Tiguex War fought between the Pueblo tribes and Coronado's men along the Rio Grande.

Spanish ambitions are exalted by the elaborate portrait of King Philip II shown above, which Gastaldi placed in the center of the full map. The monarch is seated on a throne with a gaze that follows an outstretched hand pointing toward North America. The king's outsized presence on the high seas, together with the conquistadores that dot the landscape, remind us that in the mid-sixteenth century, the Spanish were actively expanding their realm in the New World.

THE SPANISH ASSERTION OF AMERICA

Diego Gutiérrez and Hieronymus Cock, “Americae sive quartae orbis partis nova et exactissima descriptio,” 1562

This large, flamboyant, and detailed map of the western hemisphere is first and foremost a declaration of Spanish power. Cosmographer Diego Gutiérrez collaborated with the noted Flemish engraver Hieronymus Cock to produce the map for the Casa de Contratación, the agency responsible for creating charts and maps to administer the Spanish empire. The map—measuring approximately three feet square—remained the largest of its kind well into the next century. (It is unclear whether the author was Diego Gutiérrez senior or his son. Both went by the same name, and both worked for the Casa.)

The size and elaborate artistry of the map indicate that it was designed not as an aid to navigation but more as a symbolic statement of Spanish authority in the New World. At upper left are Spanish and French coats of arms, most likely commemorating the recent treaty that ended decades of war between the two powers. East of Argentina is the Portuguese coat of arms, acknowledging that country’s sphere of influence over eastern South America as set out in the Treaty of Tordesillas. Around the entire hemisphere, fantasy and reality inhabit the same space: large vessels navigate rough waters, battle for control of the South Atlantic, and sail among sea creatures. A shipwreck marks the North American coast, while the hungry cannibals in the South American interior echo the same type of detail found on Sebastian Münster’s map of 1540 (page 24).

Among the most noticeable details are the many place names along the entire coastline of South and Central America, and to a lesser extent along the North American coast. Naming these places was itself an act of Spanish control. By the time Gutiérrez drew the map, the Spanish were moving beyond the coast to the interior of what is now the southern United States. In 1540 Francisco Vázquez de Coronado headed north from northern Mexico to find the fabled cities of gold on the Great Plains, while de Soto explored the tributaries of the gulf and the Southeast. The Gutiérrez map captures the Spanish quest to name and claim the land in “La Nueva Galicia” and elsewhere. The assertion of power is underscored by the vignette in the upper middle, with King Philip as a confident Neptune sailing westward and presiding over both land and sea.

And yet, in that tentative northward reach, the Gutiérrez map inadvertently demonstrated how little was known of North America. This was the first map to name California, and properly shows it as a peninsula rather than an island (as became common in the next century). But the Atlantic coastline remains relatively confused, and the placement of large decorative elements at upper left is a tacit acknowledgment that the continental interior was poorly understood. The Gutiérrez map signaled the supremacy of the Spanish in the age of discovery, but with two crucial caveats. First, the map reminds us that North America remained a vast mystery to Europeans in the middle of the sixteenth century, at times even something of an afterthought. The heart of the map is Central and South America, with North America relegated to the corner and dominated by decorative flourishes rather than geographical detail.

Second, Spanish control in the western hemisphere would soon be challenged by other European powers. In the year this map was published, the French established Charlesfort in what is now South Carolina, and tentatively explored the coast of Florida. Though Spanish pressure drove out the French within a year, it was a sign of things to come. Within two decades, the English began to articulate their own territorial claims in North America, as shown on the next map.

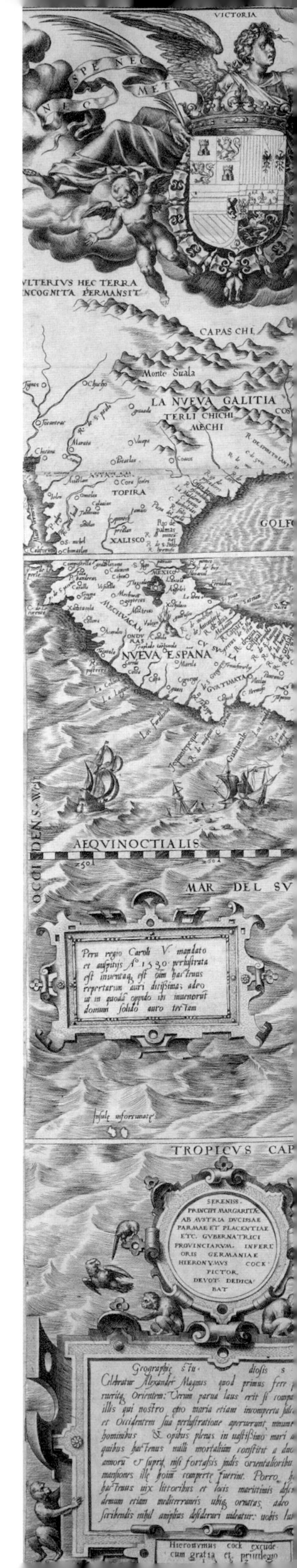

SEPTENTRIO
PARTE DI GRONLANDA
OCEANVS SEPTENTRIONALIS
PICNEMAY REG
TIERRA DE LABRADOR
CANADA PROV
TIERRA DE BACCALAOS
TIERRA FRANCISCA
TIERRA DI NORIMBERGA
AVACAL
APALCHEN
MOCOSA
TIERRA FLORIDA
ISLAS DELOS ACORES
ISLAS DE CANARIA
TROPICUS CANCRI
CVBA
OCEANVS
OCCIDENTALIS
ISLAS DE CAPO VERDE
CASTILIA DEL ORO
BENEZVELA REG
PARIA
NOVA ANDALVSIA
LA FLOTA DE PORTVGAL QVE VA PAR CALICVTE
CARTAGENA
REGIO DE BRASIL
REGIO DEL PERV
CHICA
TIERRA DE PATAGONES
MARE MAGELLANICVM SIVE PACIFICVM
GIGANTVM REGIO
Estrecho de Magallanes
OCEANVS AVSTRALIS
REGIO PATALIS
TIERRA DE MAGALLANES

THE FATHER OF THE BRITISH EMPIRE

John Dee, chart of part of the northern hemisphere, 1580

Through most of the sixteenth century, Spain was the most powerful nation on earth and the dominant force in the New World, but in the second half of the century England began to assert itself in North America. This more expansive view of England's place in America was forcefully articulated by John Dee (1527–1608), a mathematical geographer who educated some of the most influential men of the day, including Richard Hakluyt, Martin Frobisher, and Walter Raleigh. Dee also corresponded with the pivotal mapmakers Abraham Ortelius and Gerard Mercator, and by 1570 he was one of the era's most important geographers. His training in mathematics and geography made him a key adviser to almost all English navigators well into the 1580s. His relentless advocacy for a North American empire stimulated an increase in English voyages, with the expectation that settlements would follow. For decades, Dee provided the intellectual framework and justification for England's imperial vision.

In 1580 Dee drew this large map to locate English territorial claims in North America, then on the reverse side "proved" these claims through reference to historical voyages, genealogical traditions, treaties, and other evidence. From Florida to Greenland, he found precedents for English sovereignty. Some of these must have seemed extraordinary, such as his insistence that the northern reaches of the mainland belonged to England given that King Arthur had populated part of Greenland in the sixth century. Others held more weight, such as his argument that the voyages of John and Sebastian Cabot and the seafaring expeditions of the Bristol fisherman near Newfoundland gave England primacy over Spain on the North American mainland. After all, Dee argued, Columbus may have been the first to sail to the New World, but he never reached the continent itself. Dee assured the queen that these northern lands of the New World were part of the English domain.

Dee's reputation and network gave him clout in Elizabethan England. In 1579 Sir Francis Drake sailed up the Pacific coast of North America and named it "New Britain," leaving some of his few remaining men to establish a colony. The Spanish were outraged, and the queen brought Dee to London to help defend and advance Drake's claim. Dee used a map to augment Drake's assertions on behalf of the Crown, arguing that English sovereignty encompassed not just the Pacific coast but most of North America. His maps exemplified the role of geography as a tool of state power.

In 1578 Queen Elizabeth quietly granted Sir Humphrey Gilbert permission to create an English settlement in any land not controlled by a Christian prince. In doing this she was careful not to challenge Spanish sovereignty, but worried little about infringing the territorial rights of Native Americans living on the continent. Elizabeth's patent to Gilbert was the earliest English claim to territory that would become the United States. After consulting Dee in preparation for the voyage, Gilbert "granted" the geographer rights to all the land that he might discover north of the 50th parallel. Such a gift would have included most of Canada and the long-sought Northwest Passage to the Pacific Ocean, a grandiose gesture based on the assumption that the English had title to lands that they had yet to even see.

Gilbert's voyage failed, as did Sir Walter Raleigh's subsequent attempt to establish a colony at Roanoke. It would take years for English settlements to take root in North America, but the father of these ideas—the architect of the British empire as an idea and a counterweight to the Spanish Crown—was John Dee. For decades he used maps and treatises to claim for the queen a worldwide territorial dominion, convincing navigators, settlers, investors, and royalty that the future of North America was English.

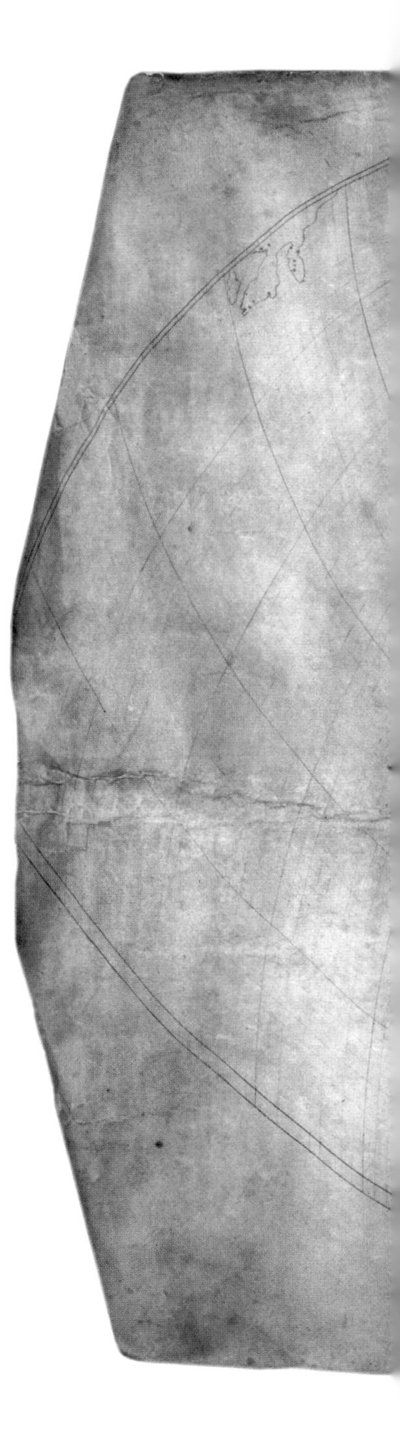

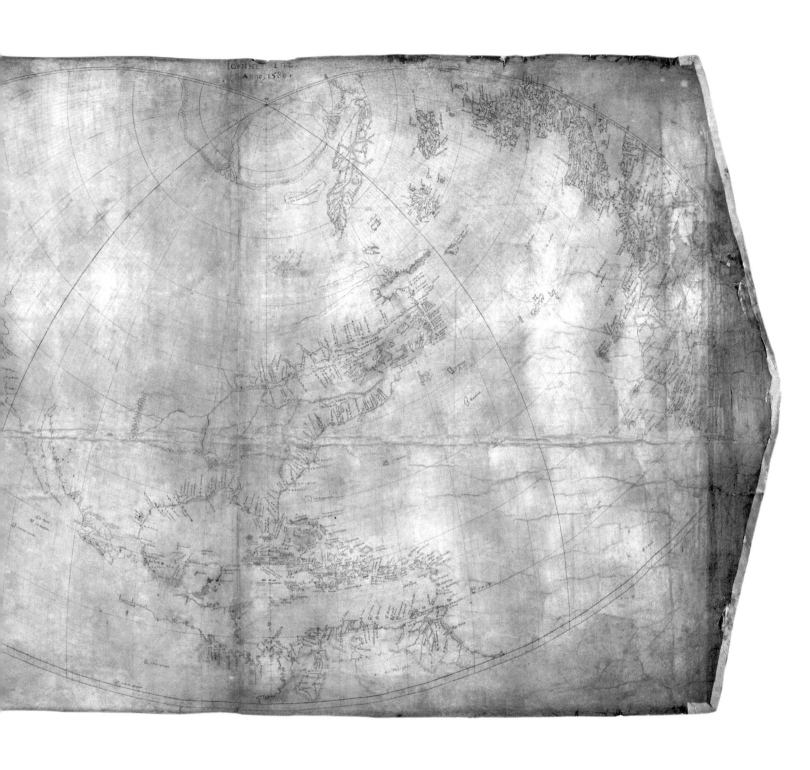
Anno, 1580.

2. 1600–1700

Early Settlement and the Northwest Passage

Diego Gutiérrez boldly claimed the Americas for the Spanish Crown in his 1562 map, as we saw in the last chapter. Yet Spain exercised little territorial control over North America. Its fort at St. Augustine in Florida was actually established in 1565 to defend against French encroachments. Sensing a similar opportunity for the English, Queen Elizabeth extended permission to Humphrey Gilbert and Walter Raleigh to establish colonies on the Atlantic coast in the 1580s. In 1583 Gilbert attempted a colony at Newfoundland; it failed and he died on the return voyage to England. The following year Raleigh explored a more hospitable climate further south, naming it Virginia in honor of Elizabeth, "the virgin queen." With one hundred soldiers he founded a settlement at Roanoke Island on North Carolina's outer banks, but the group soon abandoned the area and sailed home with Sir Francis Drake. A second settlement effort in 1587 brought men, women, and children, but the colony disappeared when it was stranded without supplies.

The failures of Gilbert and Raleigh conveyed the dangers of colonization, yet English pursuits continued. These early efforts also signaled a more general shift in seventeenth-century North America. While in the 1500s Europeans explored and then left, by the 1600s they were beginning to stay, motivated by a combination of commercial gain, religious mission, and emergent nationalism. The maps in this chapter were made as instruments of that first stage of colonization.

The earliest of these efforts were Spanish. After founding St. Augustine, the Spanish built missions and trading posts throughout the Southwest. Some missionaries lived peacefully among the Indians; more notorious was Juan de Oñate, who in 1598 led an expedition of 500 soldiers and settlers to spread Catholicism and Spanish authority. In what the Spanish named "New" Mexico, Oñate's men seized supplies from the Pueblo Indians, destroyed the village, and killed more than 800 Acoma Pueblo men, women, and children. Oñate was recalled to Spain and punished, but before that he helped to establish the first European settlement in the American Southwest, at Santa Fe.

Just as the Spanish were moving northward from Mexico, the English made a third attempt to colonize North America at a swampy spot on Chesapeake Bay that they named for King James. Within a year of founding Jamestown, Robarte Tindall sent back the first English chart of the region (page 36). This remarkable document captures Tindall's first impressions as he explored the rivers in search of a westward passage. A few years later, John Smith drew a far more detailed map of the Chesapeake, designed to promote a colony that was barely surviving because of food shortages, a failure of discipline, and deteriorating relations with the neighboring tribes of the Powhatan Confederacy (page 40). Ever the entrepreneur, Smith saw another opportunity to promote English colonization further up the Atlantic coast. Before the English had even explored (much less settled) the area, Smith claimed this as a "New England" (page 42). Smith's map shrewdly branded the area as an extension of the mother country, a familiar destination that beckoned new settlers.

While the English sought to establish a foothold in the Chesapeake and New England, the French set their sights further north. In 1608 they sent Samuel de Champlain to establish a trading outpost at Quebec on the Saint Lawrence River. Champlain's ongoing exploration of the interior laid the foundation for subsequent French claims to the Great Lakes and the Mississippi River Valley. And, while the French—like the Spanish—sought to convert souls, they spent more time building the fur trade and searching for a navigable passage across the continent. Champlain's map of "Le Canada" reflects this drive to understand the North Atlantic coast, the Saint Lawrence Seaway, and especially the Great Lakes further west (page 46).

Soon after the French established Quebec, Henry Hudson sailed up the river that now bears his name, looking for a Northwest Passage. Though he found no such passage, he used the voyage to claim enormous territory for his Dutch patrons. This paved the way for the settlement of New Amsterdam on Manhattan Island in 1624. The religious separatists sailing on the *Mayflower* in 1620 had originally planned to settle these Dutch claims along the Hudson River near Manhattan. But when their voyage was blown off course to the north, they established the Plymouth colony near Cape Cod in what Smith had named New England just a few years earlier. A "Great Migration" of religious exiles followed, extending settlement ever further into the interior and onto native lands. John Foster's 1675 map of King Philip's War records the horrific violence of those early encounters between English settlers and the Wampanoag Indians (page 54).

This rush of settlements along the Atlantic Ocean and its waterways led to a welter of imperial claims and counterclaims by the end of the seventeenth century. Jamestown, Santa Fe, and Quebec were all founded within a few years of each other, a useful reminder that there was nothing inevitable about the eventual English domination of North America. The maps in this chapter restore that contingency. The ongoing contest between the Dutch and English for control of New York is underscored by Robert Holmes' map of Manhattan in 1664 (page 52). Vincenzo Coronelli's master map of North America in 1688 captures both the state of geographical knowledge by the end of the century and the continental ambitions of the French (page 60).

This chapter also demonstrates the tenuous nature of early North American colonization, exemplified by the case of Virginia. During the desperate "starving time," Jamestown colonists even resorted to cannibalism. The colony stabilized only when its leaders introduced harsh discipline and an emphasis on agriculture, especially tobacco. Virginia's prosperity was aided by the heavy recruitment of new settlers, and the continued belief that a Northwest Passage through the continent would position the colony at the center of worldwide trade (page 44).

The demand for tobacco transformed Virginia from a struggling colony to a thriving enterprise, but at great cost. Planters, devoting ever more land and labor to the crop, came into direct conflict with the Powhatan Confederacy, which in turn prompted the Virginia Company to expel natives from the colony and to seize their lands. The increased cultivation of tobacco soon came up against a shortage of labor, which led to the introduction of slavery. When tobacco planter John Rolfe exchanged food for twenty Africans aboard a passing Dutch ship in 1619, he established a pattern that would quickly grow. Dutch and English traders soon thereafter brought Africans from the Caribbean on a regular basis, launching a practice that stimulated the early Atlantic slave trade.

In its earliest stages, slavery was a fluid practice, but in the 1660s Virginia colonists began to pass laws that defined this labor system in racial terms. These laws virtually ensured that to be black was to be enslaved, and that one born in bondage would remain so for life. The map of West Africa on page 50 marks an early moment in that evolution, when Dutch (and later English) merchants began to send slaves to the Americas in order to meet a growing demand for tobacco and sugar in Europe. Without the displacement of Indians and the introduction of slaves, Virginia could not have prospered.

Such accounts force us to reckon with our own national myths. Schoolchildren learn that, while Captain John Smith saved the Jamestown colony, he in turn was saved by Pocahontas. The New England Puritans are enshrined as religious refugees who gave thanks for their first fall harvest in a harsh and unforgiving environment. William Penn is exalted as a champion of religious freedom who forged peaceful treaties with the Lenape Indians in what would become Pennsylvania (pages 56 and 58). The maps in this chapter help us both to understand the origin of these myths and to reach beyond them to a more complex and contingent account. They enable us to see through the eyes of those early settlers, and to ask how they understood the geography of North America. They reveal the aspirations to settlement, the interests at stake, and the dynamics at work in these early years of colonization.

THE ORIGINS OF THE VIRGINIA COLONY

Robarte Tindall, a colored chart of the entrance of Chesapeake Bay, 1608

At first glance this map is a complete mystery: the geography is unrecognizable, the place names are nearly indecipherable, and the larger picture looks more than a little like a large intestine. But give it a second look, for it rewards patience.

In 1606 King James gave the newly formed Virginia Company the rights to establish another colony in North America. Having learned from earlier failures, the company adopted a joint-stock model to aggregate capital and mitigate the risk to individual investors. Just before the end of the year, the company sent 144 men and boys in three ships to the New World. After making landfall in May 1607, the party traveled fifty miles up a river they renamed in honor of their king. The settlers chose a spot with a deep-water shoreline that could be protected from Spanish attack. But, despite abundant game and protective woods, "Jamestown" was also marshy and full of mosquitoes, and thus malaria.

Two of these colonists, George Percy and Robarte Tindall, immediately set off to explore the area. Percy kept a diary, and Tindall prepared a map. Percy's diary gives a firsthand account of these early interactions between the English and Native Americans, describing both friendly and hostile encounters. For this reason, historians have paid close attention to his diary. Far less attention has been given to Tindall's map of the James and York rivers, drawn upon his return to Jamestown in April 1608. Tindall sent his map to Prince Henry along with Percy's journal, hoping that the geographically curious young prince would be pleased to share with the royal family an account of places "where never Christian before hathe been."

Tindall's map is difficult to read, and slightly disorienting. But as the first English chart of the James and York rivers, it records crucial information. The chart is oriented with west at right, where the James (at the top) and the York (below) flow from right to left. This itself is revealing: Tindall focused on the rivers rather than the adjoining land because these early settlers sought to follow them to their source, hoping to find a portage that would take them *beyond* the continent to Asia.

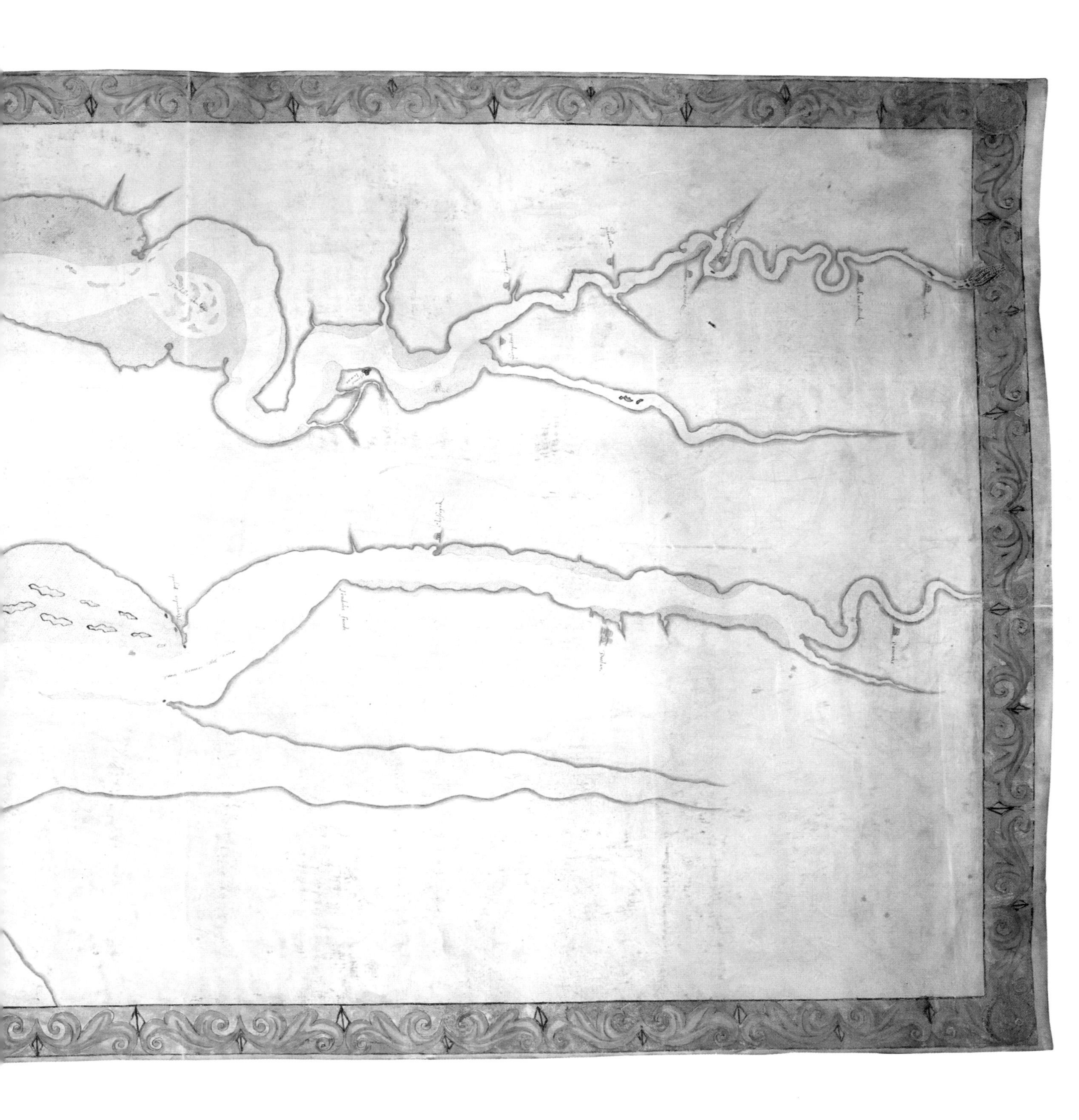

This hope of a river passage was abruptly dashed when Tindall and his party came up against the falls of the James River, near present-day Richmond. These falls are represented at the upper right corner where the chart meets its decorative border. Prevented from going any further, the men nonetheless staked a large cross at the falls and claimed the river and surrounding land for the Crown. For this reason, the falls form the far western edge of Tindall's chart, the limits of what these men were able to explore. Far downriver to the left, Tindall marked "King James his River" as well as the newfound settlement of Jamestown.

On the lower half of the map, Tindall similarly outlined the course of the York River, originally named for Prince Henry. The rivers—rather than the land—occupy the center of the map because the colonists sought to navigate around Chesapeake Bay. Tindall's attention to channels and rocky shoals further reveals the contemporary concern with the navigability of the rivers rather than the lie of the land.

The map lacks any formal elements, such as a coat of arms, a cartouche, or even a title. This suggests that it was intended for intelligence purposes rather than public consumption. Tindall also precisely identified the tribes he met on his expeditions of 1607 and 1608, for each was potentially an ally or a threat. The detail at right faintly records twelve of these native villages, carefully named and located. This makes the chart a priceless ethnographic record of the earliest stage of the Virginia colony, made before there was any guarantee that it would endure.

Indeed, conditions at Jamestown deteriorated within a few months of its founding. Men died rapidly over the summer of 1607, turning Percy's journal into a chronicle of death. As he recorded, "There were never Englishmen left in a forreigne Countrey in such miserie as wee were in this new discovered Virginia."

Tindall located the primary residence of Chief Powhatan in the village of "Poetan" at the far upper right, near the falls. The tribe that had been the "mortal enemies" of the settlers saved them with gifts of bread, corn, fish, and game. Without this help, Percy wrote, the colony would have perished. Even so, when provisions finally arrived from England in January 1608, only thirty-eight of the original settlers were still alive. The colony managed to continue with the arrival of new recruits and the imposition of harsh rules to enforce discipline and labor. Jamestown might just as easily have been swept away like Roanoke. This faint picture of the land mirrors the fragility of the colony itself.

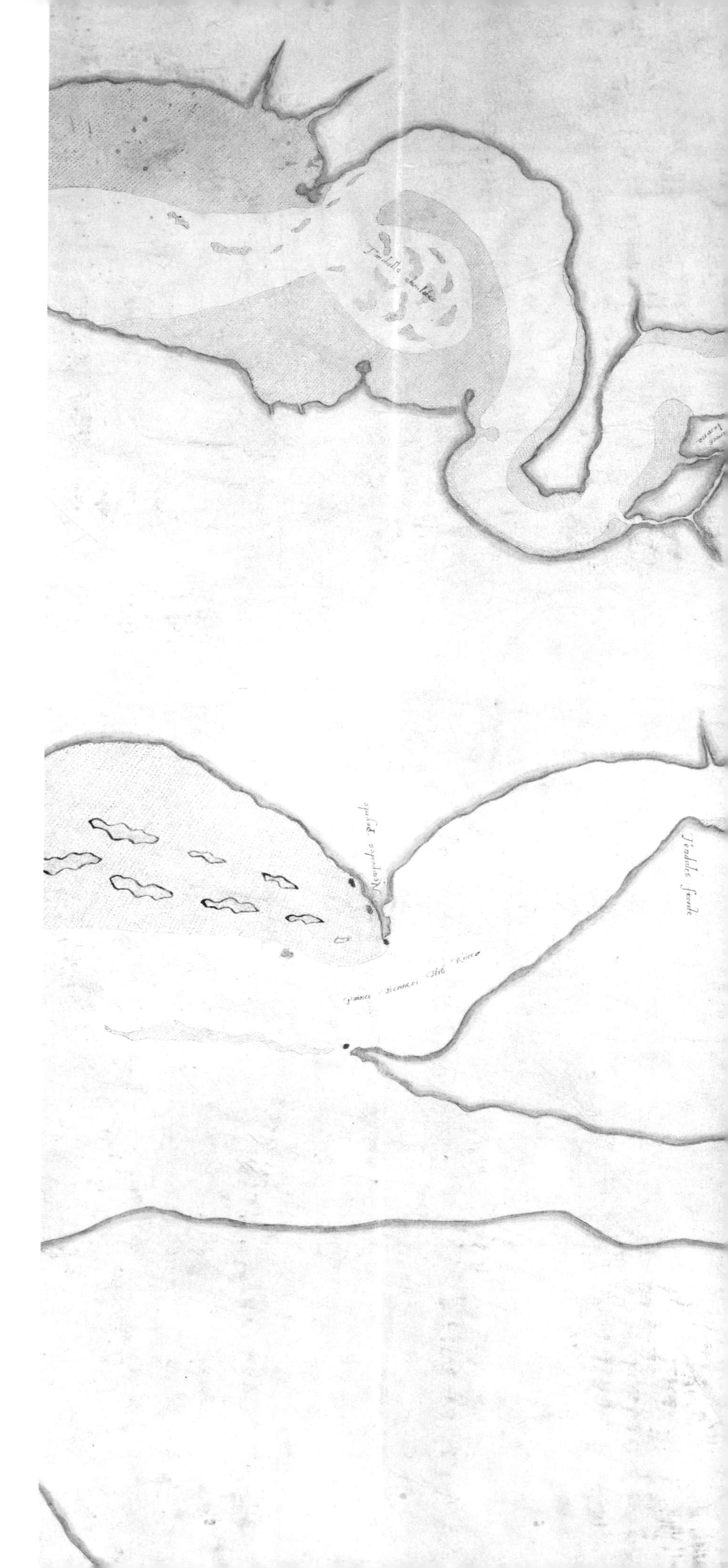

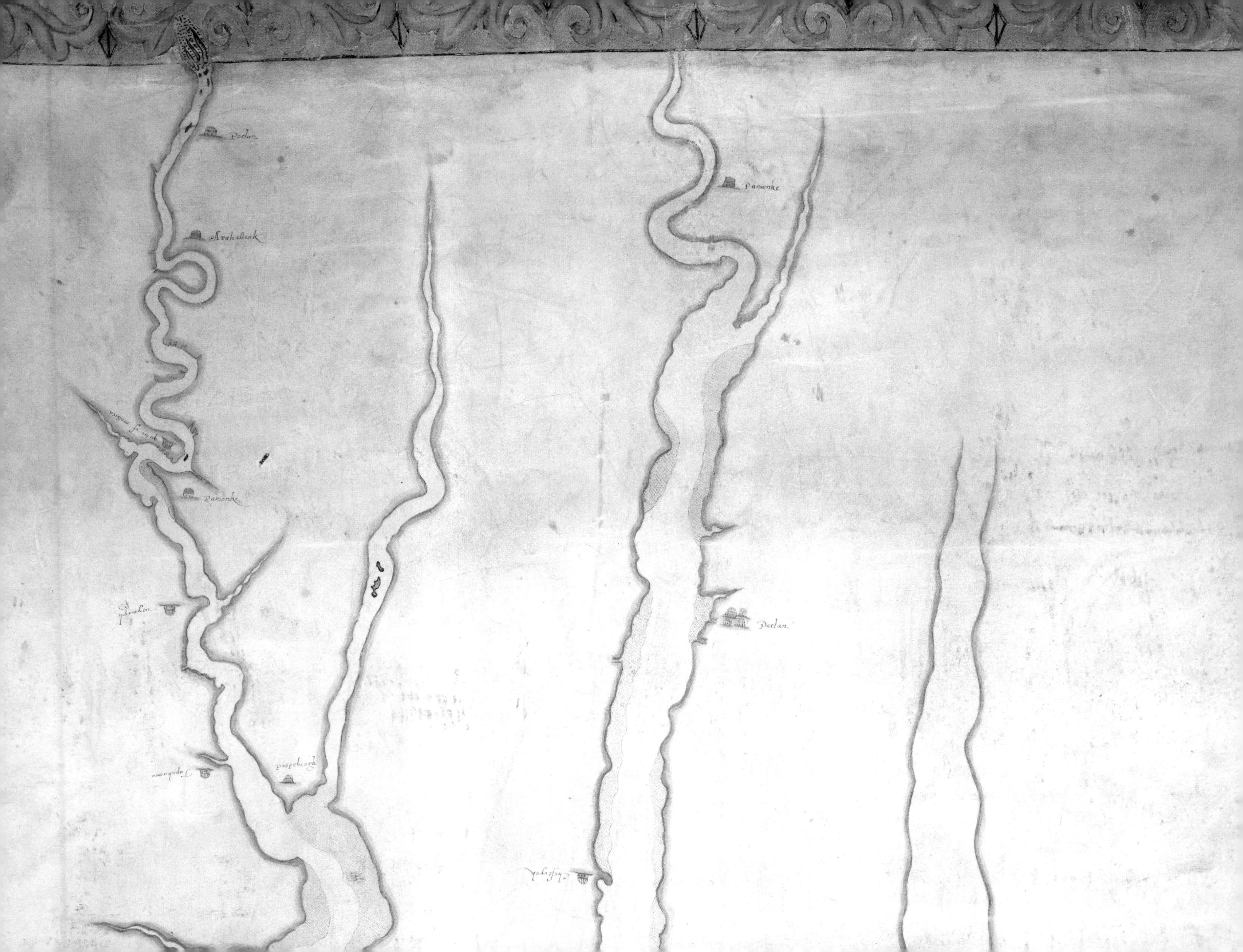
Arahatteak
Pamonke
Pamonke

VIRGINIA
Massaw:
MONACANS
MANN AHOACKS
POWHATAN
Held this state & fashion when Capt. Smith was deliuered to him prisoner
MAN GOAGS
CHA WONS
P O W H A T A N
Powhatan
Jamestowne
Point comfort
Cape Henry
Cape Charle
Smyths Iles
KUSKARAWAOKS
TOCKWOGHS
Scale of Leagues
and halfe
Leagues
Chickahokin
THE VIRGINIAN SEA
Discovered and Discribed by Captayn John Smith
Grauen by William Hole

THE SURVIVAL OF VIRGINIA

John Smith, "Virginia," 1612

No name is more closely connected with the Virginia colony than that of Captain John Smith. An early associate of the Virginia Company, he crossed the Atlantic Ocean along with Robarte Tindall and the other founders of Jamestown. Smith was a soldier and an explorer, and in his brief time as leader of the colony he imposed the harsh discipline that contributed to its survival.

Upon his arrival in Jamestown, Smith—like Tindall—was charged with exploring England's territorial claims in Virginia. Of course, in light of the Company's limited geographical knowledge of the region, any attempt to delimit the colony seemed more than a little speculative. From 1607 to 1609 Smith conducted several reconnaissance missions in the Chesapeake Bay, making contact with tribes both friendly and hostile. From Smith's own journals we have an account of his capture by a Powhatan hunting party in December 1607. When he was brought before Chief Opechancanough, he used his compass to present information about the rotation of the earth around the sun, hoping to save his life by impressing the Powhatan with this scientific instrument.

The following summer Smith took another extended trip through the bay to gather geographical intelligence from the local inhabitants. He used this information to draw his influential map of Virginia, published in 1612 and the most comprehensive picture of the region for decades. Oriented with west at the top, Smith's map presents a pleasing, holistic, and even inviting view of the Chesapeake. The decorative figures at upper left and right were likely added by someone else, but, together with the elaborate compass rose and coats of arms, these details suggest a stable and harmonious settlement. The map is far more detailed than Tindall's earlier chart of the York and James Rivers on page 36.

Both Smith and Tindall hoped to find a passage to the Pacific Ocean within the Chesapeake. But, while Tindall primarily focused on rivers and navigation, Smith paid closer attention to the adjoining land. This difference marks the shift in the nature of the colony itself: while in 1608 the colonists were primarily interested in extracting resources to send back to England, by the time Smith published his map the colony had become one of settlement and farming. Smith's careful rendering of the complex shoreline, and his identification of more than 200 Indian villages, presents a place that the English intended to make their own.

While navigating the rivers and exploring the land, Smith and his companions marked the limits of their travels by carving crosses into trees. On the map, he used a similar system of Maltese crosses to indicate the limits of his own knowledge, beyond which he relied upon information from Native Americans. These crosses appear throughout the map, forming a ring around the Chesapeake. They are visual indicators of Smith's reliance on local tribes for geographical intelligence, much as the colony itself depended upon local support for its survival.

Smith's debt to these tribes is embodied by the large native figures at upper left and right, which remind us that the bay was home to an estimated 15,000 to 25,000 natives when Jamestown was founded. It was their guidance and knowledge that enabled Smith to navigate through the Chesapeake and then compile this larger picture of the region.

Moreover, it was Powhatan—pictured at upper left—who ultimately ensured the survival of the Jamestown colony. Smith's own journal recounts the "starving time" of the winter of 1609–10, when the colony was reduced to a population of sixty, surviving on roots, herbs, acorns, berries, and a little fish. "So great was our famine," Smith wrote, that some resorted to cannibalism, but he blamed this not on the land but on the lack of planning and industry on the part of the colonists themselves.

The disastrous early years at Jamestown forced its investors to recruit more aggressively. In 1610 the Virginia Company marketed the colony as a virtual paradise that required only a measure of human labor to flourish. But the reality was much different: Jamestown survived only with gifts of food from native tribes, the arrival of new settlers, and the imposition of strict new rules by Smith and other leaders. And the colony of Virginia prospered only with the introduction of tobacco—and the importation of slave labor.

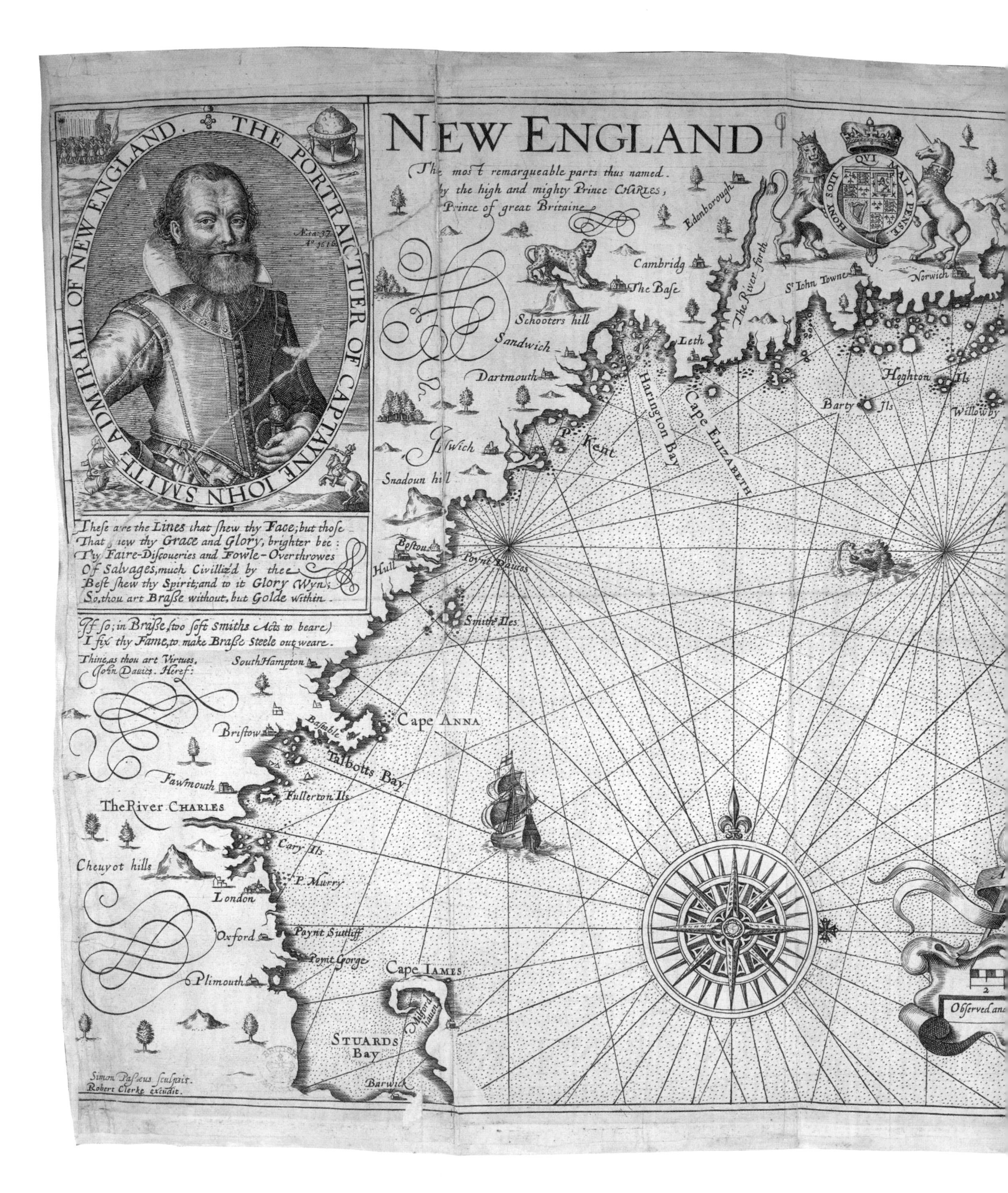

NEW ENGLAND
The most remarqueable parts thus named
by the high and mighty Prince CHARLES,
Prince of great Britaine
THE PORTRAICTUER OF CAPTAYNE IOHN SMITH ADMIRALL OF NEW ENGLAND
These are the Lines that shew thy Face; but those
That shew thy Grace and Glory, brighter bee:
Thy Faire-Discoueries and Fowle-Overthrowes
Of Salvages, much Civilliz'd by thee
Best shew thy Spirit; and to it Glory Wyn;
So, thou art Brasse without, but Golde within.
If so; in Brasse (too soft Smiths Acts to beare)
I fix thy Fame, to make Brasse Steele out weare.
Thine, as thou art Virtues,
John Davies. Heref:
Simon Pasæus sculpsit.
Robert Clerke excudit.
Edenborough
Cambridg
The Base
Schooters hill
Sandwich
Leth
The River forth
St John Towne
Norwich
Dartmouth
Harington Bay
Cape Elizabeth
Hoghton Ils
Barty Ils
Willowby
P. Kent
Ipswich
Snadoun hill
Boston
Hull
Poynt Dawes
Smiths Iles
SouthHampton
Cape ANNA
Bristow
Bastable
Talbotts Bay
Fawmouth
Fullerton Ils
The River CHARLES
Cary Ils
Cheuyot hills
P. Murry
London
Oxford
Paynt Suttliff
Poynt Gorge
Cape IAMES
Plimouth
Milford haven
STUARDS Bay
Barwick
Observed and

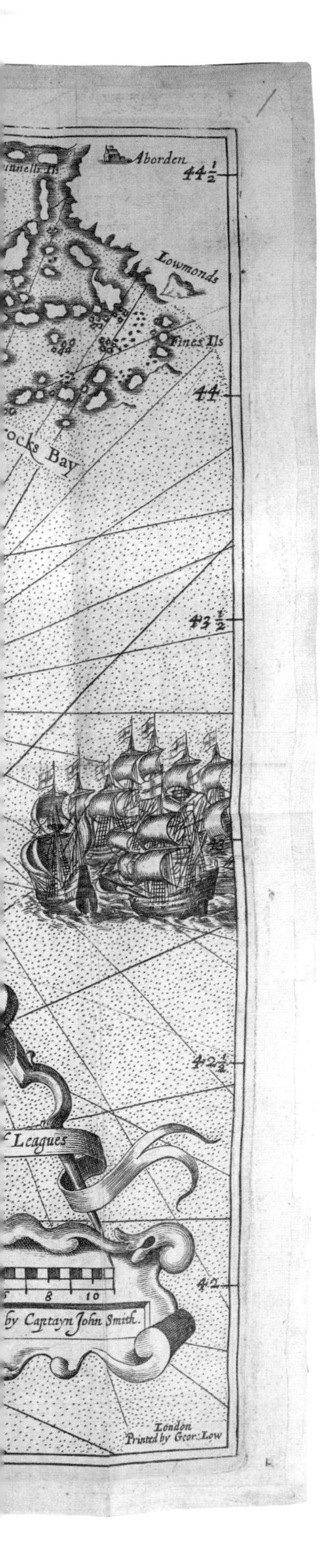

THE INVENTION OF NEW ENGLAND

John Smith, "New England," 1616

Early English attempts to settle the Chesapeake were risky affairs. The failures at Roanoke, followed by the desperation at Jamestown, chastened investors and dampened the enthusiasm of potential emigrants. One of the leaders of the Jamestown colony, Captain John Smith, returned to England in 1609 with a mixed record of success, and for a time actually distanced himself from the entire enterprise in Virginia.

Yet the commercial lure of the New World remained. In 1614 Smith joined a whaling voyage in search of gold and copper mines along the northeastern coast of America. While the sailors found neither mines nor whales, the experience convinced Smith of the potential for settlement in what was then referred to by some as "North Virginia," the region stretching from the Hudson River to Penobscot Bay. More specifically, he wondered whether he could market this forbidding and frozen coast as a hospitable emigrant destination and a worthwhile investment.

Back home in England, Smith began to brand the region in familiar and inviting terms. There was already a New Spain and a New France, so Smith designated this "New England." He then asked the heir to the throne, Prince Charles, to propose English place names to replace the indigenous ones. Through naming and mapping, Smith took the first step in creating a regional identity that endures to this day, woven through not just the geography of New England but also its cultural landscapes.

Smith's map announced that regional coherence. His portrait looms over a coastline dotted with English names, depicting an established and known landscape when in fact no English settlements existed at all. Of these names, the Charles River and Cape Ann(a) were among the few that endured, a reminder of just how fragile this venture really was. An English flotilla at right suggests that the emigration is underway, while the absence of natives frames the land as vacant. Along with Smith's large portrait and the royal coat of arms, the map conveys settlement as a sure bet. His map invited Englishmen to see this land as an extension of their own and one that they could similarly own.

Smith wrote a pamphlet to accompany the map that similarly extolled the commercial potential of "New England": fish and game would sustain the colonists, and ideal growing conditions welcomed those with energy but little money. Smith rushed his pamphlet and map to print by June 1616 to capitalize on the unexpected visit of Pocahontas to England. The map and the pamphlet circulated widely, advertising the colony as both a destination and an investment. A year later, a smallpox epidemic, most likely introduced by European traders or fishermen, ravaged the local tribes. Though natives in New England numbered as many as 100,000 when the Puritans arrived, smallpox vastly reduced the population along the coast. Some subsequent migrants even read this plague as providential, a gift from God designed to clear the way for the English.

In 1620 King James issued a patent for settlement in this "New England," confirming the name that Smith had coined. The same year, a group of religious separatists living in Leiden petitioned to settle further south, along the Hudson River. Blown off course, they landed in Plymouth, at the lower left edge of the map. There they made their home and paved the way for thousands more who arrived in the "Great Migration" of Puritans from 1630 to 1642. John Winthrop was among the first of these, and for the next twelve years he served as governor of the Massachusetts Bay colony. In that time over 12,000 traveled across the Atlantic to settle in Winthrop's "city upon a hill," driven less by Smith's vision of commercial profit than by the hope of exercising religious freedom and escaping the repression that had worsened under King James' successor, Charles.

The Puritan plan was ambitious: to establish a "Christian commonwealth" that would purify the Church of England and set an example for Englishmen back home. The Puritans' sense of themselves as a chosen people determined the structure and organization of the colony, which suppressed dissent, limited voting rights, and disapproved of extravagance. In many ways New England life was rigid, but because most of these emigrants came as families their numbers grew quickly. Though the adoption of tobacco made Virginia a wealthier colony, New England's prosperity was more evenly distributed among a much larger population. The growth of both of these colonies would have profound consequences for the indigenous peoples.

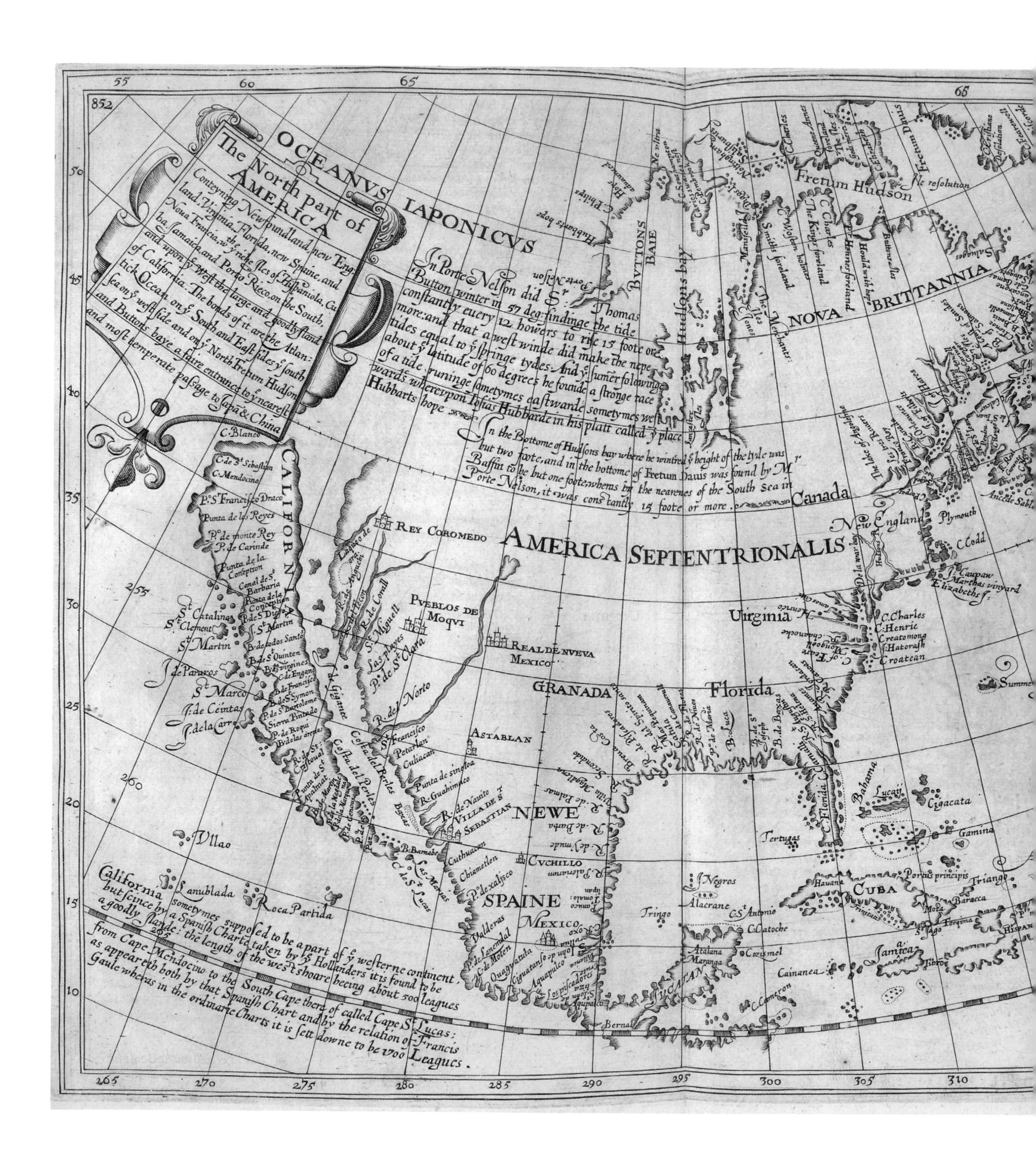

OCEANVS IAPONICVS
The North part of AMERICA
Conteyning Newfoundland, new England, Virginia, Florida, new Spaine, and Noua Francia, wth ye riche Iles of Hispaniola, Cuba, Jamaica and Porto Rico, on the South, and upon ye West the large and goodly Iland of California. The bonds of it are the Atlantick Ocean on ye South and East sides ye south sea on ye west side and on ye North Fretum Hudson and Buttons baye a faire entrance to ye nearest and most temperate passage to Japā & China
In Porte Nelson did Sr Thomas Button winter in 57 deg: findinge the tide constantly euery 12 howers to rise 15 foote or more: and that a west winde did make the nepe tides equal to ye springe tydes. And ye sumer folowinge about ye latitude of 60 degrees he founde a stronge race of a tide runinge sometymes eastwarde sometymes westwards wherevpon Iosias Hubbarde in his platt called ye place Hubbarts hope
In the Bottome of Hudsons bay where he wintred ye height of the tyde was but two foote, and in the bottome of Fretum Dauis was found by Mr Baffin to be but one foote, wheras in Porte Nelson, it was constantly 15 foote or more.
California sometymes supposed to be a part of ye westerne continent, but scince by a Spanish Charte taken by ye Hollanders it is found to be a goodly Islande: the length of the west shoare beeing about 500 leagues from Cape Mendocino to the South Cape there of called Cape St Lucas: as appeareth both by that Spanish Chart and by the relation of Francis Gaule wheras in the ordinarie Charts it is sett downe to be 1700 Leagues.
AMERICA SEPTENTRIONALIS
CALIFORNIA
BUTTONS BAIE
Hudsons bay
Fretum Hudson
NOVA BRITTANNIA
Canada
New England
Plymouth
Virginia
Florida
GRANADA
REY COROMEDO
PUEBLOS DE MOQUI
REAL DE NUEVA MEXICO
ASTABLAN
NEWE SPAINE
Mexico
CUBA
Bahama
Jamica
Hubbarts hope
Porte Nelson
C. Blanco
C. Mendocino
Roca Partida
Vllao

THE LURE OF A NORTHWEST PASSAGE

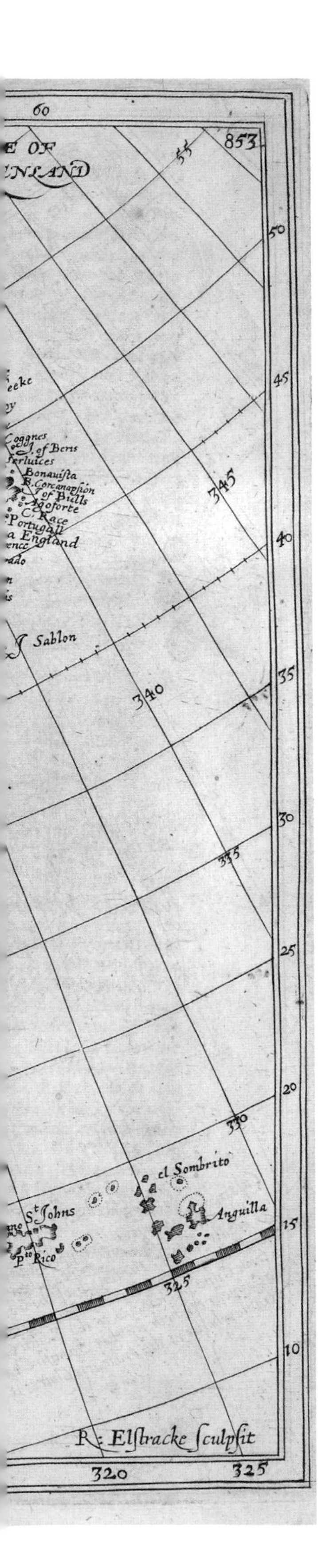

Henry Briggs, "The North Part of America," 1625

The early Jamestown settlers struggled to survive. But within a decade of its founding the Virginia colony had begun to stabilize, in part as a result of the cultivation of tobacco. Relations with the Powhatan Confederacy, however, remained hostile, and in 1622 Opechancanough (brother of Chief Powhatan) led an attack that killed more than 300 colonists along the James River. In response, the Virginia Company sought to strengthen the colony by advertising the region's climate and geography, and offering land to those willing to make the passage. Virginia was framed as a guaranteed investment for anyone willing to exert even a bit of effort, yet the deteriorating relationships with Native Americans remained an obstacle to growth.

Among the investors in the Virginia Company was the mathematician Henry Briggs, who made audacious claims in order to stimulate both investment and emigration to the colony. In 1622 he published descriptions of a navigable waterway that linked Hudson Bay to the Pacific Ocean, renewing dreams of a Northwest Passage. But geographical knowledge of the continental interior remained extremely limited. The recent expeditions of Samuel de Champlain and others up the Saint Lawrence Seaway had begun to reveal the enormous hydrographic system between the Atlantic Ocean and the Great Lakes. Yet the geography to the south and west of the Great Lakes remained a mystery, leaving plenty of room for both wishful thinking and outright fabrication.

Briggs imagined a geography that would link the Virginia colony to the Pacific Ocean. He knew that the headwaters of rivers flowing into the Chesapeake Bay were located in the mountains west of Virginia. For this reason, he carefully marked each of the rivers that flow into the Chesapeake, believing that a short portage from those headwaters over the mountains would lead to others flowing north to the Great Lakes and Hudson Bay. From there he suggested that several rivers ran west to the Pacific. In 1625 he sketched this geographical vision on the first printed English map of North America, shown at left.

What led Briggs to believe that North America could be so easily traversed? The news of Champlain's travels might have stimulated his imagination, for the Great Lakes extended far into the western interior. Briggs also claimed that Native Americans in the east had reported seeing European ships in western waters. All of this generated a picture in his mind of a relatively narrow continent.

This rosy geographical view explains why Briggs named the imaginary river flowing west to the Pacific "Hubbard's Hope": travelers navigating Hubbard's Hope west out of Hudson Bay would find themselves "very near as far toward the west as the Cape of California, which is now found to be an island." Briggs was the first to depict California as an island, propelling a myth that continued for much of the seventeenth century. While he had a stake in the Virginia colony, it is not clear whether Briggs was unwittingly or deliberately deceptive in presenting these geographical fables.

Briggs' map is best seen as a measure of contemporary geographical knowledge and emerging imperial rivalries. His goal was to find a passage to the Pacific and to "all those rich countries bordering upon the South Sea". Such a discovery through British territory would enrich and empower the English, not to mention Briggs himself. To his mind, that journey across North America ought to avoid "Newe Spain" and the missions of the Southwest, including the recently founded town of Santa Fe, which is marked as "Real de Nueva Mexico" on the map. Instead, Briggs recommended that the English cross the continent further north, taking a more temperate and "wholesome" journey "through the continent of Virginia," then via Hudson Bay and "Nova Brittania." The absence of information about the interior and the far West gave Briggs plenty of room to make claims that advanced English interests in the New World.

THE FRENCH EXPLORE AMERICA

Samuel de Champlain, "Le Canada," 1653 [1616]

The map of "The North Part of America" by Henry Briggs on the previous page framed the hope of a Northwest Passage through an English imperial lens. The French were equally committed to finding a route across the continent to the Far East. The first wave of French explorers in the sixteenth century focused on the Saint Lawrence Seaway, which carried them well into the interior. The second wave was led by Samuel de Champlain, who crossed the Atlantic dozens of times between 1598 and 1633. He sailed down the Atlantic coast to Cape Cod in 1605 and 1606, well before Henry Hudson explored New York Harbor or the Puritans settled Plymouth.

The French goals in the New World were commercial networks rather than settled colonies. To this end, Champlain was sent back to North America in 1608 to found and fortify Quebec as a trading post. With only twenty-eight men, he had no choice but to develop alliances with the local tribes. In 1609 he ventured south to explore what would become upstate New York and Vermont, including the large and narrow lake between those states that later bore his name. In his most challenging trek of 1615, he traveled 700 miles up the Ottawa River, which took him well into Iroquois territory.

All these travels brought him into regular and sustained contact with the tribes of the Great Lakes, and he formed especially strong relationships with the Algonquin, Huron, and Montagnai tribes. These alliances gave him unrivaled access to indigenous geographical knowledge, but they also necessarily pitted him against the Iroquois Confederacy. Once he returned to Paris in September 1616, Champlain combined this indigenous knowledge with his own reconnaissance to draw one of the earliest European maps of the Upper Midwest and Far Northeast. The level of detail on Champlain's map far exceeds that of Henry Briggs, and is just one indication of the superiority of French geographical knowledge at that time.

Champlain intended to publish the map in a 1619 volume detailing his voyages. But the map was not published, and the engraving lay unused until it was rediscovered in the 1650s by Pierre DuVal. After adding new information that had been accumulated in the interim, DuVal printed the map in 1653. The result is a picture of contemporary geographical knowledge that also captured emerging imperial claims. A portion of that map is enlarged here, with the entire map reproduced on the next page.

Champlain's geographical contribution grew from his extensive exploration of the Saint Lawrence Seaway and its tributaries, as well as the information he gathered from Native Americans. In 1615, he set out upriver from Quebec, marked as a fort in red along the Seaway. At the Isle of Montreal, numbered 32 on the detail here, he moved up the delta marked "R. des Prairies ou des Algonquins." From there he continued—slowly—up what would be named the Ottawa River, identified with the number 8 on the detail. This trek ultimately brought Champlain southwest to Lake Huron, named "Mer Douce" on the left edge of this image. Champlain's was the earliest European record of this lake, as well as one of the earliest efforts to map Lake Ontario just to the south, named here as "Lac St. Louis."

Champlain's extensive expeditions formed the foundation—and the limits—of this map. On the next page, notice that Champlain ends the map in the west in the middle of a lake, leaving open the possibility of a passage to the Pacific Ocean. This also reveals that he had no knowledge of Lake Michigan. Moreover, to map other areas that he had not seen firsthand, Champlain relied on Native American knowledge, such as the area around St. James Bay and portions of the Great Lakes. More generally, the presence of native tribes is prominently acknowledged throughout the map. From the "Nations du Nort" to the Iroquois "Nation du Chat" just south of Lake Erie, Champlain took pains to identify the many tribes that inhabited the greater region, and shaped his own experience in North America.

Champlain served as the de facto governor of New France until his death in 1635, when the population of Quebec remained small, at about 300. In that time, great changes had taken place further south along the Atlantic. The depiction of the Chesapeake Bay closely follows that of John Smith's map of Virginia (page 40). To finish the map, DuVal identified settlements that had developed since Champlain's original engraving: New "Angleterre," New "Hollande," Virginie, Spanish Florida, and of course "New France." Soon after the map was published, the next generation of French explorers—Louis Jolliet, Jacques Marquette, and Robert de La Salle—extended the French realm even further into what would become the United States, moving south from the Great Lakes and, simultaneously, north from the Gulf of Mexico up the Mississippi River and its tributaries.

LE CANADA
faict par le Sr de Champlain.
Mansfeld
King James his forland
Destroit de Hud
ou de Christian
Hop Aduanced
Prince Henri Cap
Baye ou ont hiuerné les Anglois
S. Thomas
terres fort haultes
Smith Forland
de Cortereal
Merchäts I.
OUTVVALLES
Vestons I.
la grande Baye ou deliurance
Cap Henriette Marie
Nations du Nort
Lac St Iean ou Piagouagami
Montagnais
James his Bay
Nations du Porc Espy
Saguenay
Kiristinous P.
Bersiamites
Tavandeköate
Port neuf
Atticamegues ou Poissons Blancs
Nouvelle France ou CANADA
Grande R. de Canada
Nadouessis
Erigoüechkak Lac St Thomas ou Kisakami
Beau Port
Montz nostre dame
Chadanaquiois
Trois Riuieres
R. de l'Assompton
algommequains
S. Esprit
R. des Prairies ou des Algonguins
Ste Croix
quebec
Socoquiois
Almouchiquois
NORUMBEGA
ACCADIE
Algonquins
Grand chasse de Cerfs et de Caribous
Pemtegouet
Baye Francoise
R. de St Laurens
Saintonge
Richelieu
R. des Iroquois
N. ANGLETERRE
Abnaqui ou
Hurons
Lac St Louis ou Ontario
Socoquiois
N. Plymue
C. Blanc ou Cod
malle barre
Hirocois auec 5 Bourgades
Nation des Loups
NLLE HOLANDE
Carantouën
SUEDE
FRANCE
lieu ou sont les Flamans appellé Nau Amsterdam
Mer Oceanne
Andastaoe
R. du Sud
du Chat
Baye de Chesapeack
R. Pauuhatan
Les Virgines
VIRGINIE
Terre non Cobnüe
Iamestan lieu des Anglois.

degrez de Latitude
Ocean Septemtrional ou Glacial
Mer du costé du Nort dite Glaciale
Nouveau Danemarq
Port de Munck
Hopecheck
Buttons Bay
Briggs bis bay
Port Nelson
Cap Philippe
Mer glacialle
Nevtra
New North Walle
Cap Southa
Mer de Hudson ou Christiane
Isle de Mansfeld
Baye ou ont
New Severne R.
New Soutvvalles
Vestons I.
Cap Henriette Marie
James his Bay
Kiristinous P.
Irini Nadous P.
Nadouessis
Les puans
Paoutig8ejenhac
les Borciers
Biserens ou Nepiciriniens et Sorciers
S. Esprit
algommequains des Prairies ou
Lac des Puants ainsi appellés pour estre venus des costes d'une Mer Salée
Sault S. Pierre
Tarontorai
S. Ioseph
Mer douce ou grand Lac des Hurons, et Atig8atan lequel a flux et reflux
Algon quins
Grand chasse de Cerfs et de Caribous
Tekari endiondi
Nation du Feu
Lac Kandechio
Les cheueux releuez
Gens de petum
Hurons
Lac S. Louis ou Onta 110
Hirocois auec 5 Bourgades
Nation Neutre
Erie Lac
Asfista gueronons
Carantouën
antou=honorons
Nation du Chat
Baye de Chesapeack
Nation ou il y a des Buffles
Virginie
Terre non encores bien descouuerte Continente a la Floride.
Floride
Pome
BRITISH MUSEUM 12 JU 1871
Explication des lettres et Chifres de la Carte
A C breton
B C de raye
C C de raze
D C S. laurens
F antiscosty
E C de gaspay
G baye de chaleur
H petit passage
I mantane
K le bic
L lesquemain
M tadoucac
N R du saguenay
O quebec
P la grande baye
R passage du nort
S le grandbanq
T banquereaux
V Ille de sable
X illes S. Pierre
Y Canpzeau
Z illes ramees
2 lac de champlain
3 Ille dorleans
4 lac S. Pierre
5 trois riuieres
6 Sault S. Louis
7 lac des bicerenis
8 riuiere de mont merency
9 sesembre
10 la heue
11 Cap de sables
12 port Royal
13 Cap des deux baye
14 R S. Iehan
15 R S.te Croix
17 R de pemetegoit
18 R de quimibequy
19 R de chouacoit
20 beau port
21 menane
22 ille per see
23 ou les sauuages uoient la mer
24 ou dautres sauuages uont a la mer
25 ille ou il y a une mine de cuiure
26 ille aux oiseaux
27 illes des bacallos
28 Croix blanches
29 belleisle
30 Cap de S.te Marie
31 Iamestan lieu des Anglois
32 lille de uilmenon
33 R de reuillon
34 R destarjon

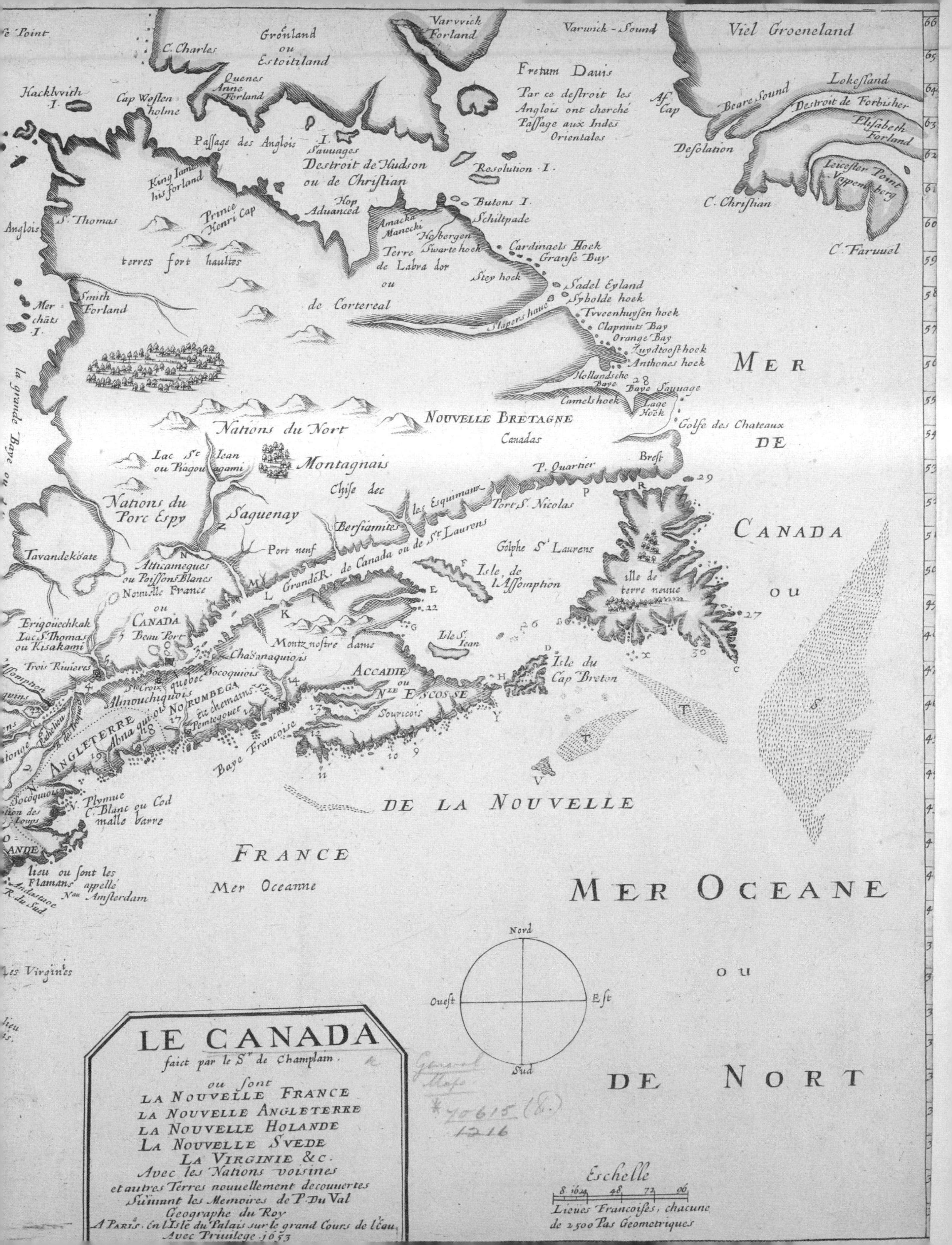
LE CANADA
faict par le Sr de Champlain
ou sont
LA NOUVELLE FRANCE
LA NOUVELLE ANGLETERRE
LA NOUVELLE HOLANDE
LA NOUVELLE SVEDE
LA VIRGINIE &c.
Avec les Nations voisines
et autres Terres nouuellement decouuertes
Suiuant les Memoires de P. Du Val
Geographe du Roy
A Paris. en l'Isle du Palais sur le grand Cours de l'eau
Avec Priuilege 1653
Eschelle
Lieues Francoises, chacune de 2500 Pas Geometriques
MER DE CANADA OU
MER OCEANE OU DE NORT
DE LA NOUVELLE FRANCE
Mer Oceanne
NOUVELLE BRETAGNE
Canadas
Nations du Nort
Montagnais
Saguenay
Golphe St Laurens
Grande R. de Canada ou de St Laurens
Isle de l'Assomption
Isle de terre neuve
Isle du Cap Breton
ACCADIE ou Nle ESCOSSE
ANGLETERRE
Nouuelle France ou CANADA
Fretum Dauis
Par ce destroit les Anglois ont cherché Passage aux Indes Orientales
Destroit de Hudson ou de Christian
Viel Groeneland
Destroit de Forbisher
Terre de Labra dor ou de Cortereal
terres fort haultes
Nord
Sud
Ouest
Est

THE ORIGINS OF THE ATLANTIC SLAVE TRADE

Hugo Allardt, "Effigies ampli Regni auriferi Guineae in Africa siti," circa 1650

It may seem odd to include a map of West Africa in a history of North America, but the connection is inextricable. The survival of Virginia came with the cultivation of tobacco, which depended upon the importation of slave labor. This places slavery at the very heart of the American experiment. The early tobacco farmers used slaves as early as 1619, and in the 1620s the Dutch founders of New Netherland brought slaves with them. By the time the slave trade ended in 1867, over 12 million Africans had been forcibly removed from their native land and most of them sent to the Americas. The peak of the trade occurred in the eighteenth century, when Africans shipped to the Caribbean and South America were sold into slavery in North America.

Though the African slave trade was launched by the Portuguese in the fifteenth century, it was the subsequent entry of the Dutch that coincided with the early settlement of America. The Dutch West India Company was founded in 1621 and initially focused on ivory, gold, and pepper. Soon the company began to export slaves to supply the growing demand for labor on the Dutch sugar plantations of northeastern Brazil. That slave trade concentrated along the southern coast of Guinea, here colored in pink at right. In 1637, the Dutch captured the Portuguese trading post at Elmina, near the cape marked "C. Corco" on the map. For the next two decades, the Dutch dominated the Atlantic slave trade. This human traffic formed one leg of the profitable "triangle trade" between Africa, America, and Europe.

The elegance and beauty of this map belies its deadly serious intent. It was published about 1650 by the engraver Hugo Allardt to celebrate the Dutch victory over Portugal in West Africa. Indeed, it was based on a Portuguese map made fifty years earlier. Large pictorial insets depict a native dance and a procession of leaders. These insets also displace the interior of the continent and draw attention to the African coast, where Dutch commercial interests were peaking. Allardt used color to distinguish the discernible "nations" of the region, reproducing divisions that had appeared on earlier European maps of Africa. He also identified the enclaves and villages of the interior, perhaps indicating the potential for further commercial growth and trade networks. Dutch ships sail along the coast, while two small Portuguese flags mark the forts that were now under Dutch control.

By the 1660s, the Dutch were transporting between 5,000 and 7,000 Africans across the Atlantic each year. Most were taken from the area east of the Volta River, near the "Costa Adra" that borders Benin. These slaves were used to build and cultivate enormous sugar plantations in Brazil and the West Indies. In North America, slavery grew more slowly because of its high cost and the continued reliance on indentured servants from Europe. In the tobacco fields of the Chesapeake, African slaves formed a minority of the labor force until 1680. Thereafter, Virginia planters began to rely more upon slavery and the English began to displace the Dutch in Africa, just as the Dutch had displaced the Portuguese decades earlier.

The increased dependence upon slavery led the colonists to pass laws to codify the practice. In 1662 the Virginia General Assembly determined that slave status would be defined by one's racial identity, and "that all children borne in this country shall be held bond or free only according to the condition of the mother." Such a condition ensured that slavery would reproduce itself in the colonies, and that bondage would be defined by race. Five years later, the General Assembly established that Christian baptism would not free children from bondage, reassuring masters that that they could evangelize without fear of losing their slaves. By 1700 slaves had entirely replaced indentured servants on the tobacco plantations of Virginia. But slaves were not confined to the Virginia colony: by 1650 there were more slaves in Dutch New Netherland than in the Chesapeake. Slavery continued to grow after the English rechristened the colony "New York," and by the 1740s black slaves made up 20 percent of its population.

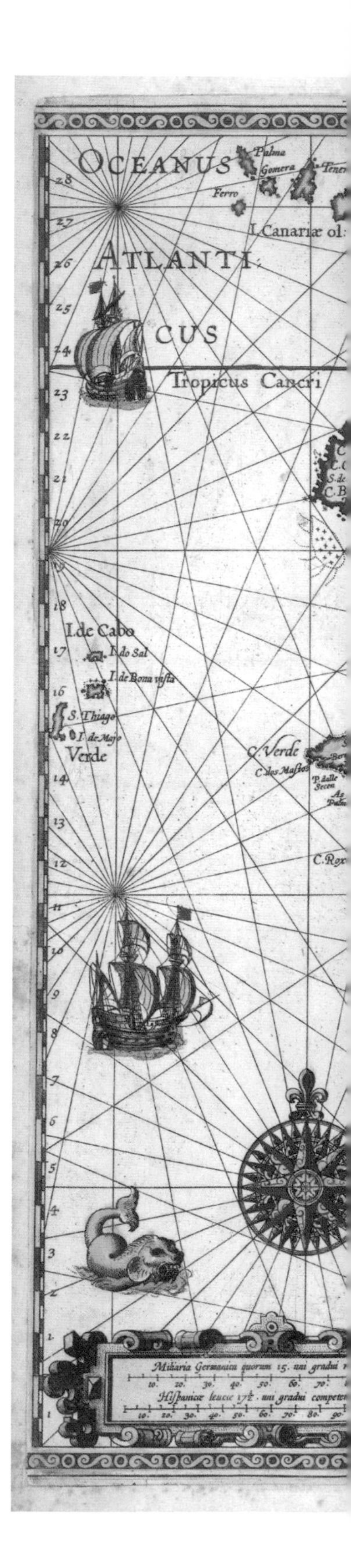

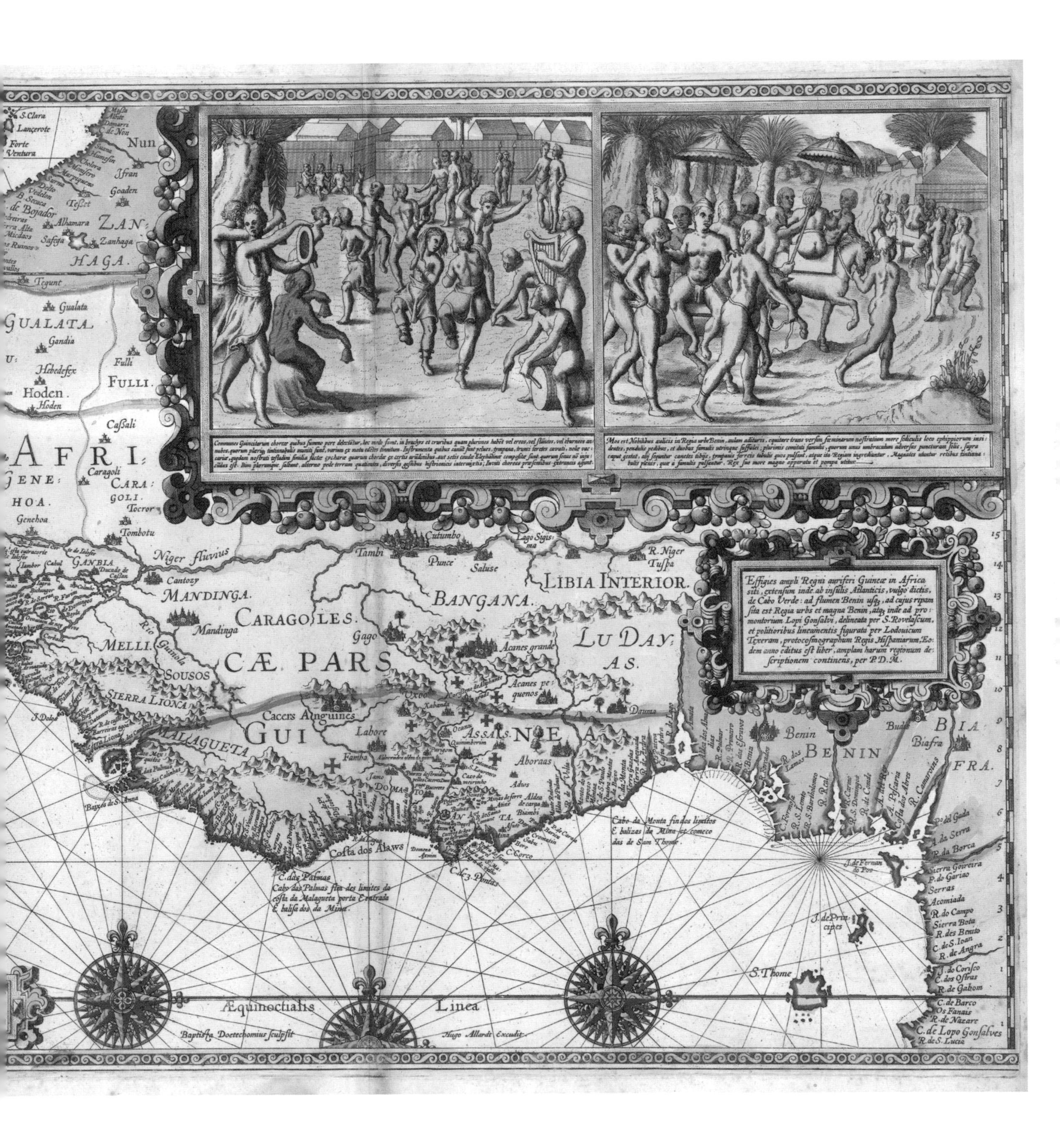
LIBIA INTERIOR
BANGANA
CARAGOLES
CÆ PARS
GUI NEA
MALAGUETA
SIERRA LIONA
MANDINGA
MELLI
BENIN
Niger fluvius
Æquinoctialis Linea
Baptista Doetechomius sculpsit
Hugo Allardt Excudit

I'LL TAKE MANHATTAN

Robert Holmes, "A Description of the Towne of Mannados or New Amsterdam as it was in September 1661," 1664

We generally think of New York as an English settlement, but this overlooks its earlier Dutch history. In 1609 the Englishman Henry Hudson contracted with the Dutch to search for a Northwest Passage by sailing up the river that later bore his name. The New Netherland Company sent another explorer a few years later to claim all territory between Virginia and New England as "New Netherland." In 1624 the Dutch began to settle a small part of that as "New Amsterdam." Yet the English never recognized these Dutch claims.

The claims of the Dutch were indeed tenuous, for New Netherland was sandwiched between more densely populated English colonies to the north and south. Moreover, the colony was primarily geared to the fur trade, making it less densely settled than other coastal colonies. The short supply of Dutch settlers drove the West India Company to invite Belgians, English, French, Germans, and Scandinavians into the colony. This also enhanced its tolerant culture, which was home to Puritans as well as Catholics. To make up for a labor shortage, the Dutch also began to import slaves. By the 1660s, free and enslaved blacks made up nearly a quarter of the local population of 1,500 Europeans, 300 slaves, and 75 free blacks.

Ongoing conflict with the English on the seas led to the "conquest" of New Amsterdam by the English. Once restored to the throne, King Charles II granted New Netherland—and the island of Manhattan—to his brother James. This exuberant map was drawn by Commodore Robert Holmes to commemorate England's seizure of the island. The map is oriented horizontally, with north at lower left. Holmes used lively color and a bird's-eye approach to celebrate this new English colony, now renamed in honor of James, duke of York. He crowded the rivers and harbors with English warships to mark the moment in September 1664 when his and other British squadrons forced the Dutch governor Peter Stuyvesant to surrender to Colonel Richard Nicolls. Holmes himself arrived in the harbor from the African coast, where he had been fighting the Dutch for control of the slave trade. Soon the English would displace the Dutch in West Africa, just as they had in North America.

With its selective detail and pictorial appearance, the map reveals little about the colony itself. But there are a few clues to life on Manhattan: the governor's house is marked at the southern tip of the island. Made of white stone, it eventually gave name to "Whitehall" Street, which persists to this day. In 1643, the Dutch West India Company retreated to the southern end of the island and erected a wall of protection against Native American attacks. This barricade eventually became "Wall Street," matched by a protective "Battery" to the right, which would become Battery Park. In 1664 these Dutch defenses made little difference, for the island was taken without a shot. The river along the lower (west) side of the island bore twenty different names before English control entrenched the name as the Hudson.

By the end of the seventeenth century the influx of English and French settlers had changed the character of New York. Yet, like Wall Street and the Battery, the Dutch persisted, reasserting their ethnic identity rather than being absorbed into the larger English culture. In fact, their presence—like that of a sizable African American community—made for an unusual level of colonial diversity. English toleration of Dutch culture included respect for religious practices as well as for property claims. Stuyvesant himself chose to remain in New York until his death in 1672, and the quaintly rendered windmill at the lower right corner of Manhattan anticipates this enduring Dutch presence. If the bold narrative of the map symbolizes the larger shift of imperial power toward the British, it also reveals the underlying pluralism and limited ethnic tolerance that would come to define the colony.

VIOLENCE AND DEVASTATION IN EARLY NEW ENGLAND

William Hubbard and John Foster, "A Map of New-England, Being the first that ever was here cut," 1677

In June 1675 English settlers in Plymouth hanged three Wampanoag Indians who were suspected of murdering a Christian Indian earlier that year. This event came after waves of European immigrants had swelled the population of New England. In response, the Wampanoag leader King Philip (also known as Metacomet) allied with the Narragansett Indians to attack English settlements, launching a vicious war that gripped southern New England for fourteen months in 1676–7.

King Philip's War remains the most destructive conflict in American history relative to the population. As such, it spawned several contemporary accounts, including that of William Hubbard, a minister from Ipswich. His "Narrative of the Troubles with the Indians in New-England" circulated widely, and distinguished itself with a woodcut map that was the first to be printed in English America. This remains one of the only surviving images from seventeenth-century New England.

Hubbard's map both documents and unwittingly explains the causes of this brutal war. It is oriented with north at the right, so that the Connecticut River flows horizontally along the top of the page. That in itself is revealing, for the westward growth of English settlement to the river significantly encroached on native lands. The new towns—Hartford, Springfield, and Northampton—indicated just how much was changing in seventeenth-century New England. By mapping the expansion of the Massachusetts Bay colony, Hubbard captured the territorial tensions that exploded in King Philip's War. Two dark vertical lines mark the northern and southern boundary of the Massachusetts Bay colony, while the lighter angled line separates out the Plymouth Colony. These boundaries also demonstrate that colonists brought their own conceptions of land—to be parceled, surveyed, and owned—to America.

Hubbard designed the map to document the Indian attacks on English settlements, numbering each of the towns to correspond to notes in his narrative. Yet he omitted those places where the English attacked the Indians, which left readers with a curiously one-sided view of the conflict. Just as revealing is the way Hubbard mapped human geography. English villages are identified by churches or houses, icons that signify civilization. By contrast, native settlements are represented by trees, reflecting an assumption that they were an extension of nature and the landscape itself. Ironically, there is evidence that Indian knowledge influenced Hubbard's map, for the stylized and oversized rendering of Lake Winnipesaukee—littered with islands—evokes native techniques of representing the landscape in a way that does not always correspond to scale.

Hubbard's larger purpose in the narrative was to argue that the war was caused not by the declining faith of the Puritans, but by the failure of the Indians to embrace Christianity. His map shows us a colonist's perspective of the conflict, wherein Christian settlers lived in constant fear of attack. Yet it was the very success of the colonies—edging westward into the wilderness—that put them in tension with indigenous tribes. Natives found themselves vulnerable and increasingly unable to protect their land. With little aid from England, the colonists began to forge a new identity, one grounded in their particular geography and circumstances. They became, in other words, less English and more American.

The terror represented by this map also contrasted sharply with the relatively peaceful settlement led by William Penn in the 1680s (see pages 56–59). The Indian tribes to the south, however, had been weakened for years by disease and prior European contact, which made it easier for Penn to enter into treaties that essentially vacated the area around Philadelphia. Hubbard's map of New England recorded the fundamental conflict and displacement of natives that accompanied every instance of European settlement in the New World. By the nineteenth century, Indians would be hard to find on American maps: they had been removed to reservations or completely erased. It seems entirely fitting, then, that the first map ever made in the British colonies would document the contest between natives and whites over control of the land: the first map made in America was a map of war.

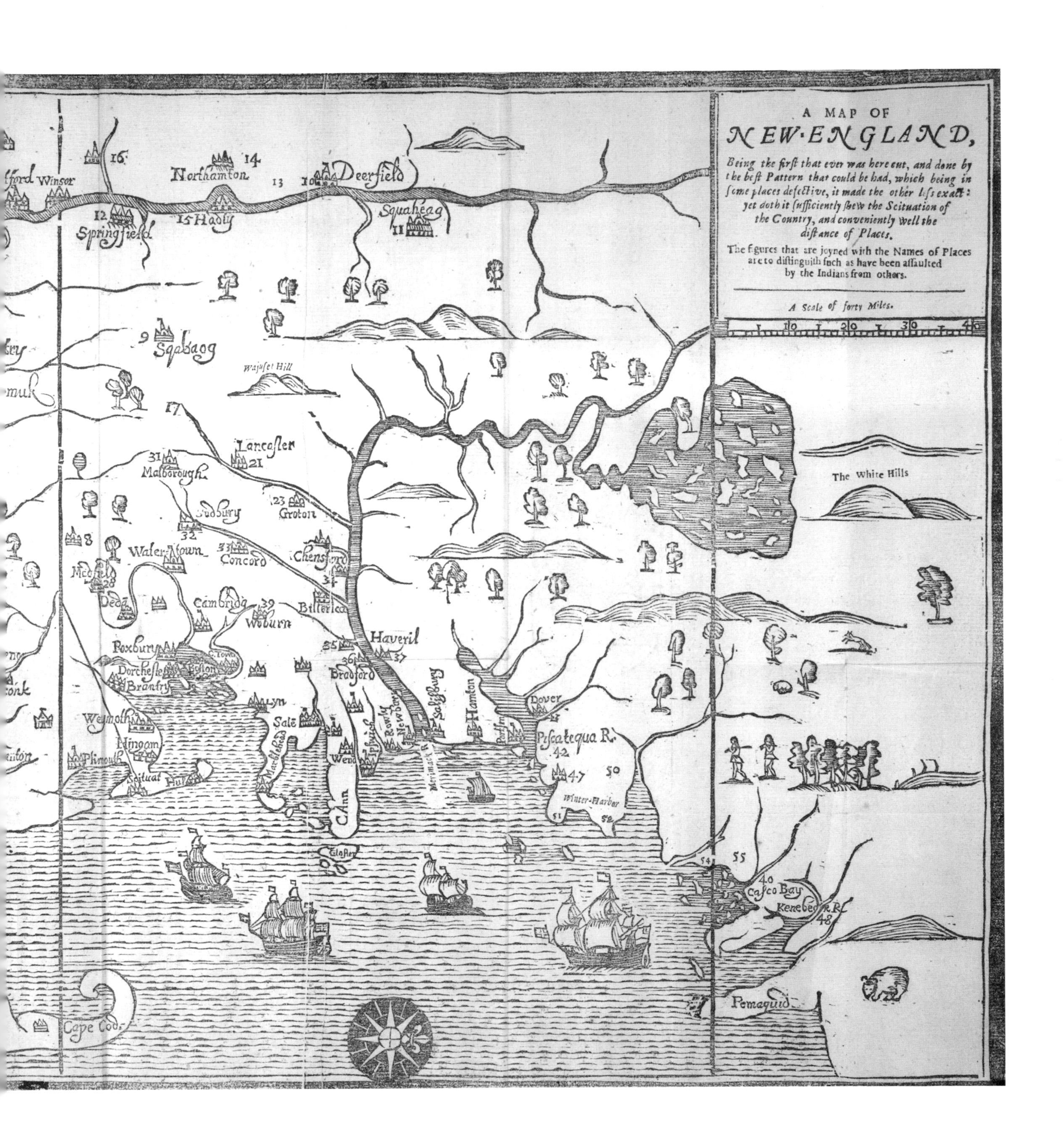
A MAP OF
NEW·ENGLAND,
Being the first that ever was here cut, and done by the best Pattern that could be had, which being in some places defective, it made the other less exact: yet doth it sufficiently shew the Scituation of the Country, and conveniently well the distance of Places.
The figures that are joyned with the Names of Places are to distinguish such as have been assaulted by the Indians from others.
A Scale of forty Miles.
10 20 30 40
Winsor
16
Northamton
14
13
10
Deerfield
12
Springfield
15 Hadly
Squaheag
11
9
Wajuset Hill
17
Lancaster
21
31
Malborough
23
Groton
Sudbury
32
8
Water-Town
33
Concord
Chensford
34
Medfield
Cambridg
39
Woburn
Billerica
Dedham
Haveril
35
37
36
Roxbury
Dorchester
Boston
Brantry
Bradford
Lyn
Salem
Weymoth
Hingam
Plimouth
Situat
Hull
Marblehead
Ipswich
Rowly
Newbery
Salisbury
Hamton
Dover
Piscataqua R.
42
47
50
Winter-Harbor
51
52
C. Ann
Merimack R.
The White Hills
54
55
40
Casco Bay
Kenebec R.
48
Pemaquid
Cape Cod

PENN'S HOLY EXPERIMENT

Thomas Holme, "A Portraiture of the City of Philadelphia," 1683, and "A Map of the Province of Pennsilvania," circa 1687

To pay off an old debt, in 1681 King Charles II made William Penn the sole proprietor of 45,000 acres of land north of Maryland and west of the Delaware River. This astonishingly powerful charter gave Penn tremendous influence over the organization and settlement of a region nearly the size of England itself. Penn was known as an advocate of religious freedom as well as a successful real-estate developer, and these two experiences directly shaped his vision of an expansive Quaker colony in America.

Penn envisioned a "Holy Experiment" that would embody democracy, economic opportunity, and religious freedom. Within months he had distributed land to about 250 "first purchasers," and that fall he sent commissioners to organize the colony, allot the grants, and site the capital city. Soon thereafter, he dispatched his fellow Quaker Thomas Holme as surveyor general of the new colony. Holme arrived in April 1682 to survey and map the new city of Philadelphia on an orderly grid, enthusing to investors back home that "such a Scituation is scarce to be parallel'd." Located between the Delaware and the Schuylkill rivers, the site was already home to a few natives and white settlers, but for the most part it remained a forest.

By the end of 1682, Holme had finalized the new street plan shown at right. His orderly grid—"two Miles in Length and one in Breadth"—incorporated Penn's ideals: even modest parcels would have space for a garden and a small orchard, and access to one of two rivers. Four squares would anchor the city's public life, while a fifth at the center would provide space for a meeting house. The numbers on the map refer to the lots granted to those first purchasers. Even the street names reflected Penn's Quaker sensibility: upon Holme's suggestion, he named them for trees rather than after illustrious leaders. Such a harmonious and composed plan must have held particular appeal for Londoners, whose ancient city had recently been ravaged by the fire of 1666. Here was a new and rationally organized town, the first European urban plan in the colonies.

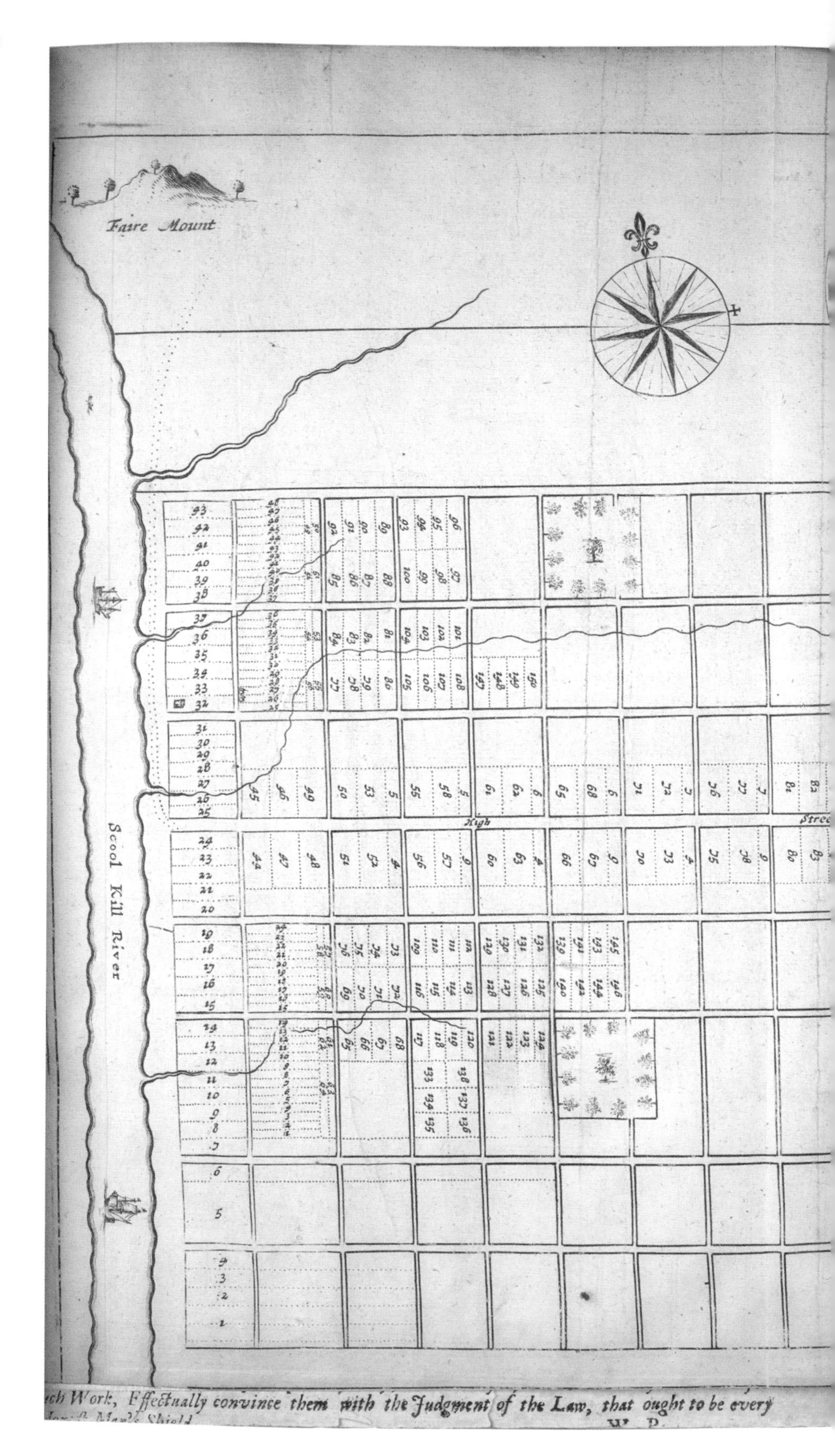

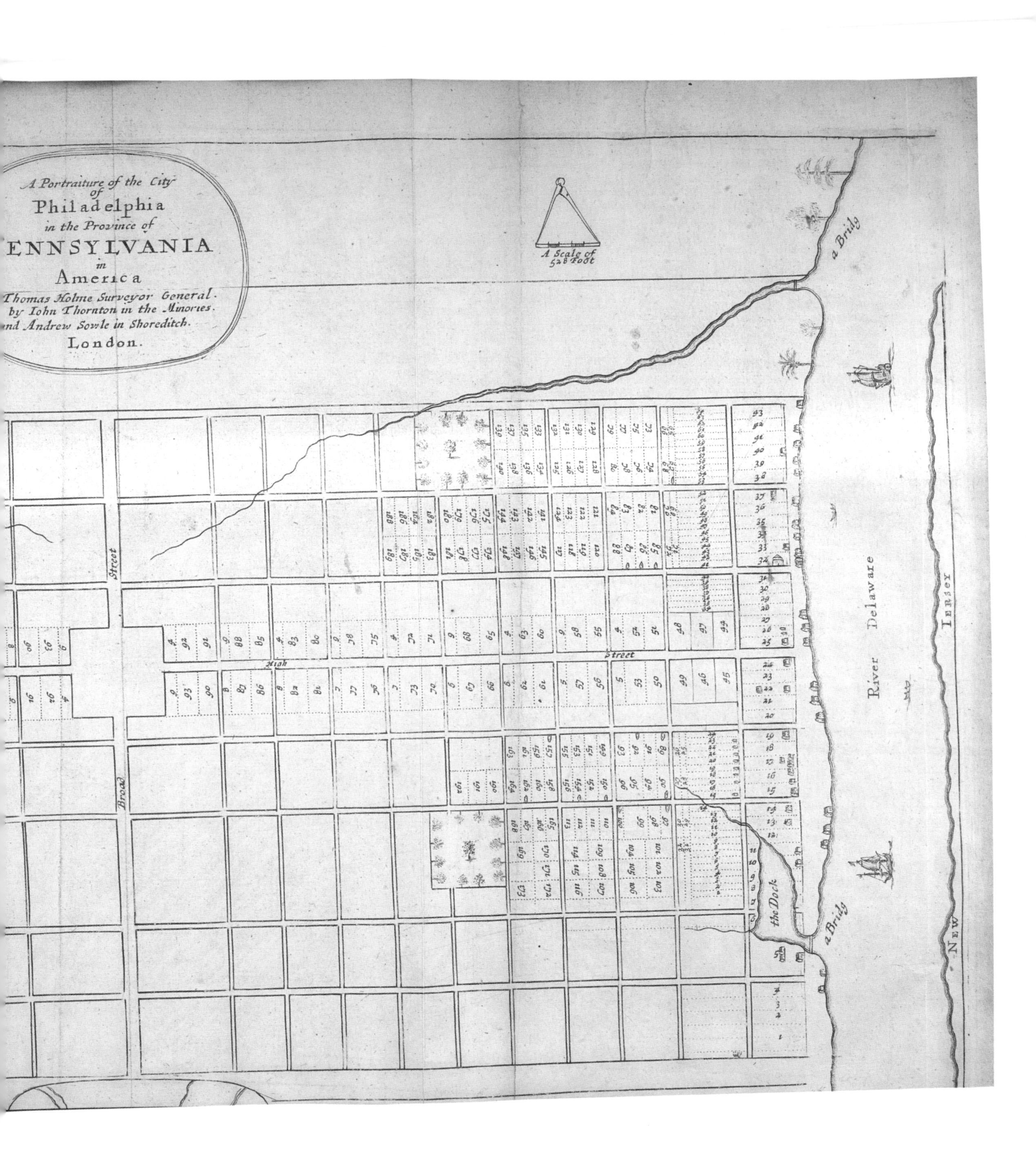
A Portraiture of the City of
Philadelphia
in the Province of
ENNSYLVANIA
in
America
Thomas Holme Surveyor General.
by Iohn Thornton in the Minories.
nd Andrew Sowle in Shoreditch.
London.
A Scale of 528 Foot
a Bridg
Broad
Street
High
Street
the Dock
a Bridg
River Delaware
New
Iersey

Upon his return to London in 1683, Penn used the Philadelphia map to advertise the colony in a pamphlet that was published in English, German, Dutch, and French. This signaled the ethnic tolerance that had already begun to shape the settlement, in sharp contrast to the relative homogeneity of the early New England towns. The following year, Penn asked Holme to create a map of the entire colony so that potential buyers—and Penn himself—could see the location of available lands.

This was easier said than done, for the pace of land grants created a welter of conflicting claims and surveys. Holme spent years sorting through these accounts. By the fall of 1686 Penn had grown impatient, writing from London with exasperation: "we want a map to the degree that I am ashamed here; ... all cry out, where is your map, what, no map of your Settlements." The surveyor responded with equal frustration, explaining that he could hardly make a map when his deputies produced such inaccurate property surveys, if they gave him surveys at all.

Holme probably finished the map in early 1687, and he gave the copy reproduced at right to Penn himself; at the time it was the most detailed map of any of the American colonies. Yet the map brought Holme even more headaches, for landholders immediately began to squabble about boundary claims and insufficient land grants. No doubt the surveyor was at the end of his rope, having been forced to produce a map with insufficient information, which only compounded property disputes in a region that had been settled so rapidly.

The map itself also captures the character of the new colony. Holme provided little topographic detail, focusing instead on the property divisions of human settlement. This included some 670 settlers with individual land grants that ranged from 125 to 5,000 acres. Where Penn had envisioned an orderly set of communities organized around central villages, within five years an avalanche of grants had overwhelmed his plan. Towns grew ad hoc, emerging out of local needs rather than following Penn's master plan. Yet the overall pattern—one that rejects the European tradition of agricultural villages and the power of central churches—also reflected Penn's ideals of tolerance, diversity, and the entrepreneurial spirit. The "Welch Tract," "Dutch Township," and "German Township" indicate the ethnic diversity that would characterize Pennsylvania. In different ways, the Pennsylvania map and the Philadelphia grid capture the sense of potential that made Penn such a successful father of the colony that bears his name.

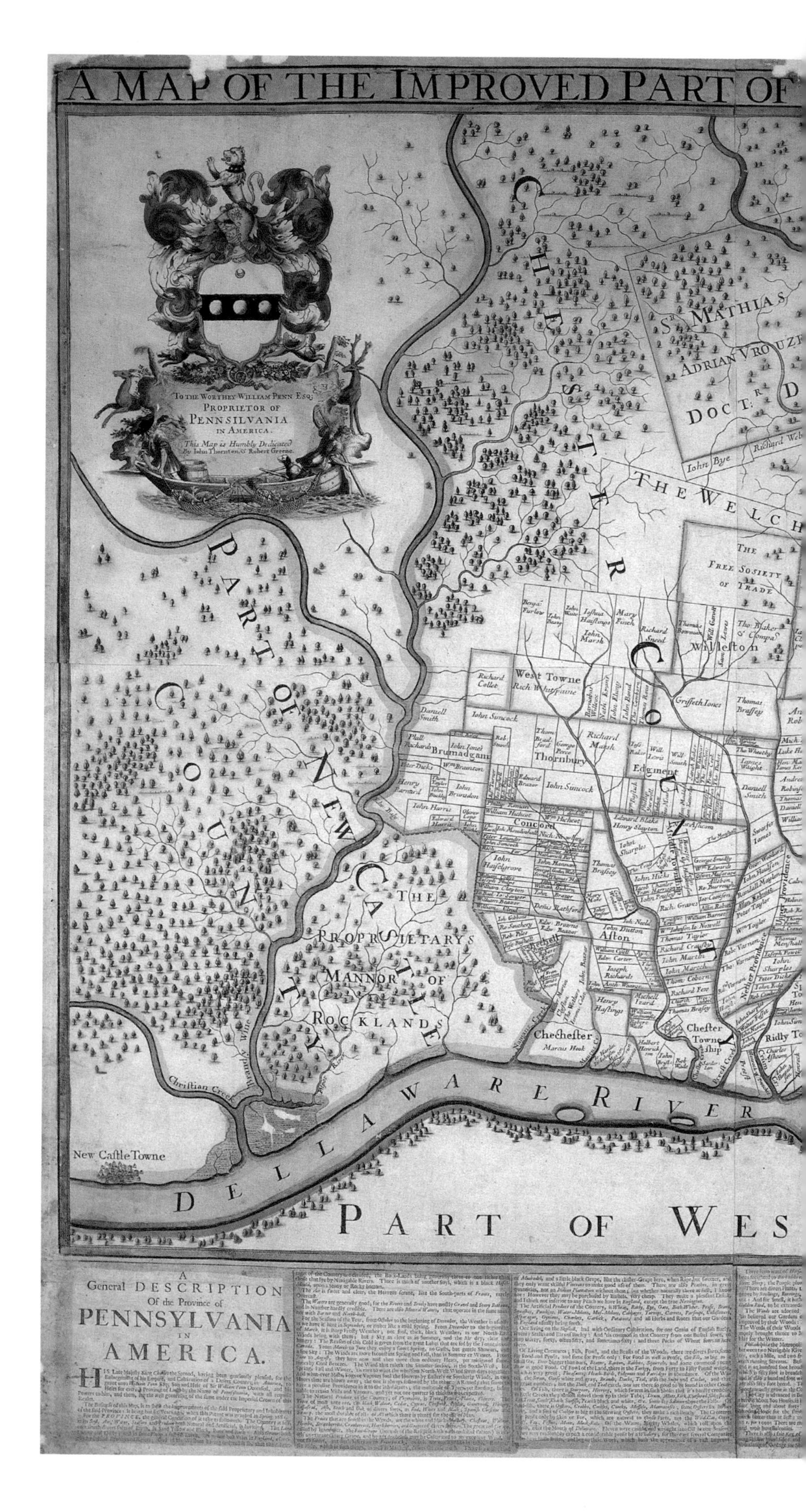

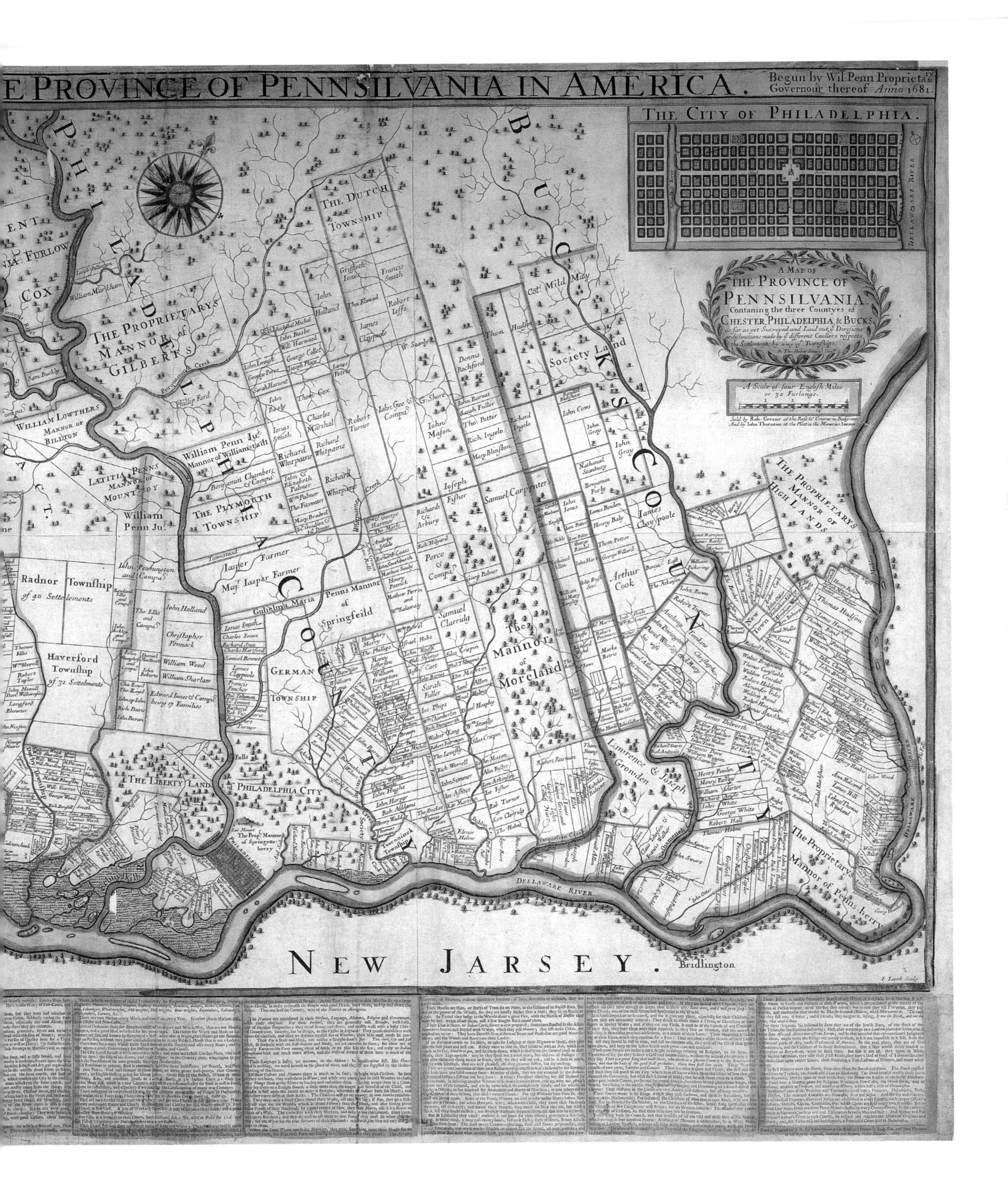

E PROVINCE OF PENNSILVANIA IN AMERICA.
Begun by Wil: Penn Proprieta.ry & Governour thereof Anno 1681.
THE CITY OF PHILADELPHIA.
DELAWARE RIVER
A MAP OF THE PROVINCE OF PENNSILVANIA.
Containing the three Countyes of CHESTER, PHILADELPHIA, & BUCKS.
A Scale of four English Miles or 32 Furlongs.
THE DUTCH TOWNSHIP
THE PROPRIETARYS MANNOR of GILBERTS
WILLIAM LOWTHERS MANNOR OF BILLTON
LÆTITIA PENNS MANNOR OF MOUNT JOY
THE PLYMOUTH TOWNSHIP
Radnor Township of 40 Settelements
Haverford Township of 32 Settelements
GERMAN TOWNSHIP
The Mannor of Moreland
Society Land
THE PROPRIETARYS MANNOR OF HIGH LANDS
The Proprietarys Mannor of Penns berry
THE LIBERTY LAND
PHILADELPHIA CITY
Toaconinck Township
DELLAWARE RIVER
NEW JARSEY.
Bridlington

FRENCH EXPANSION IN AMERICA

Vincenzo Coronelli, "America Settentrionale," in *Atlante Veneto*, circa 1688

Among the most important mapmakers of the late seventeenth century was a Franciscan friar in service to the French Crown. Vincenzo Coronelli was best known for his expertly constructed globes, the most famous of which were a pair of enormous terrestrial and celestial globes measuring over twelve feet in diameter that he crafted for Louis XIV. Just after that, he set to work on an equally ambitious atlas of the world, which was published in the 1690s. We close this chapter with the two-sheet map of North America from that atlas: it combines cutting-edge geographical knowledge with a more general assertion of French power.

As shown earlier in this chapter, English colonial efforts in North America largely focused on the Atlantic coast. By contrast, the French sought to explore rather than settle, and to press toward the interior rather than restrict themselves to the seaboard. French exploration of the Saint Lawrence Seaway and the Great Lakes by Champlain was extended to the Mississippi River and its tributaries by Louis Jolliet and Jacques Marquette in the 1670s, then Robert de La Salle in the 1680s. Though the French occasionally established outposts and small settlements, their primary goal was to develop commercial trade networks by uncovering the geography of the interior. That exploration is shown in Coronelli's confident depiction of the Great Lakes and Hudson Bay. And while he noted the presence of English settlements to the east, they remain secondary to his focus on French exploration along the Mississippi River.

These latest expeditions, however, still left much to the imagination. As is shown on the next page, Coronelli placed the mouth of the Mississippi River hundreds of miles west of its actual location, and only vaguely grasped the extent of its tributaries. Guillaume de L'Isle's's subsequent "La Louisiane" (page 66) indicates how much more of the Mississippi River drainage system would be discovered over the next three decades. To the west, Coronelli rendered California as an island, as so many mapmakers had done since Henry Briggs first made the claim more than half a century earlier. His map also substantially widened North America, revealing how much Europeans had learned about the sheer size of the continent over the course of the seventeenth century.

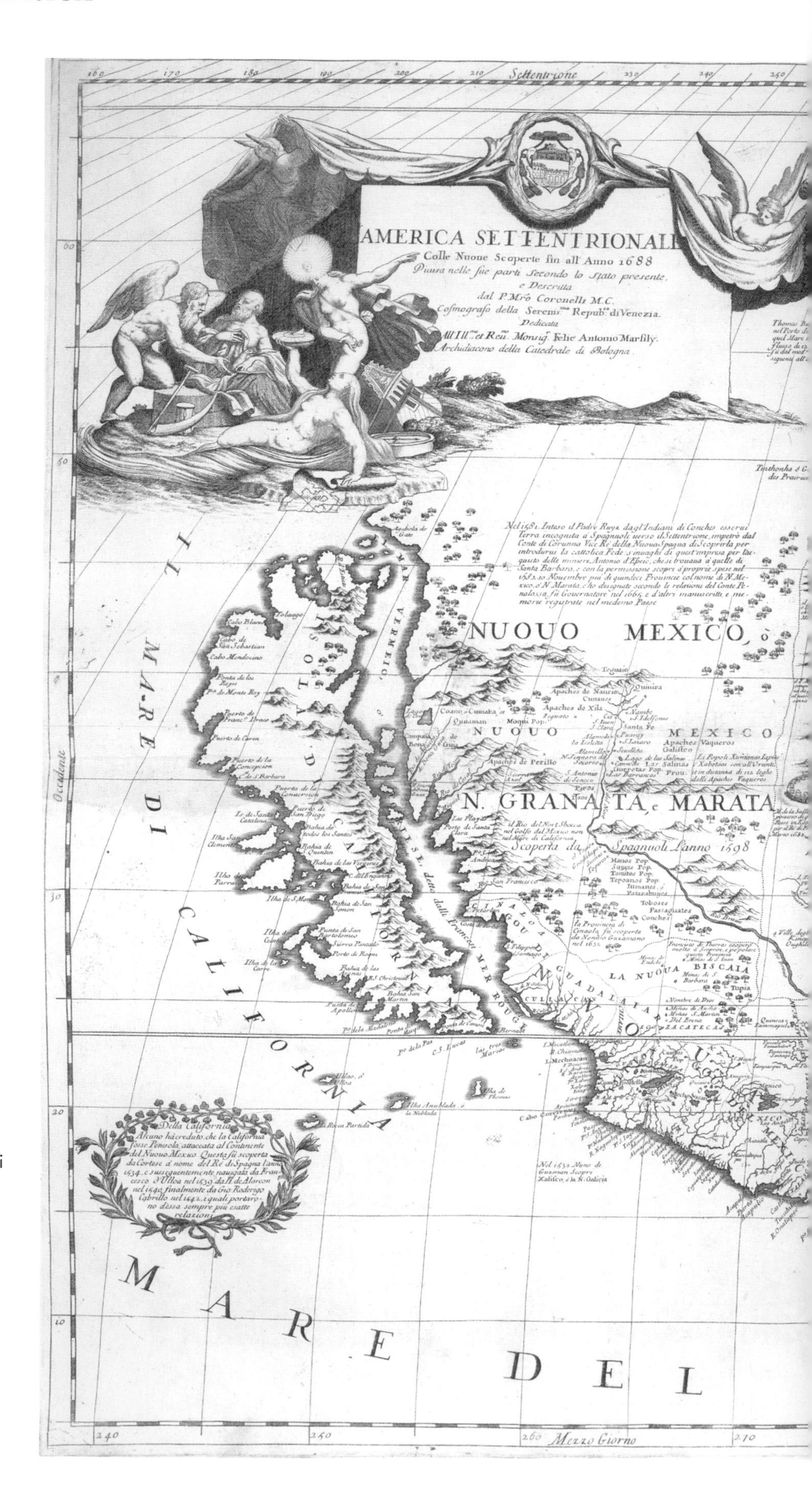

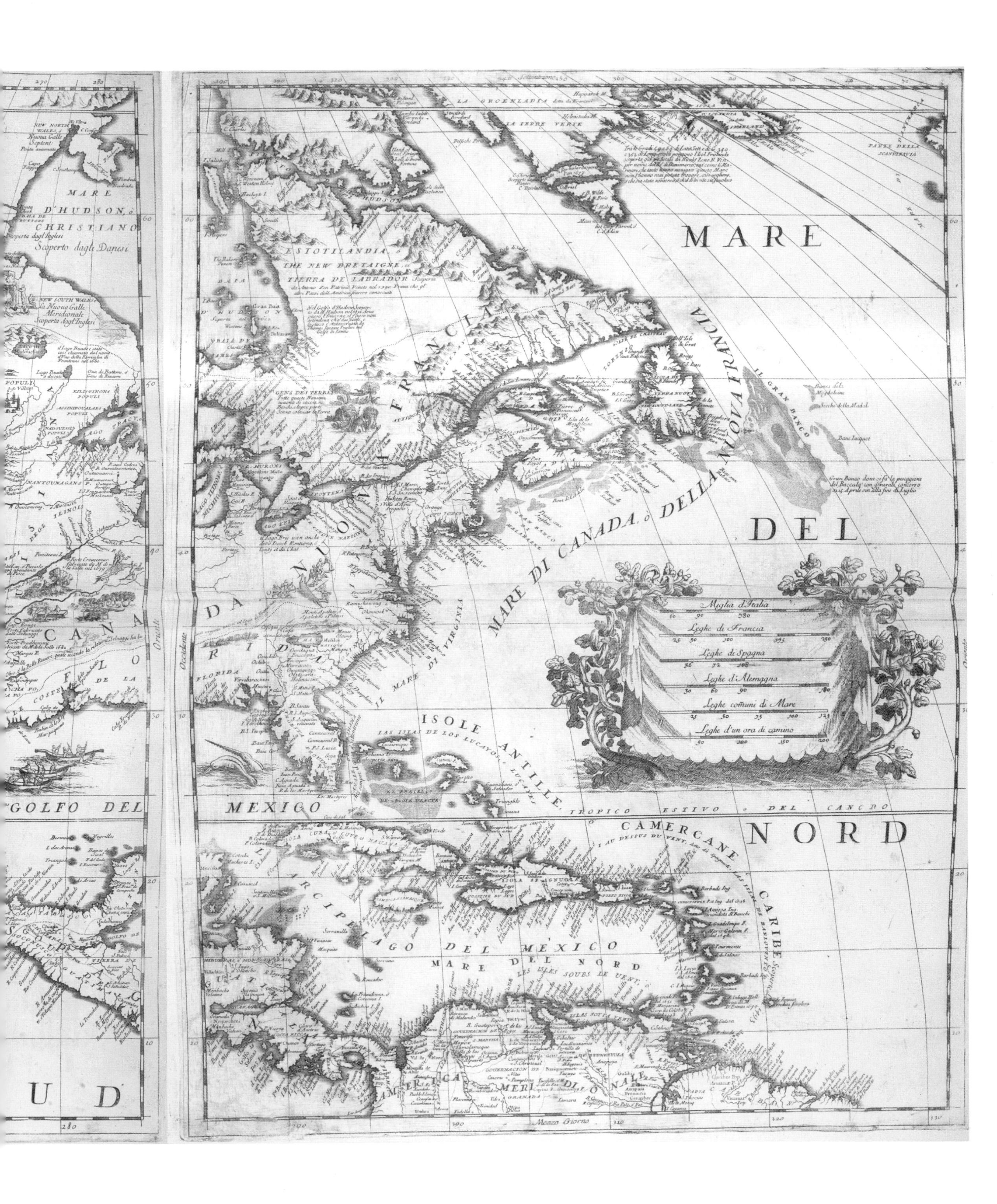

MARE D'HUDSON, ò CHRISTIANO
Scoperto dagli Danesi
NEW SOUTH WALES, ò la Nuoua Galle Meridionale Scoperta dagl'Inglesi
GOLFO DEL
U D
LA GROENLADIA
LA TERRE VERTE
ESTOTILANDIA, ò THE NEW BRETAIGNE, ò TIERRA DI LABRADOR
MARE
DEL
NORD
MARE DI CANADA, ò DELLA NUOVA FRANCIA
IL GRAN BANCO
IL MARE DI VIRGINIA
FLORIDA
ISOLE ANTILLE
MEXICO
TROPICO ESTIVO DEL CANCRO
ARCIPELAGO DEL MEXICO
MARE DEL NORD
Miglia d'Italia
Leghe di Francia
Leghe di Spagna
Leghe d'Alemagna
Leghe comuni di Mare
Leghe d'un ora di camino
Occidente
Oriente
Mezzo Giorno

The heart of the map is the interior shown at right, where Coronelli described the discoveries and forts of the French explorers. Perhaps most tantalizing was his suggestion of a short portage between the newly named "Chekagou" (Chicago) in the upper right corner and a nearby tributary of the Mississippi River, marked as "Chekagou R." The possibility of linking the Great Lakes to the Mississippi River watershed—and the Gulf of Mexico beyond—preoccupied the French throughout this period. More generally, La Salle's "discoveries" all along the Mississippi River would be cited by de L'Isle and others as evidence of French sovereignty. Coronelli advocated the expansion of French influence in North America, and his treatment of the English colonies reflected that imperial agenda. On the prior page, the Atlantic is named a sea of "Nuova Francia," which dwarfs the English presence in the "Mare di Virginia" to the south. English colonies are hemmed in by dotted lines, sharply contrasting with the expansive "Canada Nuovo" and "Louisiana," which stretch across the center of the continent.

In the west shown at right, Coronelli acknowledged the presence of the Spanish, marking the claim of Juan de Oñate to "New" Mexico with the phrase "Scoperta da Spagnuoli L'anno 1598." In his lengthy annotation at the top of this page, Coronelli also gestured toward the intrigue around the former governor of New Spain, Diego de Peñalosa. Exiled to France, Peñalosa gave valuable geographical information to Coronelli regarding the Spanish Southwest. He even suggested the possibility of aiding the French in attacking New Spain, underscoring the more general rivalry between the two empires.

Even the cartouche shown on the previous page expresses Coronelli's ambitions on behalf of the French. At the top, the winged head of Inquiry struggles to reveal America by pulling back heavy drapes, suggestively symbolizing the way in which French explorers had uncovered the geography of the continent. Beneath the figure of Inquiry stands the symbol of Truth, with sunlight radiating from her head as she points toward the geographical revelations on the map itself. This assertion of French power on the continent would directly clash with England's own imperial agenda throughout the early eighteenth century, as detailed in the next chapter.

ASSINIPOUALAES POPULI
TRACY
LAGO
NADOUESSIA POPULI
L. de Prouidenza
Sault de S. Ant. de Padoue
le Tombeau
Siou
L. des Pleurs
R. des Boeufs, ó Ouatebanga
NANTOUNAGANS
R. Noire, e Nadouessions
Outaouami
R. aux Cedres
R. Oumalouminec
il P. Marquette Gesuita, e Iolliet furono li primi ch'entrarono nel Fiume Mechissipi li 17 Giugno 1673.
R. Ouisconsing, ó Misconsin
MEXICO, ò
ILINOIS
DEGL
PAESI
Chekagou R.
Aramoni R.
Pimiteoui L.
Forte Creuecoeur fabricato da M. de la Salle nel 1679
Petit Lac, ó Picciolo lago abondante di Pesce
il F. Mechissipi conserua sempre la med: larghezza da questo * segno s'in al mare, ch'e' nauigabile * senza rapidita
Quiuira
S. Idelfonso
MEXICO
Apaches Vaqueros
Galisteo
Li Popoli Xumanas, Iapies e Xabotaos son all'Oriente, et in distanza di 112 leghe dalli Apaches Vaqueros
R. des Ilinois, ó Segnelay
Missourites
Pot a Beurre
Maroa
C. S. Antoine
R. Ouabache
Cap. Rouge
Forte fabricato dalli Seluaggi
A, e MARATA
C A N A
Forte Prudhomme fabricato da M. de la Salle 1682
M. de la Salle Piglia possesso di questo Paese in Kappaha per il Re' di Francia, 16 Marzo 1682
Kappaha
Margoi R.
P. a l'Anguille
R. Ohio, ó la Belle Riuiere, quale secondo la relatione des Seluaggi ha la
L O
Imaha
Kaskinompas
CHIKACHA PO.
TOUNICA PO.
F
DE LA
Rio de lo Spiritu Santo
Natche
I. Taenza 8 Ville di Taenza
Kotoa
COSTE
Cabo de la Cruz
C. de Hondo
Oumats
Toboses
Passaguates
Conches
Rio Brauo
Riuiere Inconnue
R. du Risque
Bahia de lo Espiritu Santo
Mar pequeno
R. de Laso
Cabo Escondido
4 Ville degli Popoli Kinepisa, ó Oughilousa
Francisco de Ibarras coopero molto a' Scoprire, e popolare questa Prouincia
Miñas de S. Iuan
Minas de Endehe
NUOUA BISCAIA
Minas de S. Barbara
Topia
KINEPISA P.
Rio Brauo
Nombre de Dios
Miñas di Aurño
Miñas S. Martin
Del Erena
ZACATECAS
Quineca
Tammapul
Lago Salso
R. Montanhas
R. Hermoso
GOLFO DEL
Tancala
Tamahalon

3. 1700-1783

Imperialism and Independence

In 1700 the French, Spanish, and British all competed for the upper hand in North America. The French and British sought access to the Ohio River, a crucial artery with tributaries stretching north and east toward the Great Lakes and draining south into the Mississippi River and the Gulf of Mexico. The French signaled their determination to control this region by founding New Orleans in 1718 and then establishing a chain of forts south from the Gulf of Saint Lawrence into the Ohio Valley. Together, these outposts formed an arc that effectively hemmed in the British along the Atlantic seaboard.

While the French forged networks on the ground, they also asserted their position through maps. Guillaume de L'Isle's authoritative and comprehensive map of North America audaciously appropriated the continental interior for the French (page 66). The British responded with maps that replaced the French geographical vision with an equally strident one of their own (pages 68–71). Further west, the Spanish began to launch expeditions north from Mexico into the region they claimed as "New Mexico." The Spanish had been chastened by the Pueblo Revolt of 1680, but in the eighteenth century they began to reclaim the Southwest and resume their search for a river passage west to the missions along the Pacific coast (page 88).

Spain's territorial claims in the Southeast were increasingly challenged by the British and French, as is shown in Mark Catesby's map on page 76. The entire Southeast was in flux: the British and Spanish clashed over northern Florida, while French traders extended their reach from the Mississippi River east toward Georgia and the Carolinas. To a limited degree, this imperial maneuvering placed Native Americans in a strategic position. A booming deerskin trade in the Southeast brought colonists into direct and sustained contact with the Cherokee and other tribes. The map on page 72 was designed to navigate the complex commercial networks and rivalries which characterized that trade.

Further north, the British strengthened their alliance with the Iroquois Confederacy in order to stave off French encroachments. Cadwallader Colden's map on page 78 conveys the geopolitical character of British–Iroquois diplomacy in western New York. Meanwhile, the French built competing alliances and trade networks with other tribes to consolidate their hold on the Ohio Valley and the Lower Mississippi. The resulting treaties were often ignored when tribes no longer served European purposes, as when the South Carolina economy shifted from the deerskin trade to the cultivation of rice. But in an era when European powers vied for control of the continent, indigenous peoples exercised a certain level of power.

The British quest to map its colonies in order to contain French expansion intensified at mid-century. In 1754 the governor of Virginia sent a young George Washington west over the Allegheny Mountains to halt French military incursions along the southern edge of Lake Erie. The information gained on this mission proved crucial to British strategy when war broke out with the French a few months later (page 84). Lewis Evans subsequently designed a map of the middle British colonies in order to encourage British migration into the trans-Allegheny West and thereby repel the French (page 86). Evans' map was just one of several British efforts to advance geographical knowledge of the colonies; the most comprehensive was Joshua Fry and Peter Jefferson's profile of Virginia (page 80). This was the first map to reveal the entire Chesapeake river system, a geographical advantage that helped to make tobacco the most important export of the eighteenth century, and Virginia the largest and wealthiest of Britain's American colonies.

The growth of tobacco could not have occurred without the expansion of the slave trade. Malachy Postlethwayt designed the map on page 74 to advance the British role in that trade and the particular interests of the new Royal African Company. The growth of the slave population mirrored a more general population increase throughout the colonies. In 1713 about 360,000 European colonists were living in North America. By the time war broke out between the French and the British in 1755, that figure was closer to 1.5 million, a fourfold increase in just four decades.That population began to expand beyond the seaboard settlements of Connecticut and Massachusetts into Pennsylvania, Maine, New Hampshire, and Vermont. Similarly, by the 1730s Virginians were moving beyond the Tidewater and the Piedmont into the western valleys.

Victory in the French and Indian War gave the British unrivaled control over North America east of the Mississippi

River, and at the same time sowed the seeds of American independence. The colonists believed that the ejection of the French would lead to more freedom—particularly for trans-Allegheny settlement—while the British treated the victory as an opportunity to tighten governance over the colonies and to extract more wealth to help pay for the late war. These British regulations and taxes generated colonial discontent that ultimately led to the revolution in the 1770s.

The English colonists saw themselves as competent self-regulators, and crackdowns on governance only heightened their awareness of their political rights. Among the most consequential of those restrictions is one documented in the 1775 map of Boston on page 90. In mid-April, British forces were sent west to confiscate weapons that were rumored to be stored in the town of Concord. This British map illustrated the ensuing skirmishes at Lexington and Concord, where the "shot heard round the world" opened a war for independence. As one of the earliest maps of the Revolutionary War, it conveys the individual maneuvers and operations that led to the Siege of Boston. In 1775, however, there was little indication that these skirmishes would lead to a civil war throughout the colonies, much less a movement for independence that would influence the arc of world history.

For the next two centuries, Americans were taught to see the Revolution as a war to secure liberties violated by the Crown, even as slavery grew more entrenched. What, then, did the revolutionary rallying cry of liberty actually mean? In August 1775 King George declared the colonies to be in open rebellion. A few months later, the governor of Virginia offered to liberate slaves who remained loyal to the Crown. The proclamation outraged Southern slaveholders, who believed that it manipulated slaves in order to suppress the Revolution. But it also reminds us how central slavery was to the colonial experience. Indeed, it was the widespread practice of slavery that sensitized Southern colonists to the violations of their liberties. In other words, notions of "freedom" were inextricably connected to slavery.

American independence was won in 1782 when General George Washington—with tremendous French military support—defeated the British at the Battle of Yorktown. Sebastian Bauman's eyewitness map of that event on page 92 became a symbol of colonial emancipation. The final map of this chapter was used by the British to negotiate the boundaries of this new nation, a palpable reminder of the sheer contingency of history. How would a continued French presence in the Ohio Valley have shaped that region, and would it have limited British settlements to the seaboard? Might negotiations—rather than rebellion—have kept the colonies part of the British empire? This chapter demonstrates the degree to which economic and diplomatic decisions—many of which were made through maps—had far-reaching consequences in the eighteenth century.

THE WAR OF THE MAPS

Guillaume de L'Isle, "Carte de la Louisiane et du Cours du Mississipi," 1718

Vincenzo Coronelli's map of North America on page 60 documents the contemporary confusion regarding the path of the Mississippi River and its tributaries. That mystery was largely solved by Guillaume de L'Isle, whose maps of the interior represent a quantum leap of geographical accuracy. In recognition of these skills he was named cartographer to the French king, and his maps remained influential throughout the eighteenth century. For our purposes, de L'Isle is central not only because of his geographical precision, but also because of his ability to advance French territorial claims through his masterful map of "La Louisiane" in 1718.

The conflict between the French and the British was rooted in opposing views of territorial sovereignty. The French claimed that the explorations by Robert de La Salle up the Mississippi River gave them rights not only to the river, but also to its tributaries. Conversely, the British asserted that the Treaty of Utrecht (1713) awarded them control of the same region via their relationship with the Iroquois. These competing interpretations of sovereignty set up an unresolvable conflict. In this context, de L'Isle was asked to make a map of North America that would ground and defend French claims to the continent. Rather than solely relying on past maps, he sought new and firsthand information such as field reports from trappers and traders. The result was both geographically precise and full of propaganda.

De L'Isle used the map to stretch the French sphere of influence to its limits. He centered the map on the French claim to the Mississippi River, elegantly presented as the axis of the continent. The tributaries extend that claim both east and west.

By marking the routes of La Salle, Bienville, and St. Denis across the southern portion of the map, de L'Isle sought to buttress French claims to the lower Mississippi River and its tributaries. The map became an instrument of strategy and diplomacy, an indication that the French sought to amplify, rather than diminish, their presence in the region. These imperial designs are apparent elsewhere on the map. De L'Isle took care to identify former and current native villages in order to present the interior as an inhabited space. The French had limited interest in settling this territory, and instead focused on its mineral wealth. In this context, the native inhabitants were not rivals; instead, they were diplomatic allies, trade partners, geographical informants and, potentially, even a source of labor.

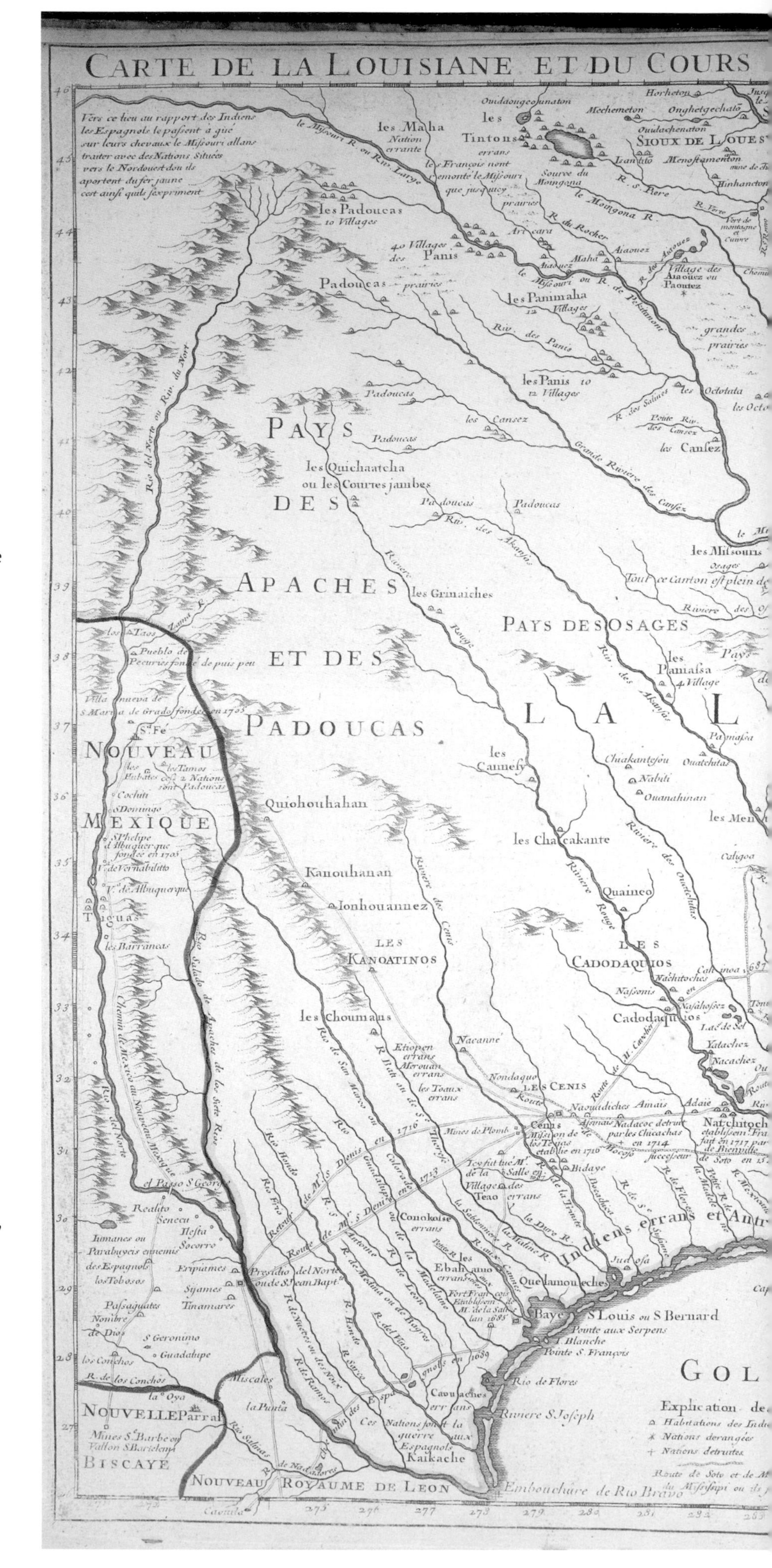

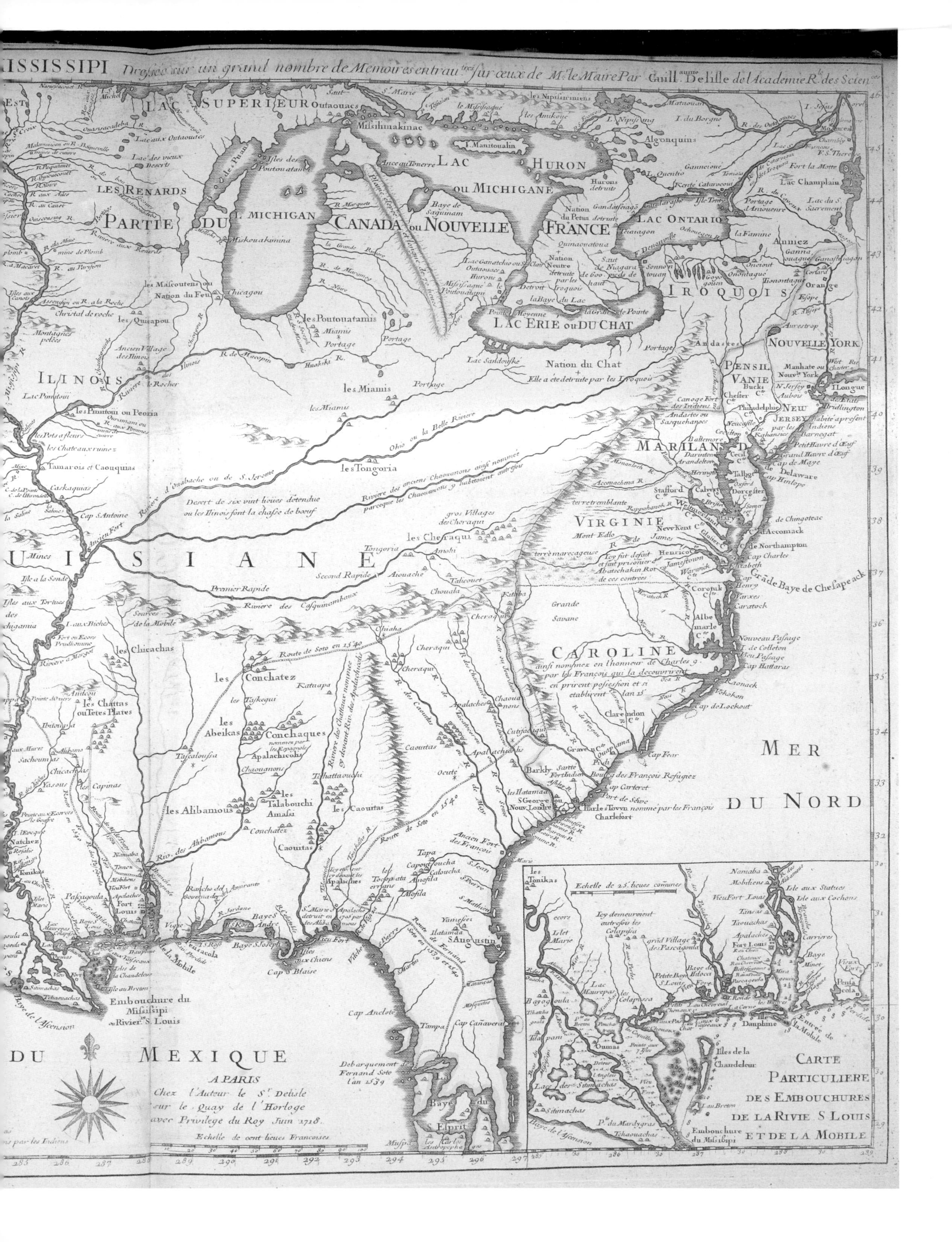

ISSISSIPI Dressée sur un grand nombre de Memoires entr'autres sur ceux de Mr. le Maire Par Guill.aume Del'Isle de l'Academie R. des Scien
LAC SUPERIEUR
LAC HURON
LAC ou MICHIGANE
I. MICHIGAN
LAC ONTARIO
LAC ERIE ou DU CHAT
LAC Champlain
PARTIE DU CANADA ou NOUVELLE FRANCE
LES RENARDS
IROQUOIS
NOUVELLE YORK
PENSILVANIE
NEW JERSEY
MARILAND
VIRGINIE
CAROLINE
ILINOIS
LOUISIANE
Ohio ou la Belle Riviere
Desert de six vint lieües detendue ou les Ilinois font la chasse de boeuf
Grâde Baye de Chesapeack
MER DU NORD
DU MEXIQUE
Embouchure du Missisipi ou Rivier. S. Louis
Route de Soto en 1540
A PARIS
Chez l'Auteur le Sr. Delisle sur le Quay de l'Horloge avec Privilege du Roy Juin 1718.
Echelle de cent lieues Francoises
CARTE PARTICULIERE DES EMBOUCHURES DE LA RIVIE S. LOUIS ET DE LA MOBILE
Echelle de 25 lieues communes

Herman Moll, "A New Map of the North Parts of America Claimed by France," 1720 (pages 70–1)

The primary goal of de L'Isle's map—shown here and on the prior page—was to limit the British presence in North America. The main area of contention was the Ohio Valley shown at right, which he mapped as both devoid of British settlers and home to native tribes. To assert French power, he boldly and expansively marked the interior as "La Louisiane," effectively trapping the British colonies along the eastern seaboard and relegating Spain's "Nouveau Mexique" to the continent's western edge. To add insult to injury, de L'Isle's annotations suggested that Charleston and Carolina were named for French—rather than English—royalty.

De L'Isle published the map just as the French founded the port of New Orleans in 1718. The geography reinforces the political message: in de L'Isle's rendering the Mississippi River watershed "naturally" aligns with the extent of French territory. Given how unclear imperial claims remained after the Treaty of Utrecht, it is difficult to imagine a more strategic use of cartography. While the British were impressed by de L'Isle's cartographic skill and command of geography, they were outraged by his overt attempt to encroach upon British claims. The map also confirmed British fears that French knowledge of geography was far superior to their own. De L'Isle's map raised the stakes of the geopolitical struggle between Britain and France over control of the American interior.

Herman Moll responded in kind with a map of his own. Moll was a leading map publisher of the early eighteenth century. In 1715 he published a map that depicted the British dominions along the eastern seaboard as strategically positioned relative to the rest of the continent. Moll was thus particularly troubled by de L'Isle's 1718 map, which attempted to limit Britain's territorial sphere. He responded with the pointedly titled "New Map of the North Parts of America Claimed by France" (next page).

The very title of the map hints at Moll's sense of disbelief: how could de L'Isle claim that the "adjoining territories" of England and Spain were marginal to the French center? His map dripped with sarcasm, sniping that "all within the Blew Colour" is what is "claim'd" by France, while "The Yellow Colour what they allow ye English." In characterizing de L'Isle's map as propaganda, Moll aimed both to challenge French claims and to fortify British settlements beyond the seaboard.

To this end, he labeled French territorial claims as "incroachments," which were all the more outlandish to him given how few of them had actually settled North America. He even urged the British to preserve alliances with the Iroquois and Cherokee as a way of containing the French. Over the next few decades the British became increasingly aware of the threat posed by the French around the Great Lakes, which they considered an extension of British territory in light of their relationship with the Iroquois or "Five Nations." Moll repeatedly referenced these alliances on his map. He also reached back into history, invoking John Cabot's arrival in the New World in 1498 as evidence of British sovereignty in North America.

Though Moll criticized de L'Isle's map as a source of French imperialism, he also acknowledged its importance as a geographical document; in fact, Moll's map directly relied on de L'Isle. But Moll made geographical contributions of his own, particularly in his representation of the topography, road system, and Indian settlements of the Carolina backcountry. His larger purpose with this map, however, was not to uncover geographical knowledge but to limit French power. Note that he claimed Newfoundland to the Carolinas as British land. With this wider geographical scope, he visually reduced French claims. The coat of arms at left projected British power further, not just in the east but across the continent. Noticeable as well are the notations of "good pasture ground" and "good ground," and especially the general absence of an Appalachian range, which de L'Isle had used on his map as a natural western barrier to the British colonies.

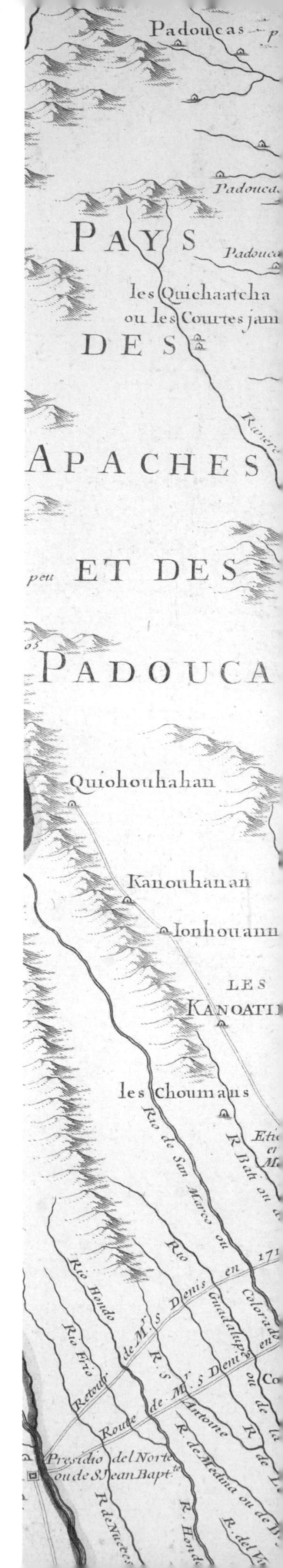

le Missouri ou R. de Pekitan oui
Aiaoüez ou Paoutez
Chemin des Voyageurs
Riviere aux Renards
Miskouakimina
les Panimaha
12 Villages
Riv. des Panis
grandes prairies
mine de plomb
R. a la Mine
mine de Plomb
R. a Macaret
R. au Parisien
les Mascoutens ou Nation du Feu
Chicagou
les Panis 10
12 Villages
Isles aux Canots
Assenisipi ou R. a la Roche
Christal de roche
les Octotata
R. des Salines
les Octotata
les Quicapou
Chicagou R.
R. S. Joseph ou des Miamis
les Cansez
Petite Riv. des Cansez
Montagnes pelées
les Cansez
Ancien Village des Ilinois
R. de Macopin
Huakiki
Grande Riviere des Cansez
Misisipi R.
ILINOIS
Riviere des Ilinois
le Rocher
Padoucas
Lac Pimitoüi
les Pimitoüi ou Peoria
Ouramani ou R. aux Pommes
mine de cuivre
des Akansas
le Missouri R.
les Missouris
Salines
les Pots a fleurs
Osages
les Chateaux ruinez
Tout ce Canton est plein de Mines
Tamarois et Caouquias
Riviere des Osages
R. Fourchue
Maramec R.
C. de la Pointe
C. de l'Hirondelle
Caskaquias
Riviere d'Ouabache ou de S. Jerome
PAYS DES OSAGES
Desert de six vint lieües d'etendue ou les Ilinois font la chasse de boeuf
Riv. des Akansas
les Paniassa
4 Village
Pays plein de Mines
R. a la Saline
la Saline
Cap S. Antoine
Ancien Fort
LA LOUISIANE
Mines
Paniassa
Isle a la Sonde
les Cannesy
Chiakantesou
Ouatchitas
Nabiti
Ouanahinan
Isles aux Tortues
Premier Rapide
Riviere des Casquinas
Lac des Mitchigamia
Kappa
Sources de la Mobile
les Menitous
I. aux Biches
Fort ou Ecors Prudhomme
les Chacakante
Riviere des Ouatchitas
les Chicachas
Route de Soto
Riviere a Margot
Caligoa
Riviere Rouge
Route de Soto
les Conchatez
Quaineo
Kappa
Pointe d'Oziers
Anitcou
les Chattas ou Tetes Plates
les Taskegui
Ibitoupa
LES CADODAQUIOS
les Akansa
les Abeikas
Conchaques
nommez par les Espagnols
Cahinoa 1687
Isles aux Mures
Nachitoches
Sachoumas
Alibamo
Apalachicolis
Tascaloussa
Nassonis
Ance percee
Chicachas
Chaouanons
Nasahossez
Tonicas
Cadodaquios
Lac de Sel
Route de Soto
R. des Yasous
Isles aux chevreuils
Yasous
les Capinas
les Talabouchi
Amasi
Yatachez
les Alibamous
Nacachez
Cours de Soto
Taensas
Pointe aux Ecorces
le Gouffre
Ouachita
Route de M. de Bienville
I. l'Evesque
Route de M. de Tonty en 1702
Route de Soto en 1540
Conchatez
Nondaquo
LES CENIS
Route
Route de M. Cavelier
Natchez
Rosalie
Caouitas
Naouadiches
Ainais
Adaie
Riv. Rouge
Riv. a Lalleman
Riv. des Alibamous
Assinais
Nadacoc detruit par les Chicachas en 1714
Natchitoches etablissem.t François fait en 1717 par M.r de Bienville
Guachoia ou mourut Soto
Tonikas
Naniaba
Tomez
Mobiliens
Vieu Fort
Apalaches
Rancho del Governador
R. del Amirante
Cenis
Mission de los Tejas etablie en 1716
Mocoso successeur de Soto en 1542
Bidaye
Ecu fut tué M.r de la Salle en 1687
R. de la Trinite
R. des Ouachas
Pascagoula
Fort Louis
Village des Teao errans
R. de Flores
la Madele
Petite R. de S.
R. Mexicano
R. Jordano
Bayes Andre
S.t Marie detruit en les Aliba
Bocachia
Loupeloussa
Islet Marie
Bayogoula
Tchatchagoula
Lac Maurepas
Colapissa
Baye S. Louis
Chataux
Vigie
I. S.te Rose
Pensacola
Baye S. Joseph
Sablonniere
la Dure R.
Indiens errans et Antropophages
Tilapani
Lac Pontchartrain
Oumas
I. au Chat
Isle aux Vaisseaux
I. Dauphine
Riv. Perdide
R. de la Mobile
Isles aux Chiens
la Maline R.
R. aux Cannes
Judosa
Isles de la Chandeleur
Vieu Fort
Cap S.t Blaise
Quelamoueches
Cap du Nord
Sitimachas
Tchaouachas
Isle au Breton
Embouchure du Misisipi ou Rivier.re S. Louis
Fort François etablissem.t de M. de la Salle en 1685
Baye S. Louis ou S. Bernard
Pointe aux Serpens
I. Blanche
Baye de l'Ascension

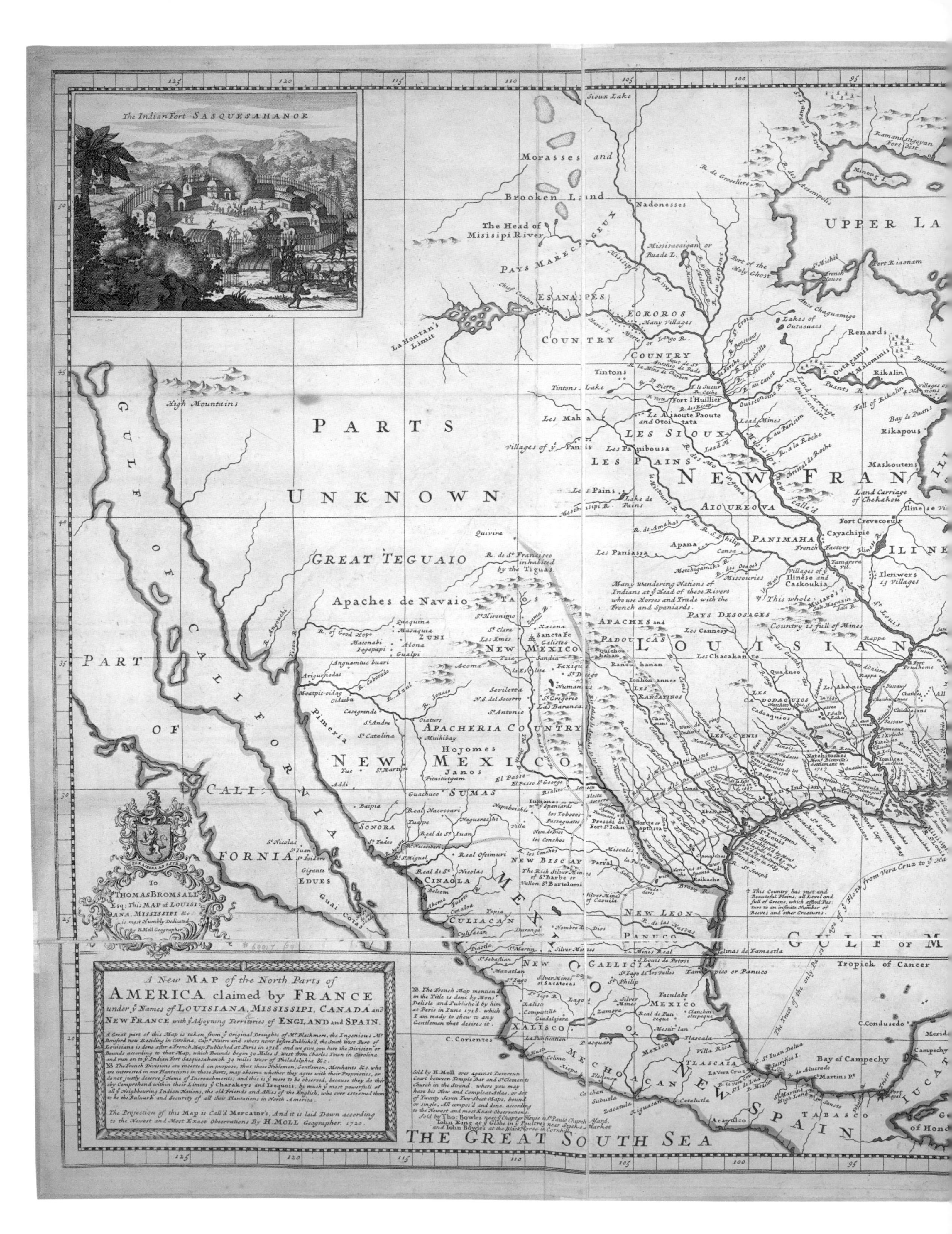

The Indian Fort SASQUESAHANOK
To
THOMAS BROMSALL
Esq; This MAP of LOUISIANA, MISSISSIPI &c.
is most Humbly Dedicated
By H. Moll Geographer
A New MAP of the North Parts of
AMERICA claimed by FRANCE
under yͤ Names of LOUISIANA, MISSISSIPI, CANADA and
NEW FRANCE with yͤ Adjoyning Territories of ENGLAND and SPAIN.
A Great part of this Map is taken from yͤ Original Draughts of Mr. Blackmore, the Ingenious Mr. Beresford now Residing in Carolina, Capt. Nairn and others never before Publishe'd, the South West Part of Louisiana is done after a French Map Published at Paris in 1718. and we give you here the Division or Bounds according to that Map, which Bounds begin 30 Miles S. West from Charles Town in Carolina and run on to yͤ Indian Fort Sasquesahanok 30 miles West of Philadelphia &c.
NB The French Divisions are inserted on purpose, that those Noblemen, Gentlemen, Merchants &c. who are interested in our Plantations in those Parts, may observe whether they agree with their Proprieties, or do not justly deserve yͤ Name of Incroachments; and this is yͤ more to be observed, because they do thereby Comprehend within their Limits yͤ Charakeys and Iroquois, by much yͤ most powerfull of all yͤ Neighbouring Indian Nations, the old Friends and Allies of the English, who ever esteemed them to be the Bulwark and Security of all their Plantations in North America.
The Projection of this Map is Call'd Mercator's, And it is laid Down according to the Newest and Most Exact Observations By H. MOLL Geographer. 1720.
NB. The French Map mention'd in the Title is done by Monsr. Delisle and Publishe'd by him at Paris in June 1718. which I am ready to shew to any Gentlemen that desires it.
Sold by H. Moll over against Devereux Court between Temple Bar and St. Clements Church in the Strand. where you may have his New and Compleat Atlas, or Set of Twenty Seven Two-Sheet Maps, bound or single, All compos'd and done, according to the Newest and most Exact Observations.
Sold by Tho: Bowles next yͤ Chapter House in St. Pauls Church Yard, John King at yͤ Globe in yͤ Poultrey near Stocks Market and John Bowles at the Black Horse in Cornhill.
THE GREAT SOUTH SEA
PARTS UNKNOWN
GREAT TEGUAIO
NEW FRANCE
LOUISIANA
NEW MEXICO
GULF OF CALIFORNIA
PART OF CALIFORNIA
NEW SPAIN
GULF OF MEXICO
Tropick of Cancer
Bay of Campechy
UPPER LAKE
Morasses and Brooken Land
The Head of Misisipi River
LES SIOUX
APACHERIA COUNTRY
Apaches de Navaio
NEW BISCAY
NEW LEON
NEW GALLICIA
XALISCO
MECHOACAN
TLASCALA
SONORA
CINAOLA
CULIACAN
PANUCO
C. Corientes
High Mountains
Many Wandering Nations of Indians at yͤ Head of these Rivers who use Horses and Trade with the French and Spaniards.
This Country has vast and Beautifull Plains, all Level and full of Greens, which afford Pastures to an infinite Number of Beeves and other Creatures.
Country is full of Mines

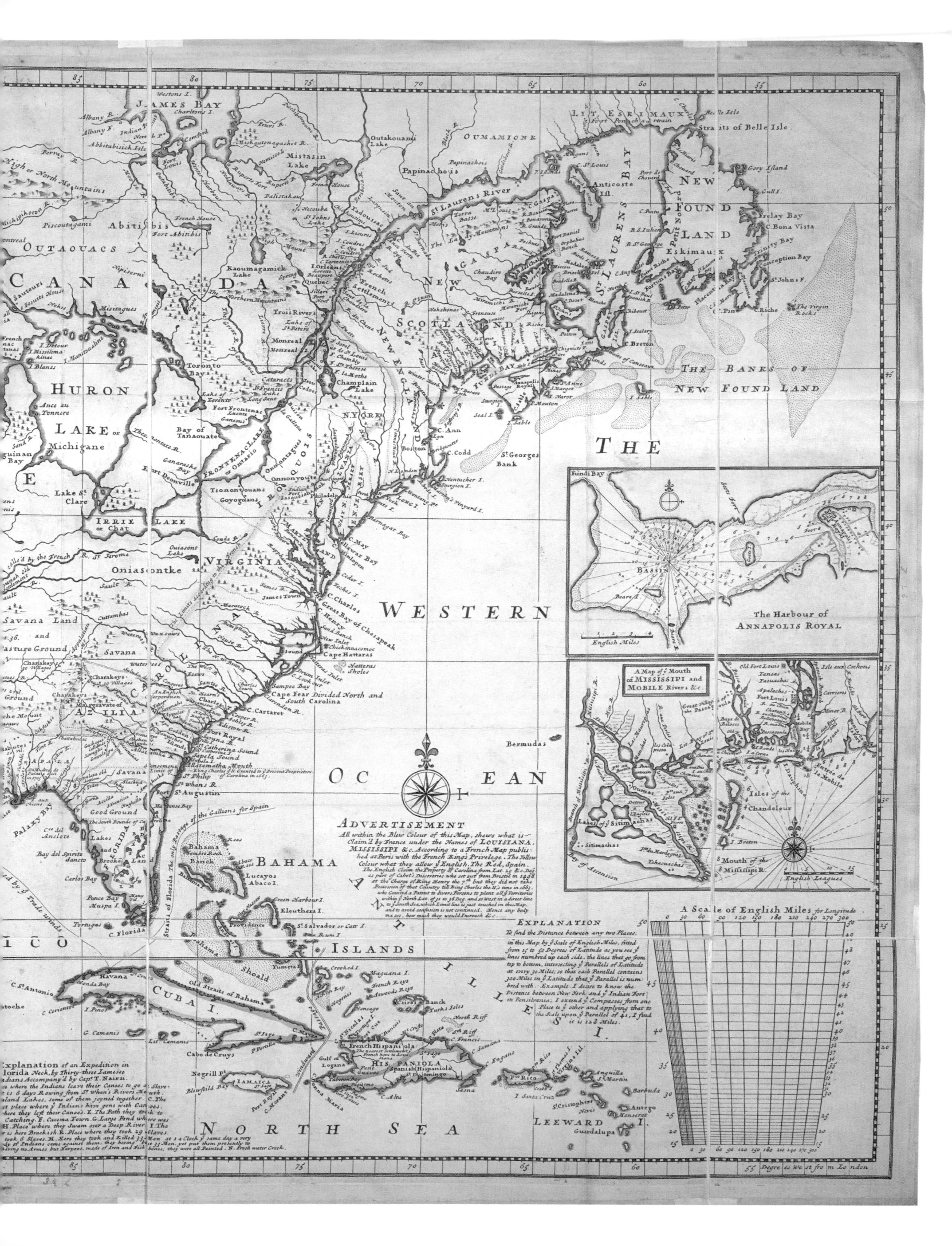

JAMES BAY
CANADA
OUTAOUACS
HURON
LAKE or Michigane
IRRIE LAKE or Chat
FRONTENAC LAKE or Ontario
IROQUOIS
NEW ENGLAND
N. YORK
NEW SCOTLAND
VIRGINIA
CAROLINA
FLORIDA
St Laurens River
St LAURENS BAY
NEW FOUND LAND
Straits of Belle Isle
THE BANKS OF NEW FOUND LAND
THE WESTERN OCEAN
BAHAMA
ANTILLE ISLANDS
CUBA I.
HISPANIOLA
JAMAICA
NORTH SEA
LEEWARD I.
Bermudas
Cape Hatteras
Great Bay of Chesapeak
Fort St Augustin
The Harbour of ANNAPOLIS ROYAL
A Map of ye Mouth of MISSISSIPI and MOBILE Rivers &c.
English Miles
English Leagues
A Scale of English Miles for Longitude
Degrees West from London
ADVERTISEMENT
All within the Blew Colour of this Map, shews what is Claim'd by France under the Names of LOUISIANA, MISSISSIPI &c. According to a French Map publis'hed at Paris with the French King's Privelege. The Yellow Colour what they allow ye English. The Red, Spain. The English Claim the Property of Carolina from Lat. 29 &c. Deg. as part of Cabot's Discoveries who set out from Bristol in 1498 at the Charge of King Henry the 7th but they did not take Possession of that Country till King Charles the II's time in 1663, who Granted a Patent to divers Persons to plant all ye Territories within ye North Lat. of 31 to 36 Deg. and so West in a direct line to ye South Sea, which Limit line is just touched in this Map, and to avoid confusion is not continued. Hence any body may see, how much they would Encroach &c.
EXPLANATION
To find the Distance between any two Places in this Map by ye Scale of English Miles, fitted from 15 to 50 Degrees of Latitude as you see ye lines numbred up each side. the lines that go from top to bottom, intersecting ye Parallels of Latitude at every 30 Miles; so that each Parallel contains 300 Miles in ye Latitude that ye Parallel is numbred with. Exampl. I desire to know the Distance between New York and ye Indian Fort in Pensilvania; I extend ye Compasses from one Place to ye other and applying that to the Scale upon ye Parallel of 41, I find it is 128 Miles.

NATIVE AMERICANS NAVIGATE THE DEERSKIN TRADE

Circa 1721

Though it lacks a formal title, the lengthy description of this map tells us much: "This map describing the situation of the Several Nations of Indians to the NW of South Carolina was copyed from a Draught drawn & painted on a Deer skin by an Indian Cacique and presented to Francis Nicholson Esq. Governour of South Carolina by whom it is most humbly Dedicated To His Royal Highness George Prince of Wales."

At first glance this may look less like a map than a slightly confusing organizational chart. With the right context, however, the diagram becomes a sophisticated guide to the trade war among Native Americans and European settlers in the early eighteenth century. To decode the map, first consider that it depicts space in terms of networks and relationships rather than absolute physical distance. At left is an angular grid of streets that represents "Charlestown," the bustling port of the newly named royal colony of South Carolina. The thirteen circles of varied size represent the relative power of southeastern tribes. Many, but not all, of these circles are connected by lines indicating trade networks and alliances. At lower right, the colony of Virginia is connected primarily to the Nasaw (or Catawba), while lines connect the port town of Charlestown to several other tribes.

Though the geography may not be obvious at first, rotating the map to the left makes it easier to see Charlestown and Virginia along the Atlantic coast, flanked at left by the tribes of the interior. The map captures an intensely competitive era in the deerskin trade. While tribes angled for advantage with the British colonies, Virginia and South Carolina vied to corner the market at their respective ports. The Carolina colony had been founded only a few years earlier as a geographical buffer between Virginia and Spanish Florida. Carolina's economy initially depended on fur and deerskins, and it very quickly surpassed Virginia as the most important center of Indian trade. As those hides were profitably exported to England, Carolina traders penetrated ever further into the backcountry, eventually challenging Virginia's own territorial claims and trade networks.

The explosion of the deerskin trade soon depleted animal numbers. This was just one of several grievances among southeastern Indian tribes that led to the Yamasee War in 1715. Another was the growth of rice plantations, which displaced the tribes from their historical lands. The ensuing war decimated the deerskin trade even further. In the wake of this collapse, the Crown appointed Francis Nicholson governor of the new South Carolina colony. Nicholson was instructed to build new forts in the interior to guard against French encroachments, and to strengthen trade with the tribes. In pursuit of the latter, he summoned the Cherokee, Catawba, and Creek Indians for a meeting, where he was given this map.

The author of the map is not definitively known, but it was most likely a Cherokee leader, or cacique, who sought to strengthen trade relationships with the Carolina colony and its port at Charlestown. Note the path alongside the top of the map, which suggestively circumvented the Catawba to directly link the Cherokee and Charlestown. Moreover, there is no path that connects the Cherokee to Virginia, an acknowledgment that the deerskin trade between the two had recently ended. In light of that, the map may have been a Cherokee strategy designed to cultivate trade with South Carolina.

Whether drawn by a Cherokee or a Catawba, this map was designed to influence Nicholson as he navigated the complex network of commercial and diplomatic relationships in the wake of a violent and deadly war. Just a few years later Nicholson received a very similar map from the Chickasaw, one designed to strengthen their trade relationship with the Carolina colony at the expense of the Catawba. Both are superb examples of cartography from a non-European perspective, and they underscore the dramatic and fluid relationships at work on a continent full of shifting alliances, networks, and rivalries. They also remind us that it was the Europeans who created the category of "Indian," lumping together groups who saw themselves as distinct—and sometimes competing—peoples. Finally, a chart such as this highlights the larger point of this book: that maps are simultaneously reflections of reality and instruments of persuasion.

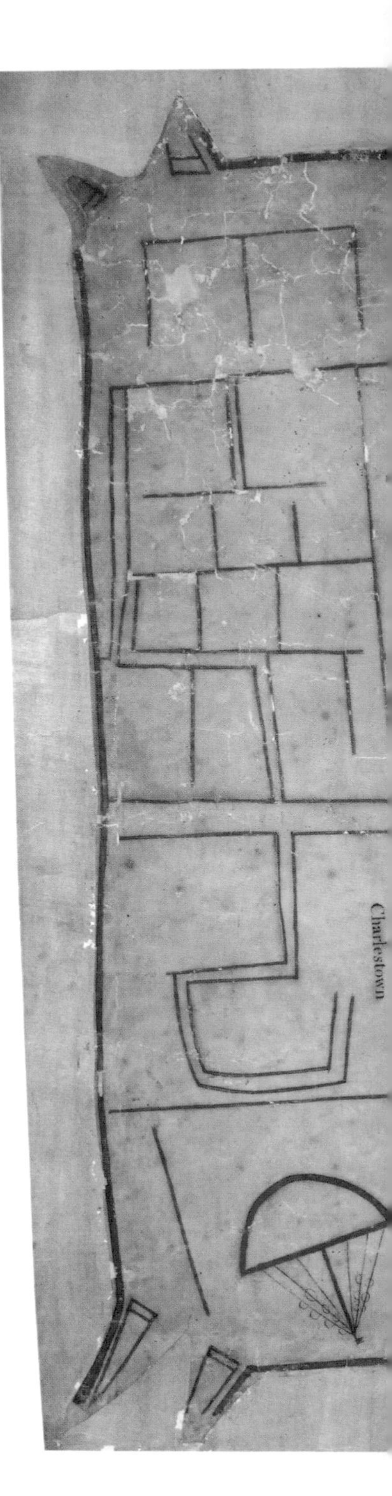

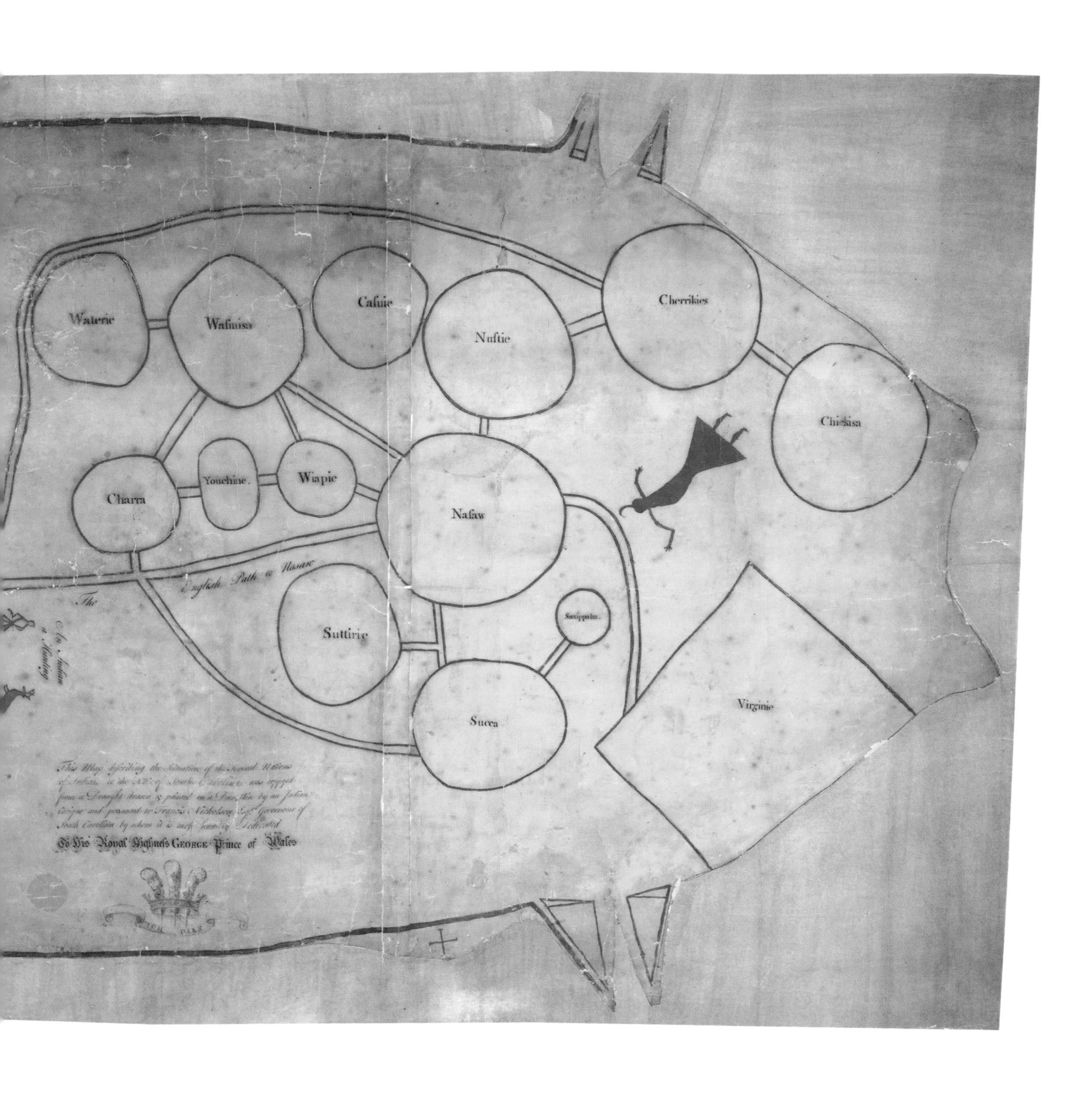
Waterie
Wafmisa
Cafuie
Nuftie
Cherrikies
Chickisa
Charra
Youchine.
Wiapie
Nafaw
Suttirie
Saxippaha.
Succa
Virginie
To His Royal Highnefs GEORGE Prince of Wales

ENGLAND AND THE SLAVE TRADE

Malachy Postlethwayt, "A New and Correct Map of the Coast of Africa," 1757

In 1713 the Spanish granted the British the exclusive right to supply their New World colonies with slaves. This grant of the "Asiento" in South America also coincided with an increased demand for slavery in the British colonies. In fact, the growth of labor-intensive crops—Caribbean sugar, Virginia tobacco, and Carolina rice—directly stimulated the slave trade. By 1750 the British were importing 50,000 to 60,000 Africans each year to their colonies, nearly half of whom came through South Carolina's booming port of Charleston.

Among the strongest advocates of the British slave trade was the political economist Malachy Posthlethwayt, who sought to make West Africa a center of the British empire in the 1730s and 1740s. But this future, he warned, was threatened by the current state of trade along the West African coast. Postlethwayt saw British merchants undercutting one another, which in turn made it easier for the French, Portuguese, and Dutch to enlarge their own share of the trade. In his mind, the best way to strengthen the British position in West Africa was to consolidate all commercial exchange within the newly established Royal African Company. This would also advance a mercantilist system whereby African slaves would be paid for by British products and East Indian commodities.

Postlethwayt drew a map to promote this mercantilist vision and Britain's position in the slave trade more generally. The lengthy annotations at left detail the imperial rivalries at work along the crowded Cape Coast. To highlight the even more frenzied activity on the Gold Coast, he drew an inset map that used national flags to identify the rival imperial interests at each port or point of entry. He first published his map in 1746 in a full-throated defense of the Atlantic slave trade in which he called for the British to fortify and protect their ports along the coast.

Postlethwayt's map reminds us how crucial slavery was to the survival and prosperity of the colonies. The Atlantic slave trade, he wrote, affords "our Planters a constant Supply of Negroe-Servants for the Culture of their Lands in the Produce of Sugars, Tobacco, Rice, Rum." That "Negroe-Trade" provided the British with an "inexhaustible Fund of Wealth and Naval Power." For this reason, Postlethwayt urged that British positions on the Slave Coast be reinforced, for these were the key to extending the empire into the interior. Along the coast of the map, he reminds readers that the British had opportunities to expand commerce into the interior, though they had been rebuffed by the French along the far western coast, north of the River Gambia.

Even as early as 1750, the slave trade had prompted criticism, which Postlethwayt countered by arguing that slaves were far better off in service to British planters than subject to ongoing warfare at home in Africa. But by 1757 Postlethwayt himself had become an ardent critic of the slave trade. He remained, however, a firm advocate of the expansion of imperial interests in West Africa, and insisted that the way for the British to end the trade was through expanded commerce with interior African kingdoms.

All told, from 1607 to 1807 over 3 million Africans were sent to the Americas through the British slave trade.

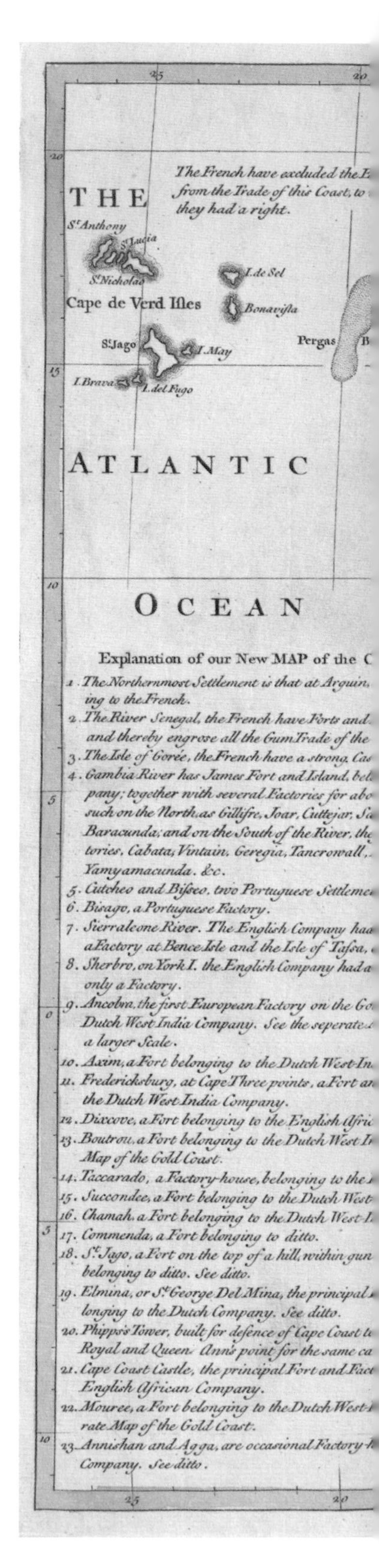

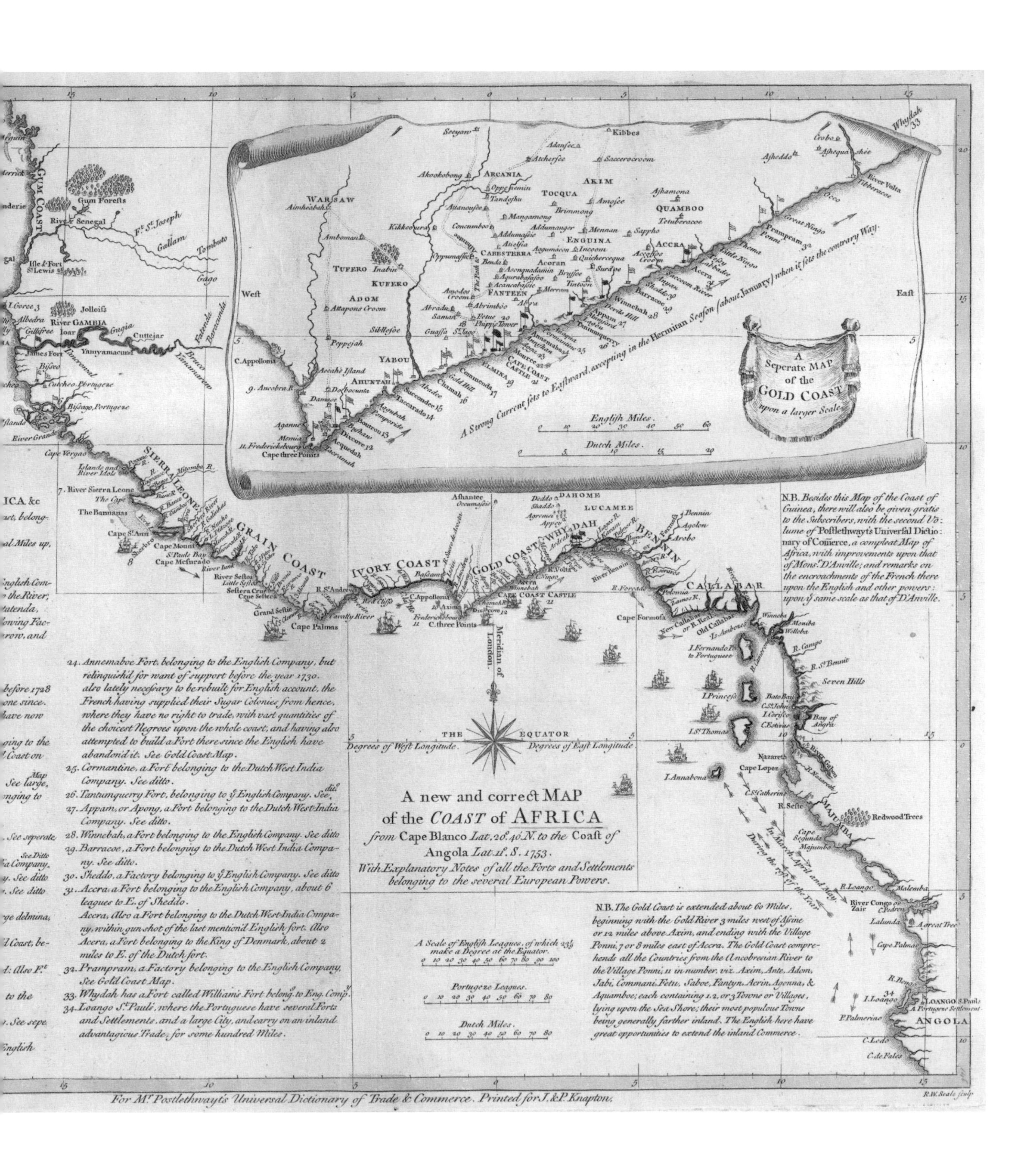

A new and correct MAP of the COAST of AFRICA from Cape Blanco Lat. 20°. 46. N. to the Coast of Angola Lat. 11°. S. 1753.
With Explanatory Notes of all the Forts and Settlements belonging to the several European Powers.
A Seperate MAP of the GOLD COAST upon a larger Scale.
A Strong Current sets to Eastward, excepting in the Hermitan Season (about January) when it sets the contrary Way.
English Miles.
Dutch Miles.
N.B. Besides this Map of the Coast of Guinea, there will also be given gratis to the Subscribers, with the second Volume of Postlethwayt's Universal Dictionary of Comerce, a compleat Map of Africa, with improvements upon that of Mons. D'Anville; and remarks on the encroachments of the French there upon the English and other powers: upon ye same scale as that of D'Anville.
24. Annemaboe Fort, belonging to the English Company, but relinquish'd for want of support before the year 1730. also lately necessary to be rebuilt for English account, the French having supplied their Sugar Colonies from hence, where they have no right to trade, with vast quantities of the choicest Negroes upon the whole coast, and having also attempted to build a Fort there since the English have abandon'd it. See Gold Coast Map.
25. Cormantine, a Fort belonging to the Dutch West India Company. See ditto.
26. Tantumquerry Fort, belonging to ye English Company. See ditto.
27. Appam, or Apong, a Fort belonging to the Dutch West-India Company. See ditto.
28. Winnebah, a Fort belonging to the English Company. See ditto
29. Barracoe, a Fort belonging to the Dutch West India Company. See ditto.
30. Sheddo, a Factory belonging to ye English Company. See ditto
31. Accra, a Fort belonging to the English Company, about 6 leagues to E. of Sheddo.
Accra, Also a Fort belonging to the Dutch West-India Company, within gun-shot of the last mention'd English fort. Also Accra, a Fort belonging to the King of Denmark, about 2 miles to E. of the Dutch fort.
32. Prampram, a Factory belonging to the English Company, See Gold Coast Map.
33. Whydah has a Fort called William's Fort belong. to Eng. Comp.
34. Loango St. Paul's, where the Portuguese have several Forts and Settlements, and a large City, and carry on an inland advantagious Trade, for some hundred Miles.
A Scale of English Leagues, of which 25 make a Degree at the Equator.
Portugeze Leagues.
Dutch Miles.
N.B. The Gold Coast is extended about 60 Miles, beginning with the Gold River 3 miles west of Asine or 12 miles above Axim, and ending with the Village Ponni, 7 or 8 miles east of Accra. The Gold Coast comprehends all the Countries from the Ancobresian River to the Village Ponni; 11 in number, viz. Axim, Ante, Adom, Jabi, Commani, Fetu, Saboe, Fantyn, Acrin, Agonna, & Aquamboe; each containing 1, 2, or 3 Towns or Villages, lying upon the Sea Shore; their most populous Towns being generally farther inland. The English here have great opportunities to extend the inland Commerce.
THE EQUATOR
Degrees of West Longitude.
Degrees of East Longitude
Meridian of London.
GUM COAST
SIERRA LEONE
GRAIN COAST
IVORY COAST
GOLD COAST
WHYDAH
BENNIN
CALLABAR
DAHOME
LUCAMEE
ANGOLA
MAJUMBA
For Mr. Postlethwayt's Universal Dictionary of Trade & Commerce. Printed for J. & P. Knapton.
R.W. Seale sculp

WAR AND DISCORD IN THE SOUTHEAST

Mark Catesby, “A Map of Carolina, Florida and the Bahama Islands,” in *The Natural History of Carolina, Florida and the Bahama Islands*, 1731–43

A century before John James Audubon thrilled Americans with his *Birds of America*, the Englishman Mark Catesby published the equally stunning *Natural History of Carolina, Florida and the Bahama Islands*. Catesby first visited Virginia and the West Indies in 1712, spending seven years studying plants and animals in their native habitats. He returned to London just long enough to realize that he needed to know more, and in 1722 he visited South Carolina to continue his work.

Catesby could not have known that his American visits would coincide with a period of profound upheaval. At first glance the elegant map he created to guide readers through his *Natural History* reveals little of this disruption. But, on closer inspection, we can see some clues to the imperial rivalries at work in the Southeast. Note that Catesby used color to identify the presence of the British (pink), the French (brown), and the Spanish (yellow). However neatly colored on the map, these colored spheres were far less stable on the ground. Catesby noted that the British charter for Carolina extended well south of St. Augustine. The Spanish rejected this claim, which led to several skirmishes, raids, and wars between the two on this contested borderland. In 1702 and again in 1740 the British raided St. Augustine to protest the Spanish practice of welcoming escaped slaves from the Carolinas and offering to liberate them in exchange for conversion to Catholicism.

The most important of these conflicts occurred at Stono, near Charleston in South Carolina. In 1739, a group of armed slaves rebelled, heading south toward promises of liberty in Spanish Florida. As they marched, the group grew to a critical mass of one hundred. A battle with the British killed many of these rebels, and those who survived were immediately

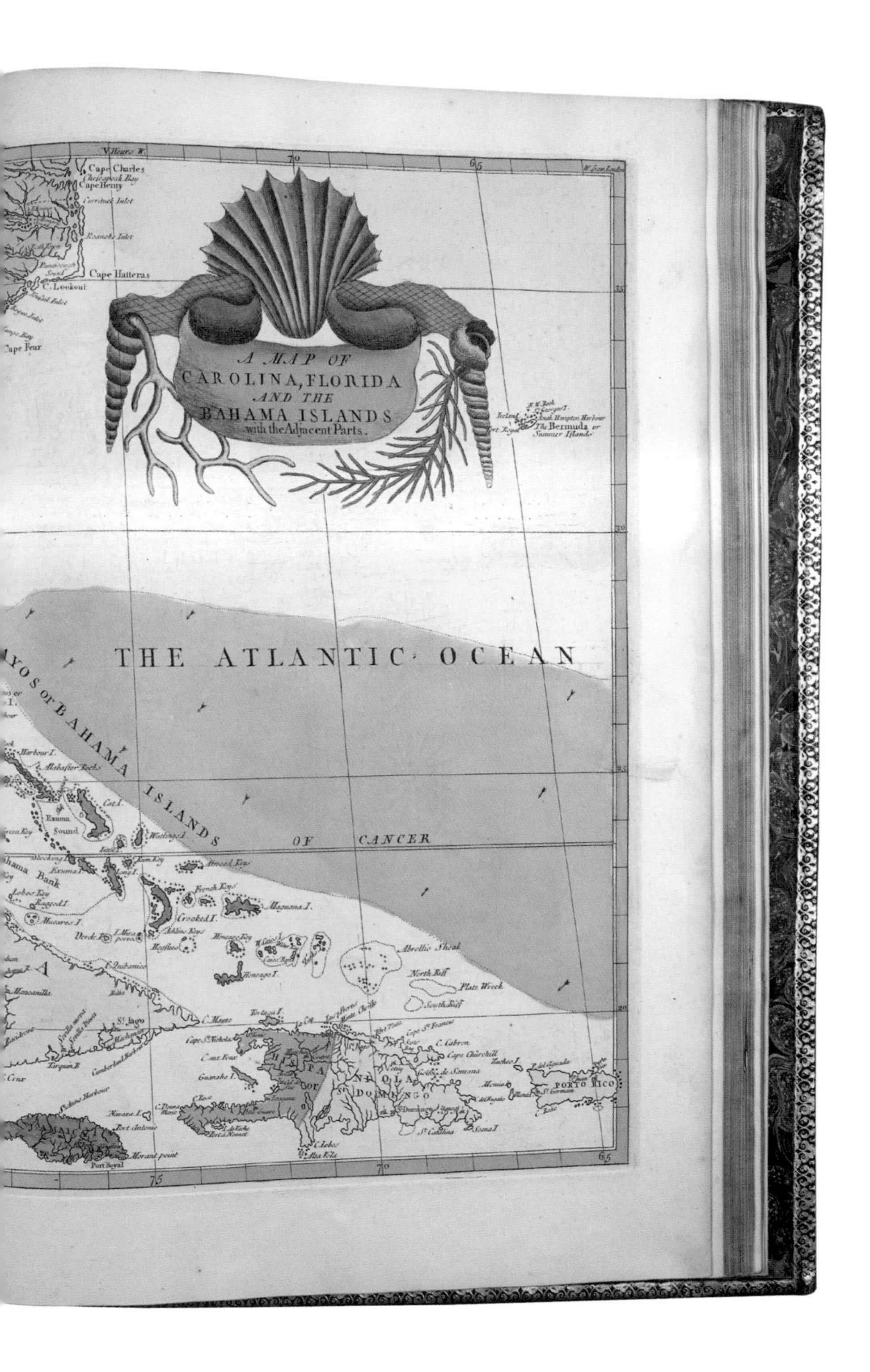

armed by the Spanish to help protect St. Augustine. The Stono Rebellion terrified British slaveholders and led them to tighten slave codes in South Carolina and to suspend the slave trade temporarily. Just as British colonists gradually asserted their rights in the eighteenth century, so too did slaves protest their own bondage.

Complicating matters further was the Yamasee War of 1715–7, in which British settlers drove several native tribes south into Florida. The destructiveness of that war challenged the very viability of the Carolina colony, which was saved only by an alliance between the Cherokee and the British.

Catesby's map is also one of the first to identify the new colony of Georgia. In 1732 James Oglethorpe petitioned the Crown to create this new colony as a haven for the "worthy poor" of England. Oglethorpe's high-minded goals included a prohibition against slavery, but the law was repealed once Georgians realized how profitable the plantations of Virginia and the Carolinas were. In fact, by 1740, well over half of South Carolina's population were slaves, almost all of whom were put to work in the rice fields that had grown up almost overnight in the Lowcountry.

Florida was riven by its own internal discord. As indicated by the map, the Spanish controlled much of the colony through the early eighteenth century. Yet the borders were continually contested. The dotted horizontal line across Florida marks the British claim for the southern border of the Carolina Colony. This territorial conflict drove years of border warfare between the Spanish and the British. With the defeat of the French in the Seven Years' War in 1763, the British took control of the entire colony, only to watch it revert to the Spanish when Americans achieved independence in 1783.

Despite these divisions and rivalries, Catesby presented a coherent geographical region in the eighteenth century. With its increasing dependence upon large-scale plantation agriculture, the Southeast was drawn into a transatlantic system of commercial exchange stretching thousands of miles to Africa and Europe. The profitability of that system in turn intensified British investment in the colonies.

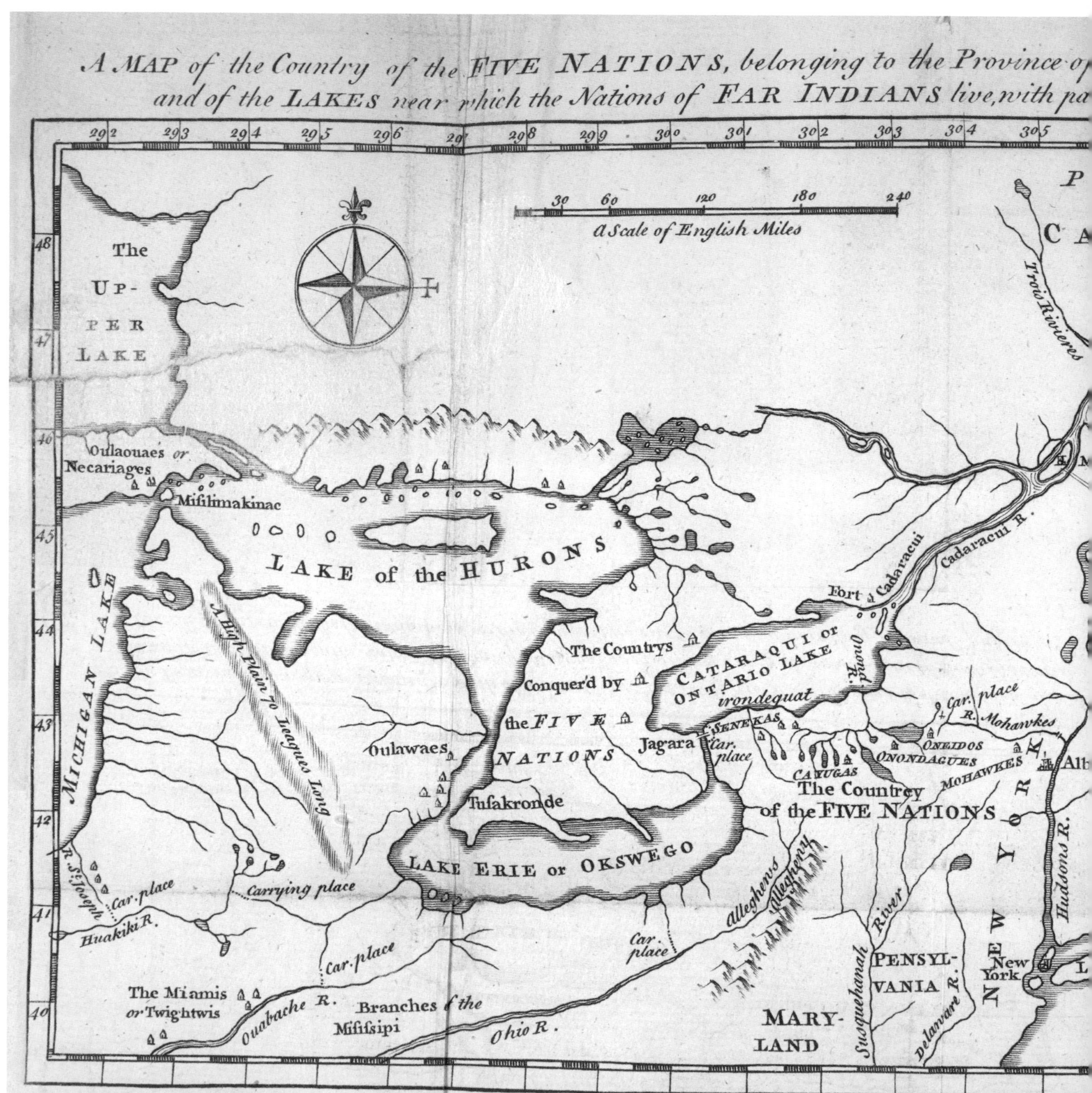

A MAP of the Country of the FIVE NATIONS, belonging to the Province
and of the LAKES near which the Nations of FAR INDIANS live, with
A Scale of English Miles
30 60 120 180 240
The UPPER LAKE
Oulaouaes or Necariages
Misilimakinac
LAKE of the HURONS
MICHIGAN LAKE
A High Plain 70 Leagues Long
The Countrys Conquer'd by the FIVE NATIONS
Oulawaes
Tusakronde
Jagara
Car. place
CATARAQUI or ONTARIO LAKE
irondequat
Fort Cadaracui
Cadaracui R.
Onond R.
SENEKAS
CAYUGAS
ONONDAGUES
ONEIDOS
MOHAWKES
R. Mohawkes
The Countrey of the FIVE NATIONS
NEW YORK
Hudsons R.
Trois Rivieres
LAKE ERIE or OKSWEGO
Carrying place
Car. place
Huakiki R.
The Miamis or Twightwis
Ouabache R.
Branches of the Missisipi
Ohio R.
Alleghenys
Allegheny
Susquehanah River
PENSYLVANIA
Delaware R.
MARYLAND
New York
N.B. The Tuscaroras are now reckon'd a sixth Nation, & live between the Onondagues & Oneidos; & the
received to be the seventh Nation at Albany, May 30th. 1723; at their own desire, 80 Men of that Nation
The chief Trade with the far Indians is at the Onondagues rivers mouth where they must all pass to go

IROQUOIS DIPLOMACY

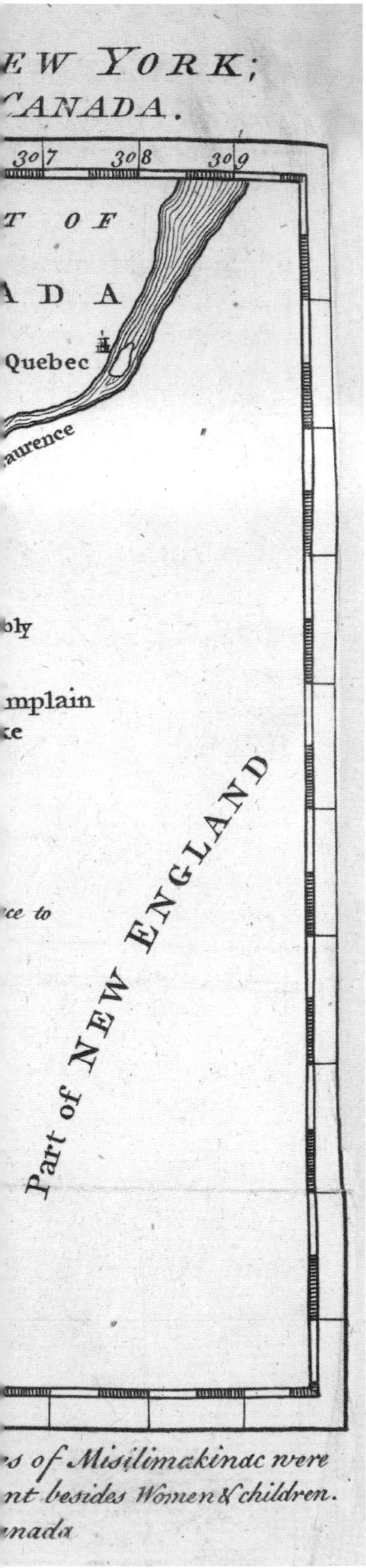

Cadwallader Colden, "A Map of the Country of the Five Nations, belonging to the Province of New York; and of the Lakes near which the Nations of Far Indians live, with part of Canada," 1755

If the Cherokee map on page 72 showed us a native perspective on trade and diplomacy, this gives us a colonial view of the same. Cadwallader Colden was surveyor general of the New York colony, and one of the first Europeans to chronicle Indian life and history in North America. His 1723 map of the Five Nations of the Iroquois captures the geopolitics of the early eighteenth century. In fact, Colden's map is a snapshot of the diplomatic maneuvering in the Ohio Valley that eventually led to the French and Indian War.

The map's history begins in 1720, when New York welcomed William Burnet as the new colonial governor. Burnet sought to improve trade with the Algonquian-speaking tribes to the west, which were closely aligned with French traders. To that end, Colden traveled to the frontier outpost of Albany in the fall of 1721 to negotiate an agreement with the Iroquois. His aim was to ensure that tribes further west could safely travel across Iroquois lands in order to trade with the British in New York. The continued growth of Indian trade was crucial to the commercial success of New York, both internally and relative to the other colonies and across the Atlantic.

After he returned from Albany, Colden wrote a long history of the Five Nations that was designed to challenge French power in the Great Lakes and Upper Mississippi River. The map itself reflects a British goal, but, given the limited contemporary knowledge, Colden was forced to rely upon Guillaume de L'Isle's French map (page 66) for the geographical detail. A closer look reveals that Colden and Burnet had other ambitions as well: several places marked "carrying place" or "car. place" indicate their belief that only a short distance separated the Great Lakes and the headwaters of the Ohio River. With the Ohio draining into the Mississippi, this was indeed an important claim. These notations on the map underscore the British aim of building a network of communication and transportation that would ultimately reach the Gulf of Mexico.

In these years, the Iroquois Confederacy had emerged as a powerful force of its own, concluding treaties with both the British and the French in 1701. Colden was keenly aware of this situation. As he put it, the Iroquois "used" the French Jesuits as hostages and could easily have destroyed the emerging colony of Quebec, shown at upper right. Their diplomatic skills made them a feared adversary, particularly given how "extreamly Revengeful the Indians naturally are." In response, the British proceeded deliberately and carefully to cultivate an alliance with the Iroquois that might facilitate westward trade and settlement.

Like the maps by Herman Moll and Guillaume de L'Isle, this one captures the intense rivalry between the French and the British for control of the Ohio Valley and the Mississippi River. While the French controlled the waterways, the British allied with the Iroquois Confederacy and used those alliances to claim control even where they had no settlers. Firsthand British knowledge of this region was also sorely limited, placing them even more at the mercy of the Iroquois. This also reminds us that British ambitions in the West were just beginning to gain momentum in the 1720s and 1730s.

Colden himself warned that the French could use the Saint Lawrence River and the Great Lakes to penetrate the interior and to corner trade with native tribes. In response, Governor Burnet—with Colden's support—urged Great Britain to restrict trade between the colonists and the French. Though the plan failed, the map indicates the emerging British designs on the trans-Allegheny West. In fact, this 1755 reissue of the map was published just after the French and British had gone to war.

TOBACCO AND VIRGINIA

Joshua Fry and Peter Jefferson, “A Map of the Most Inhabited Part of Virginia containing the whole Province of Maryland with Part of Pensilvania, New Jersey and North Carolina,” 1755

The British were disturbed by the growing French presence in the Ohio Valley during the 1740s, and particularly concerned by the vulnerability of western Pennsylvania and Virginia. In 1748 the British Board of Trade and Plantations solicited new and more detailed maps in order to secure those regions. To that end, the governor of Virginia commissioned the surveyors Joshua Fry and Peter Jefferson—father of Thomas Jefferson—to develop a new and detailed topographic map of the colony. Fry and Jefferson delivered their original draft to London in 1752; geographical errors near Lake Erie led to the substantially revised edition of 1755 that is reproduced here and on the next page. By that time, tensions between the French and British had led to war.

The power of this map lay primarily in its detailed depiction of the entire river system of Virginia, much of which is shown at right. For the first time, the four principal rivers as well as their tributaries were shown together, creating a larger picture of a colony that was at the height of its success in tobacco cultivation. Virginia was home to more extensive and navigable rivers than any other colony on the seaboard, leading one contemporary to remark that most every tobacco farmer had “a river at his door.” The James, York, Rappahannock, and Potomac rivers brought ships into the heart of the colony, and directly connected many of these tobacco planters to European markets.

Tobacco grew quickly in the Virginia colony with the corresponding growth of slavery. From the early settlements near Jamestown, the crop expanded north across the York and Rappahannock until it dominated the Chesapeake. By 1750 tobacco was North America’s most valuable export, enriching the colony and making it the population center of British America. The region covered by this map was home to 400,000 by 1740 (the population of Virginia alone was 260,000). Fully one-quarter of those were slaves. The profits from tobacco and slavery enabled Virginia to replicate the English model of a rural gentry and a landed elite. This in turn fostered a social, economic, and political hierarchy that profoundly influenced everything from the institutionalization of slavery to emerging colonial notions of liberty.

The artistic cartouche at lower right on the next page exemplifies the influence of tobacco farming in mid-eighteenth-century Virginia. Created by two noted London artists, it features the commercial exchange around tobacco: in the foreground two planters negotiate with a ship’s captain, while to the right an accountant with his back turned carefully records the profits. The four slaves, minimally dressed, each undertake a different aspect of the labor required for the success of tobacco.

That commerce would decline considerably during the Revolution. A credit crisis in 1772 fueled discontent among planters and traders. Once the war for independence began, tobacco cultivation dropped sharply, forcing Virginians to diversify into foodstuffs and other products. Virginia remained the pre-eminent British colony, and its gentry became the leaders of the Revolution and the early national period. The irony here is crucial, for it was the slaveholding elite that embraced the revolutionary spirit of liberty, no doubt in part because those men understood firsthand how fragile this liberty really was.

The Fry–Jefferson map was reprinted in eight states, and remained the most authoritative picture of Virginia and the adjacent area for forty years.

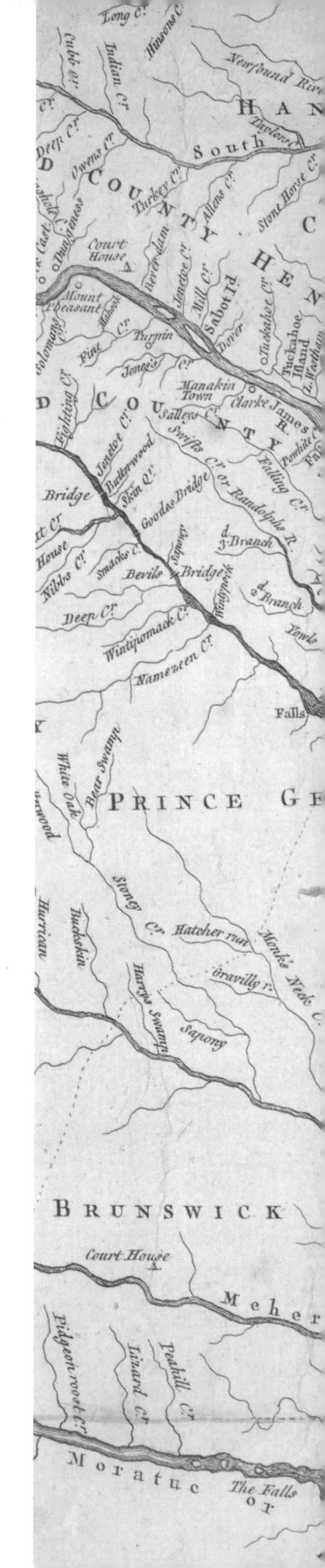

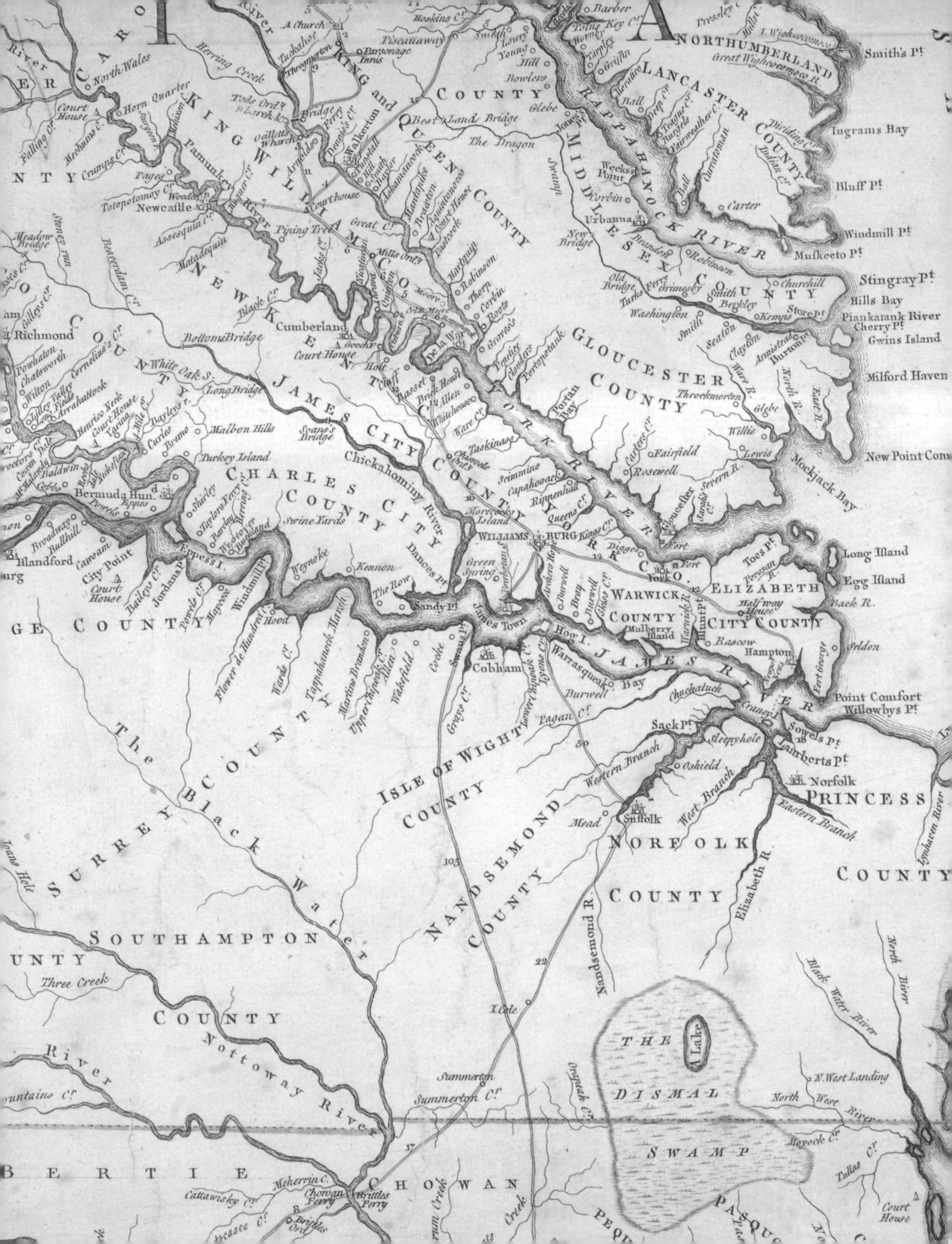

NORTHUMBERLAND
LANCASTER COUNTY
RAPPAHANOCK RIVER
MIDDLESEX COUNTY
KING and QUEEN COUNTY
KING WILLIAM CO.
NEW KENT C.
GLOUCESTER COUNTY
YORK RIVER
JAMES CITY COUNTY
CHARLES CITY COUNTY
YORK CO.
WARWICK COUNTY
ELIZABETH CITY COUNTY
JAMES RIVER
ISLE OF WIGHT COUNTY
NANDSEMOND COUNTY
NORFOLK COUNTY
PRINCESS
COUNTY
SURREY COUNTY
SOUTHAMPTON
COUNTY
BERTIE
CHOWAN
THE DISMAL SWAMP
A Lake
The Black Water
Nottoway River
Chickahominy River
Pamunkey River
Richmond
Williamsburg
James Town
York
Hampton
Norfolk
Suffolk
Urbanna
Newcastle
Cumberland
Blandford
City Point
Bermuda Hun.
Court House
Walkerton
Smith's Pt.
Ingrams Bay
Bluff Pt.
Windmill Pt.
Muskeeto Pt.
Stingray Pt.
Hills Bay
Piankatank River
Cherry Pt.
Gwins Island
Milford Haven
New Point Com
Mockjack Bay
Long Island
Egg Island
Back R.
Point Comfort
Willowbys Pt.
Sowels Pt.
Lamberts Pt.
Elizabeth R.
Eastern Branch
West Branch
Western Branch
Nandsemond R.
Black Water River
North River
North West River
N. West Landing
Summerton
Summerton Cr.
Meherrin C.
Chowan Ferry
Brittles Ferry
Three Creek
Cobham
Hog I.
Warrasqueak Bay
Mulberry Island
Sandy Pt.
Green Spring
Swine Yards
Weynoke
Kennon
Eppes I.
Turkey Island
Malbon Hills
Bottoms Bridge
Long Bridge
White Oak S.
Piping Tree
Best Land Bridge
The Dragon
Herring Creek
Piscattaway
Hoskins Cr.
Tuckahoe
Throgmorton
Parsonage
Innis
North Wales
Horn Quarter
Totopotomoy Cr.
Falling Cr.
Flower de Hundred
Tappahanock Marsh
Martins Brandon
Wakefield
Chuckatuck
Sack Pt.
Sleepyhole
Oshield
Craney I.
Mead
Fort George
Seldon
Halfway House
Rascow

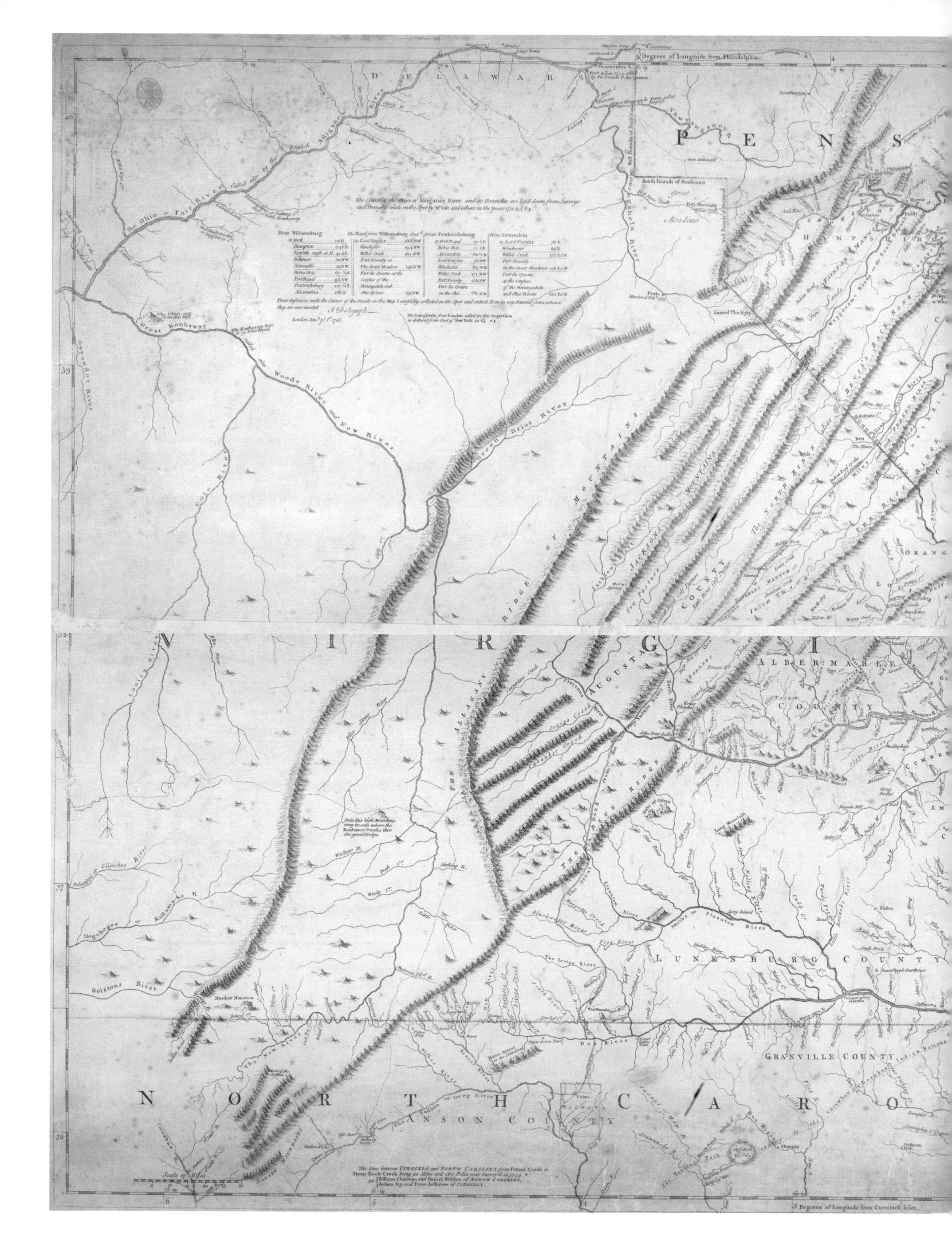

Degrees of Longitude from Philadelphia
D E L A W A R S
P E N S
The Course of the Ohio or Alliganey River and its Branches are laid down from Surveys and Draughts made on the Spot by Mr. Gist and others in the years 1751 2, 3 & 4.
From Williamsburg
to York 12 E.
Hampton 24.5 E.
Norfolk cross at H. 32 S.E.
Delawar 30 NW.
Newcastle 56 NW.
Hobss Hole 67
Port Royal 93 NW
Fredericksburg 107
Alexandria 168 N
The Road from Williamsburg Contd
to Lord Fairfax 186 NW
Winchester 194 NW
Wills's Creek 261 NW
Fort Necesity in The Great Meadow 346 NW
Fort du Quesne at the Conflux of the Monongahela and Ohio Rivers 391 NW
From Fredericksburg
to Port Royal 20 S.E.
Hobss Hole 52 S.E.
Alexandria
Lord Fairfax 76 NW
Winchester 84 NW
Wills's Creek 171 NW
Fort Necesity 238 NW
Fort du Quesne on the Ohio 284 NW
From Alexandria
to Lord Fairfax 78 E
Winchester 86 E
Wills's Creek 173
Fort Necesity in the Great Meadows
Fort du Quesne at the Conflux of the Monongahela and Ohio Rivers
These Distances with the Course of the Roads on the Map I carefully collected on the Spot and entred them in my Journal from whence they are now inserted. J. Dalrymple
London Jany ye 1st 1755
The Longitude from London added in this Impression is deduced from that of New York in 6h 4' 0.4
the Ohio or Fair River Called also by the English Alliganey
Monongahela River
South Bounds of Pensilvania
Great Meadows
Laurel Thicket
Great Konhaway
Woods River and New River
Green Briar River
Guyondot River
Cole River
Louisa River
Clinches River
Hogehegee or Kallamako R.
Holstons River
Meadow Mountain
The New River
Walkers R.
Sinking R.
Blue Stone River
from this high Mountain may be seen where the Konhaway breaks thro the great Ridge
RIDGE OF MOUNTAINS
JACKSONS MOUNTAINS
THE NORTH RIDGE
DEVIL'S BACK BONE
THE BLUE RIDGE
THE ALLAGANY
V I R G I
AUGUSTA
ALBEMARLE COUNTY
Staunton River
Smith's or Staunton River
Blackwater River
Pigg River
The Irvine River
LUNENBURG COUNTY
Dan River
GRANVILLE COUNTY
N O R T H C A R O
The Yadkin or Deep River
ANSON COUNTY
The Line between VIRGINIA and NORTH CAROLINA, from Peters Creek to Steep Rock Creek being 90 Miles and 280 Poles, was Survey'd in 1749.
By William Churton, and Daniel Weldon of NORTH CAROLINA, Joshua Fry, and Peter Jefferson of VIRGINIA.
Scale of Miles
3 Degrees of Longitude from Currituck Inlet

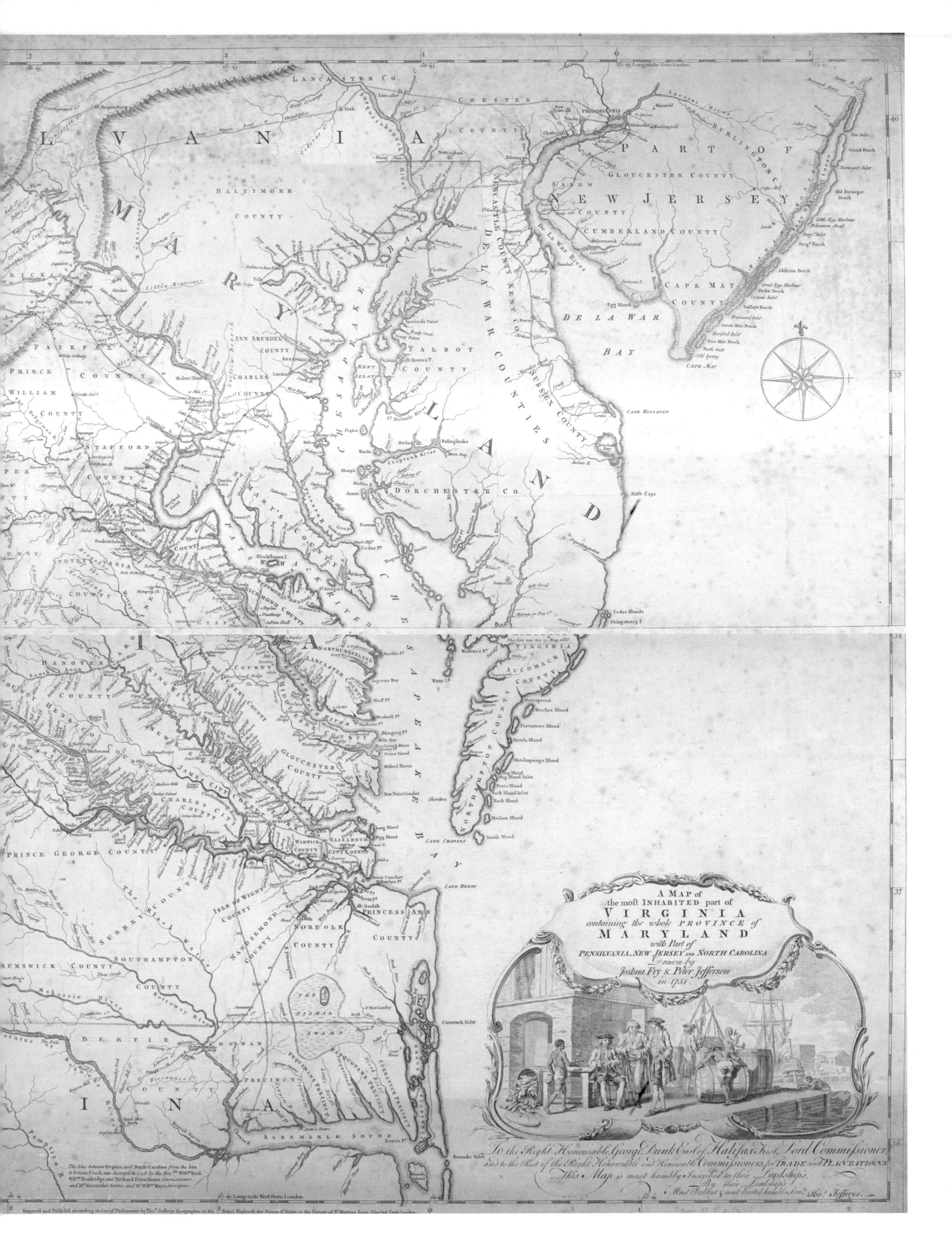

A MAP of
the most INHABITED part of
VIRGINIA
containing the whole PROVINCE of
MARYLAND
with Part of
PENSILVANIA, NEW JERSEY AND NORTH CAROLINA
Drawn by
Joshua Fry & Peter Jefferson
in 1751.
To the Right Honourable George Dunk Earl of Halifax First Lord Commissioner,
and to the Rest of the Right Honourable and Honourable Commissioners for TRADE and PLANTATIONS
This Map is most humbly Inscribed to their Lordships,
By their Lordships
Most Obedient & most devoted humble Servt. Thos. Jefferys.
The Line between Virginia and North Carolina from the Sea
to Peters Creek, was Surveyd in 1728 by the Honble Willm. Byrd,
Willm. Dandridge and Richard Fitzwilliams, Commissioners;
and Mr. Alexander Irvine and Mr. Willm. Mayo, Surveyors.
Longitude West from London
Engraved and Published according to Act of Parliament by Thos. Jefferys Geographer to His Royal Highness the Prince of Wales at the Corner of St. Martins Lane, Charing Cross London.
LANCASTER CO.
CHESTER COUNTY
PHILADELPHIA
PART OF BURLINGTON C.
GLOUCESTER COUNTY
NEW JERSEY
SALEM COUNTY
CUMBERLAND COUNTY
CAPE MAY COUNTY
DE LA WAR BAY
CAPE MAY
CAPE HINLOPEN
NEWCASTLE KENT SUSSEX
DE LA WAR COUNTIES
MARYLAND
BALTIMORE COUNTY
ANN ARUNDEL COUNTY
ANNAPOLIS
KENT ISLAND
TALBOT COUNTY
DORCHESTER CO.
PRINCE GEORGE COUNTY
CHARLES COUNTY
ST. MARY'S COUNTY
CHESAPEAKE BAY
PATOWMACK RIVER
FAIRFAX
PRINCE WILLIAM COUNTY
STAFFORD COUNTY
KING GEORGE COUNTY
WESTMORLAND COUNTY
RICHMOND COUNTY
SPOTSYLVANIA COUNTY
NORTHUMBERLAND
LANCASTER COUNTY
MIDDLESEX COUNTY
ACCOMACK COUNTY
NORTHAMPTON COUNTY
CAPE CHARLES
CAPE HENRY
HANOVER COUNTY
KING WILLIAM
HENRICO COUNTY
NEW KENT COUNTY
GLOUCESTER COUNTY
JAMES CITY COUNTY
CHARLES CITY COUNTY
WARWICK COUNTY
ELIZABETH CITY COUNTY
PRINCE GEORGE COUNTY
SURREY COUNTY
ISLE OF WIGHT COUNTY
NANSEMOND COUNTY
NORFOLK COUNTY
PRINCESS ANN COUNTY
SOUTHAMPTON COUNTY
BRUNSWICK COUNTY
THE DISMAL SWAMP
CHOWAN
BERTIE COUNTY
PERQUIMANS PRECINCT
PASQUOTANK PRECINCT
CURRATUCK PRECINCT
ALBEMARLE SOUND
Currituck Inlet
Roanoke Inlet
Roanoke Island

A YOUNG GEORGE WASHINGTON MAPS THE CLASH OF EMPIRES

"A map of the country between Will's Creek and Lake Erie, shewing the designs of the French for erecting forts southward of the lake; drawn ... before the erection of Fort Duquesne," circa 1754

Having established control over the Great Lakes in the early eighteenth century, the French began to push south into the headwaters of the Ohio and Mississippi rivers. Their ambition was to create a north–south axis of forts that would act as a continuous line of communication from Canada to Louisiana while also preventing the westward expansion of the British across the mountains of Virginia. Both European powers believed that they had a rightful claim to the region. The British argued that their alliance with the Six Nations gave them territorial control over the Ohio Valley, while the French countered that the discoveries of La Salle and Marquette gave them rights to the Upper Mississippi Valley.

In response to the French movement into the Ohio Valley, Virginia's lieutenant governor, Robert Dinwiddie, sent George Washington—just twenty-one at the time—with a forceful message for the French commander, who had erected a fort near Lake Erie, at the upper right of the map. Dinwiddie explained to Washington that "The Lands upon the River *Ohio*" were the property of the British Crown, and that the construction of French forts in this region would be considered acts of hostility. On October 31, 1753, Washington set off from Williamsburg to deliver the message across hundreds of miles. His route is recorded on this manuscript map, which was likely drawn by Washington himself, or copied from his original.

Facing difficult and unknown terrain, Washington pressed through mountains and forests to the headwaters of the Allegheny River. Beyond the Forks of the Ohio, he met with Indian chiefs at Logstown to shore up alliances and gain information about French activity. Soon thereafter he spent an evening at the Indian village of Venango with French soldiers. Having had a bit too much to drink, Washington explained, the French soldiers "gave license to their Tongues to reveal their sentiments more freely. They told me, That it was their absolute Design to take Possession of the *Ohio*, and by G— they would do it."

The French calculated that, though they were far outnumbered by the British, the latter were "too slow and dilatory" to prevent them from taking what was rightfully theirs. When Washington finally conveyed Dinwiddie's warning to the French commander at the newly erected Fort Le Boeuf along French Creek (at the upper right corner of the map), he was politely rebuffed. Given the expanding French presence near the Ohio River, the British were not really in a position to demand anything. When Washington insisted that the French stop taking British prisoners in the region, he was told that "no *Englishman* had a Right to trade upon those Waters; and that he had Orders to make every Person Prisoner who attempted it on the *Ohio*, or the Waters of it."

Washington returned to Williamsburg in the middle of January 1754. He may have failed to limit French expansion, but he produced a map and detailed notes that showed the British exactly what they were up against. Moreover, as a result of his mission Washington knew more about the Ohio Country than anyone else in the colonies. On the map he warned that "The French are now coming down ... to prevent our Settlements." He urged the British to respond by constructing "a fort near Shanapins Town" at the Forks of the Ohio (at the center of the map). He also confidently asserted that the French presence was insufficient to defend the forts they had built near Lake Erie. While at Fort Le Boeuf, he had surreptitiously counted the men and canoes he saw passing by on French Creek, and from that number extrapolated the general number of French soldiers in the region.

Dinwiddie promptly published Washington's map and journal. He then enlisted 200 men to march into the valley to erect a fort at the confluence of the Monongahela and Allegheny rivers, for Washington recommended this site as giving "absolute Command of both Rivers." Soon the French seized the site and renamed it Fort Duquesne, using it as a strategic outpost during the nine-year fight against the British known as the French and Indian War. Britain's ultimate victory in the war led to the nearby construction of Fort Pitt, which eventually became Pittsburgh.

Washington's unassuming sketch map outlines a remote region that drew the two most powerful empires in the world into war, even though both had questionable claims to the territory. The result was extraordinary: the war, which lasted until 1763, ended French claims in North America. Within a decade, the British colonists were again led into war by George Washington, this time against their own country.

A Scale of Miles

10	20	30	40	50	60	70	80	90	100

The French are now coming from their Forts on, and near the Lake Erie, to Venango to Erect another Fort, – and from thence they design to the Forks of Monongehele and to the Logs Town and so to continue down the River building at the most convenient places in order to prevent our Settlements &c.

NB: A little below Shanapins Town in the Forks, is the place where we are going immediately to Build a Fort, as it Commands the Ohio and Monongehele

Fort
Fort
Cassawaga
Venango
Logs Town
Mingo Town
Ohio
River
Shanapins Town
Turtle Creek
Mr. Fraziers
Queen Aligueppas
Monongehele
Mr. Gists new Settlemt.
Turkey Foot
Aligany
Mountains
Savage River
Wills Creek
Potomack River
Wappacomo

THE PRESENT CONJUNCTURE OF AFFAIRS IN AMERICA

Lewis Evans, "A General Map of the Middle British Colonies, in America," 1755

The previous map captured the trek of young George Washington into the West, where he witnessed French efforts to move toward the confluence of the Monongahela and the Allegheny rivers at present-day Pittsburgh. This spot became the flashpoint between the French and the British as they went to war over claims to the larger "Ohio Country." Recently it had become clear that the Ohio River and its tributaries potentially connected the interior to the established colonies, the Great Lakes, and the Mississippi River down to New Orleans.

These geographical revelations raised the stakes between Britain and France in 1754. In response, the Pennsylvania legislature asked the surveyor Lewis Evans to map the relationship between the bustling seaboard colonies and this promising interior. Evans complied, issuing an authoritative map that detailed "the present Conjuncture of Affairs in America," meaning the full-scale war that was raging between France and Britain over the Ohio Country. Along with a pamphlet printed by Benjamin Franklin, the map urgently called for the British to confront the growing French presence in the trans-Allegheny West.

Evans' small map was designed to be tipped into a short pamphlet promoting western settlement. This size limited the amount of topographic detail that he could offer, yet the map is bursting with information. The complex topography of the Allegheny Mountains is detailed, but more important was the inviting picture of the Ohio Country. "The English have several Ways to Ohio," he wrote, "but far the best is by Potomack." He carefully annotated the river systems of the Ohio Valley, distinguishing gentle flows from rapids and falls and pointing out lands where British settlers might form a bulwark against the French. He identified waterways navigable by canoes, boats, and larger vessels, and enthusiastically marked short portages that offered the possibility of inland transportation between different watersheds.

Note the way Evans also carefully delineated Indian lands. At first glance, this strikes the modern reader as an acknowledgment of the indigenous peoples who greeted European settlers in North

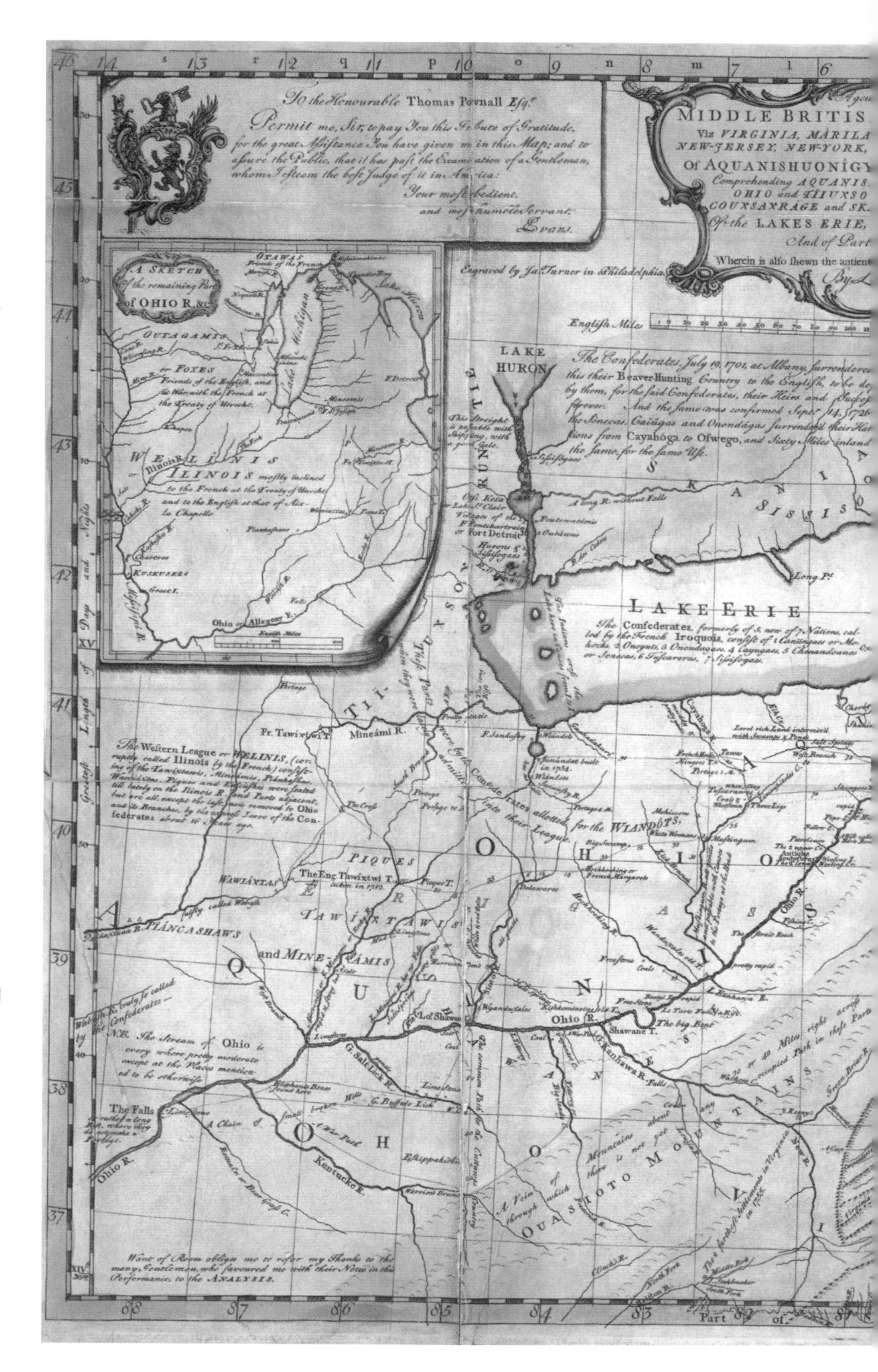

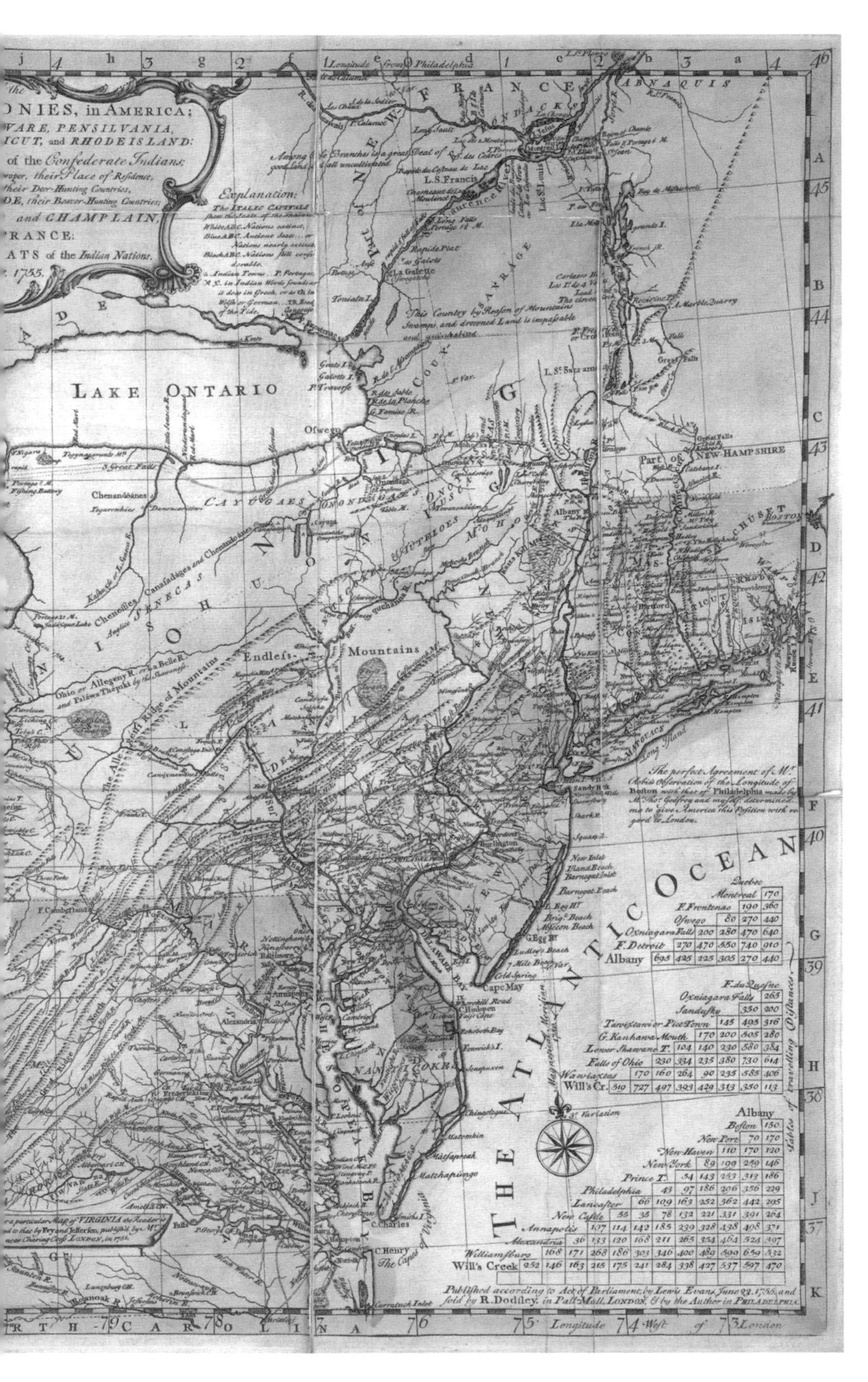

America. Yet this too was a strategic decision, for the British considered the lands of the Five Nations—which had then expanded to six— to be an extension of their own. As Evans put it, "whatever is theirs, is expressly acceded to the English by Treaty with the French." To the extent that they could map these tribes, the British envisioned their own empire stretching not just along the seaboard but also into the interior.

The map was used by General Edward Braddock during the war against the French, and was reprinted and pirated for decades. Perhaps most intriguing is Evans' closing statement in the pamphlet, where he entertained the possibility that the colonies themselves might eventually seek independence. This was two decades before the Revolutionary War, a response to the incipient voices of independence voiced in Massachusetts. But Evans dismissed this as the "Height of Madness," arguing that the colonies were far better off with Britain than independently facing "French power" in the west.

The very fact that Evans' map was printed in the colonies is suggestive. Benjamin Franklin's press was just one of several in British North America that produced newspapers and pamphlets; by contrast, no printing press existed in Spain's northern colonies of Florida and New Mexico, nor was there a press operating in New France. The British colonies also boasted a number of libraries and bookshops in New York, Boston, and Philadelphia. These would foster a high level of literacy and a dynamic print culture, which proved instrumental in spreading the anti-British sentiment that led to the Revolution. In other words, it would not be long before Lewis Evans' map took on a very different meaning from what he had intended.

SPANISH GEOGRAPHICAL INTELLIGENCE IN THE SOUTHWEST

Don Bernardo de Miera y Pacheco, "Plano geographico de la tierra descubierta y demarcada," 1778

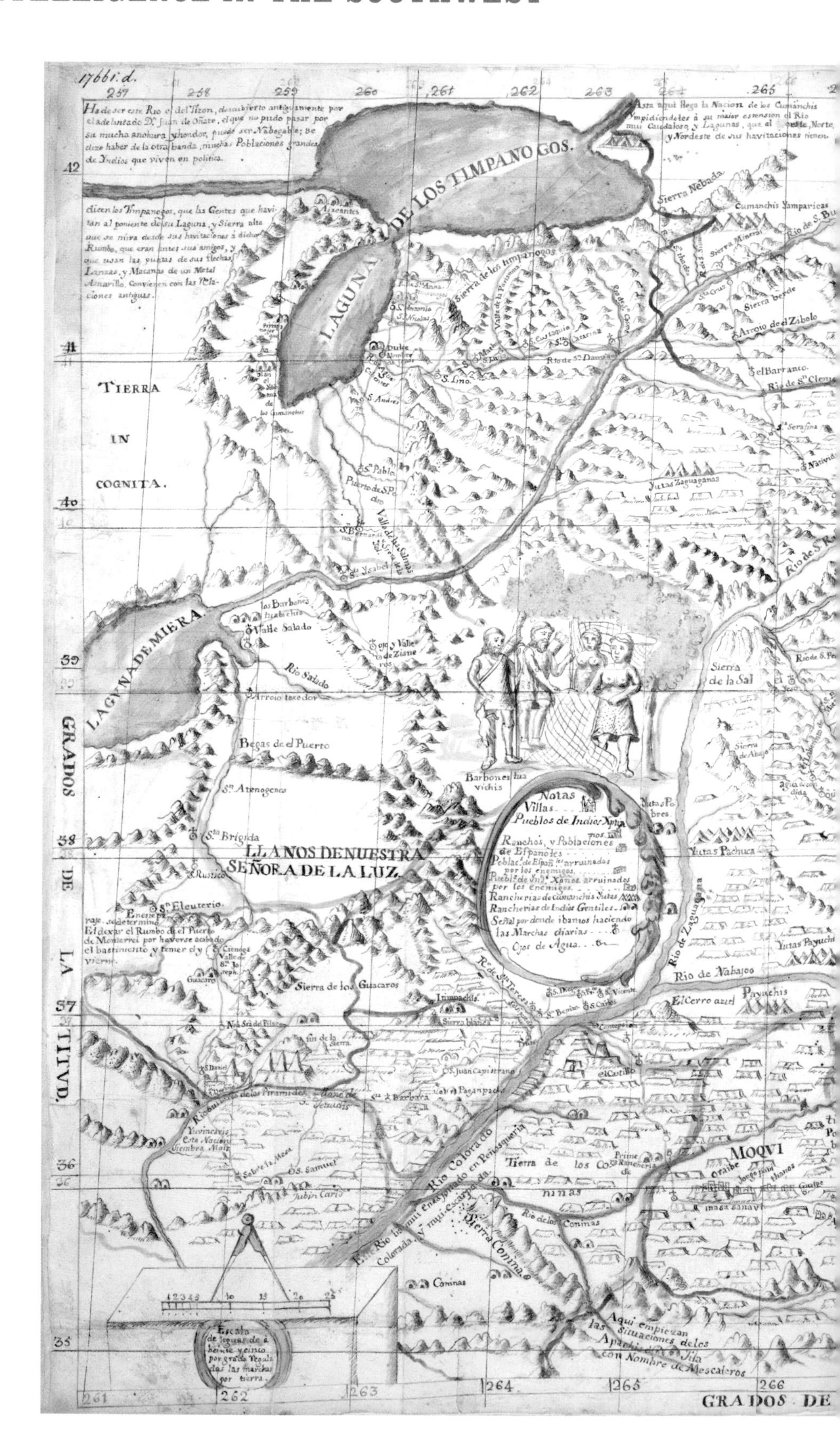

July 1776 is enshrined as a moment of unrivaled importance in American history, when the Declaration of Independence announced emancipation from the British Crown. It was also when, 3,000 miles to the west of Philadelphia, a band of Spanish missionaries and explorers left Santa Fe on a reconnaissance mission through what is now northern New Mexico, western Colorado, eastern Utah, and northern Arizona.

The explorers had two goals: first, to find an overland passage to the missions of Alta California, thereby solidifying Spain's northern frontier; second, to demystify the area now known as the Four Corners, a region that had remained largely unexplored by Europeans. The Spanish faced a host of hostile neighbors to their north: Russians and British bent on settling the northern Pacific coast, French exerting pressure from the east, and native tribes such as the Pueblo who had historically resisted Spanish control.

The expedition of Francisco Atanasio Dominguez and Silvestre Vélez de Escalante included the skilled cartographer and artist Bernardo de Miera y Pacheco. But no sooner had they set off than they began to encounter problems. Traversing some of the most complex and difficult terrain of the Southern Rockies, they became lost, and several members fell ill. If not for the aid of a young Ute Indian guide, the entire expedition might very well have perished. Upon their successful return in January 1777, Miera y Pacheco produced one of the most comprehensive maps of the Southwest yet made. This gave Spanish missionaries their first comprehensive profile of the region based on actual observation.

The symbolism of the map is hard to miss: a papal chariot in the upper right corner marks the power of the Church over Spanish North America. The very sight would have confirmed British fears of popery. To extend Spanish influence, Miera y Pacheco advocated settlements near Salt Lake, on the San Juan River, and at the confluence of the Gila and Colorado rivers. His advice was not taken, and Spain's control over the region weakened in subsequent decades.

A close look at the map also reveals Spanish priorities and perspectives. The depiction of "bearded Indians" at the center of the map testifies to Miera's

ethnographic interest in the mysterious native tribes that were rumored to live in this area.

The expedition also convinced him that the Rocky Mountains were the backbone of North America, with headwaters of rivers that flowed eastward and westward into two different oceans. The most geographically accurate aspect of the map was the depiction of the upper Colorado River Basin, as well as the San Juan and Dolores rivers. On the map, the San Juan River is marked "Rio de Nabajoo," fed by several tributaries to the north.

But there were also serious errors, most of which grew out of Miera's hope of finding a navigable river flowing west from the Rocky Mountains through the Great Basin to the Pacific Ocean. At upper left, he asserted a large river flowing west from Lake Timpanogos. Miera invested similar hope in the Rio de S. Buenabentura to the east (later named the Green River), which he believed would drain to "Laguna de Miera" and the Pacific beyond.

Miera's map influenced geographic knowledge for decades, as is apparent in maps compiled by Alexander von Humboldt and Zebulon Pike in about 1810. But perhaps just as intriguing is its diplomatic influence. When the Americans completed the Louisiana Purchase in 1803, they argued that the southwestern border included all the land to the Rio Grande, including much of present-day Texas. The Spanish countered that the border lay much further east, and used the topographic detail on this map to demonstrate their superior knowledge of the terrain and their historical rights to the region. That western border of Louisiana Territory would remain contested for years, first with Spain and later with Mexico. The entire area mapped here became part of the United States in 1848, though its rich and complex Spanish and Mexican heritage endures.

THE SHOT HEARD ROUND THE WORLD

"A plan of the Town and Harbour of Boston ... Shewing ... the late Engagement between the King's Troops & the Provincials ...," 1775

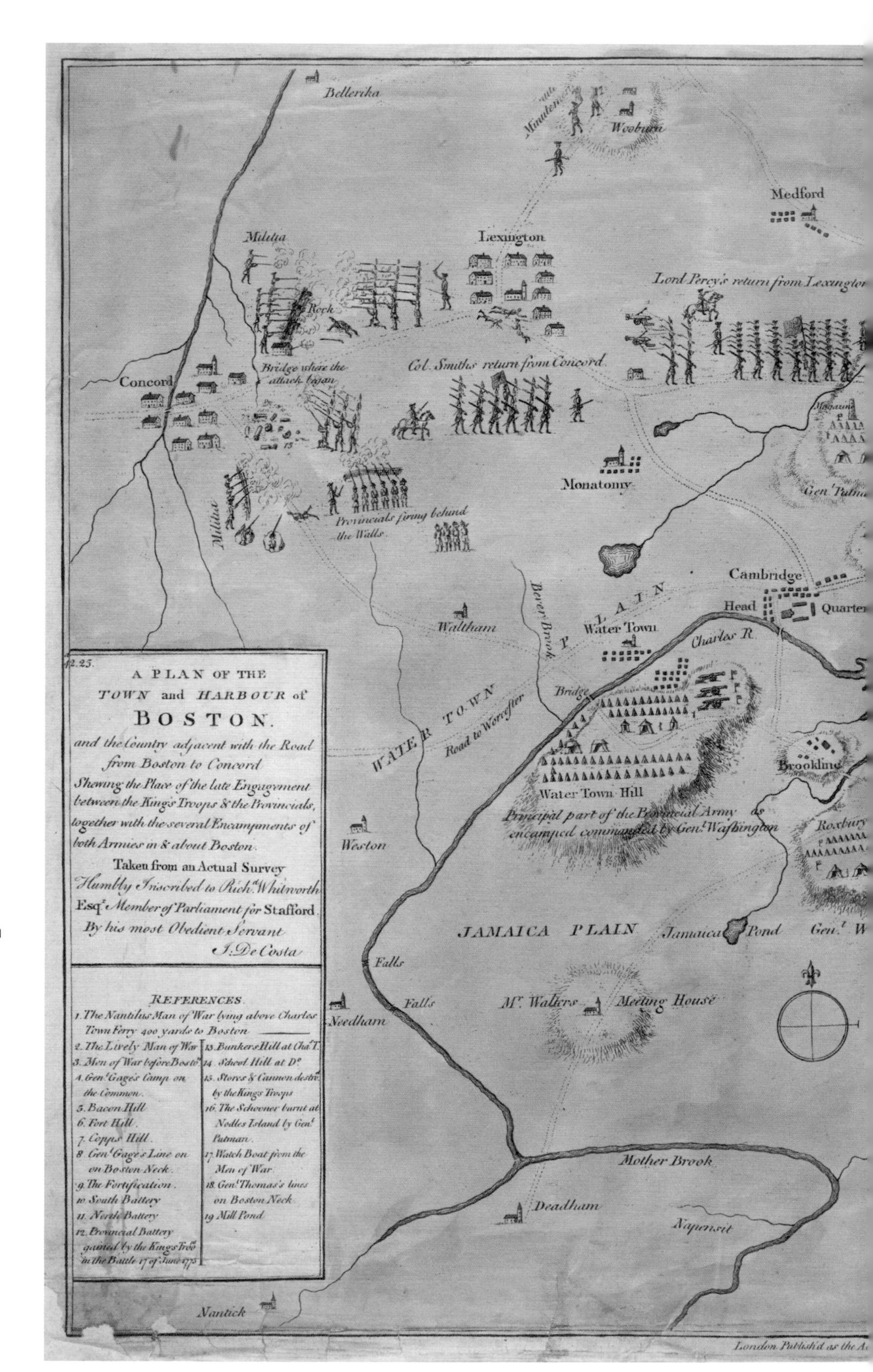

After the British victory over the French in 1763, the colonists expected a reprieve from war and a general expansion of liberties. Instead, the British tightened control over the colonies and demanded greater taxes to help pay for the late war. This led to defiance, which was initially sporadic and centered in Massachusetts. In 1770 a confrontation between colonists and royal troops was quickly dubbed the Boston Massacre. Three years later Bostonians dumped stores of British tea into the harbor, prompting Parliament to close the port and to crack down on town halls and other political activity. This only convinced more colonists that the British posed a direct threat to their liberties.

To coordinate this growing resistance, twelve of the thirteen colonies sent representatives to the First Continental Congress in October 1774. Such a meeting generated solidarity among the colonists and helped them to see their common plight. They agreed to reconvene the following May if their demands were not met. In the meantime, Committees of Safety began to organize local governments in order to wrest political power away from Britain and its colonial governance structure. In March 1775 Patrick Henry cried "Give me liberty, or give me death," urging the colonists to move from political mobilization to armed opposition.

Within a month these tensions exploded in Massachusetts. The conflict began when the British commander in the colonies, Thomas Gage, sent forces stationed in Boston to confiscate gunpowder and weapons stored in the nearby town of Concord. On the night of April 18, Lieutenant Colonel Francis Smith led his men across the Charles River through Cambridge and Menotomy toward Lexington. Paul Revere alerted nearby towns that the British were advancing, which drew Minutemen toward Lexington, at the upper edge of the map. The British met the rebels on Lexington Common at dawn, and killed eight Americans.

The British marched on to Concord, only to find that the weapons had been moved and hundreds

of rebels had taken up positions along the North Bridge. This map—among the earliest records of the Revolutionary War—depicts a British perspective on the fighting that ensued. The Americans are shown camped behind walls and rocks, firing on the British from three sides. Describing the "Bridge where the attack began," the map implies that the Americans ambushed the British regulars. Smith and his men were forced to retreat, dogged by rebel sniper fire that drove the men—along with Lieutenant General Hugh Percy's troops—back to the safety of Boston.

Just weeks later, the colonists reconvened at the Second Continental Congress, authorizing an army to fight the British and naming General George Washington as its commander. The skirmishes at Lexington and Concord became the "shot heard round the world," drawing reinforcements on both sides. British warships surrounded Boston, while newly formed American armies camped along the Mystic and Charles rivers. Though unrecorded on the map, the Siege of Boston coincided with an outbreak of smallpox that ravaged those trapped in the city. The outbreak continued until 1782, killing more Americans than died in the Revolutionary War itself.

Given Washington's heroic service in the French and Indian War (page 84), it must have been particularly poignant for the British to learn that he would command the "Provincial Army" shown here. Both sides raced to hold the high ground of Dorchester and Charlestown, which led to the Battle of Bunker Hill in June. This map was published immediately after that battle, before anyone could have known the war would ultimately end British control over the colonies.

WHERE THE BRITISH LAID DOWN THEIR ARMS

Sebastian Bauman, plan of the investment of York and Gloucester, 1781

The rebellion around Boston shown on the previous page quickly turned into a civil war, and ultimately became a struggle for national liberation. Today, we assume that the colonists uniformly rejected British control, but before the war about a third of them remained loyal to the Crown, while another third hoped for reconciliation. Even as the war galvanized more supporters for independence, the great military question was how the poorly trained and ill-funded rebels could defeat one of the world's strongest armies. In part it was the persistence of colonial soldiers and civilians that forced the British to abandon the war. Equally important was the intervention of the French, who allied with the Americans and critically aided them at the Battle of Yorktown in October 1781.

For much of this six-year war the outcome was in doubt. With a stalemate prevailing in the northern colonies by 1778, the British turned south to capitalize on that region's strong loyalist sentiment. After conquering Savannah and Charleston, Lord Cornwallis, commander of the British forces in the South, moved up the coast in 1781 to attack American supply and training bases in Virginia.

The commander of the Continental Army—George Washington—sought to fight the British in New York. But French forces under the Comte de Rochambeau stressed the importance of striking in the Chesapeake, by both land and sea. Washington accordingly sent men under the leadership of the Marquis de Lafayette to confront the British in Virginia. From April to August, Lafayette drove Cornwallis back, forcing him to retreat to Yorktown and await support from the British Navy.

Cornwallis chose Yorktown for its strategic location on a narrow spot along the York River. Across the river, Gloucester Point offered a potential escape route for the British soldiers. But ultimately that geography worked against Cornwallis by trapping the British against the river, in a twenty-day siege that became the climax of the Revolutionary War. The map at right recounts the critical moments of that campaign.

In August 1781, Cornwallis built enclosed fortifications—known as redoubts—that ringed the southern edge of Yorktown. The British and Hessian forces in those fortifications are marked in pink. Within a month, however, 8,300 American and French troops had arrived at Yorktown, and an additional 17,500 were camped nearby at Williamsburg. With these numbers, the allies far outnumbered the British and Hessian forces at Yorktown and across the river at the tip of Gloucester Point. Absent from this early copy of the map is the French fleet on the York River, which supplied crucial aid to the Americans by blockading the British in a way that prevented both reinforcements and evacuation.

To the left in yellow are Rochambeau's French forces, while the American military under General Washington is shown at right in blue. The turning point was on October 11, when the Americans and the French advanced toward Yorktown to build a series of trenches and earthworks (marked as blue and yellow lines to the south of the town) that surrounded British defenses. With the blockade on the river in force, the British were left largely stranded, without necessary supplies and reinforcements.

Over the next few days, the British fired against the allies from their positions at Redoubts 9 and 10, the southeastern corner of the British fortifications shown near the river. On October 14, Commander Alexander Hamilton and his men stormed Redoubt 10, a move that many considered terribly risky but which ultimately moved the allies closer to the British, where they intensified their bombardment.

In response, the British attempted to retreat across the river, though a sudden windstorm prevented either a successful crossing to Gloucester Point or a safe return to Yorktown. The allies closed in even tighter, shown on the second line of trenches, and within three days Cornwallis signaled his willingness to surrender. By this time, the Continental Army had taken over 7,000 British prisoners of war.

The Battle of Yorktown forced the British to abandon the war effort and to begin negotiating for peace. This spare and elegantly colored map captured that turning point. It was drawn by Sebastian Bauman, a forty-two-year-old major in the Second Continental Artillery Regiment who had been trained as an engineer in the Austrian Imperial Army. He drew the map just after the British surrendered, and it became the blueprint for all subsequent renderings of the battle. In the foreground are the quarters of Rochambeau and Washington, near "The Field where the British laid down their Arms." Those words were truer than Bauman realized, for this was the last major land battle of the war.

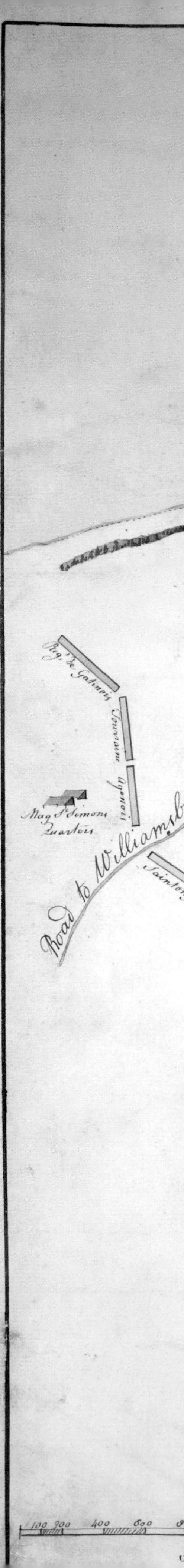

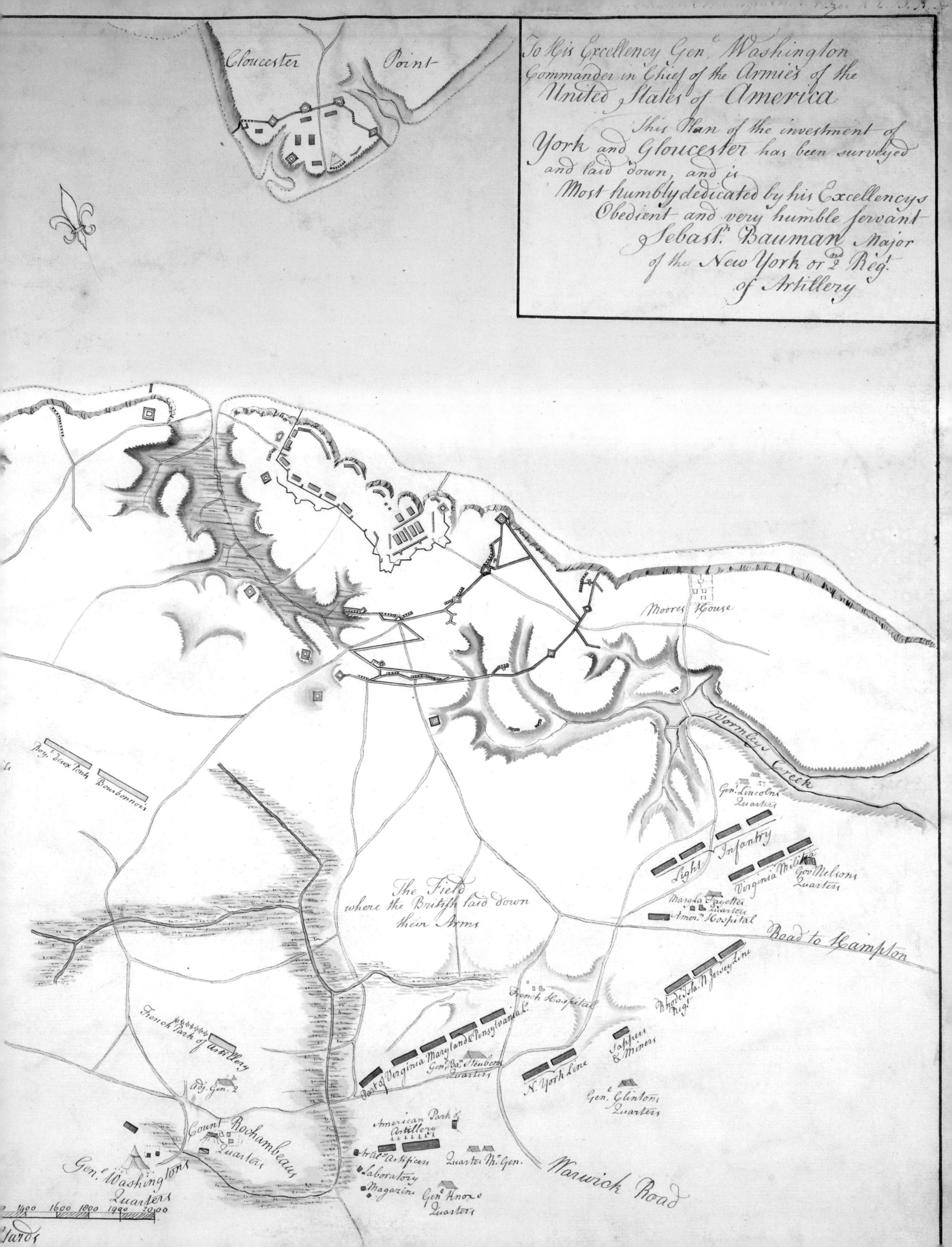

To His Excellency Genl. Washington
Commander in Chief of the Armies of the
United States of America
This Plan of the investment of
York and Gloucester has been surveyed
and laid down, and is
Most humbly dedicated by his Excellencys
Obedient and very humble servant
Sebastn. Bauman Major
of the New York or 2nd Regt.
of Artillery
Gloucester Point
Moores House
Wormley Creek
Roy. deux Ponts
Bourbonnois
Genl. Lincolns Quarters
Light Infantry
Virginia Militia
Gov. Nelsons Quarters
Marqs. La Fayette Quarters
Amern. Hospital
Road to Hampton
The Field where the British laid down their Arms
French Hospital
Rhode Isla. N Jersey Line Regt.
French Park of Artillery
Part of Virginia Maryland & Pensylvania Ls.
Genl. Bn. Steubens Quarters
N. York Line
Sappers & Miners
Genl. Clintons Quarters
Adj. Genl.
Count Rochambeaus Quarters
Genl. Washingtons Quarters
American Park of Artillery
Artiller Artificers
Laboratory
Magazine
Quarter Mr. Genl.
Genl. Knoxs Quarters
Warwick Road
1400 1600 1800 1900 2000
Yards

John Mitchell, "A Map of the British Colonies in North America ...," 1775 [1755]

The eighteenth-century geopolitical contest between the French and the British in North America produced a steady stream of maps. Among the most influential of these was John Mitchell's "Map of the British Colonies in North America." Mitchell was born in Virginia, was educated in Edinburgh, and practiced medicine in the colonies before returning to Britain in 1746. Thereafter he developed a keen interest in North American geography, and especially the strategic position of the Ohio Valley.

In the early 1750s the earl of Halifax, who presided over the British Board of Trade, asked Mitchell to compile a map that would help defend British claims to the Ohio Valley. Mitchell responded with a massive map measuring 4½ feet by 6½ feet. He detailed both physical and human geography in a way that was comprehensive and thoroughly British in its perspective.

A bold red line follows the Mississippi River before turning northeast through Lake Michigan and then east toward the Saint Lawrence Seaway. This line marked the British interpretation of their territorial borders, as laid down in the Treaty of Utrecht in 1713. With this line, the British claimed all of the Ohio Territory and much of the Great Lakes. Mitchell substantiated this British claim on the map itself, as shown on the next page. Just east of the Mississippi River, he identified several longstanding English settlements in order to buttress English sovereignty. He then reinforced those territorial claims by pointing to the presence in the same region of Native American tribes that had either allied with the British or signed treaties with them. All of Mitchell's annotations were designed to limit French claims to the area east of the Mississippi River.

The French, however, asserted a boundary far to the east, marked on the map by a thick yellow line. The vast region between the red and yellow boundaries formed the heart of the conflict between the French and the British. In fact, by the time Mitchell issued the first edition of his map in 1755, the two nations had gone to war over this territory. The British consulted the map throughout the French and Indian War, and their victory ended the French presence in North America altogether. Yet this British dominance would not last. By the 1770s the British were again at war, this time with the colonists themselves. At the end of the war this very copy of Mitchell's map was used by Britain's chief negotiator, Sir Richard Oswald, to establish the boundaries of the new United States at the Treaty of Paris.

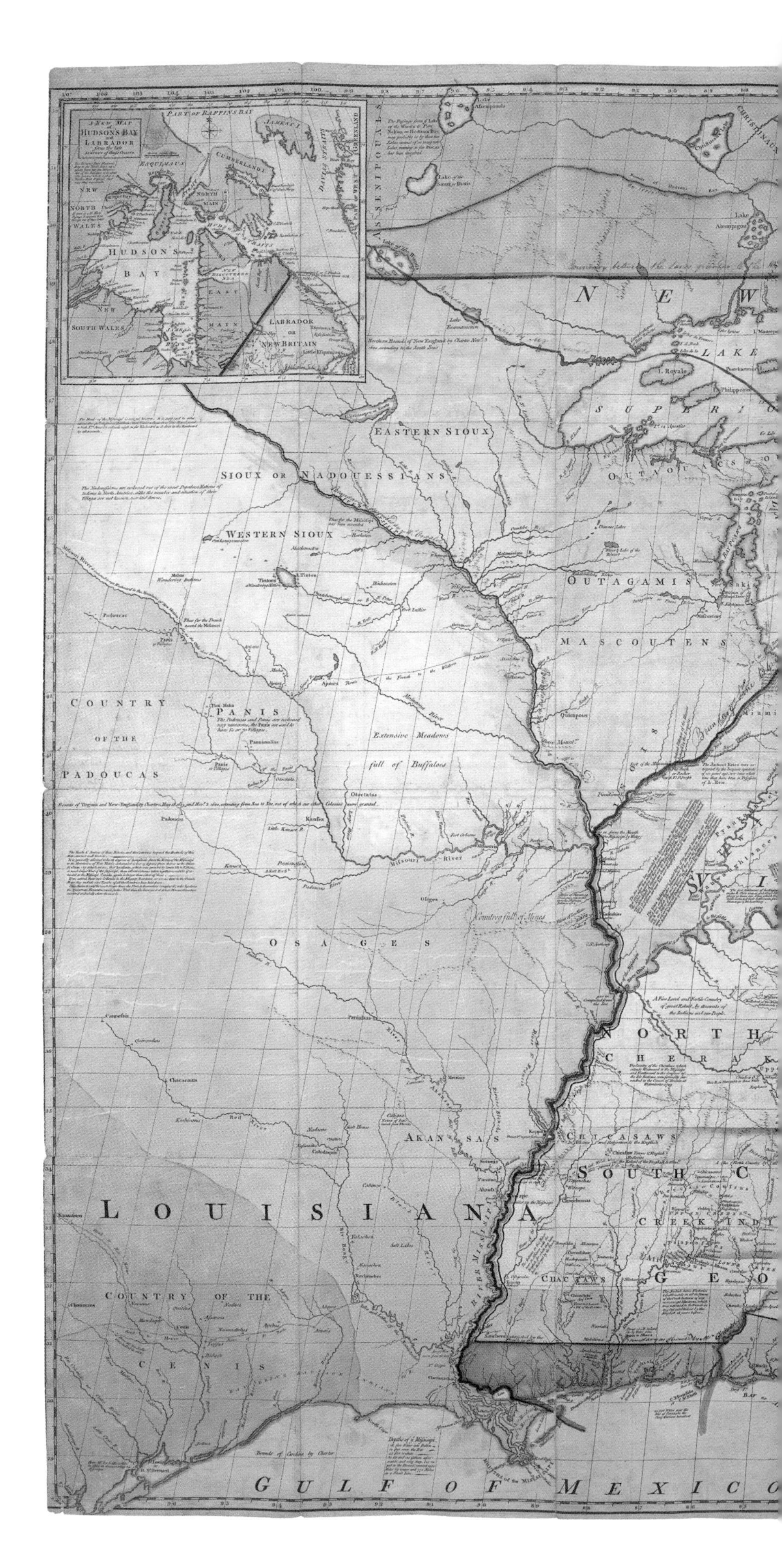

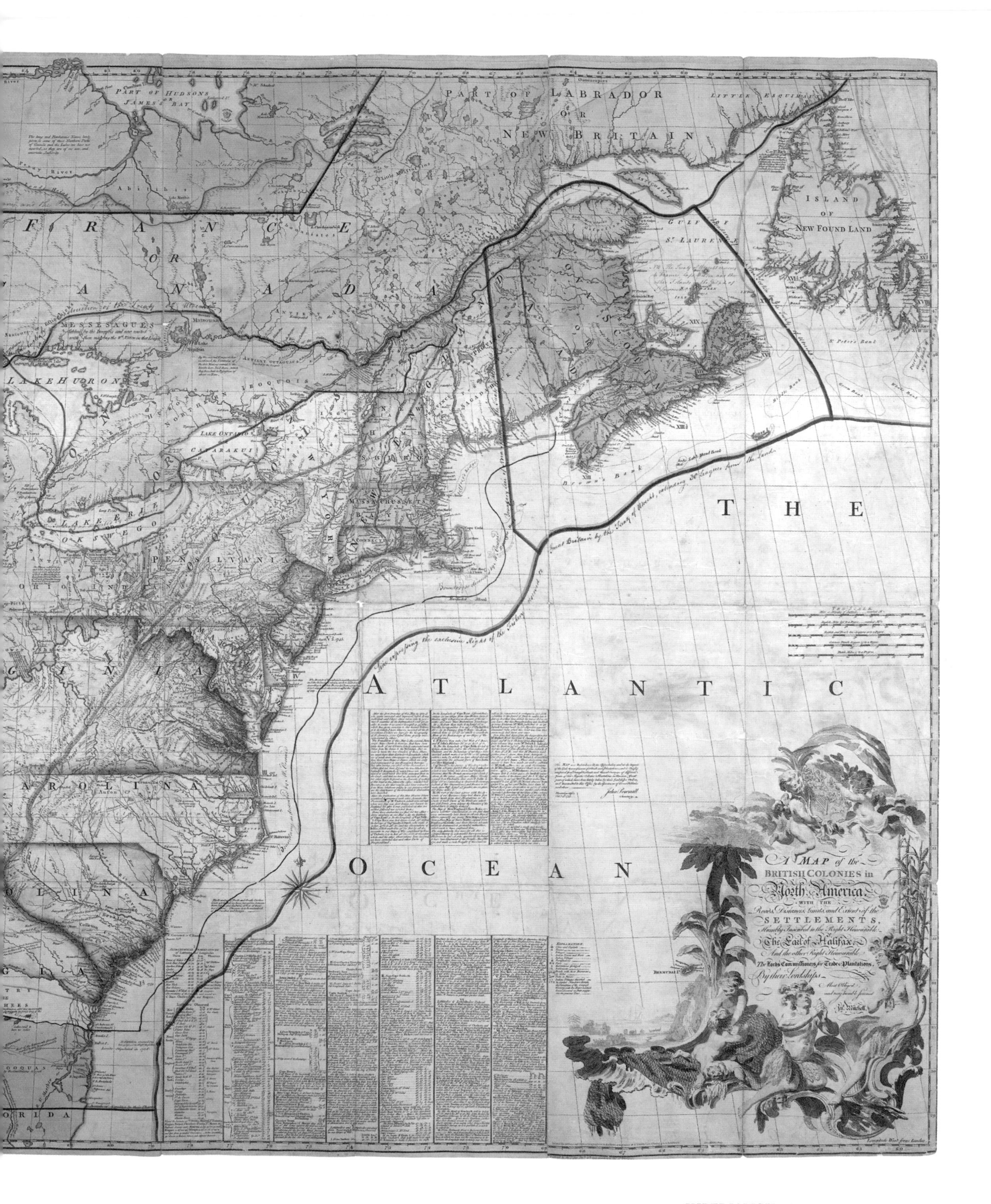
PART OF HUDSONS BAY
JAMES BAY
PART OF LABRADOR
OR
NEW BRITAIN
ISLAND
OF
NEW FOUND LAND
GULF OF ST. LAURENCE
F R A N C E
OR
C A N A D A
MESSESAGUES
LAKE HURON
IROQUOIS
LAKE ONTARIO
LAKE ERIE
PENSILVANIA
MASSACHUSETS BAY
Brown's Bank
THE
ATLANTIC
OCEAN
BERMUDAS I.
A MAP of the
BRITISH COLONIES in
North America
WITH THE
Roads, Distances, Limits, and Extent of the
SETTLEMENTS,
Humbly Inscribed to the Right Honourable
The Earl of Halifax
And the other Right Honourable
The Lords Commissioners for Trade & Plantations,
By their Lordships
Most Obliged
and very humble Servant
Jno. Mitchell

The Antient Eriez were extirpated by the Iroquois upwards of 100 years ago, ever since which time they have been in Possession of L. Erie.
Fort of the Miamis
The Rock or Rocher
Pimitiou L.
Pimiteoui
Copper Mine
R. Emicouen
R. Cahoki
Cahokies
Tamaroas
Kaskaskies
C. St. Anthony
Iroquois R.
River Theakiki
Miamis
Portage
Half way Cross
Wabache R.
Pyankashees
Wauwaughtanees
Great Wiaut
Little Wiaut
Pickawillanees or Picts
Delaware T.
Hockhocken or Margaret's T.
Delawares
Lower Shawnoes
Bever Ponds
Sandoski
The Places called Licks, and Lick Creeks are Salt water, which afford plenty of Salt to Man & Beast in those Inland Parts. The Resort of all sorts of Game, Huntsmen, Traders & Warriors, especially the Salt Ponds.
OHIO
The first Settlement of the English on the R. Ohio was at and about Allegheny 30 Years ago. Since which they have extended their Settlements from Shenango to Pickawillany
to the falls with a very gentle Current
River Ohio about 1 m. broad & 5 or 6 fathom deep
Forks of the Misisipi
920 m. Computed from the Sea
Cuttawa or Catawba R.
Fredericks R.
Naked Cr.
Cumberland R.
Walkers
the Extent of the English Settlements 1750.
A Fine Level and Fertile Country of great Extent, by Accounts of the Indians and our People.
River of the Cherakees
Hogohege R.
Pelisipi River
Clinch's R.
Holston R.
Augusta
NORTH CARO
CHERAKEES
The Country of the Cherakees which extends Westward to the Misisipi and Northward to the Confines of the Six Nations was formally surrendered to the Crown of Britain at Westminster 1729
Upper Cherakees
Chastoe O.T.
This R. is Navigable to those Falls
Euphasee
Gr. Tellique an English Factory
Chatuga
Quanessee English Factory
Watoga
Naquassee
Keowee
Fredrick
Yadkin R.
Catawba R.
CATAWBAS
CHICASAWS
In Alliance and Subjection to the English
Chicasaw Towns & English Factories
the Extent of the English Settlem.ts
SOUTH CAROLI
A fine Fertile Country by all Accounts
Lower Cherakees
Chicasaws
Coussa
Abecochee
Abecas or Coussas
Puckantalla
Wokukay
Hallebee
Ockfuskee
Euphalaw
UPPER CREEKS
CREEK INDIANS
Chatahoochee
Coweta O.T.
Cowetas
Euchees
Westoes
Talapoosees
Tomeche
Cussetas
Ocmulgee
The Little Mountains
Little River
Saluda R.
Congarees
Saxegotha Township
Amelia Township
Orangeburgh Township
New Windsor
Fort Moore
Augusta
Colleton
Granville C.
Ogeechee R.
Mount Pleasant
Ebenezer
On these Rivers the lower Creek Indians formerly dwelt, before the War with Carolina in 1715, when they removed to Chatahoochee River.
Chachumas
Wiloupo
Tapouchas
Cowachitow
Tombecbe
Ayanabe
Alibamous
Savannas

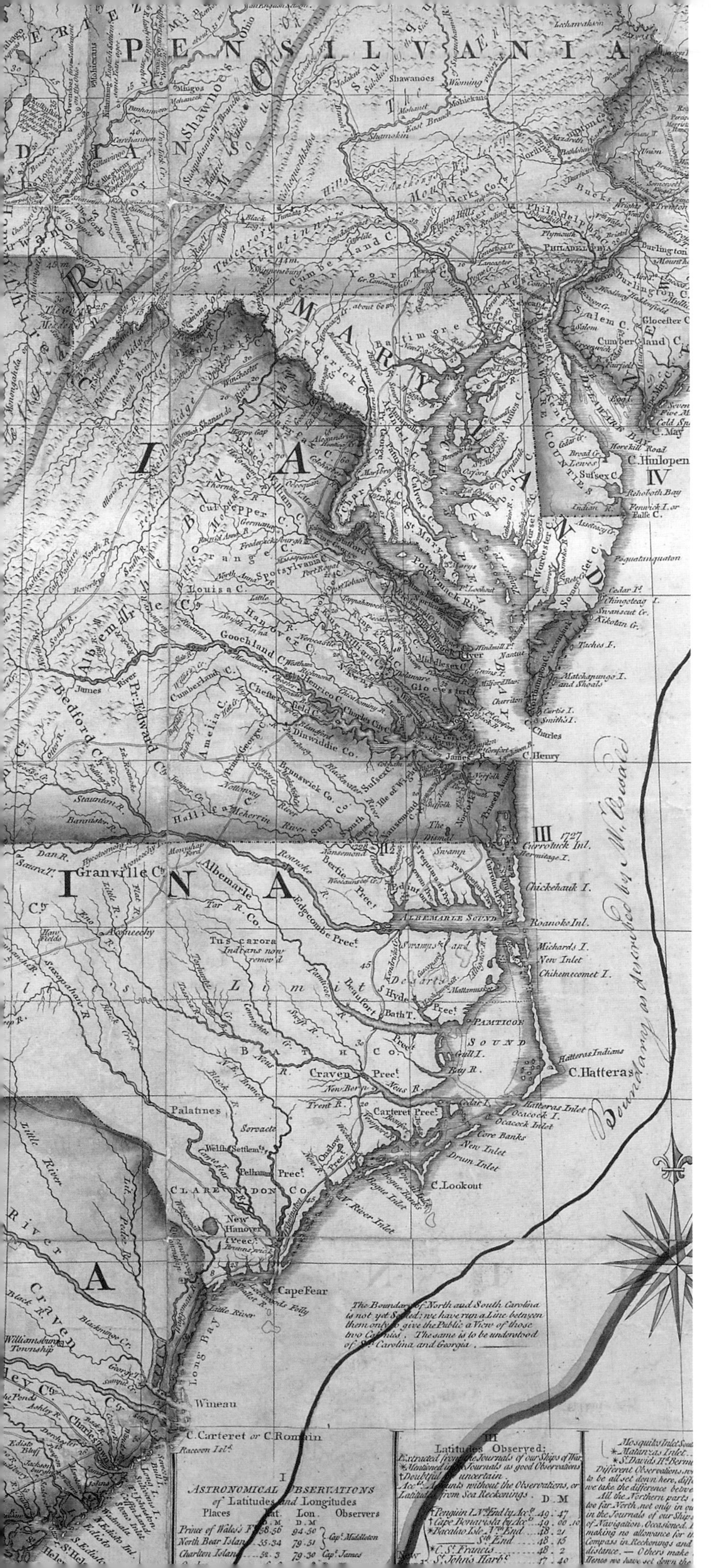

Oswald heavily annotated his copy of the map to make sense of the competing claims of the British and the Americans. The most important of these annotations are his thin red lines, which mark the boundaries of the new United States. But Oswald also included the thick yellow and red lines to reference French and British territorial claims prior to the end of the French and Indian War. He included these historical lines to note the last legal definition of boundaries in North America, a point of reference for subsequent negotiations. In the same vein, he prominently emblazoned "Six Nations" across the western part of Virginia and north through the Great Lakes. This was a reference to earlier treaties that "gave" the British suzerainty over the Iroquois lands (the Iroquois might have interpreted that arrangement differently). By attributing sovereignty to the "Six Nations," Mitchell was in fact laying claim for the British.

Ironically, this British claim gave the new United States a boundary much further west of the original colonies. The American delegation to the Paris peace negotiations included John Jay, Benjamin Franklin, and John Adams. They successfully pressed the British to relinquish claims of territorial sovereignty all the way west to the Mississippi River, considerably enlarging the new nation.

This "red line" copy of the map used in the peace negotiations was given to King George III, a record of lost colonies. As Matthew Edney has observed, there is no small irony in the fact that a map designed in 1755 to protect and extend the British empire in North America was ultimately used to dismember it. In fact, one wonders whether the map—in depicting geographical coherence—might have implicitly suggested a nation long before one materialized in the Revolution. In this respect, maps have the power to suggest what might be as well as what is.

By 1791 the Mitchell map had been reprinted twenty-one times in four languages, and pirated many times more; it circulated widely in Europe and North America, and was used to negotiate boundary disputes into the early twentieth century. Just as powerful is the symbolic influence it has exerted down to our own day, for it remains the first recognizable picture of the nation. With its articulation of emerging states along the seaboard, as well as its geographical reach into the trans-Mississippi West, the map looks familiar to us. This gives it a particular hold over our imagination, and is perhaps yet another reason why one early map scholar called it the most important map in American history.

4. 1783–1835

A Nation Realized

The Declaration of Independence remains one of the most powerful political documents in human history, an assertion of rights and equality that continues to inspire democratic movements worldwide. But while it galvanized a rebellion against tyranny, at bottom the declaration was a statement of principles rather than a blueprint for governance. It did little to establish an alternative to British monarchy, nor did it answer thorny questions of administration or state power. A stable political framework was all the more important given the astonishing and unexpected geographical changes brought by the Treaty of Paris in 1783. The American delegation to those treaty negotiations won a western boundary for the nation at the Mississippi River. This made the United States one of the largest nations in the world at its founding.

Those western territories were home to Native Americans who had no voice in the transfer of power from the British to the Americans. Several tribes northwest of the Ohio River actively resisted American rule for several years after the Revolution. The conflict was settled—if temporarily—by the 1795 Treaty of Greenville, which acknowledged Indian title to lands west of the Appalachians while simultaneously reaffirming US sovereignty in the region. This tenuous arrangement did little to stabilize the larger relationship between American settlers and Native Americans. Tribes further east also discovered that American independence brought negative consequences; the Iroquois, for example, found that their wartime alliance with the British cost them control over lands in New York and Pennsylvania.

The problems facing the new nation extended far beyond native–white relations. The resolution of wartime debts, jurisdiction over western lands, and the admission of new states were just some of the issues that American independence left unresolved. Driving all of this uncertainty were more fundamental questions about the nature of political authority. With the British no longer in control, how would Americans govern themselves? What did these former colonists have in common with one another, and how would they forge a larger national identity?

Many of the maps in this chapter were designed to confront those challenges. We open with Abel Buell's map of the new nation, a rare picture that both captures the flush of victory and anticipates the problems ahead. Only a few copies of Buell's map were made, but its influence was amplified when it was incorporated into Jedidiah Morse's bestselling geography textbook. With this wide circulation, Buell's map became a fixture in homes and schools across the country and exposed Americans to a new common geographical identity. Like rituals such as the Fourth of July and Washington's birthday, maps had the power to cultivate a sense of nationhood. For this reason, geography became an essential element of the American curriculum after the Revolution. The first generation of girls to be formally educated was widely taught to replicate maps of their nation with great care and artistry, as shown on page 118.

Yet even as a national identity began to coalesce, Americans struggled to forge a stable administrative state. The first attempt came with the Articles of Confederation, formed during the Revolutionary War. Suspicious of centralized power and monarchy, the framers constructed a weak government that was based on a contract between the states rather than a binding union. But the limits of this government soon became apparent when it was unable to levy taxes or raise an army. Tasked with improving this system, the framers instead drafted an entirely new constitution that invested the federal government with more power. To ensure ratification, they made several compromises. They included a Bill of Rights and a bicameral legislature, but also the notorious "three-fifths" compromise, which enlarged the population of slave states to strengthen their representation in Congress.

The location of the national capital was itself a compromise, as detailed on page 106. Maps of Washington, D.C. became recognizable symbols of this unprecedented experiment in representative government. The iconic power of the plan of Washington is apparent in Edward Savage's portrait of the president and his family (page 6). The first family gathers around a table to examine a map of the proposed national capital, drawing attention to the future that it represents. At far right a black servant stands inconspicuously to attention, his role in the scene—and in the nation—left unclear. The elegance, serenity, and optimism of Savage's painting was entirely at odds with the bitter partisanship of the early national era. Satirical maps of redistricting in Massachusetts—captured on page 116—struck a chord with Federalists. They also introduced the American term of the "gerrymander" to capture a troublesome yet ubiquitous feature of representative democracy that has only grown over time (page 254).

Though the Articles of Confederation lasted less than a decade, in that time Congress fundamentally shaped the nation's geography by passing the Northwest Ordinance. That legislation provided for the survey, dispensation, and settlement of land north of the Ohio River, which in turn laid the foundation for the states of Wisconsin, Michigan, Illinois, Indiana, and Ohio, and for portions of Minnesota. Further south and east, new settlers streamed into upstate New York, Kentucky, and Tennessee. This land rush was guided by John Filson's map of "Kentucke," which portrayed the western lands as fertile and free for the taking. In this respect, Filson was part of a long tradition dating back to John Smith and Henry Briggs, who designed maps to encourage frontier settlement (page 102).

That migration also created administrative challenges. While Filson was beckoning Americans west, the newly created Post Office faced the daunting task of delivering mail across this expansive national territory. American independence directly stimulated the need to deliver mail *between* the colonists; previously, most correspondence had been conducted with Britain. Increased demand prompted Congress to establish the mail as one of the federal government's first permanent responsibilities. Abraham Bradley's large map of the United States on page 108 captures the challenge of building a network of communication over such sprawling geography. Yet Bradley's maps also implied that without a reliable postal network this *country* could not become a *nation*. Similarly, it was the inefficient delivery of mail across the Atlantic that led Benjamin Franklin to investigate and map the Gulf Stream in the 1780s (page 104).

When Bradley completed his first postal map in 1796, the Mississippi River formed the nation's western boundary. For American settlers along the Ohio River, the Mississippi was not just a national boundary but a crucial artery for transportation and trade. Early in Thomas Jefferson's presidency, the Spanish and French limited American access to the port of New Orleans. When Jefferson in response sought to purchase New Orleans from the French, he was unexpectedly offered all of Louisiana. Though the president was uncertain whether he had the constitutional authority to acquire foreign territory, he took the opportunity and instantly doubled the size of the nation. But Americans knew little about this land, so Jefferson proposed an expedition up the Missouri River then west to the Pacific Ocean. By comparing maps of the West before and after the expedition of Meriwether Lewis and William Clark (pages 112–115), we can see how much information was gained.

The Louisiana Purchase was an unexpected windfall, but many Americans wondered whether a healthy republic could be sustained over such an immense region. For decades thereafter American maps dismissed much of the western plains and the Southwest as a "Great American Desert." East of the Mississippi, however, steamboats and canals were swiftly transforming national geography in the 1820s and 1830s. Cadwallader Colden's map on page 120 celebrated the commercial power of the Erie Canal. This engineering marvel linked the old Northwest and the Great Lakes to New York City in a way that would have been unimaginable even a few decades earlier. In a sprawling nation, the importance of these technologies is hard to exaggerate: canals, steamboats, and later railroads created new regions and networks, integrated the population, and accelerated the circulation of goods and information. The result was economic growth that had far-reaching consequences. The emergence of the textile industry in the Northeast, for instance, generated a demand for cotton that entrenched and expanded slave labor into new lands and profoundly shaped the southern economy.

We close this chapter with a map that both encapsulates this era and anticipates the next. John Melish's "Map of the United States" (page 122) was published in the flush of victory after the war of 1812. By including the information brought back by western expeditions, Melish gave Americans a more accurate sense of continental geography. The map was also a tool of statecraft that was used to settle international boundaries. But it was the suggestive picture of a nation extending to the Pacific that made the map so striking. Long before the phrase "Manifest Destiny" was coined, Melish's map anticipated the rapid territorial expansion of the 1840s.

NATIONAL ASPIRATIONS

Abel Buell, "A New and Correct Map of the United States of North America," 1784

After declaring independence, the American patriots faced the more difficult task of establishing a government of their own. As a reaction against the British monarchy, they initially rejected a strong central power. Instead, the Articles of Confederation invested sovereignty in the individual states, which in turn formed a "firm league of friendship" with one another. Each state sent representatives to a Congress that had limited jurisdiction beyond conducting foreign policy, maintaining national defense, and arbitrating interstate disputes. Yet this weak government lacked the power to enforce laws or even collect taxes from a geographically dispersed population that shared little more than a common wartime enemy.

This fragile period is captured on the first map of the country published in the United States. The map was created by Abel Buell, a Connecticut engraver who was not above using his skills to counterfeit currency. The geography on Buell's map was not terribly original, for he largely drew on earlier maps by Lewis Evans and John Mitchell shown in Chapter 3 (pages 86 and 94). More striking is the elaborate cartouche designed to commemorate independence. A radiant sun illuminates a brightly colored flag; at right a young man holds a small globe labeled "America," with a Phrygian cap—the symbol of liberty—atop his staff and the date of independence inscribed at his feet. Buell even embedded his patriotism in the geography of the map itself by fixing the prime meridian at Philadelphia rather than relying on the standard of Greenwich in England.

Only a few copies of Buell's map were made, but its influence was amplified when Jedidiah Morse hired Amos Doolittle to adapt the map for his bestselling schoolbook *Geography Made Easy*. Morse proudly introduced the map by reminding students that the "tyranny of Britain" compelled the colonists to declare "themselves free, sovereign, and independent States ... after a long, unnatural and destructive war." Indeed, Morse's geography text was itself an intellectual declaration of independence, for he insisted that students learn their subjects through American authors such as himself rather than Europeans.

Geography Made Easy went through dozens of editions from 1784 through the early nineteenth century, which meant that thousands of children were first exposed to American geography through Doolittle's map. The importance of Buell's map—and Doolittle's adaptation—is compounded by its timing: in the immediate aftermath of the Revolution, nationhood was not a self-evident concept, but rather one that had to be articulated and accepted. The colonists had to *learn* to define themselves as Americans, and to identify those in other states as their countrymen. Symbolic maps such as Buell's, and school maps such as Doolittle's, taught Americans young and old to see the nation as an extension of themselves.

A national identity could not exist, however, without some kind of national *authority*. By the time Buell published this map, the Articles of Confederation faced heavy criticism. In 1786 a rebellion of indebted former Revolutionary War soldiers convinced many that a stronger central government was necessary to maintain order and to foster growth. Alexander Hamilton and James Madison advocated a national government that could levy and collect taxes and regulate interstate trade. Above all, the "Federalists" argued for a government of three branches, one with a dominant legislative branch supplemented by a judiciary and an independent executive.

The ratification of the Constitution in 1789 settled some of these debates, and laid down the political structure that survives down to our own day. Yet Buell's map reminds us that at its founding the nation was an aspiration more than anything else.

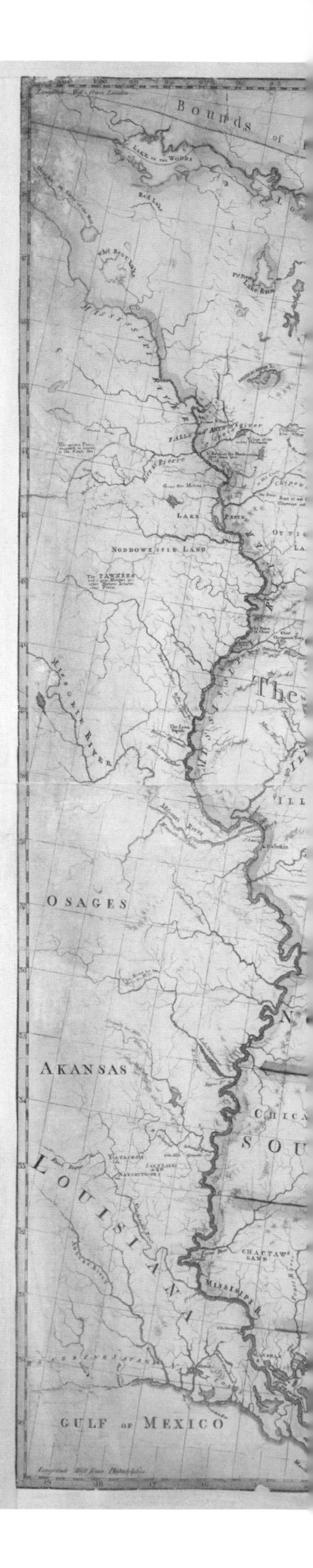

JAMES BAY
PART of LABRADOR
GULF OF ST LAURENCE
Bay by the Treaty of Utrecht
C A N A D A
ANTIENT ALGONQUINS
ANTIENT OUTAOUAS
RIVER ST LAURENCE
NOVA SCOTIA
BAY OF FUNDY
LAKE SUPERIOR
LAKE HURON
LAKE MICHIGAN
LAKE ONTARIO
LAKE ERIE
NORTHERN IROQUOIS
MASSACHUSETS
S T A T E
PENSILVANIA
BALTIMORE
CHESAPEAK
C. CHARLES
C. HENRY
V I R G I N I A
The Laurel Mountains
C A R O L I N A
C A R O L I N A
CREEKS LAND
G I A
APALACHEES
East FLORIDA
Bay Appalache
THE ATLANTIC OCEAN
A NEW and correct MAP of the UNITED STATES of North America
Layd down from the latest Observations and best Authorities agreeable to the PEACE of 1783
Humbly Inscribed to his Excellency the Governor and Company of the STATE of CONNECTICUT
By their Most Obedient and very humble Servant
Abel Buell

AN INVITATION TO SETTLEMENT

John Filson, "Map of Kentucke," 1784

Daniel Boone embodies the myth of the American frontier. He became a folk hero in his own lifetime, one of many trappers, hunters, and explorers who traveled into the new "Kentucky Country" in the 1760s and 1770s. These men brought back tales of an abundant land, bounded by the Blue Ridge Mountains and the Ohio River, which was then home to the Shawnee Indians. Given the limited knowledge of this region, speculation ran rampant about its potential for settlement. A trickle of migration at the end of the Revolutionary War grew into a flood, as thousands of Americans and Europeans, gripped by Kentucky fever, followed Boone's path west through the Cumberland Gap.

This explosive growth—like the celebrity of Boone himself—was no accident. The chief promoter of Kentucky was John Filson, a schoolteacher who acquired 12,000 acres in the territory in 1782. Realizing that the value of his land depended on the prospects of settlement in the region, he befriended Boone and other frontiersmen to learn the geography, geology, river systems, and native tribes of his new home. The result was Filson's enthusiastic description of Kentucky, which included this map as well as a dramatic account of "The Adventures of Daniel Boon." Published on Boone's fiftieth birthday, Filson's account was a hit in the United States as well as in Britain, France, and Germany.

Filson's map and narrative stimulated the rush to Kentucky. His map presented an inviting territory easily accessed through roads and rivers. He described a fine and well-watered land, confining the presence of Native Americans to wigwams north of the Ohio River. Filson also named forts, towns, and roads for the early white settlers in the territory. By doing this he established the territory's recent past in order to claim its future. For instance, in Fayette County Filson identified the "bloody battle" of Blue Licks, fought ten months after the "final" battle of Yorktown. There, Boone squared off against a much larger force of British and Indians to defend the emerging settlements of the territory. In the lower right corner of the map we see Boone's more lasting legacy, the Wilderness Road he established from Virginia through the Cumberland Gap. In both examples, Filson took care to name—and thereby claim—the land for settlement and development.

Through his publication, Filson helped to establish Boone as an American hero and Kentucky as a land of opportunity. The map seamlessly integrates information with promotion, and its wide circulation shaped early perceptions of this region. Its pleasing and balanced appearance focuses not on the entire extent of Kentucky but on its central territory, one hundred square miles of "the most extraordinary country that the sun enlightens with his celestial beams." Navigable rivers stretch across abundant and fertile land, draining into the (slightly misplaced) Ohio River. Even the cartouche—dedicated to Congress and to George Washington, commander of the Continental Army—underscores Filson's confidence in Kentucky's future. He closed his narrative in the same manner, proposing a new settlement on the Lower Mississippi River that would siphon trade from Spanish-controlled New Orleans and eventually make America the commercial rival of Europe.

Filson's map is a remarkable and influential document of the early frontier. It captures the optimism of the new nation just as Americans began to turn their attention to the trans-Appalachian West. His general predictions of growth were realized: by 1784, 30,000 migrants had arrived, a number that had more than doubled by the end of the decade. In 1800 Kentucky was home to 221,000 settlers, nearly twenty percent of whom were enslaved.

This rapid migration to Kentucky created a degree of social fluidity and instability that characterized several frontier settlements in this era. Like Kentucky, many of these communities were home to the earliest religious revivals of the Second Great Awakening. Just northwest of Lexington, where the map reads "Abundance of Cane," was the first of these camp meetings. In 1801 over 25,000 Protestants converged on the area that Boone himself had named Cane Ridge, an enormous number considering that Lexington was home to just 2,000 people at the time. The revival that began at Cane Ridge spread through the frontier and then across the country, lasting for decades and bringing mass conversions and new practices of worship to Christianity. The Second Great Awakening challenged established church authority and transformed Protestant theology. In this respect, the frontier was at the center, rather than the periphery, of American life.

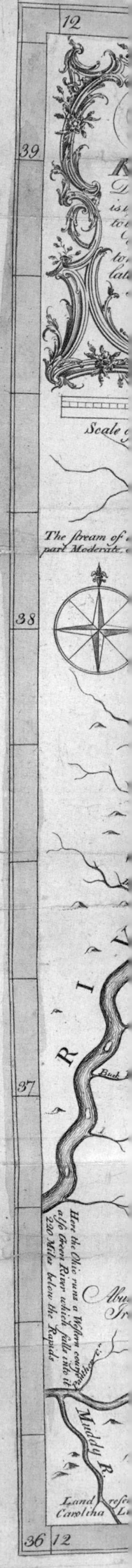

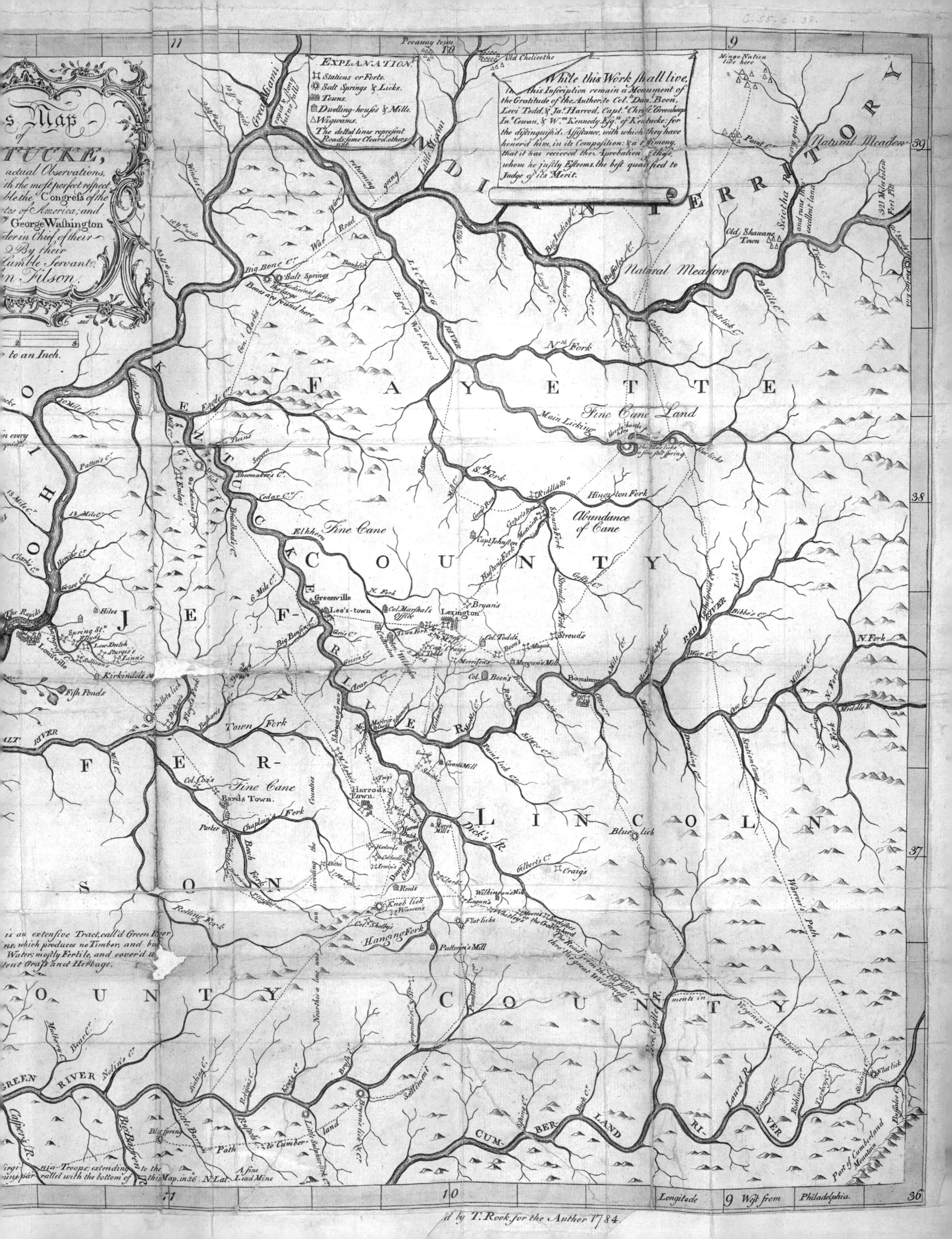

EXPLANATION
Stations or Forts.
Salt Springs & Licks.
Towns.
Dwelling-houses & Mills.
Wigwams.
The dotted lines represent Roads, some Cleard, others not
While this Work shall live, let this Inscription remain a Monument of the Gratitude of the Author, to Col.^s Danl Boon, Levi Todd, & Ja.^s Harrod, Captn Chris.^r Greenup, In.^o Cowan, & Wm Kennedy Esq.^rs of Kentucke; for the distinguish'd Assistance, with which they have honor'd him, in its Composition: & a Testimony, that it has received the Aprobation of those, whom he justly Esteems, the best qualified to Judge of its Merit.
s Map
of
TUCKE,
actual Observations,
ith the most perfect respect
ble the Congress of the
tes of America; and
George Washington
der in Chief of their
By their
umble Servant,
n Filson.
to an Inch.
INDIAN TERRITORY
FAYETTE
COUNTY
JEFFERSON
LINCOLN
COUNTY
KENTUCKE RIVER
LICKING RIVER
Great Miami
Little Miami
Old Chelicothe
Pecaway town
Mingo Nation live here
Old Shawane Town
Natural Meadow
Sciotha R.
Big Bone C.^k
Salt Springs
Medicinal Spring
The large Bones are found here
War Road
Bird's War Road
N.^th Fork
Main Licking
Fine Cane Land
the Blue licks a fine salt spring
St Fork
Riddle's St.^n
Hinkston Fork
Abundance of Cane
Fine Cane
Elkhorn
Capt. Johnson
Greenville
Lee's-town
Col. Marshal's Office
Lexington
Bryan's
Strouds
Col. Todd's
Boon's
Morgan's Mill
Col. Boon's
Boonsburg
RED RIVER
The Rapids
Louisville
Hites
Spring St.^n
Low Dutch
Sturgis's
Linn's
Kirkindols
Fish Ponds
Bullets lick
Salt River
Town Fork
Col. Cox's
Fine Cane
Bards Town.
Parker
Chaplain's Fork
Beech Fork
Harrods Town.
Grants Mill
Danville
Knob lick
Warren's
Col. Shelby's
Hanging Fork
Patterson's Mill
Flat licks
Logan's
Whitley's
Craig's
Gilbert's C.^k
Dick's R.
Blue lick
Warrior's Path
is an extensive Tract, call'd Green River
ns, which produces no Timber, and but
Waters, mostly Fertile, and cover'd with
lent Grass and Herbage.
GREEN RIVER
Big Barren R.
CUMBERLAND RIVER
Laurel R.
Part of Cumberland Mountain
Virginia to Kentucke
Path to Cumberland
Big Spring
Virginia Troops extending to the
parallel with the bottom of this Map in 36 N. Lat.
A fine Lead Mine
Longitude 9 West from Philadelphia.
11
10
9
39
38
37
36
d by T. Rook for the Author 1784

ATLANTIC OCEAN
LABRADOR
Hudson's Str.
Davis Str.
R. S. Lawrence
N. England
Nova Scotia
Virginia
NORWAY
Baltick
Ireland
Britain
France
B. Biscay
Spain
Portugal
Part of BARARY
LAND of the ESKIMAUX's or LABRADOR
GULF of St. LAURENCE
R. S. Lawrence
NOVA SCOTIA
NEW ENGLAND
NEW HAMPSHIRE
MASSACHUSETS
CONNECTICUT
Cape Cod
Bay of Fundy
St. Georges Bank
Nantucket I.
Long I.
NEW YORK
NEW JERSEY
LAKE MICHIGAN
LAKE HURON
LAKE ONTARIO
LAKE ERIE
Fort Detroit
Niagara falls
PENNSYLVANIA
Philadel.
MARY LAND
Delaware Bay
VIRGINIA
OHIO RIVER
NORTH CAROLINA
Chesapeak Bay
C. Hatteras
SOUTH CAROLINA
GEORGIA
P. Royal
WEST FLORIDA
EAST FLORIDA
PART OF THE GULF of MEXICO
GULF of FLORIDA
BAHAMA I.
ATLANTIC OCEAN
Bermuda I.
4 Minutes
3½ Minutes
3 Minutes
2½ Minutes
facing page 315.

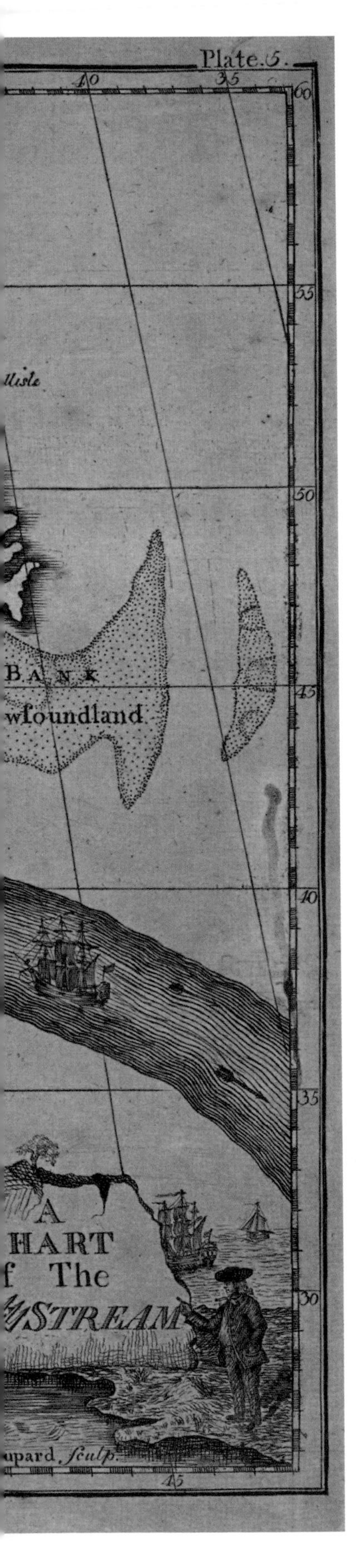

THE CURRENTS OF THE ATLANTIC WORLD

James Poupard and Benjamin Franklin, “A Chart of the Gulf Stream,” 1786

In the summer of 1726 Benjamin Franklin was a young man of twenty, sailing home to the colonies after his first visit to London. As his ship approached the North American coast, a wet hot wind picked up and the water changed color to show an abundance of grass and other marine life. At the same time, the ship’s pace slowed considerably, though Franklin was unable to account for any of these abrupt changes.

Twenty years later Franklin observed a similar puzzle: ships sailing east seemed to move more quickly across the Atlantic than those bound for the American colonies. Franklin was drawn to this problem for a third time another twenty years later, but with greater urgency. In his capacity as Deputy Postmaster General for the American colonies, he heard from customs commissioners in Boston that mail packets traveling from Falmouth in England to New York were taking two weeks longer to arrive than those sailing from London to Rhode Island, even though the former trip was a shorter distance. Franklin consulted his cousin, the experienced navigator Timothy Folger, who speculated that the packet ships heading to Rhode Island must have been piloted by captains who understood the Gulf Stream. Sailors had long known about this current even though it had not been mapped or documented in navigation or maritime guides.

Folger had learned of the Gulf Stream through his experience among New England whalers. They described a current that flowed up the coast from Florida and then turned east, one powerful enough to separate whaleboats from their larger ships. British captains heading toward America were fearful of the rocky shoals of George’s Bank (just east of Nantucket Island on the map), so they often sailed further south and thereby placed themselves directly against the Gulf Stream, which was moving east. This could add days to their journey, and though they were often warned as much by American whalers, their advice was usually ignored.

Folger charted the basic dynamics of the current for Franklin, showing him how it broadened and narrowed. Franklin printed and distributed Folger’s chart among British sea captains in 1769 or 1770 in the hopes that they might use it to their advantage, but the advice was (again) ignored. During the Revolutionary War this maritime knowledge became sensitive information, leading Franklin to stop distributing the chart to British sailors. Once the war ended, he published his own picture of the Gulf Stream, shown here, in the *Transactions of the American Philosophical Society*. The main map focuses on the varied strengths of the current along the Atlantic coast, while the inset depicts its entire course. In a small cartouche at lower right Franklin stands on a spit of land, sharing the map with Neptune. Engraved by James Poupard, Franklin’s map was a hit, and by the end of the 1780s the “Gulph Stream” had entered the American lexicon.

While Franklin and Folger were researching this current, William Gerard de Brahm was working in the same vein. A migrant to the Georgia colony, de Brahm noticed a current near the future site of Miami while surveying the shoreline. In 1771 he sailed up the coast to Newfoundland Bank, and then east across the Atlantic. De Brahm published his own chart of the Gulf Stream, which traced a current moving along the North American coast before joining others flowing south out of Hudson Bay and the Saint Lawrence Seaway. Similarly, on this map Folger and Franklin show the current moving along the American coast before shifting south of George’s Bank. De Brahm and Franklin were thus developing the idea of the Gulf Stream at the same time, each drawing on his own experience. Their initial charts were published simultaneously about 1770, a remarkable coincidence given that the current had shaped European exploration, settlement, and trade for over two centuries.

ENGINEERING THE NATION'S CAPITAL

Andrew Ellicott and Pierre Charles L'Enfant, "Plan of the City of Washington, in the Territory of Columbia," 1792

The success of the American Revolution made clear that the colonists had many different visions of their political future. Their first attempt to establish an administrative framework came in the Articles of Confederation, which produced a weak central government as a reaction against monarchy. The shortcomings of the Articles led to fierce debates about how to proceed. Those advocating an entirely new Constitution and a stronger central government dubbed themselves Federalists. They met passionate resistance from Antifederalists who feared that concentrated national power directly threatened their liberties. As a result, the new Constitution included ten amendments that prioritized individual rights and placed limits on national authority.

Just after the states ratified the Constitution, the newly empowered government faced its first serious test. Alexander Hamilton had long argued that the nation should assume the state debts incurred during the Revolutionary War. But many Southerners remained skeptical, concerned that this would enrich the wealthy at the expense of ordinary Americans. In 1790 Hamilton forged a compromise with James Madison and the newly appointed Secretary of State, Thomas Jefferson: Southerners would not object to the federal assumption of war debts, and in return the nation's permanent capital would be located in the South.

The new Congress codified this compromise in the Act of 1790: after a ten-year stint in Philadelphia, the capital would move to a location between Maryland and Virginia, two Southern and slaveholding states. This gesture was designed to affirm both the influence and interests of Southerners. President George Washington selected the site for the new national capital at the confluence of the Potomac and the Anacostia rivers (then known as the Eastern Branch), not far from his Mount Vernon home. The Potomac seemed particularly appealing given its central location in the nation. Moreover, with headwaters that lay near the Ohio River, the Potomac held out the promise of linking the seaboard to the interior. On the detail at right, depth soundings are shown along the Potomac River to indicate its navigability.

To design the capital, President Washington appointed Pierre Charles L'Enfant, a French architect who had fought alongside the patriots in the Revolutionary War. Two hundred years later, L'Enfant's plan still frames the city. Like most eighteenth-century American towns, Washington was organized on a grid, though L'Enfant added a series of diagonals to facilitate movement across the city. As shown on the detail below at right, he established a central axis along Pennsylvania Avenue to connect the president's home to the legislature. Jefferson suggested wide boulevards and limited building heights to ensure both light and air, and to lend the capital a stately atmosphere. In a similar vein, L'Enfant reserved a few elevated sites around the city for anticipated monuments to commemorate the Revolution. The main avenues would be named after the fifteen states, while a series of central squares would provide space for relaxation and yet more national memorials.

L'Enfant finished his plan in the summer of 1791, though before it was approved he was fired for his inability to compromise. Andrew Ellicott stepped in to formally submit the plan, aided in these final stages by the freeborn African American Benjamin Banneker. Once approved, the plan was engraved in Philadelphia by Thackara & Vallance in late 1792. This was just weeks after the cornerstone of the White House—marked as "President's House"—was laid by slaves.

Multiple editions of the layout were published in the United States, Paris, and London, indicating the widespread interest in this new national capital. The diamond-shaped plan of the city became a familiar image in the early republic. In the 1790s Edward Savage painted the Washington family seated around a large copy of L'Enfant's plan (page 6). Savage knew that viewers would recognize the map even with just a portion exposed. For the next several decades, in fact, American schoolgirls commonly copied, painted, and embroidered the L'Enfant plan as part of their civic education. The map of this new planned city was a fixture of popular culture, and a symbol of national independence itself.

If the capital appears slightly unfamiliar here, that is because of changes made after 1900. The mall was extended, and wider streets reflect the City Beautiful movement that had taken the nation by storm. Yet L'Enfant's original plan largely survives—a national capital designed from scratch. Unlike Paris and London, Washington, D.C. would be primarily a center of political power, geographically separate from its cultural and financial capitals. If that is still true, it is also by design.

PLAN
of the CITY of
Washington
in the Territory of Columbia,
ceded by the States of
VIRGINIA and MARYLAND
to the
United States of America,
and by them established as the
SEAT of their GOVERNMENT,
after the Year
MDCCC.
Lat. Capitol, 38: 53, N.
Long. 0: 0.
GEORGE TOWN
Presidents House
Capitol
POTOMAK RIVER
EASTERN BRANCH
PART OF VIRGINIA WITHIN THE TERRITORY OF COLUMBIA.
PART OF MARYLAND WITHIN THE TERRITORY OF COLUMBIA.
OBSERVATIONS
explanatory of the
Plan.
THE positions for the different Edifices, and for the several Squares or Areas of different shapes, as they are laid own, were first determined on the most advantageous ground, ommanding the most extensive prospects, and the better susceptible f such improvements, as either use or ornament may hereafter all for.
LINES or Avenues of direct communication have been devised, connect the separate and most distant objects with the principal, d to preserve through the whole a reciprocity of sight at the same time. ttention has been paid to the passing of those leading Avenues over the ost favorable ground for prospect and convenience.
NORTH and South lines intersected by others running due East and est, make the distribution of the City into Streets, Squares, &c. and those es have been so combined as to meet at certain given points with these ergent Avenues, so as to form on the Spaces "first determined," the different uares or Areas.
SCALE OF POLES.
Breadth of the Streets.
THE grand Avenues, and such Streets as lead immediately to public places, are from 130 to 160 feet wide, and may be conveniently divided into foot ways, walks of trees, and a carriage way. The other Streets are from 90 to 110 feet wide.
IN order to execute this plan, Mr. ELLICOTT drew a true Meridional line by celestial observation, which passes through the Area intended for the Capitol; this line be crossed by another due East and West, which passes through the same Area. These lines were accurately measured, and made the basis on which the whole plan was executed. He ran all the lines by a Transit Instrument, and determined the Acute Angles by actual measurement, and left nothing to the uncertainty of the Compass.
Presidents House
Pennsylvania
Virginia
Capitol.
New
York
chusetts
North
South
East

FORGING A NATIONAL NETWORK

Abraham Bradley Jr., "A Map of the United States, Exhibiting the Post-Roads ...," 1796

The Revolutionary War demonstrated the need for an infrastructure that would both stimulate internal growth and protect against external threats. Among the first attempts to address these challenges was the Postal Service Act of 1792, which transformed the operation and scope of domestic mail. The act authorized new post offices in the nation's remote but growing frontier regions, an acknowledgment that the mail was a federal obligation that extended throughout its domain. By launching a commitment to regular, scheduled delivery, Congress also endorsed—and even prioritized—public access to information. Yet this overhaul of the mail was not prompted or even facilitated by transportation innovations; mail continued to be delivered on horseback, with the subsequent introduction of stagecoaches to accommodate the increasing volume and weight of newsprint.

The Postal Service Act rapidly expanded the circulation of mail and made the Post Office the largest and most important federal agency in the early republic. In effect, the act created not just an infrastructure but also a market for information. This surge of mail turned the Post Office into a hive of activity from the 1790s to the 1820s, as successive postmasters worked to create, manage, and above all coordinate a massive and continuously developing network of communication. At the center of the action was Abraham Bradley, Jr., who joined the Post Office in 1791 as a clerk. With this large map, Bradley ambitiously set out to visualize the entire postal operation in both practical and symbolic terms. In both its general structure and its many details, the map reflects Bradley's vision of nationhood.

Some of the map's specific features reflect Bradley's own identity as a staunch Federalist. To honor the first president, he identified Washington, D.C. as the national capital, even though the seat of government would not move south from Philadelphia for another four years. Bradley also measured longitude not from Greenwich, England—as was customary—but from Washington, D.C. His was among the first American maps to assert this new prime meridian.

More generally, Bradley's was the most comprehensive and detailed map of the nation up to that point. Issued in two parts, it emphasized post and stagecoach roads, branch post offices, and ports of entry. The result was an advertisement for the mail system itself, with an established north–south corridor and inroads into western New York, the hinterlands of Maine and Vermont, and the more sparsely settled South. While most contemporary maps focused on towns, boundaries, rivers, and topography, Bradley omitted these details in order to foreground the connections *between* places. This was a map designed to convey the lived experience of space rather than just the measure of distance. As shown on the next page, regions crowded with postal roads reveal comparatively dense areas of settlement, while sparsely connected areas reflect smaller populations with a lower demand for mail delivery.

Among the most notable elements of the map is the chart of the mail schedule at the far right, and enlarged on the next page. Here Bradley synthesized a tremendous amount of information in order to track the physical path of the post down to the hour. With a single line, he followed the mail from northern Maine to Georgia, with branch routes listed at the bottom of the chart. At a glance, viewers could see the entire seaboard network in both space and time. Bradley even differentiated the summer and winter schedules to allow for changes in weather and navigability.

The very decision to compile and publish such a schedule implied an assumption of regular service. And such a schedule could be created only through a systematic observation of delivery that was then aggregated and averaged. The chart confidently assumed a predictable rhythm of information. With the map and schedule, Bradley announced one of the largest and earliest commitments of the federal government, one that American citizens would quickly come to expect as a basic function of the state. He continually revised his maps to reflect the expansion of the network.

The Post Office quietly and continuously integrated the far-flung reaches of the physical territory into a coherent national space. As John Calhoun remarked in a speech in 1817, "the mail and the press are the nerves of the body politic. By them, the slightest impression made on the most remote parts is communicated to the whole system." Bradley's maps both reflected and advanced that goal, making the United States a nation not just in name but also in operation.

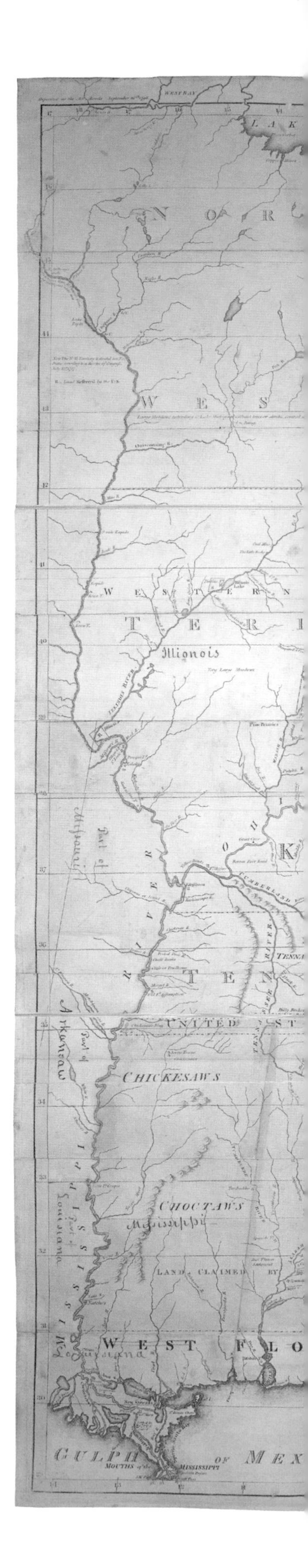

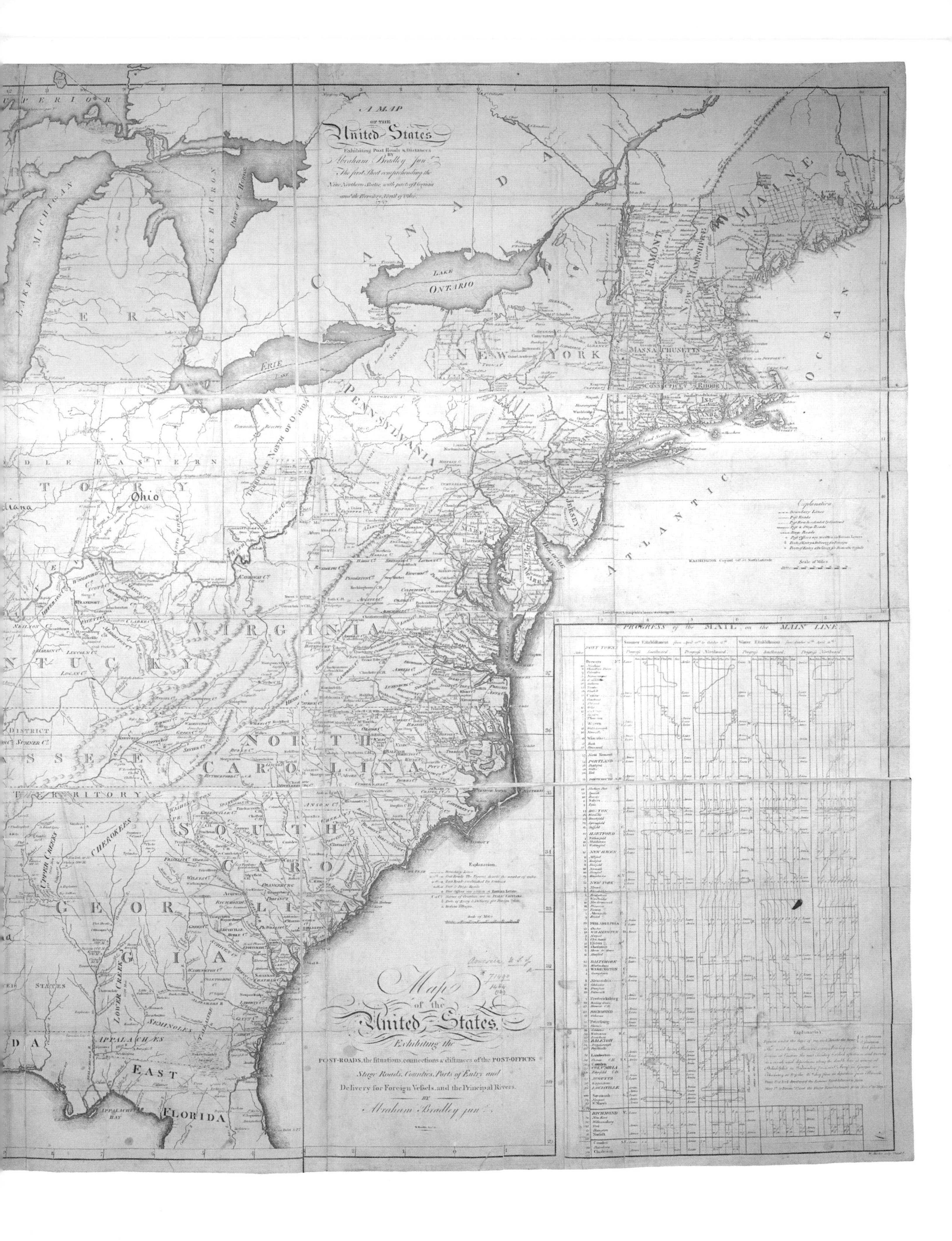
A MAP
OF THE
United States
Exhibiting Post Roads & Distances
BY
Abraham Bradley Jun.
Map
of the
United States,
Exhibiting the
POST-ROADS, the situations, connections & distances of the POST-OFFICES
Stage Roads, Counties, Ports of Entry and
Delivery for Foreign Vessels, and the Principal Rivers.
BY
Abraham Bradley jun.
PROGRESS of the MAIL on the MAIN LINE
Explanation
LAKE ONTARIO
ERIE LAKE
LAKE HURON
LAKE MICHIGAN
CANADA
NEW YORK
PENNSYLVANIA
VERMONT
NEW HAMPSHIRE
MAINE
MASSACHUSETTS
CONNECTICUT
RHODE
JERSEY
DELAWARE
VIRGINIA
KENTUCKY
NORTH CAROLINA
SOUTH CAROLINA
GEORGIA
EAST FLORIDA
APPALACHES
SEMINOLES
CHEROKEES
UPPER CREEKS
LOWER CREEK
Ohio
ATLANTIC OCEAN
APPALACHIE BAY

ONTARIO
Lake St. Clair
Detroit
ERIE
LAKE
Niagara F.t
FALLS
Newark
Queenston
Buffaloe C.
Long Point
Presque Isle
Lands of the SIX NATION
Kanandaigua
Williamsburg
PENNSYLVANIA
LYCOMING C.
Connecticut Reserve
TERRITORY NORTH OF OHIO
EASTERN
Ohio
Sandusky R.
Miami R.
Ft. Defiance
Ft. Wayne
Ft. Adams
Ft. Recovery
Ft. Hamilton
Cincinnatus
Columbia
SCIOTO RIVER
Judge Simmes Purchase
Marietta
Belville
Gallipolis
Conveyed to Settlers at Gallipolis
KANHAWAY Co.
Pittsburg
Greensburg
Washington
Wheeling
Brownsville
Union
FAYETTE C.
Berlin
Bedford
BEDFORD C.
HUNTINGDON C.
Huntington
Lewistown
MIFFLIN C.
Northumberland
CUMBERLAND
Carlisle
Chambersburg
Cumberland
Hagers T.
Williamsport
Martinsburg
Winchester
Moorfields
HARDIE C.
Strasburg
Woodstock
New Market
Rockingham C.H.
Staunton
AUGUSTA Co.
Bath C.H.
Hot Springs
Sweet Springs
Greenbrier C.H.
RANDOLPH Co.
PENDLETON Co.
HARDY Co.
SHENANDO Co.
LOUDOUN Co.
Leesburg
ALEXANDRIA
Colchester
Dumfries
Fredericksburg
Falmouth
CULPEPPER Co.
ORANGE Co.
STAFFORD Co.
ALBEMARLE Co.
Charlottesville
Lexington
VIRGINIA
Columbia
Richmond
RICHMOND
Manchester
Petersburg
Cabbinpoint
Liberty
Lynchburg
New London
Prince Edward C.H.
Charlotte C.H.
AMELIA Co.
DINWIDDIE Co.
BEDFORD Co.
CAMPBELL Co.
Rockymount or Franklin C.H.
Martinsville
Pittsylvania C.H.
Halifax
Mecklenburg
LUNENBURG Co.
BRUNSWICK Co.
SUSSEX Co.
Southampton C.H.
HENRY Co.
PATRICK Co.
Montgomery C.H.
Wythe C.H.
Abingdon
WASHINGTON Co.
RUSSEL Co.
KY
Lexington
Bourbonton
Boonesboro
CLARKE Co.
MADISON Co.
Crab Orchard
MASON Co.
BOURBON Co.
Washington
Springfield
Preston
NORTH
CAROLINA
KNOXVILLE
KNOX Co.
JEFFERSON Co.
SEVIER Co.
GREEN Co.
Greenville
Jonesboro
HAWKINS Co.
WASHINGTON Co.
BUNCOMB Co.
BURKE Co.
Morganton
Lincolnton
LINCOLN
RUTHERFORD Co.
Iredell
Salisbury
Wilkes
WILKES Co.
Rockford
Huntsville
Germanton
Bethania
Salem
Lexington
Randolph
Hillsboro
ORANGE Co.
Chapel Hill
RALEIGH
JOHNSTON C.
Smithfield
WAYNE Co.
Chatham C.H.
MOORE
Cabarras
Montgomery C.H.
Averysboro
Waynesboro
DOBBS Co.
Kinston
GLASGOW Co.
PITT Co.
Greenville
Tarboro
NASH C.
Louisburg
WARREN Co.
HALIFAX
Warrenton
Oxford
GRANVILLE Co.
Person C.H.
Caswell C.H.
Rockingham C.H.
Williamsboro
NORTHAMPTON Co.
Princeton
Murfreesboro
Winton
BERTIE Co.
Windsor
Williamston
Washington
Ft. Lowdon
Chota

YORK
MASSACHUSETTS
CONNECTICUT
RHODE
ISLAND
JERSEY
DELAWARE
DELAWARE BAY
Long Island Sound
ATLANTIC
OCEAN
CHESAPEAK BAY
Albany
Schenectady
Cooperstown
Catskill
Kingston
Pittsfield
Stockbridge
Northampton
Springfield
Worcester
Concord
BOSTON is in SUFFOLK C.
Salem
Marblehead
Gloucester
Newburyport
PORTSMOUTH
PORTLAND
Providence
Newport
New London
HARTFORD
New Haven
Litchfield
NEW YORK
Newark
Elizabethtown
Brunswick
Princeton
TRENTON
Burlington
PHILADELPHIA
Chester
Wilmington
Lancaster
Reading
Easton
Bethlehem
Dover
Cape May
Sandy Hook
Cape Cod
Nantucket or Sherburn
Martha's Vineyard
Holmes Hole
Edgartown
Montauk Point
Sagharbour
Explanation
Boundary Lines
Post Roads
Post Roads extended by Contract
Post & Stage Roads
Stage Roads
Post Offices are written in Roman Letters
Ports of Entry & Delivery for Foreign
Ports of Entry & Delivery for Domestic Vessels
WASHINGTON Capitol 38° 53' North Latitude
Scale of Miles
Longitude Computed from Washington
PROGRESS of the MAIL on the MAIN LINE.
POST TOWNS
Miles
Summer Establishment from April 15th to October 15th
Winter Establishment from October 15th April 15th
Progress Southward.
Progress Northward.
Sun. Mon. Tue. Wed. Thu. Fri. Sat.
Leave
Arrive
Brewers
Machias
Chandler's River
Columbia
Narraguagus
Gouldsboro
Sullivan
Trenton
Bluehill
Castine
Penobscot
Prospect
Belfast
Camden
Thomaston
Warren
Waldoborough
Newcastle
Wiscasset
Bath
Brunswick
North Yarmouth
PORTLAND
Biddeford
Welles
York

55.
LOUISIANA
Drawn by S.Lewis
Tanner Sc.
NEW NORTH WALES
HUDSON
BAY
NEW SOUTH WALES
BRITISH POSSESSIONS
NORTH
NEW MEXICO
NEW ALBION
NEW NAVARRE
CALIFORNIA
NEW BISCAY
PACIFIC OCEAN
GULF OF MEXICO
W. FLORIDA

BEFORE LEWIS AND CLARK

Samuel Lewis, "Louisiana," in Aaron Arrowsmith, *A New and Elegant General Atlas*, 1804

President Thomas Jefferson came into office with an agrarian vision for his country; it required more land to accommodate a nation of farmers. He also inherited a belief that North America offered a passage to India—with all its attendant commercial rewards. As seen in earlier chapters, this hope stretched back to the fifteenth century. These two assumptions help to explain the president's enthusiasm for the Louisiana Purchase and his desire to understand the geography of the Far West.

In large part Jefferson's interest in a Northwest Passage was driven by geopolitics. By the time he became president, Spain had established control of the Southwest and had built presidios up the Pacific coast to San Francisco. Russian traders were actively extending their networks further up the coast, while British explorers continued their search for a transcontinental portage from Hudson Bay to the Pacific. Finally, the French, under Napoleon, flirted with the hope of rekindling their interior empire of trade along the Mississippi River. In 1800 North America was alive with imperial competition.

Spain continued to claim much of the continent west of the Mississippi River, and became increasingly wary of American encroachment after the Revolution. Late in 1802, Spanish authorities abruptly closed the port of New Orleans to American trade, which gravely threatened farming in the Ohio Valley. Almost simultaneously, Spain was negotiating a secret treaty to turn over the Louisiana Territory to France. Jefferson was aware of this development, and sought to negotiate with the French to reopen access to New Orleans. To his great surprise the French foreign minister was instructed to offer all of Louisiana to the Americans for $15 million, mostly through the forgiveness of debts incurred during the French Revolution. In April 1803 the deal was complete, and on Independence Day Jefferson announced the Louisiana Purchase to the public.

Before the land transfer was finalized, Jefferson sent a secret message to Congress requesting support for an expedition to the Pacific to be undertaken by his personal secretary, Meriwether Lewis. Once authorized, the president instructed Lewis to "explore the Missouri river, & such principal stream of it" in order to discover "the most direct & practicable water communication across this continent, for the purposes of commerce." Jefferson was keenly aware of the multiple imperial interests at play in the Far West, and hoped that improved geographical knowledge of the region would enable the Americans to leverage their position with Europeans as well as native tribes.

Just what did Americans know about the Far Northwest before Meriwether Lewis and William Clark conducted their expedition? Here is the first map of the Louisiana Territory after it was transferred to American control, engraved by Samuel Lewis and published in Aaron Arrowsmith's *A New and Elegant General Atlas*. Lewis took a broad view of Louisiana in order to situate it within the geography of North America. By deemphasizing national and imperial boundaries, he drew attention to the topography and river systems of the West.

Lewis based his map on one drawn by Antoine Pierre Soulard in 1795. The Spanish governor of Louisiana in St. Louis instructed Soulard to survey the interior in order to gather geographical intelligence about the Mississippi and Missouri river basins. Soulard's map included a few errors that were replicated on Lewis' 1803 map here, and it was those errors which led Jefferson to believe that there was a viable water route to the Pacific. The most striking of these is the nearly uninterrupted chain of relatively low mountains from Canada down through New Mexico. This suggested that the "Stoney Mountains" were narrow and easily traversed, which in turn fueled Jefferson's hope of a transcontinental passage. The map also suggests that the headwaters of the Missouri River were extremely close to those of the rivers flowing west to the Pacific Ocean. This vastly underestimated the course and reach of the Missouri, as well as the size of the mountains. But the errors on the map show us contemporary views of the continent, and by extension Jefferson's motives for the expedition.

AFTER LEWIS AND CLARK

"A Map of Lewis and Clark's Track, across the Western Portion of North America," copied by Samuel Lewis from the original drawings of William Clark, 1814

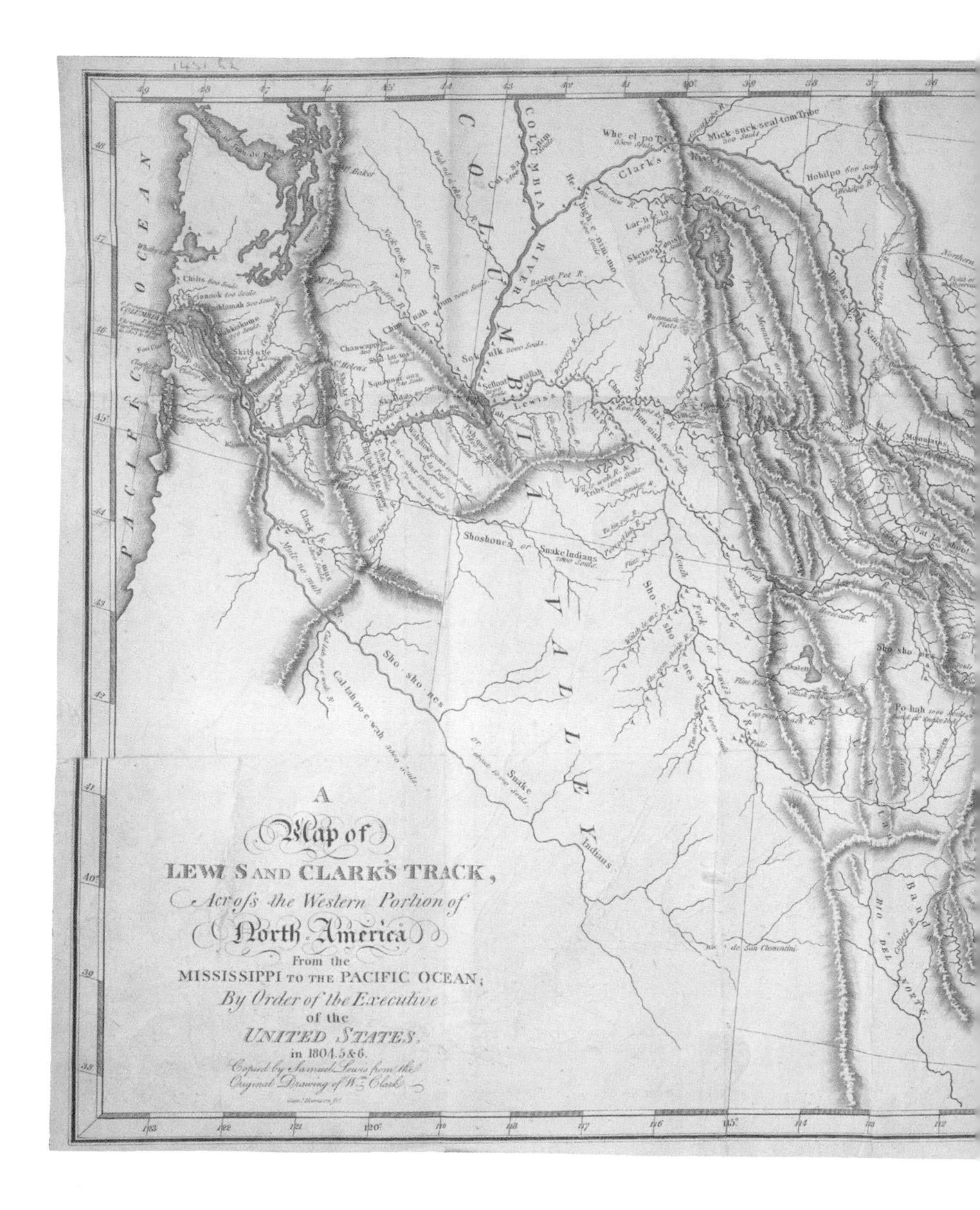

As the previous map reveals, it was the relative *lack* of geographical knowledge that led President Jefferson to sponsor an expedition beyond the nation's western boundary to the Pacific Northwest. In his official request to Congress, Jefferson argued that such an expedition would clarify the geography of North America. Like so many before him, he also sought to advance trade by discovering a Northwest Passage to the Pacific and Asia beyond. Finally, he hoped to strengthen the American fur trade relative to the British by establishing trading posts along the Mississippi River that would receive goods from the Missouri and its tributaries. In sum, Jefferson had several reasons to send Meriwether Lewis and William Clark up the Missouri River in the spring of 1804. Two years later, the expedition returned to St. Louis having traversed more than 7,000 miles. The information they brought back utterly reshaped the map of North America.

When they set off, Lewis and Clark knew comparatively little about the continental interior, and what they thought they knew was often wrong. There was no navigable passage to the Pacific Ocean, nor was there a short portage in the Rocky Mountains between the headwaters of rivers running east and west. Rather than a single low, narrow chain of mountains as depicted on the previous page, Lewis and Clark found multiple and complex ranges of much higher elevations. This immediately ended any hope of an easy overland route. After crossing the mountains in both directions and from multiple approaches, the expedition brought back a far *wider* picture of the West.

It took more than ten years for the reconnaissance knowledge gained by Lewis and Clark to circulate. Once William Clark returned to St. Louis in 1806, he spent years compiling a map of the expedition to accompany the publication of his journals. Late in 1810, the map was redrawn by Samuel Lewis, who made the map on page 112 as well. Published in 1814 alongside Clark's journals, it remains one of the most important American contributions to nineteenth-century geography.

First and foremost, the map captures the extent of the Columbia River system in the west and the Missouri River system in the east. The great bend of the Missouri River was brought into focus, as were the proper courses of the Clearwater, Columbia, and Snake rivers. Mountains and valleys were elaborated in a way that must have awed contemporary viewers. The expedition also ended the assumption that western rivers remained wide and navigable near their headwaters. Instead, Clark described rivers that were narrow at their source and that broadened as

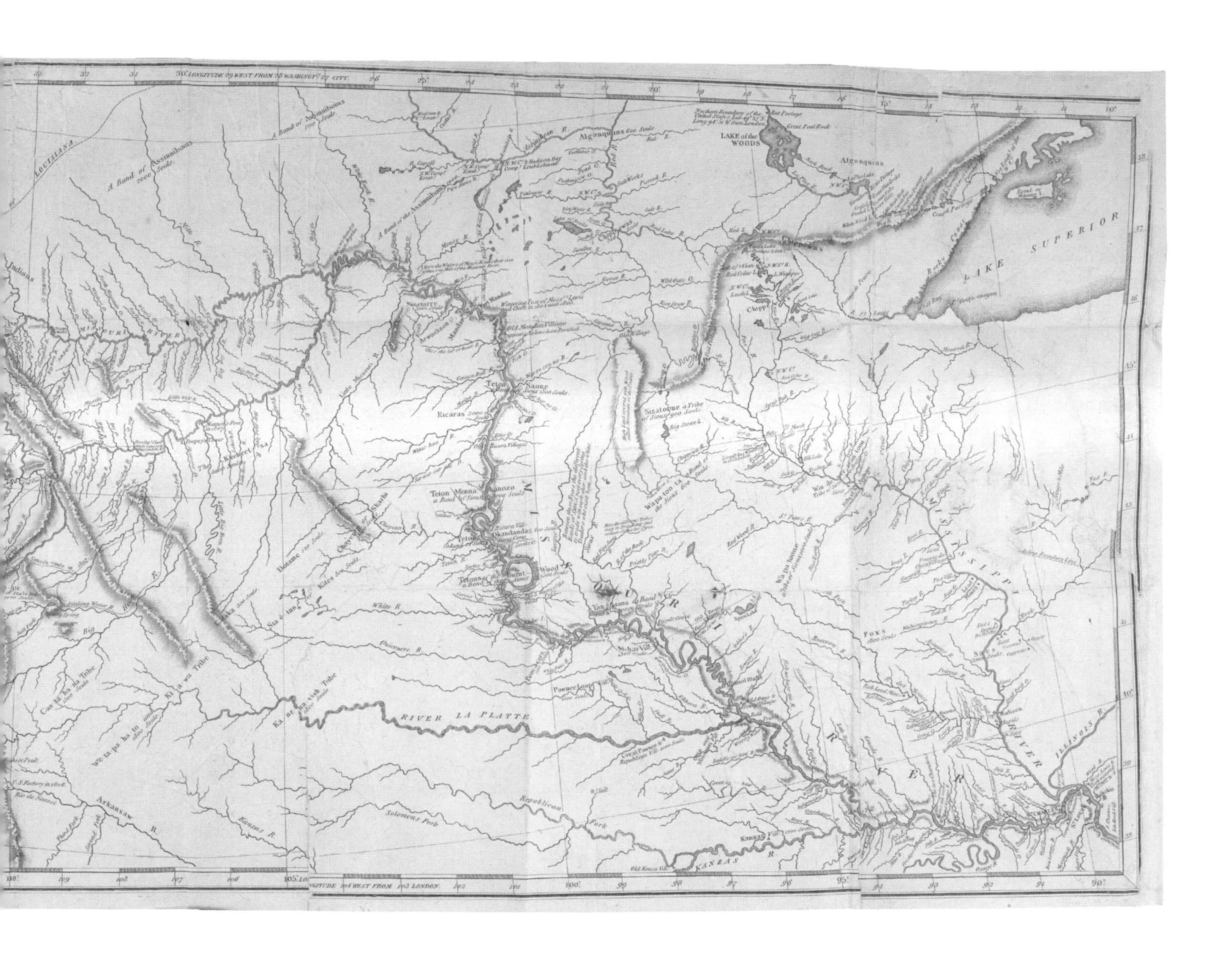

they flowed circuitously toward the Pacific Ocean or the Mississippi River.

Clark's map was not without errors, of course. Because the expedition did not venture south, the depiction of the Southern Rockies remained vague at best. Moreover, like his contemporaries, Clark believed that the headwaters of some of the great western rivers—the Colorado, the Rio Grande, the Big Horn, and the Yellowstone—converged. And yet these limitations detract little from Clark's achievement in drawing a far more complex picture of the American Northwest, one that would remain authoritative until the expeditions of John Fremont and Charles Wilkes in the 1840s. Equally important is the way that the expedition—and its map—drew attention to the West as a region of its own rather than a pathway to Asia. In their voluminous notes, Lewis and Clark detailed the West's Indian populations, resources, and potential for settlement. In this sense, Jefferson's venture was eminently successful: though it definitively ended the search for a Northwest Passage, it opened an era of exploration on the continent itself.

DEMOCRACY SUBVERTED

"The Gerry-Mander, or Essex South District Formed into a Monster!," *Salem Gazette*, April 2, 1813

Those who bemoan the political polarization of our own day may take comfort in knowing that this is nothing new. Periods of intense partisanship appear throughout American history, and in fact the party system itself emerged out of bitter political rivalries in the 1790s. Soon after the Americans won independence from Britain, intense disagreements over federal power, personal liberties, and the nation's future drove the birth of the Federalists and the Republicans.

Moreover, the roughly equal power exercised by the two parties in the early republic drove each of them to seek any competitive edge they could find. In 1811 Massachusetts—which had for years been dominated by the Federalists—suddenly turned toward the Jeffersonian Republicans. With control of both the governorship and the state legislature, the Republicans acted to maintain their advantage by passing a redistricting bill that coincided with the most recent census. Governor Elbridge Gerry signed the bill, and Republicans immediately drew electoral boundaries that confined Federalists to Essex County's interior in order to minimize their statewide voting power. The gambit worked: in the next election cycle, with *fewer* votes the Republicans managed to elect *more* representatives than the Federalists.

The Federalists were understandably outraged, and accused the Republicans of fraud. At a Boston dinner party, a Federalist examining a map of one of the newly configured districts observed that its contorted shape resembled that of a salamander. Another quipped that the district ought to be deemed a "Gerry-Mander," a jab at the governor who had signed the bill. The joke morphed into Elkanah Tisdale's satirical map of the district, first published in the Boston *Gazette* in 1812, and soon reprinted in other Federalist papers. The map is not just named for Governor Gerry, but also mocks him personally: his profile is caricatured with a hooked nose at Middleton and a jutting chin at Lynnfield. Gerry was a signer of the Declaration of Independence and vice president under James Madison. Despite this sterling reputation, however, his name would be forever associated with political manipulation.

The Federalist *Salem Gazette* published this version of the map as part of a series that excoriated the opposition and its tactics. The scandal coincided with the outbreak of war with Britain in 1812, which Federalists blamed on the Republican president James Madison. This explains the *Gazette's* urgent call for "Federalists! Followers of Washington!" to turn out on election day and outvote the Republicans.

The paper accused the Republicans of "unholy party spirit," for in redrawing district boundaries they had taken advantage of representative democracy to create a permanent political advantage. The nearly hysterical rhetoric reveals the level of anger felt by the Federalists. Despite the fact that they had been the majority party in Massachusetts, shrewd measures on the part of the Republicans had effectively turned them into a minority. Hence the paper's argument that the Gerry-Mander was not just politics as usual, but a technique that "stifles the voice of the *Majority*." To the Federalists, this was nothing less than the betrayal of representative government itself.

This rhetoric mobilized the Federalists, who regained control of the Massachusetts legislature and immediately repealed the redistricting law. But the term "gerrymander" stuck, and has been invoked ever since to describe perceived unfairness in redistricting.

The practice of gerrymandering long predates the term. To those who have used it successfully, gerrymandering is a perfectly legal and democratic instrument to maximize a party's representation; to those on the losing end, it signals an absolute subversion of democracy. In our own time, we have been reminded more than once that the electoral college operates in the same way, whereby the winner of the popular vote may still lose the presidency.

Efforts to redraw boundaries to ensure a particular outcome have been reviewed by the Supreme Court for decades. Yet the court has overturned such redistricting efforts only when they effectively diminish the representation of racial minorities. In the 1990s Justice Sandra Day O'Connor pushed back against bizarrely shaped districts in Texas, arguing that they were clearly designed to dilute minority voting power. By contrast, the court is only beginning to rule efforts to limit the power of a particular *party* as unconstitutional as shown on page 254. The upshot is a nation with less competitive elections, for voters have been sorted into safer districts. If this seems undemocratic, it is as old as the republic itself.

SALEM GAZETTE.

VOLUME XXVII.] SALEM, FRIDAY, APRIL 2, 1813. [No. 27.

THE GERRY-MANDER, OR *ESSEX SOUTH DISTRICT FORMED INTO A MONSTER!*

FEDERALISTS! FOLLOWERS OF WASHINGTON!

AGAIN behold and shudder at the exhibition of this terrific Dragon, brought forth to swallow and devour your Liberties and *equal* Rights. Unholy party spirit and inordinate love of power gave it birth;—your patriotism and hatred of tyranny must by one vigorous struggle strangle it in its infancy. The iniquitous Law, which cut up and severed this Commonwealth into Districts, is kindred to the arbitrary deeds of NAPOLEON when he partitioned the territories of innocent nations to suit his sovereign will. This Law inflicted a grievous wound on the Constitution,---it in fact subverts and *changes our Form of Government*, which ceases to be *Republican* as long as an *Aristocratic* HOUSE OF LORDS under the form of a Senate tyrannizes over the People, and silences and stifles the voice of the *Majority*.

When Tyranny and arbitrary Power thus make inroads upon the Rights of the People, what becomes the duty of the citizen? Shall he submit quietly and ignominiously to the decrees of Usurpers? Are the citizens of this Republic less jealous of their rights than their ancestors? Will you, then, permit a Party to disfranchise the People,---to convert the Senate Chamber into a Fortress in which ambitious office-seekers may entrench themselves and set at defiance the frowns of the People? No,---this usurping Faction must be dislodged from its strong-hold.

Arise, then, *injured Citizens!* Turn out! turn out! let Monday Next be the day of your Emancipation---by one manful Struggle reclaim your usurped Rights---and frown into obscurity those audacious men who unblushingly boasted---"*We have secured the Senate for ten years, and should have been Fools if we had not done it.*" Prove on election day that the Folly of their men is equal to their want of honesty and contempt of the People. Elect patriots who will be loyal to the Constitution, and faithful to the interests of the State,---men who will harmonize with our illustrious STRONG,---and who will not sell our Drafted Soldiery, like Hessians, to any Foreign Power.

requisition their utmost cunning to devise means expressly to accommodate their own views and to promote their own interest, should still be unwilling to adhere to the policy prescribed by themselves.

Such a man would have to learn, that violence and subordination must be employed to accomplish what dishonesty and trick had failed to secure. It is in fact the settled maxim of some, who, if the testimony of their own professions is to be credited, have much love for the people and a sacred regard for their rights, to contemn all legal provisions, which impede their ambitious designs, however just and necessary to secure the peace and order of society; to execute by force whatever they are prohibited to do by law, and to aggrandize themselves without respect to justice or decency. Upon no other principle can we account for the conduct and activity of those, who, at a late election in this town, scrupled not to bring to the poll, not only alien residents, who were known never to have been naturalized, not only a certain clergyman (the learned Comentator of Politics for this vicinity) whom they had, in common with the rest of his brethren, disqualified to vote in the choice of town officers, and thus to involve him in the turpitude of their transgressions, and to multiply the causes of repentance, (unless he so far believes, "that the end would sanctify the means," that there would be *non causa penitentiæ*); but even to seduce citizens from *other* towns, to invest the minority with a power, which they would be sure to abuse. He, who promises freedom to the captive, when he increases the number and weight of his chains, acts as consistently, as he does, who declaims much and loud about the liberties of the people and the rights of his country, when all his actions evince, that he regards neither the one nor the other, any farther, than he discovers a direct advancement of his own interest. It is greatly to be desired, that *honest* men would sometimes suffer themselves to be influenced as well by objects of *vision*, as *hearing*, and compare *actions* with *professions*; and then let judgment decide upon the worth of character, and the safety of reposing confidence. CIVIS.

MR. PRINTER,—I am a native of *Great-Britain*, and some years ago was naturalized at *Gottenburg*; but for several months I have been engaged in trade in this State. I have seen very pressing notices of the Marshal for alien enemies to report themselves forthwith While I remain here I wish to comply with the laws and ordinances of the country, and feel no enmity towards the government or people. A peace or change of times would destroy the profits of my present speculations, and in that case I should probably soon depart. Moreover, I do not find that the U. S. of America are at war with my adopted country. Now I will thank any of your correspondents to inform we what is proper or expected from one in my circumstances; as there are many in the country, and more expected, similarly situated. A. B.

"UNDER A SIGNATURE,"

Is the correct, and was once the universally received phrase. But a spirit of innovation and hyper-criticism directed by gross misconception has engendered and brought into too general use the awkward and absurd term, "*over a signature*."—To show how mischievous tolerated abuses are, we would refer to the following from the Boston Gazette. "*The Road to Peace*.

"Messrs. Russell & Cutler—I have read a very sensible, but small tract [compose your muscles, gentle reader] *over the above title*," &c.

Now, the tract, "*over*" this title is worthy of all praise for its correctness and candor, and may be read *without offence* and certainly with advantage, *by all parties*. It is peculiarly calculated for *distribution*—it *repels no reader*, whatever be his politics or character.

Given *over* my hand and seal, March, 1813.

THE GOOD OLD WAY.

DARKNESS VISIBLE.

Mons. GALLATINI has directed the Lights near the Chesapeake and Nantucket to be extinguished, so that our Enemies may be left in the dark; and of course it will be perfectly easy to place Torpedoes under the Fleet and blow it out of our waters.—How perfectly in character is it for our present rulers to *extinguish* instead of illumining, and to scatter *darkness* where their predecessors diffused *light*. It is the sport of these Proclamation-men, like Mad Tom,

"To play at bowls with sun and moon,
"And shade them in eclipses."

It is expected that every man who does not wish to be drafted as a conscript and sent out of the State to fight the battles of the Southern Nabobs—every Farmer who wishes to prevent his farm and stock from being seized and sold to pay the enormous taxes which are shortly to be laid on them, will turn out on the first Monday of April next, and support the Peace Ticket.—New Hampshire has set a bright example; let us follow it. You all know you were prosperous and happy under Washington; you are now wretched and miserable under Madison.—You cannot, you will not give your suffrages for men who are in favor of the present war, unless you are unjust to yourselves, and unfaithful to your country.—*Augusta Herald.*

Salem,

FRIDAY, APRIL 2, 1813.

SOUTH with a Vengeance!

The Senatorial District made and called "ESSEX SOUTH *District*," by a Gerrymander Legislature, has the following *peculiarities*:

1. It contains three Towns (Salisbury, Amesbury and Haverhill) more *northerly* than any other town in the County.

2. An East and West line drawn through the centre of the County leaves Salisbury, Amesbury, Haverhill, Methuen, a large part of Andover and a small part of Middleton [six] on the *Northern* side.

3. The only towns wholly in the *southern* section of the county that the district contains, are Danvers, Lynnfield, Salem, Marblehead and Lynn [five.]

4. The *northernmost* town of *another county* is tacked on it.

☞ There neighbor, take a piece of chalk and see if it does not come out so.

Measures for Defence have been adopted in all the ports threatened by the British Squadron. At Wilmington a Committee of Safety has been appointed. The Honorable Messrs. BAYARD and HORSEY, (U. S. Senators from Delaware) and other zealous Federalists constitute this Committee. These are the Patriots on whom the country must rely when the hour of extreme peril approaches. Our democratic rulers have sufficient talents to plunge the nation in misery and calamity, but none but Federalists can rescue the nation and save its "drowning Honor." The deceased Heroes WASHINGTON, LINCOLN, KNOX, WAYNE, MORGAN and HAMILTON,—the Soldiers of Liberty who atchieved our Independence,—were zealous and decided *Federalists* to the last hour of their lives;—and even WASHINGTON was calumniated and traduced as a British Tory by the Democratic party,—which party he solemnly declared in a letter to a friend was "A CURSE TO THIS COUNTRY." This same party is now scourging the country with War, and the People *feel* the truth of the words of WASHINGTON.

The Right of Suffrage.

ALL men will agree in the importance of having the right of suffrage properly defined by law; and most men, when uninfluenced by passion, will acknowledge the necessity of enforcing the law with the utmost strictness, and can correctly calculate the immense benefits which would result from such a course.—These are the natural dictates of the understanding and of conscience. But man in *theory* and in *action* is not the same being. In the dispassionate moments of speculation, he listens to the suggestions of his reason, and approves of the admonitions of his understanding; but in practice he acknowledges no law but convenience, and feels no satisfaction for his exertions except in triumph and victory.—To a person, whose astonishment had not been so often excited, till he had become incapable of new surprise, by the moral phenomena which are continually occurring in the political world in these days of strange delusion and of disasters; it would appear incredible, that a party of men, after they had had the opportunity of putting in

A SCHOOLGIRL MAPS HER COUNTRY

"A Map of the United States by Catharine M. Cook," 1818

Educational opportunities for young women exploded during the early republic. While girls had routinely been educated in their homes before 1800, thereafter families began to enroll them in schools. Hundreds of female academies appeared to meet this demand, ranging from small and temporary enterprises to more stable institutions that endure to this day.

Girls were often exposed to the same subjects as boys, though with interesting variations. Geography was considered essential for both sexes, but for girls it often included extensive map-drawing exercises. From 1790 to the 1830s, thousands of students drew, painted, and stitched maps as part of their education. Most common were maps of the nation and its capital, shown on pages 100 and 106. Some of these were copied and traced, while others were drawn freehand, using only the grid of longitude and latitude as guides. While most reflect care and precision—with carefully composed borders and river systems—others bear the marks of more artistic freedom.

Maps and geography were considered particularly appropriate material for girls, a "useful" pathway to literacy and citizenship that also honed traditional feminine skills of "accomplishment" such as painting or needlework. As John Pinkerton wrote in 1818 in the preface to his own atlas, "[Geography] is a study so universally instructive and pleasing that it has, for nearly a century, been taught even to females." Sarah Pierce, founder of the Litchfield Female Academy in Connecticut, stressed geography and map drawing as a way to strengthen "principles of association" and "readiness of memory." For young girls in the new republic, the ability to create a map of the nation and the world was a mark of one's education.

This page is taken from the penmanship journal of Catharine Cook, who attended a well-established school for girls in Vermont. Like her classmates, Cook was taught to use penmanship as a way to practice other subjects: across several pages she wrote out lessons in history, geography, and astronomy with elaborate calligraphy. The journal concluded with a series of hand-drawn maps: first the world, then the nation shown here, and finally the individual states.

Cook's map of the United States is representative of an exercise that was undertaken in hundreds of schools across the country. She took great care to reproduce the formal elements of a map, such as the "neatline" around the edge alongside the measures of longitude and latitude. The elaborate cartouche at lower right reflects Cook's attention to calligraphy and illustration, both of which were integral to the curriculum. State names are drawn in a separate style of calligraphy, while meticulous application of color demarcates boundaries. Together, these details reveal the sustained attention that these projects required, no doubt as a way to occupy the long school days that developed in the nineteenth century. These maps took months to complete, and many were then publicly evaluated in formal competitions held at the end of the school year. The charming misspellings on Cook's map—of Pennsylvania, Lake Superior, Louisiana, and Mississippi—remind us that behind these projects were individual learners often just entering their teenage years.

This map—like so many others—also reflects something deeper at work in early nineteenth-century education. By drawing their country, these students were making the nation real, inscribing its abstract boundaries and administrative units and visualizing the topography and river systems of distant regions that most would never see firsthand. In the process, young girls and boys connected themselves to their fellow citizens, rendering the nation as a coherent and stable entity. Such an exercise was especially relevant in the 1810s, when the War of 1812 tested—and then vindicated—American independence.

Map drawing was more commonly taught to girls than boys in this era. Catherine Beecher, founder of the Hartford Female Academy, recalled (less than fondly) the emphasis on map drawing and artistic accomplishment in her own education. Many young girls who had been exposed to map work in British and American academies went on to become teachers to support themselves, and brought these exercises with them into an ever growing network of schools around the country. The practice declined by the 1840s, when inexpensive wall maps, atlases, and other cartographic materials flooded the market and made it less necessary for students to create their own learning materials. Immensely charming artifacts, these maps also reveal the daily experience of the first generation of girls to be formally educated in the new nation.

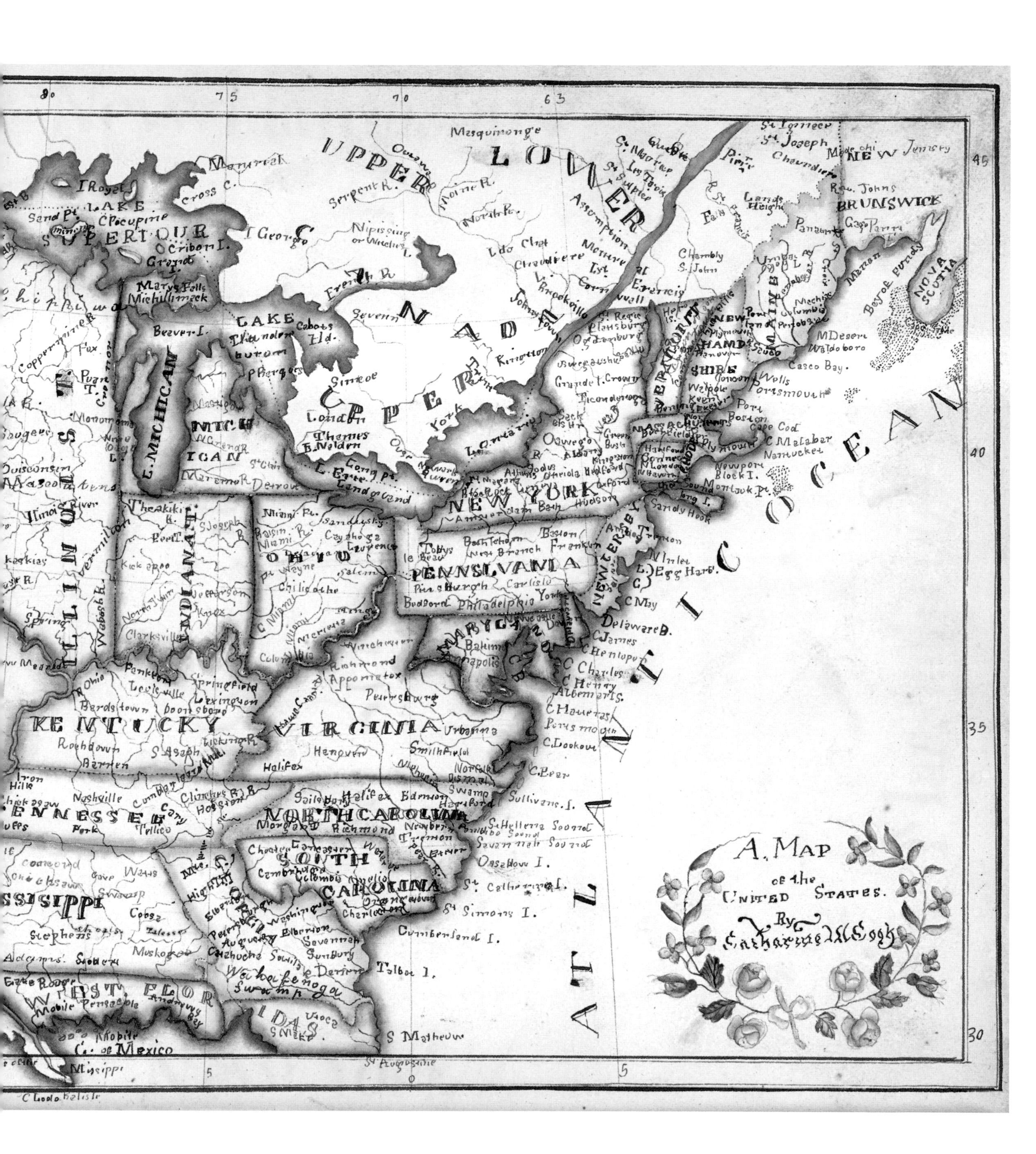
A Map
of the
United States.
By
Catharine McEogh
UPPER
LOWER
CANADA
UPPER
ATLANTIC OCEAN
NEW YORK
PENNSLVANIA
OHIO
INDIANA T.
ILLINOIS T.
MICHIGAN
L. MICHIGAN
KENTUCKY
VIRGINIA
NORTH CAROLINA
SOUTH CAROLINA
MARYLAND
NEW JERSEY
VERMONT
NEW HAMPSHIRE
MASACHUSETTS
MAINE
NEW BRUNSWICK
NOVA SCOTIA
WEST FLORIDAS
LAKE SUPERIOUR
C. of Mexico
Philadelphia
Pittsburgh
Sandy Hook
Cape Cod
Casco Bay
Delaware B.
C Henlopen
C Charles
C Henry
C Hatteras
Cumberland I.
S Matheuw
St Augustine
Nashville
Lexington
Richmond
Petersburg
Halifax
Norfolk
Mobile
Pensacola
Savannah
Charleston
Augusta
Chillicothe
Kingston
York
London
45
40
35
30
80
75
70
65
5
0
5

A LITTLE SHORT OF MADNESS

Cadwallader D. Colden, "New York," 1825

In 1807 Congress asked the Treasury secretary, Albert Gallatin, to survey the country's entire transportation network in order to set priorities for improvements. Gallatin's lengthy and comprehensive response has been called the greatest planning document in American history. He emphasized the importance of infrastructure to integrate the seaboard and the interior. A road across the Appalachian range, for instance, would draw the west closer to eastern settlements, while a network of canals would facilitate trade between seaports and the Great Lakes. Finally, Gallatin proposed a series of canals to create an inland navigation network from Massachusetts to southern Georgia. This was an outlandishly expensive project for the young nation, but the War of 1812 forcefully demonstrated the need for an internal system of transportation. This was all the more important given the sheer size of a country that extended from the Mississippi River to the Atlantic Ocean.

One of Gallatin's specific suggestions was the construction of a canal from Lake Ontario to the Hudson River. New Yorkers responded with enthusiasm, and when little federal support materialized they took matters into their own hands. The state formed a canal commission and appointed former New York City mayor DeWitt Clinton as its leader. In 1816 the commission proposed a 350-mile east–west route from the Hudson River to Lake Erie, which former President Jefferson called "a little short of madness." With signatures of support from 100,000 New Yorkers, however, construction of the canal began on Independence Day in 1817.

The Erie Canal used over fifty locks to traverse 600 feet of elevation, and was far longer than any canal in Europe at that time. The project reconfigured American geography by connecting the western interior to the coast. The old Northwest and the Great Lakes region were both brought into the orbit of New York City, enlarging its economic prospects at the expense of Boston. The canal reduced the cost of shipping by over 90 percent, and within a few years

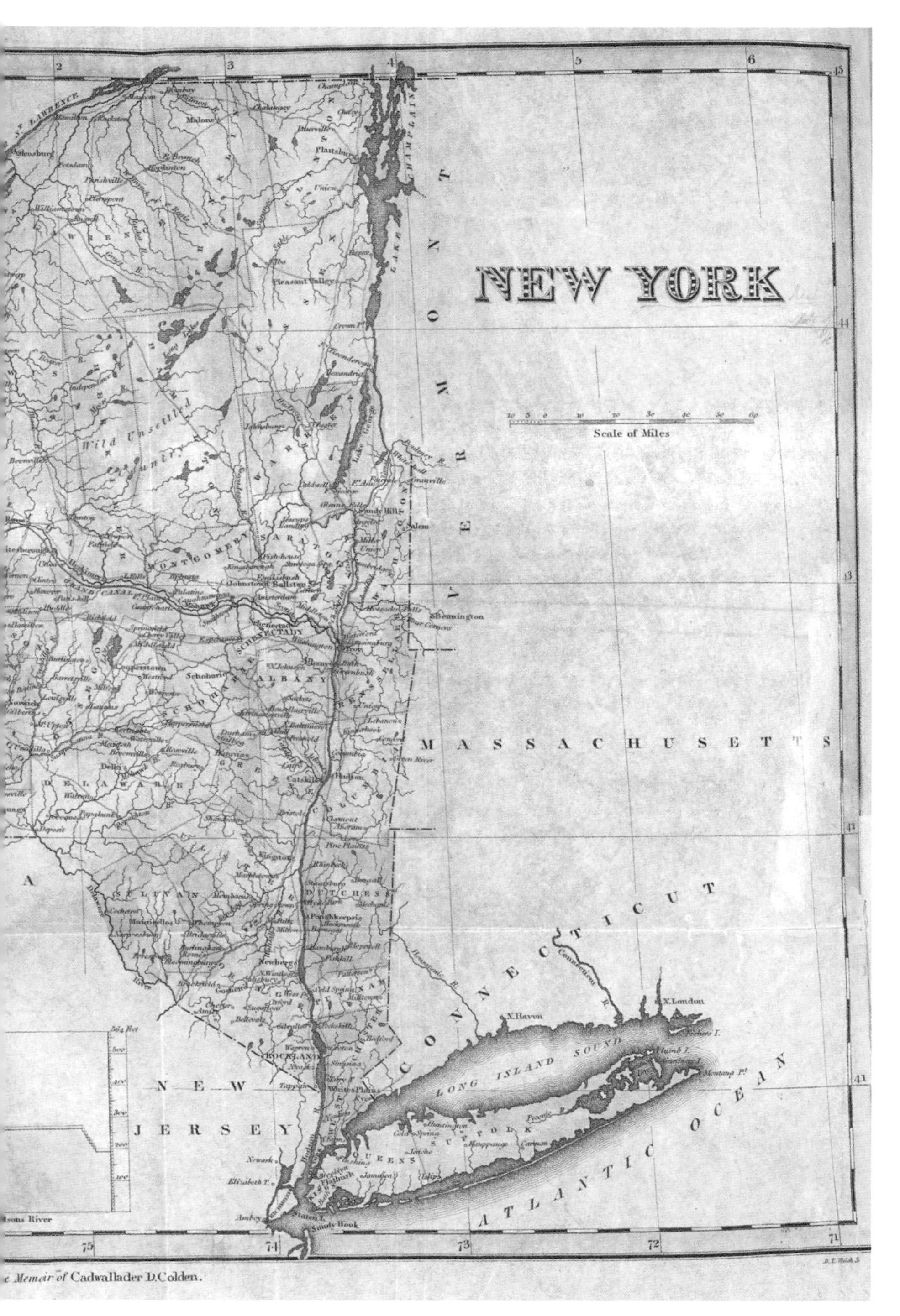

it was transporting a greater volume of material than the entire Mississippi River system. This fever of activity created new towns all along the route, and expanded existing western settlements at Buffalo and Detroit.

The Erie Canal ushered in a canal-building boom throughout the country that lasted until the advent of railroads in the 1840s. Together with steamboats, canals shrank distance, fueled migration, and fed commerce, shifting an economy centered on local transactions—and often involving barter—to one centered on cash. Like the digital revolution of our own day, canals advanced communication, realigned regional networks, and created entirely new markets.

The effects are even more striking when we recall a map made by Cadwallader Colden in the 1720s (page 87). Colden had been sent west to forge an agreement with the Iroquois Confederacy that would facilitate the safe travel of other tribes through their territory. On that map the future path of the Erie Canal is in a region characterized as "The Country of the Five Nations," an indication of Iroquois power and dominance at that time.

A century later, Colden's grandson, Cadwallader David Colden, oversaw the construction of the canal as the mayor of New York City. To commemorate this achievement, the younger Colden published a history of western New York, stressing that not a single white inhabitant lived in the area as late as the 1780s. In his history, Colden reprinted his grandfather's map to demonstrate just how much had changed over the course of a century.

While the elder Colden mapped western New York as Native American land, his grandson presented the entire state as the arena of white settlement. Moreover, here Colden celebrates the transportation improvements and their effect on New York's economic future. Along the bottom of the map, cross-sectional diagrams demonstrate the engineering achievement not just of the "Grand Canal," but an earlier canal constructed from Lake Champlain to the Hudson River. The canals not only integrated a geographically expansive state, they transformed its presence in the larger region. These projects amplified the power and prosperity of New York, and profoundly shaped the emerging territories of the Midwest.

A CONTINENTAL FUTURE

John Melish, "Map of the United States, with the contiguous British & Spanish Possessions," 1823

In 1839 newspaperman John O'Sullivan declared that the United States was divinely ordained to spread its civilization westward. O'Sullivan was not the first to claim that American expansion was providential, but he did introduce the phrase "Manifest Destiny." If there is a cartographic picture of this idea, here it is. As the first map of the United States to encompass the entire continent, it anticipates the territorial expansion of the 1840s. Yet it was designed not by an American but a Scotsman, and published more than two decades before O'Sullivan coined the phrase.

John Melish emigrated from Scotland to Philadelphia in 1810 to establish the nation's first commercial mapmaking firm. On earlier visits he had traveled widely through the country, and these experiences served him well in his new business venture. Melish settled in Philadelphia at an auspicious moment for the new nation, when relations between the United States and Britain had begun to deteriorate. Though there were multiple sources of tension, the breaking point came when the British repeatedly impressed American sailors into the Royal Navy. Such insults prompted American war hawks to demand retaliation, and in 1812 Congress declared war against Britain.

The War of 1812 was widely opposed by Federalists, who considered it little more than an elaborate maneuver to advance the aims of the Republican Party under President James Madison (page 116). In the end, no territory was gained from Britain, and the half-hearted American effort to invade Canada was easily repelled. Yet Americans had held their own against a superior British naval force, and that was enough to qualify the conflict as a second war of independence. The return of peace in 1815 sparked a sustained wave of patriotism.

In was in this moment of heightened nationalism that Melish began to compile an ambitious new map of the United States. The first edition included new geographical information brought back by the Lewis and Clark expedition (page 114). In the final edition of the map (shown here), Melish also described the western plains as the "Great Desert," a phrase that had just a year earlier appeared on Stephen Long's new map of the far west based on his own expedition of 1819.

With its elaborate detail, high quality of production, and up-to-date information, the map instantly became one of the most sought-after profiles of North America. It was used in the Adams–Onís Treaty negotiations of 1819, which transferred control of Florida from Spain to the United States. It also helped to set the boundary between Spanish California and the Oregon Territory at the forty-second parallel north.

Yet the map's enduring power derives not from its role in contemporary statecraft, but from a larger message. Melish initially designed the map to extend to the Rocky Mountains, which formed the nation's western boundary. Yet somewhere in the process of compiling it he changed his mind, and decided to extend the map far beyond the boundaries of the United States to the Pacific Ocean. Upon its publication in 1816, Melish explained that his map was a "picture," one that showed "at a glance the whole extent of the United States territory from sea to sea." Though much of the Far West was not yet part of the United States, Melish found it more aesthetically pleasing to include the entire continent, which paralleled the "expansion of the human race from east to west."

By showing the whole continent, Melish foreshadowed—and perhaps subtly influenced—the nation's westward trajectory, just as John Mitchell had with his map of 1755 (page 94). For instance, he outlined the reach of the Missouri Territory in green, intentionally leaving its western boundary undefined. Just two years later, the US and Britain negotiated an agreement to jointly occupy the territory of Oregon. The map, oddly, had anticipated the nation's westward reach.

Melish's map was updated and reissued twenty-five times. The final version included the new state of Missouri, the admission of which in 1820 involved an important and controversial compromise over slavery. Missouri was admitted to the Union as a slave state with the understanding that this institution would be barred from all future states carved out of the Louisiana Territory north of parallel 36° 30′. In documenting this ominous agreement over slavery, and encompassing the entire Far West, Melish's monumental map showed Americans not just where they were, but also where they were heading.

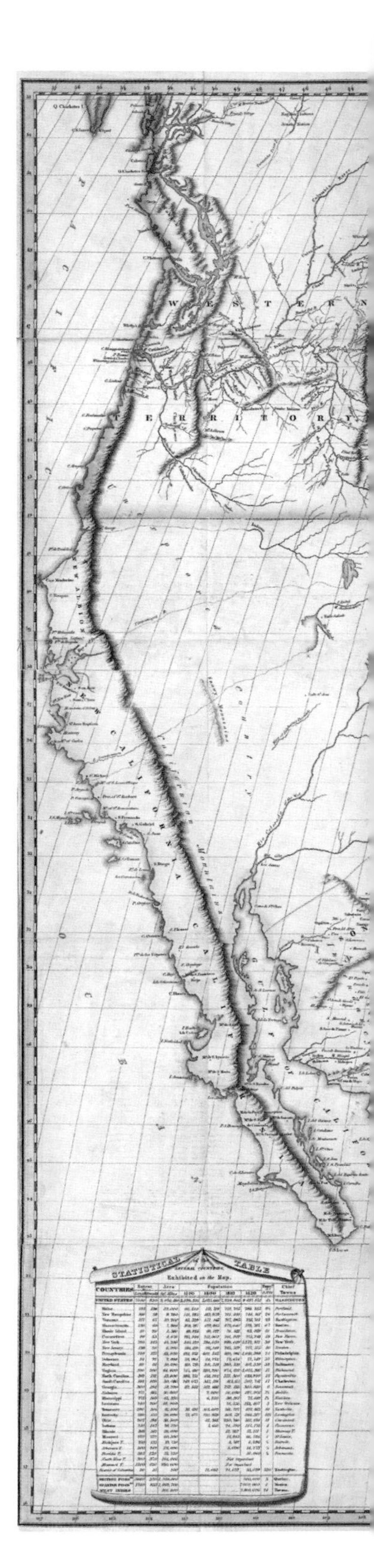

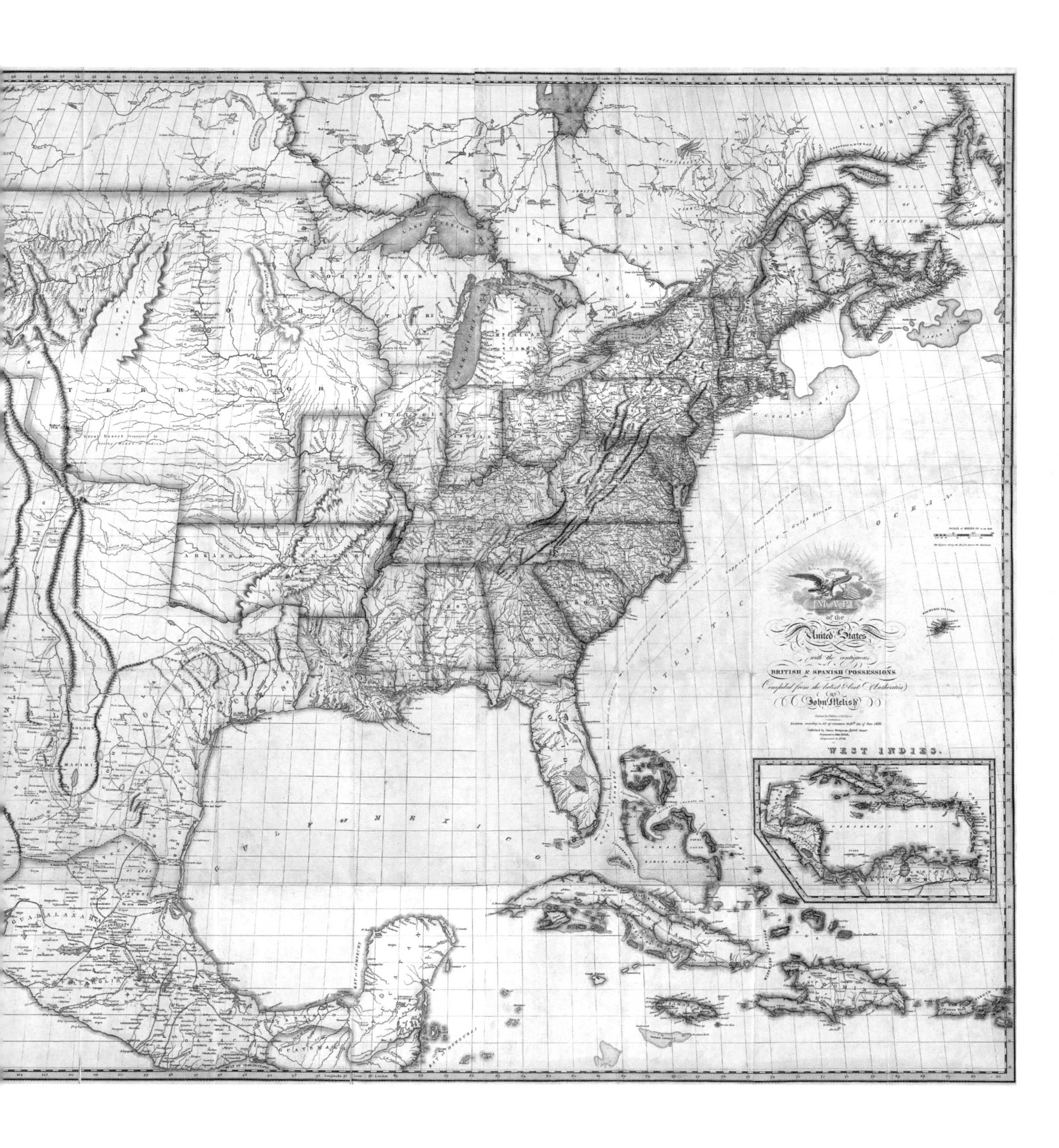
MAP
of the
United States
with the contiguous
BRITISH & SPANISH POSSESSIONS
Compiled from the latest & best Authorities
BY
John Melish
WEST INDIES
LAKE SUPERIOR
LAKE MICHIGAN
NORTH WEST
TERRITORY
ARKANSA
CAROLINA
GEORGIA
ATLANTIC
OCEAN
GULF OF MEXICO
Supposed Limits of the Gulph Stream
LABRADOR
GULF OF ST. LAURENCE
BAY of CAMPECHY
BAY of HONDURAS
GUATEMALA
CARIBBEAN SEA
GREAT BAHAMA BANK

5. 1835–1874

Expansion, Fragmentation, and Reunification

Americans were on the move in the mid-nineteenth century, in every sense of the word. Beginning in the 1830s, westering migrants began to claim land in Texas, Oregon, and Utah, well before those regions were part of the United States. A few years later, prospectors from around the world descended upon the gold fields of California, just months after that territory was acquired from Mexico. The growing demand for labor in the emerging industrial centers of the north drew thousands of Americans off the farm, while millions of Europeans joined them in search of greater opportunity.

While migration is often narrated in terms of liberation, for many it signaled a profound loss of freedom. President Andrew Jackson made room for land-hungry settlers and prospectors in Georgia by dispossessing Native Americans and sending them west. Planters establishing new cotton and sugar plantations in Mississippi, Louisiana, and Texas forced the sale and migration of hundreds of thousands of slaves from the upper to the lower South. Nearly 300,000 prospectors, farmers, and Mormons heading to the Far West dramatically disrupted Native American lands on the Great Plains.

Several of the maps in this chapter explore the relationship between nineteenth-century migration and national expansion. Much of this expansion was driven by technological innovation and the rise of a market economy. As shown on page 120, the Erie Canal enabled farmers to produce not just for their own subsistence and barter but also for distant markets. Similarly, Eli Whitney's cotton gin significantly expanded the amount of land cultivated in the South, so that by 1860 one-third of the nation's cotton crop was grown *west* of the Mississippi River. This increase in cultivation stimulated Northern textile manufacturing, which in turn made cotton the nation's largest export. In 1858 James Henry Hammond declared that "Cotton is king," meaning that not even the English monarchy could challenge the economic might of the American South. By then the US was producing 75 percent of the world's cotton supply.

Westward expansion was often led by migrants who crossed national borders to settle new land. In the 1820s Moses Austin led a group of Southerners to Tejas (the territory of today's Texas) at the invitation of Mexico, which hoped to stabilize and develop its northern frontier. In 1836 those migrants joined with Mexican-born settlers to declare themselves independent. Many in this new Republic of Texas sought annexation by the United States, but American presidents were understandably reluctant to provoke war with Mexico or antagonize Northerners who keenly opposed the entry of a new slave state into the Union.

Just north of Texas, Congress organized a territory dedicated to the "civilized" tribes displaced by the Indian Removal Act of 1830. The map on page 130 shows the geopolitics—not to mention the injustice—involved in President Jackson's policy of removal. The War Department was responsible for protecting these emigrant tribes, while also separating them from white settlers to the immediate east in Arkansas and Missouri. The result was a complex hierarchy of internal and external boundaries that continued to shift as the nation expanded westward in the 1840s.

The annexation of Texas in 1845 prompted a controversial war with Mexico, and commercial firms quickly issued maps to meet public demand (page 134). At the same time, the federal government issued detailed maps to encourage overland migration to Oregon and ultimately American control of that territory (page 132). The subsequent victory over Mexico in 1848 gave the United States control over California, the Intermountain West, and an enlarged Texas. The timing was remarkable: no sooner had the US acquired California than gold was discovered at the foothills of the Sierra Nevada Mountains. Within weeks the US military had sponsored the most detailed map yet of those mining districts (page 136). Each of these three maps was designed to advance the nation's control over the Far West.

Some Americans celebrated this dramatic territorial growth as a sign of Providence, while others worried that it only extended the life of slavery. As slavery grew, so too did Southern regional distinctiveness. In response, many Northerners began to assert a contrary identity, one grounded in the principle of "free labor." To be sure, the concept of "free labor" was more than a little ironic given that low-wage jobs proliferating in the North gave workers little mobility. However, by 1850, sectional differences were undeniable. As the map on page 138 shows, waves of German and Irish immigrants sought the harbors of Boston, New York,

Philadelphia, and San Francisco; few chose the South.

Regional differences surfaced in other ways. Evangelical revivals fostered an array of social and moral reforms in the antebellum era, the most successful of which was temperance. The map on page 126 reveals the energy that drove the contemporary war against alcohol. While temperance was a national crusade, however, most evangelical movements were concentrated in the Northern states. Samuel Gridley Howe's effort to teach geography to the blind (page 128) grew out of a community of reformers in Boston. Slavery made the South more paternalistic, less likely to support social reforms, and downright hostile to more assertive movements such as abolition.

That reform impulse found political expression in the Whig Party, which had organized against President Jackson's two-term Democratic administration from 1829 to 1837. Through much of the 1840s—even as the nation expanded to the Pacific Ocean—Whigs and Democrats maintained national constituencies. But in 1854 Stephen Douglas fractured this party system by introducing a bill that repealed the prohibition against slavery in the Louisiana Territory. The prospect of slavery extending north of the 36° 30′ line outraged many Northerners. They abandoned their Whig and Democrat homes to organize a new Republican party, committed to the principle that Congress had the obligation to prohibit slavery from the western territories. Through the 1850s, John Jay and other Republicans urgently issued maps to publicize the geographical threat of slavery (page 140). Jay's map demonstrates that it was the fate of slavery in the *West*, and not in the *South*, that drove the sectional crisis.

The Republicans lost the presidential election of 1856. Four years later, they won the electoral college and the White House without the support of a single slave state. Slaveholders in the Deep South considered the very election of Abraham Lincoln a threat to their future; they responded by leaving the Union. In his inaugural address, Lincoln made clear that he would protect slavery where it existed, his aim being to halt the momentum of secession. For a few weeks he kept the states of the Upper South in the Union. But in April 1861, a crisis at Fort Sumter led the president to call up the militia, which prompted Virginia and three other slave states to join the Confederacy.

In the first months of the Civil War, Lincoln refused to attack slavery, and overruled generals who used their military authority to issue emancipation orders in the South. The president believed that a conservative policy on slavery was needed to keep the loyalty of the border states, particularly Kentucky. But the slow progress of the Union Army led Lincoln to adopt emancipation as a military measure. Though highly limited in scope, the Emancipation Proclamation fundamentally shifted the meaning of the war. What began as an effort to suppress a rebellion and preserve the Union ultimately ended slavery and redefined American citizenship.

Maps played a crucial and often unexpected role in that conflict. Here we examine not the many maps designed for battle but, rather, those that measured the strength of the rebellion. For Lincoln, a path-breaking map of the distribution of slavery (page 142) helped him to see that the Confederacy's greatest asset was its labor system. Similarly, data-driven maps (page 146) shaped General William Tecumseh Sherman's campaign through Georgia, which ultimately accelerated Union victory and the destruction of slavery.

The Civil War ended a brutal labor system that had endured for centuries. But the liberation of four million slaves was swiftly compromised with the end of Reconstruction. In state after state, whites violently subjugated the freedmen and attacked Republican leaders in a manner that ultimately led to the collapse of Reconstruction governments. The map on page 148 shows the dynamics of this resistance in New Orleans, a pattern that extended across the South and anticipated the future of the entire region: by 1877 the nation had abandoned Reconstruction and turned its attention elsewhere.

THE GEOGRAPHY OF SIN

Reverend John Christian Wiltberger, "Temperance Map," 1838

In 1801 an evangelical camp meeting drew thousands to the Kentucky frontier, launching a religious revival that quickly spread to every corner of the country. In the early stages of this awakening, many evangelicals concentrated on individual piety and salvation, but by the 1820s they began to focus on social ills such as alcohol. It was easy to see why. Between 1800 and 1830, Americans drank so much that one historian dubbed the nation the "Alcoholic Republic." In part this increase in consumption resulted from the country's abundant grain supply, which farmers could easily distill into whiskey. But alcohol was also integral to American life: often safer than water, it was consumed at work, essential to weddings and funerals, and ubiquitous on election day.

Temperance activists—aided by influential preachers—built an astonishingly effective grassroots movement that used extended religious networks to spread the message of abstinence. The results were stunning. Within ten years of its founding in 1826, the American Temperance Society had established 5,000 branches. These branch organizations convinced young and old alike to take the "Cold Water Army" pledge and abstain from alcohol entirely. Through songs, prayers, and broadsides, they insisted that even modest consumption led to drunkenness, sin, and early death. Exaggerated as these claims may sound, temperance was one of the most successful reform movements in the nation's history. Abraham Lincoln was among its adherents, a lifelong teetotaler. The temperance movement peaked in the 1830s and 1840s; by 1855 most states had limited the production of alcohol.

Among the more creative temperance advocates was the Philadelphia minister John Wiltberger, who invited readers to navigate sin and temptation through this imaginary map. The upper territory lays out a geography of righteousness, where Industry, Improvement, Prosperity, and Plenty reward those who follow the Cold Water River of abstinence. But danger lay ahead of those who imbibed even modestly. At left, an archipelago seductively presents the occasional drink as both pleasing and innocent. But beyond the islands of Medicine and Hospitality we quickly discover that socially acceptable and modest drinking lead to the Sea of Intemperance. Along these middle latitudes, among the only ways to reach safety or redemption is through the narrow Tee Total Railroad. By contrast, how easy it is to submit to temptation and descend toward the islands of Murder, Larceny, and Poverty; beyond this, one is ever more likely to slip into the Land of Inebriation. Such a place is rife with danger, offering only False Hopes, False Comforts, and Ruin.

Allegorical maps such as this one had long been used to model ideals of love, courtship, and propriety. But even within this long tradition, Wiltberger's map stands out for its geographical realism and prurient detail. With this colorful and inviting image, he aimed to preach the gospel of abstinence to a broad audience. The result is an elaborate and compelling geography that uses fiction to teach that happiness comes only through self-denial. Even the orientation of the map reinforces the message, for redemption lies to the north—toward heaven—while to the south lie decay and disorder.

After its extraordinary victories in the 1840s, the temperance movement waned before resurging when the exploding urban population renewed concerns about the use of alcohol among young men in the Gilded Age. In the early twentieth century, temperance won its greatest triumph in the Eighteenth Amendment, which banned alcohol for over a decade and inadvertently created an underworld economy of speakeasies, black markets, and crime.

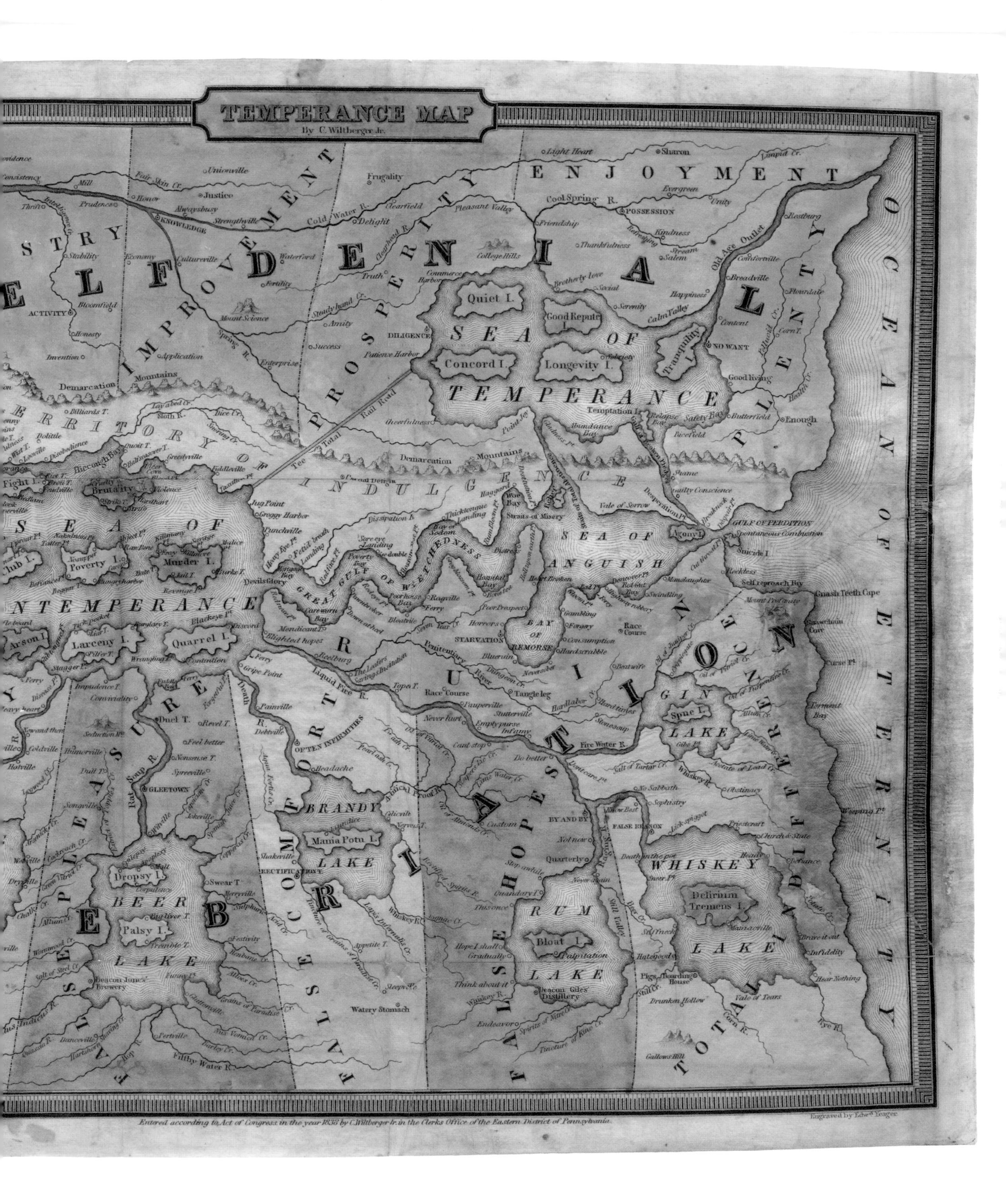
TEMPERANCE MAP
By C. Wiltberger Jr.
ENJOYMENT
SELF DENIAL
INDUSTRY
IMPROVEMENT
PROSPERITY
PLENTY
OCEAN OF ETERNITY
SEA OF TEMPERANCE
Quiet I.
Good Repute I.
Concord I.
Longevity I.
Tranquility I.
TERRITORY OF INDULGENCE
SEA OF INTEMPERANCE
Fight I.
Brutality I.
Club I.
Poverty I.
Murder I.
Arson I.
Larceny I.
Quarrel I.
GREAT GULF OF WRETCHEDNESS
SEA OF ANGUISH
Agony I.
GULF OF PERDITION
BAY OF REMORSE
RUIN
DESTRUCTION
OCEAN
GIN LAKE
Spue I.
BRANDY LAKE
Mania Potu I.
BEER LAKE
Dropsy I.
Palsy I.
RUM LAKE
Bloat I.
WHISKEY LAKE
Delirium Tremens I.
PLEASURE
FALSE COMFORT
FALSE HOPES
TOTAL INDIFFERENCE
Demarcation Mountains
Rail Road
Tee Total
Cold Water R.
Cool Spring R.
Fire Water R.
Liquid Fire R.
KNOWLEDGE
ACTIVITY
DILIGENCE
POSSESSION
NO WANT
STARVATION
GLEETOWN
Duel T.
Watery Stomach
Deacon Jones' Brewery
Deacon Giles' Distillery
Gallows Hill
Vale of Tears
Entered according to Act of Congress in the year 1838 by C. Wiltberger Jr. in the Clerks Office of the Eastern District of Pennsylvania.
Engraved by Edwd Yeager

AN ATLAS FOR THE BLIND

Samuel Gridley Howe and Samuel Ruggles, map of Vermont for the blind, 1837

The map on the previous page reflects the creative energy of antebellum reformers, who used every means at their disposal to rid the country of alcohol. The same spirit that animated temperance also infused contemporary efforts to modernize schooling. Among the leaders of this movement was Horace Mann, who broadened access to education and experimented with new methods of instruction. Typical of these reformers was the Bostonian Samuel Gridley Howe, who dedicated himself to educating blind students.

In 1832 Howe founded the Institution for the Education of the Blind, the first school of its kind in the United States and now known as the Perkins School for the Blind. Such schools had long existed in Europe, but Howe broke with tradition by insisting that blind students could direct their own education by learning to read. In the era before Braille, such a task was easier said than done. Howe hoped to cultivate literacy through innovative teaching and learning materials that improved upon French techniques of raised script. In 1835 he designed his own raised typeface, which replaced curves with streamlined narrow and angled letters to facilitate tactile identification. This Boston Line Type was used in all of the learning materials at the Institution.

Throughout the 1830s Howe developed new instructional materials, and he made geography a foundation of his institutional curriculum. Initially, those techniques relied on a tutor, who would explain the arrangement of the states and then ask the student to do the same. Yet this required a sighted interpreter, and Howe wanted blind students to learn on their own. Working with the printer Samuel Ruggles, he created a massive globe that measured almost five feet in diameter. The globe still stands in the Perkins History Museum, though blind students had difficulty grasping—both literally and figuratively—the overall geography of the world on such a large apparatus. This inspired Ruggles to produce an atlas that would enable blind students to explore geography at their own pace and through their own efforts.

This map of Vermont is from one of the few surviving copies of that atlas. Ruggles and Howe used dotted lines to mark state and national borders and solid lines to trace rivers. They added a second parallel line to denote the widening of the Connecticut River as it flowed south. Elsewhere on the map, unique symbols indicate distinct physiographical features. Small hachuring is used to mark the mountain ranges across the state, while the waters of Lake Champlain and Lake George are set off from land by subtle horizontal ridges. Individual letters throughout the map reference towns listed and described on a separate page. The spare markings on the page are designed to maximize tactile legibility for the reader.

Whether Howe's atlas was useful to students is hard to say, for some found Boston Line Type difficult to master. In spite of that, it quickly became the most widely used raised typeface in the country. By 1840 Howe and Ruggles had published forty-one books with Boston Line Type, including the New Testament and four geography texts. Tracts extolling the virtues of temperance were also published, reflecting Howe's own evangelical and moral convictions. A committed opponent of slavery, Howe married Julia Ward, who famously penned the overtly Christian "Battle Hymn of the Republic" just after the outbreak of the American Civil War.

Howe insisted that Boston Line Type was superior to coded systems such as Braille. To his mind, the latter only further segregated and isolated blind readers by using an arbitrary set of symbols instead of embossed letters that could be read by those with sight and without. But by the turn of the century, Boston Line Type and several other systems were overtaken by Braille.

Howe's atlas—like the schoolgirl map on page 118—underscores the concerted effort in the antebellum era to widen access to education and reach new segments of the population. This drive bore significant fruit by the end of the century, when nearly every state had established a system of common and public schools.

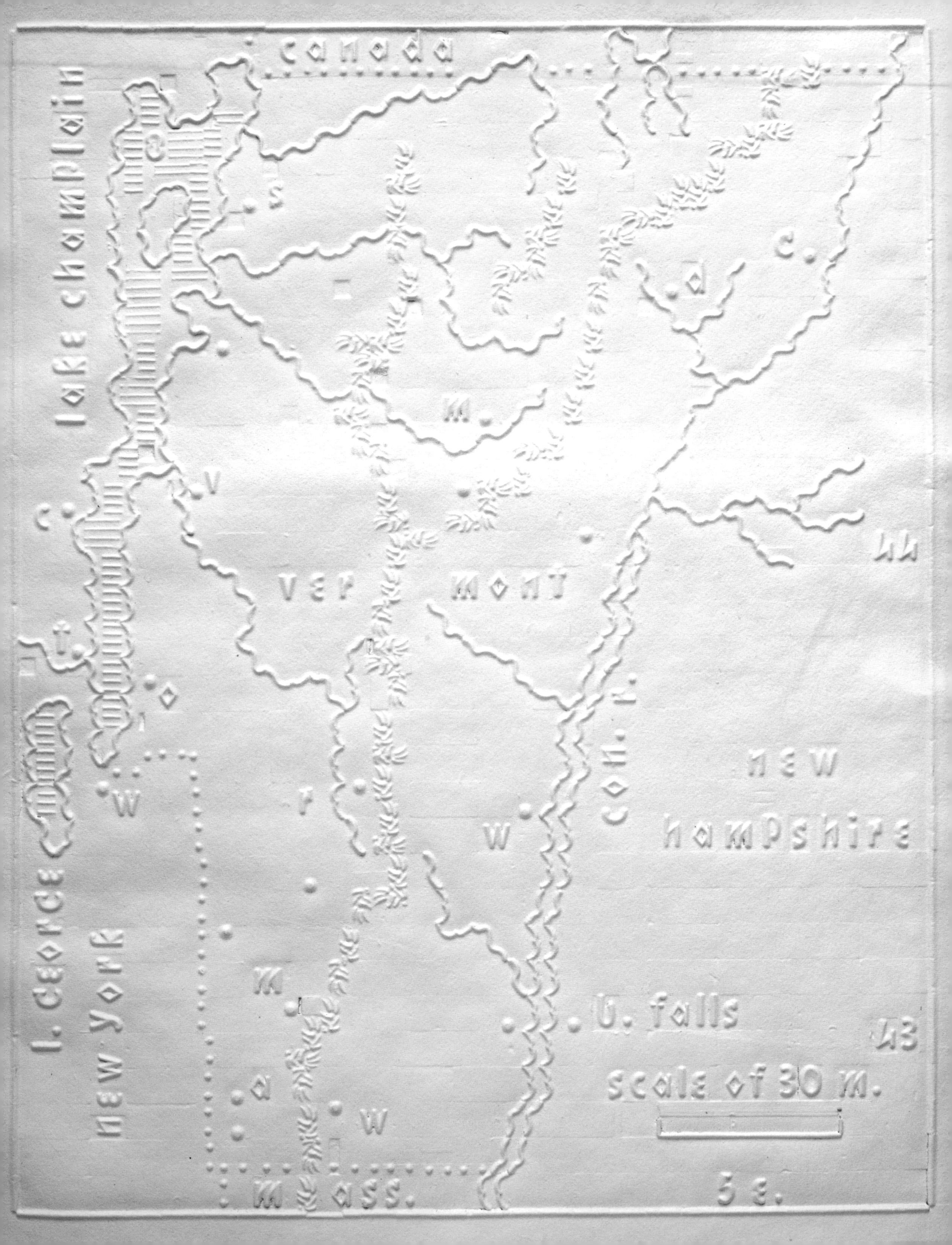

canada
lake champlain
l. george
new york
ver mont
new hampshire
con. r.
u. falls
scale of 30 m.
mass.
5 e.

THE ENEMY WITHIN

Charles Gratiot, "Map Illustrating the Plan of the Defences of the Western & North-Western Frontier," 1837

The American Southeast had long been inhabited by Native Americans when Europeans began to settle its shores in the seventeenth century. Among the largest of these tribes was the Cherokee Nation, which extended from northern Georgia into eastern Tennessee and the Carolinas. But the territorial domain of the Cherokee rapidly contracted during the eighteenth century, and by the Revolutionary War they had surrendered half of their land to the colonists.

When gold was discovered within the boundaries of the Cherokee Nation in the late 1820s, the state of Georgia claimed jurisdiction over the land. By this time, the Cherokee had established a system of education and courts, which prompted the federal government to identify them as one of the five "civilized tribes" of the Southeast, along with the Chickasaw, Creek, Chocktaw, and Seminole. Georgia's presumption of sovereignty was immediately challenged by the Cherokee, leading the matter into the political and judicial arena. President Andrew Jackson sided with the state of Georgia, and even proposed a bill to remove the tribes. Jackson argued that the removal of the Cherokee was part of a long American tradition: just as tribes further north had either been removed or eliminated "to make room for the whites," so too should southeastern tribes be relocated to the western frontier.

Several southeastern tribes resisted removal, and the Supreme Court ruled that Georgia had no jurisdiction over Cherokee land. In a rather stunning act, Jackson simply ignored the Court's decision and moved forward with removal. He even framed this as a benevolent and generous policy that would protect tribes from aggressive land seekers. Moreover, Jackson argued, removal would enable these Indians to pursue "happiness in their own way" while also making space for white settlement. Though most tribes acquiesced, the Cherokee leader, John Ross, continued to resist, pleading with Congress and the president to respect Indian sovereignty.

In 1835 military representatives negotiated with several other tribal leaders to accept this "voluntary" removal. John Quincy Adams, Jackson's successor, opposed the policy, but even his leadership was no match for land-hungry settlers in Georgia and more general anti-Indian sentiment throughout America. In 1838 the army forced the remaining 16,000 Cherokee on an arduous trek to the Indian Territory. One quarter of the tribe perished during this westward journey.

"The Trail of Tears" ended the presence of Native Americans east of the Mississippi. But forcibly moving this population west created a new challenge of defining—and securing—the nation's western border. The task of establishing a defensive frontier fell to the engineer Charles Gratiot, the son of a French trader in Spanish St. Louis who was among the first graduates of West Point. Gratiot and his colleagues in the War Department proposed a network of forts and garrisons to accomplish three related goals. The first was to protect these emigrant tribes from more powerful and aggressive tribes in the region, such as the Comanche. The second was to keep the peace between the many emigrant tribes that had been forcibly relocated to this new territory. The third was to separate the emigrant tribes from their white neighbors in Arkansas and Missouri.

Here Gratiot mapped out a new defensive frontier that would stretch from the Missouri River to the northern edge of Texas. Within that corridor, about 200 miles wide, he identified a network of forts and depots that the military could reach within days either by roads (marked in red) or by waterways (marked in blue). Leavenworth, which had been established a decade earlier as a stop on the Santa Fe Trail, now became Fort Leavenworth, a base from which the military could arbitrate disputes between tribes or between tribes and white settlers. Dozens of other fortifications formed a militarized frontier both to protect emigrant tribes and to segregate them from white settlers to the east.

The very fact that the War Department proposed such a complex plan underscores the fraught history between European settlers and Native Americans in North America. After independence, the United States had pursued several different Indian policies, some of which were more cooperative than others. Jackson forcefully implemented a vision of separation by creating an altogether new territory for the tribes of the Southeast. Though designed to be a perpetual home for the emigrant tribes, much of this "Indian Territory" would be appropriated for white settlement in the Oklahoma land rush in 1890. This was just one example of a pattern that characterized the entire history of America: permanent Indian territory was anything but permanent.

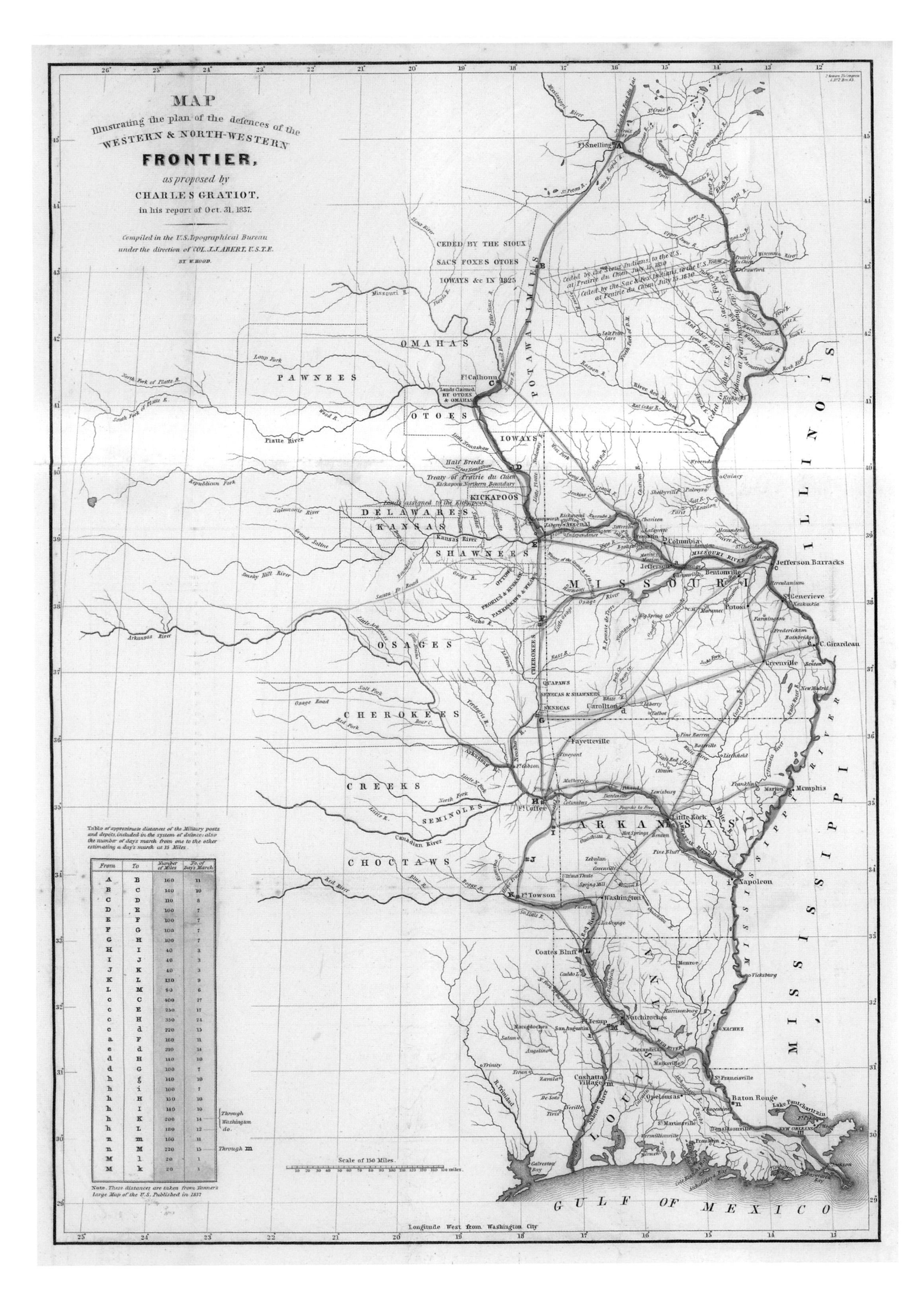

From	To	Number of Miles	No. of Day's March
A	B	160	11
B	C	140	10
C	D	110	8
D	E	100	7
E	F	100	7
F	G	100	7
G	H	100	7
H	I	40	3
I	J	40	3
J	K	40	3
K	L	130	9
L	M	90	6
c	C	400	27
c	E	250	17
c	H	350	24
c	d	220	15
a	F	160	11
e	d	210	14
d	H	140	10
d	G	100	7
h	g	140	10
h	i	100	7
h	H	150	10
h	I	140	10
h	K	200	14
h	L	180	12
n	m	160	11
n	M	220	15
M	l	20	1
M	k	20	1

Through Washington do. (h–K, h–L)
Through m (n–M)

Note. These distances are taken from Tanner's large Map of the U.S. Published in 1837

OPENING THE OREGON TRAIL

Charles Preuss, “Topographical Map of the Road from Missouri to Oregon,” from the field notes and journal of Capt. J. C. Fremont, section IV, 1846

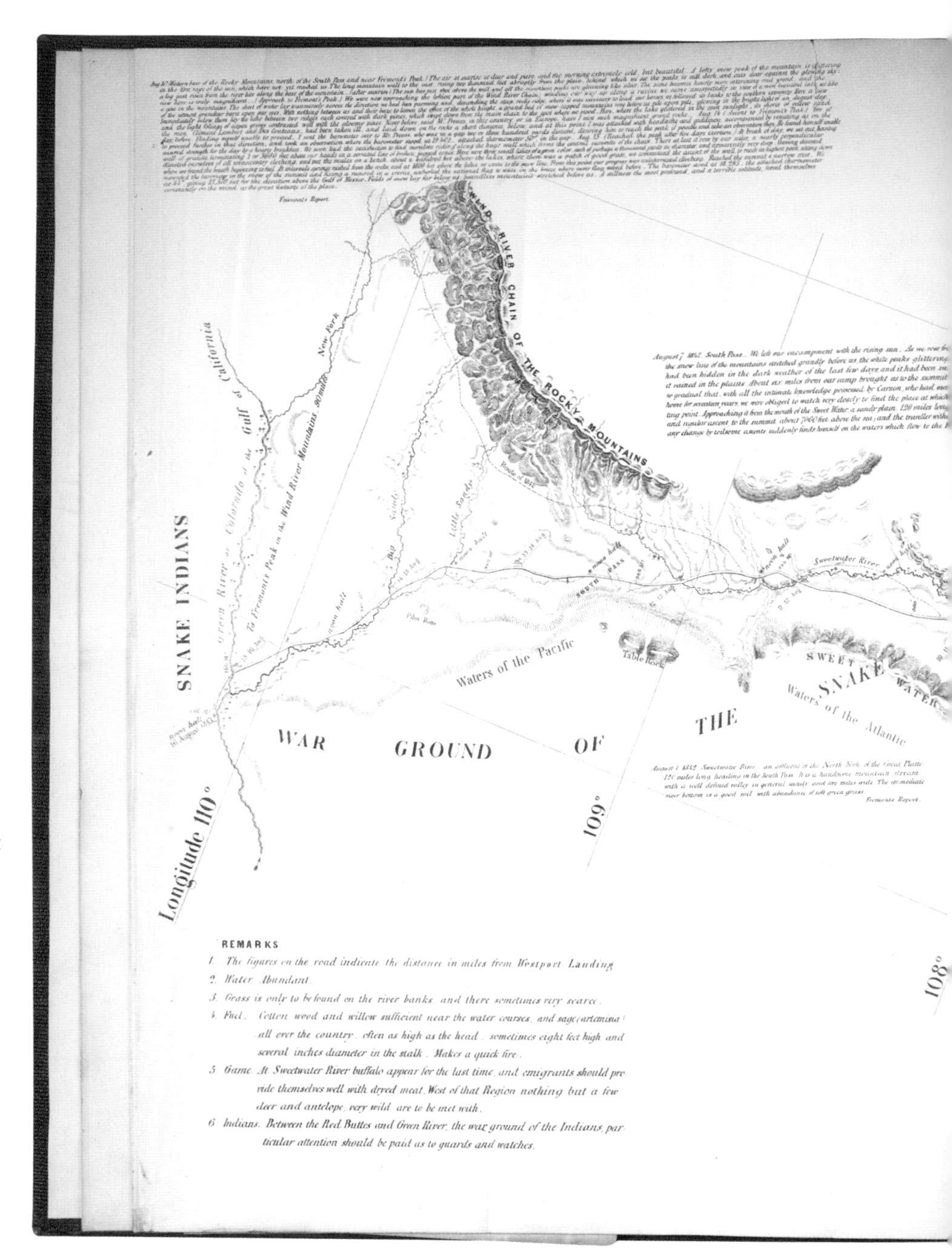

The map on the previous page captures the assumption—first articulated by Thomas Jefferson—that the Louisiana Territory would become a permanent home to Native Americans. That vision of the future, however, was complicated when thousands of Americans migrated across the Great Plains to the Far West. Farmers began heading to the fertile lands of Oregon’s Willamette Valley in the 1830s. In the 1840s, Mormons migrated west to flee persecution, followed by thousands of prospectors streaming toward the gold fields of California.

These migrants brought Oregon and California into the American imagination well before those areas were part of the national domain. At the same time, the explorer John Charles Fremont made a series of expeditions that vastly enlarged American geographical knowledge of the western interior. Fremont’s expeditions were facilitated by his marriage to Jessie Benton, daughter of Missouri senator Thomas Hart Benton. A champion of Manifest Destiny, Benton forcefully advocated the annexation of Texas, and sponsored Fremont’s western expeditions as a way to bring the Far West under American control.

Fremont’s extended journey of 1843 took him through Oregon and California. Upon his return, his wife developed his detailed field notes into a memoir that burnished his reputation and advanced his political career. Even more consequential than the memoir were the maps produced by Charles Preuss, a German immigrant who had joined two of Fremont’s expeditions. Despite a sour temperament and an abiding hatred of the outdoors, Preuss created some of the most important maps of the American West in the 1840s. Among the most impressive of these was a seven-sheet series of the Oregon Trail.

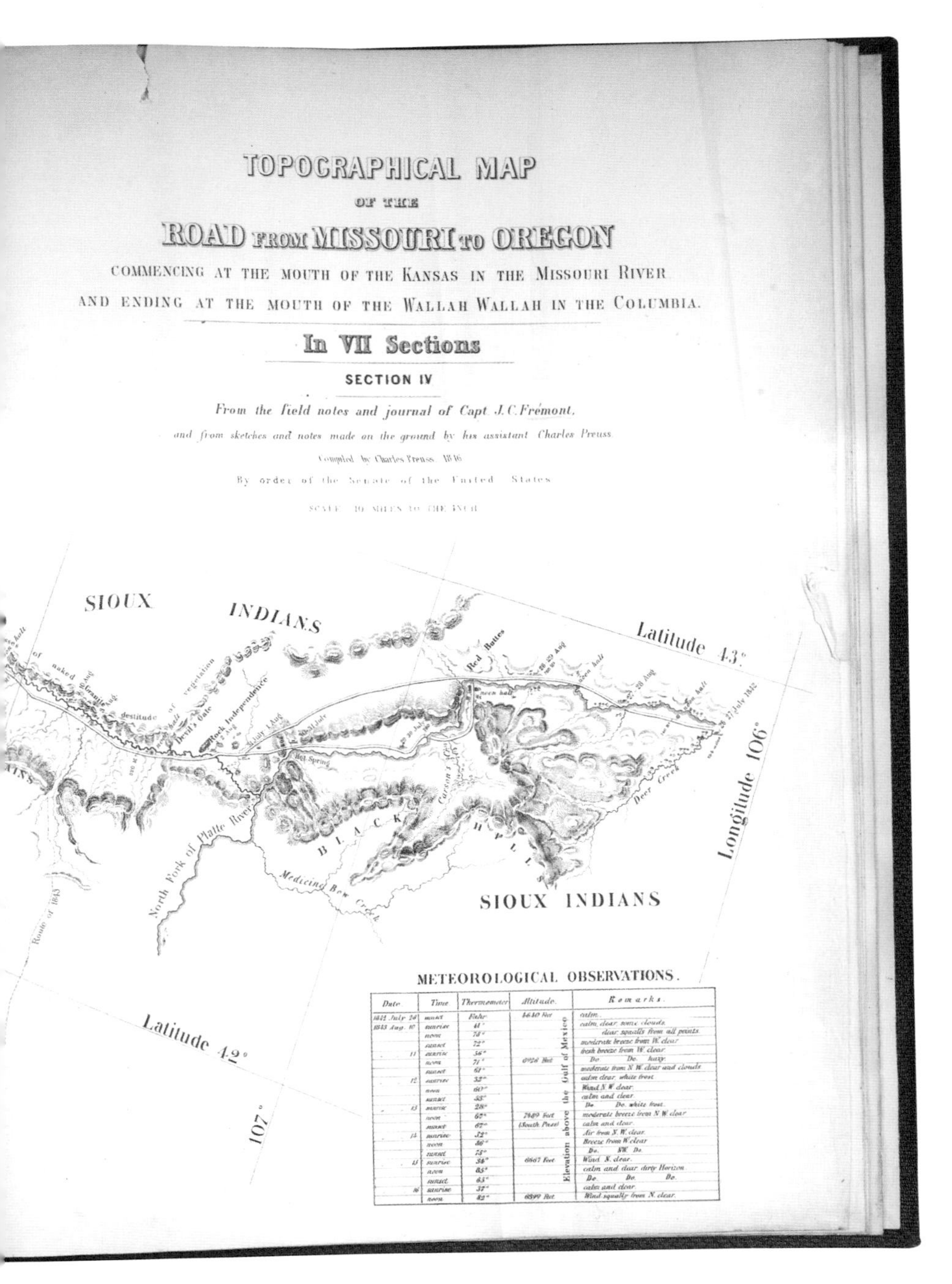

Preuss drew these maps at the request of David Atchison, Missouri's second senator and as ardent an expansionist as Benton. Atchison hoped that detailed maps of the Oregon Trail would facilitate not just migration but also the construction of strategic military forts along the way. To compile the maps, Preuss drew from Fremont's journal as well as his own field experience on these expeditions. Shown here is the fourth sheet of the series, which covers the area of Wyoming just northeast of Casper through South Pass to Jackson.

Preuss oriented the maps to show the geography that lay immediately ahead on the trail. Each sheet gave practical field information about the presence of Native Americans as well as climate, wildlife, fuel, and game. Information about temperature and rainfall was especially welcome in the 1840s, for detailed knowledge of climate in this region—long stereotyped as the "Great American Desert"—would not be widely available for another decade. For all these reasons, Preuss' maps were among the most important and accurate profiles of western geography of the time. The Mormons also used the maps, though ironically their goal was not Oregon or California, but instead the blank spaces south of the trail in what was then northern Mexico. There, they hoped to live and worship together free from the persecution that plagued them in the United States.

By the time the maps were published, the nation was at war with Mexico and negotiating with Britain over control of the Oregon Territory. Ultimately, both of these engagements enlarged the nation's reach to the Pacific Ocean. Once these territories were secured, maps such as Preuss' supported a mass migration to the Far West. From 1840 to 1860, 300,000 Americans made the overland trail to Utah, Oregon, and California. In this context his maps of the Oregon Trail are not just passive records of geographical information but active instruments of national expansion. If 1846 was indeed the "year of decision" for the United States in the West, as Bernard DeVoto has argued, Preuss and Fremont guided these decisions. Their detailed maps enabled Americans to see, understand, and take control of the West.

A CONTINENTAL NATION

Samuel Augustus Mitchell, "A New Map of Texas Oregon and California," 1846

The map of the Oregon Trail on the previous page was published immediately after the United States annexed Texas, and just before it acquired California, Oregon, and the Intermountain West. We often consider this enormous growth in the mid-1840s as an inevitable stage of American history, but it hinged on a series of highly contested events. Just as maps of the Oregon Trail stimulated the imagination of migrants and expansionists, so too did this pocket map show Americans a much larger West that would come almost entirely under American control by 1848.

The question of Texas had long bedeviled Americans. In 1821 Mexico achieved independence from Spain, thereby inheriting a vast North American empire that encompassed California, Texas, and everything in between. Officials in Mexico City struggled to control these distant territories against raids by equestrian bands of Apache and Comanche that roamed across what is now Texas and Oklahoma. Hoping to fortify this northern frontier, Mexico offered land to Americans in exchange for vague promises to convert to Catholicism. By 1835, 35,000 Americans had migrated west into Texas, many of them slaveholders intending to grow cotton. Mexico responded by outlawing slavery in 1829 and closing the border to Americans the following year. Both strategies proved fruitless: Americans continued to immigrate into Texas illegally, and the slave population increased.

Relations between Americans and Mexicans in Texas also deteriorated, the result of cultural differences and the desire among many whites to separate from Mexico. In 1836 a group of these American settlers—and a few Mexican allies—declared themselves an independent republic and immediately sought annexation by the United States. Washington demurred for fear of provoking a war with Mexico, which refused to acknowledge Texas as an independent state. Moreover, the admission of Texas to the Union would immediately upset the delicate balance between slave and free states in Congress. Texas thus remained independent for the next nine years. Then, in 1844, the Democrat James Polk campaigned for the presidency on a platform of national expansion. His victory signaled a political shift, and sitting President John Tyler immediately proposed to annex Texas.

At the end of 1845 President Polk celebrated the addition of Texas as "a bloodless achievement." He aimed to continue that momentum by pressing Congress to negotiate a favorable boundary with Great Britain in Oregon. Thousands of Americans had already migrated to the Far West, well before it was part of the country. Soon those settlers began to demand federal protection, forcing the question of whether the United States ought to annex Oregon as it had Texas. Faced with their cries of "54° 40′ or Fight," Polk advocated a boundary across the forty-ninth parallel north, marked on the map in dark ink. The British countered by proposing a border further south on the Columbia River, along with a smaller American province on the coast. The United States rejected the offer, and ultimately won a national boundary on the forty-ninth parallel.

Meanwhile, Mexico treated the American annexation of Texas as a violation of national sovereignty. Here Samuel Augustus Mitchell's highly popular pocket map claimed the Rio Grande as the southern border of Texas, though Mexico insisted that the Nueces River to the north was the proper boundary. That dispute led to a war that engulfed California as well, which in turn declared its own independence from Mexico. Though the war was brief, it remains one of the most divisive in American history, regarded by many Northerners as a naked land grab on behalf of slaveholders and expansionists. Others worried that the nation's belligerence and aggressive growth had transformed it into a corrupt and overextended empire.

Mitchell's map captures all of these geopolitical shifts of the 1840s. It was particularly valuable to Mormons, tens of thousands of whom were then fleeing persecution in the Midwest. Just weeks after the map was published, Brigham Young asked for copies to aid the exodus of the Latter-Day Saints to Deseret, later Utah. With information about travel distances and geography, Mitchell's map was a reliable, inexpensive, and portable companion for Mormons leaving the United States. Little did they realize that their refuge in northern Mexico would soon become American territory. The Treaty of Guadalupe Hidalgo ended the war with Mexico in 1848, and granted all of "Upper California," depicted in pink, to the United States. Within days, the discovery of gold in California had sparked a frenzied worldwide migration to San Francisco that is detailed on the next map.

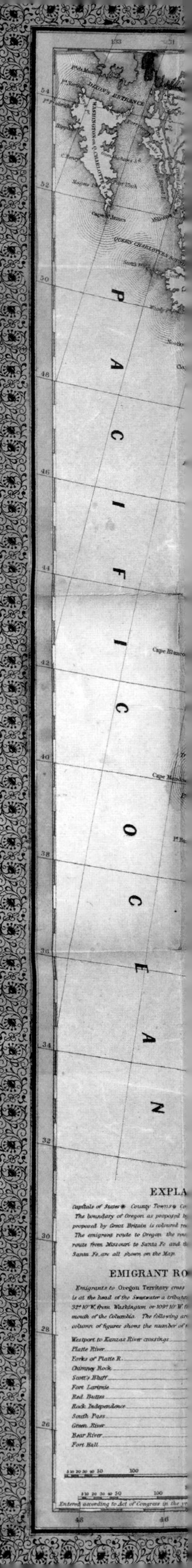

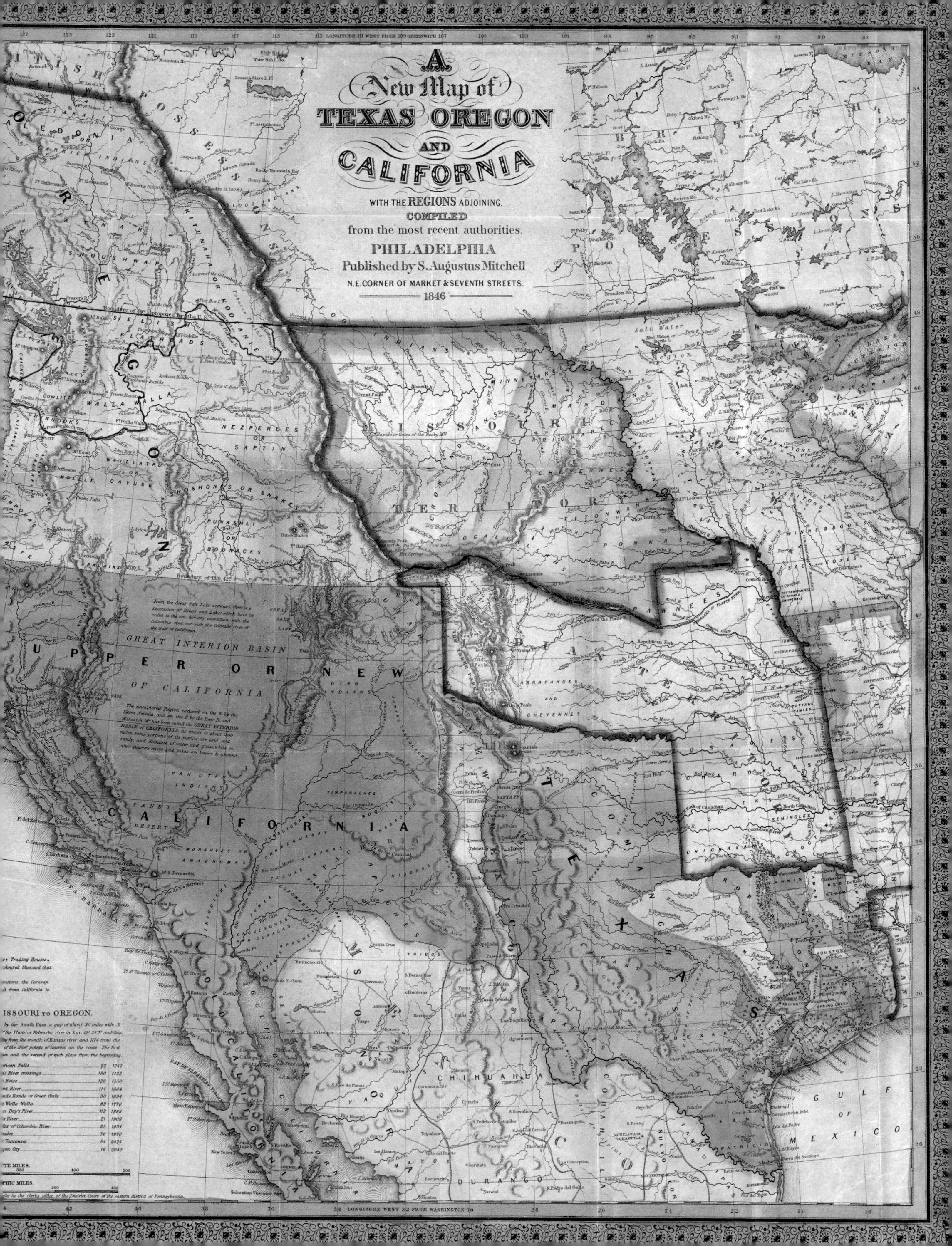

A New Map of
TEXAS OREGON
AND
CALIFORNIA
WITH THE REGIONS ADJOINING.
COMPILED
from the most recent authorities.
PHILADELPHIA
Published by S. Augustus Mitchell
N.E. CORNER OF MARKET & SEVENTH STREETS.
1846
GREAT INTERIOR BASIN
UPPER OR NEW
OF CALIFORNIA
CALIFORNIA
MISSOURI
TERRITORY
OREGON
TEXAS
GULF OF MEXICO
CHIHUAHUA
DURANGO
LONGITUDE WEST FROM WASHINGTON

GOLD IN CALIFORNIA

E. O. C. Ord, "Topographical Sketch of the Gold & Quicksilver District of California, July 25th, 1848"

On February 2, 1848, the Treaty of Guadalupe Hidalgo ended the war with Mexico and extended American territory to the Pacific Ocean. Just days before the treaty was signed, a New Jersey carpenter found a few pieces of gold on the American River in central California. He shared his discovery discreetly with the mill's owner, John Sutter, who tried in vain to keep the news quiet. After visiting the site, California's military commander Richard Mason informed his superiors that there was enough gold in the country drained by the Sacramento and San Joaquin rivers to finance the war with Mexico one hundred times over. Mason described a frenzied scene, with sailors deserting their ships, artisans abandoning their shops, and soldiers leaving their military posts from Sonoma to Monterey to seek their fortune in the Sierra Nevada foothills. His breathless report fueled a worldwide lust for gold that drew thousands to San Francisco.

Mason's report in the summer of 1848 was accompanied by this map, one of the first based on observation rather than speculation. Compiled by Lieutenant Edward Ord, it depicts the topography and river system of the Central Valley with a focus on its mineral resources. The map is centered on Sutter's Fort, but describes subsequent gold found to the north and south and the roads to access each. It promised even greater riches by locating reports of gold discoveries along several rivers flowing out of the Sierra Nevada Mountains to the east.

By focusing on the land and mineral resources, Ord presented a landscape that was relatively devoid of human settlement. He marked sites of mines, mills, and diggings, but made little mention of Native Americans and others living in the Sacramento Valley. In this respect the map suggested a vacant land of abundant wealth, with few obstacles to development. The topographic and mineral detail on the map abruptly ends at the peak of the Sierra Nevada, beyond which knowledge was limited. Ord also shrewdly marked the quicksilver mines near San Francisco Bay, which by 1851 produced half of the global supply of mercury, the chemical element needed to refine gold.

Though the map was made by an agent of the US military, in both form and content it served as an invitation to prospecting. Together with Mason's report, it was eagerly received by President Polk, who celebrated the discovery of gold as a sign of American providence and Manifest Destiny. Victory over Mexico yielded massive territorial gains, while negotiations with Britain made Oregon Territory part of the United States. In the closing months of his presidency, Polk marveled at this geographical transformation. The Mississippi River, which had originally marked the nation's western boundary, no longer reached even its midpoint. Polk predicted that California would soon be as important as the Louisiana Purchase, with San Francisco Bay rivaling the commercial reach of New Orleans.

Yet this territorial growth also brought serious strains. Conflicts with Indian tribes in California and elsewhere in the West reminded the president that this land had long been inhabited. Even more urgent were the political concerns that slavery might extend into California. The president dismissed these by arguing that the climate and soil of the West would never support large-scale agriculture. But he also defended the interests of slaveholders by acknowledging and validating the state rights arguments that had begun to emerge in the South. Polk proposed to settle the crisis by extending the Missouri Compromise line—which divided free from slave territories along the 36° 30′ parallel—to the Pacific. Instead, arguments over the fate of these new territories raged until a series of compromises strengthened protections for slavery in the South in exchange for the admission of California as a free state.

In a larger sense, the timing of these events demonstrates that there was nothing providential about the expansion of the United States between 1803 and 1848; rather, these territorial gains came through purchase, negotiation, military force, and—in the case of the California gold fields—extraordinary good fortune.

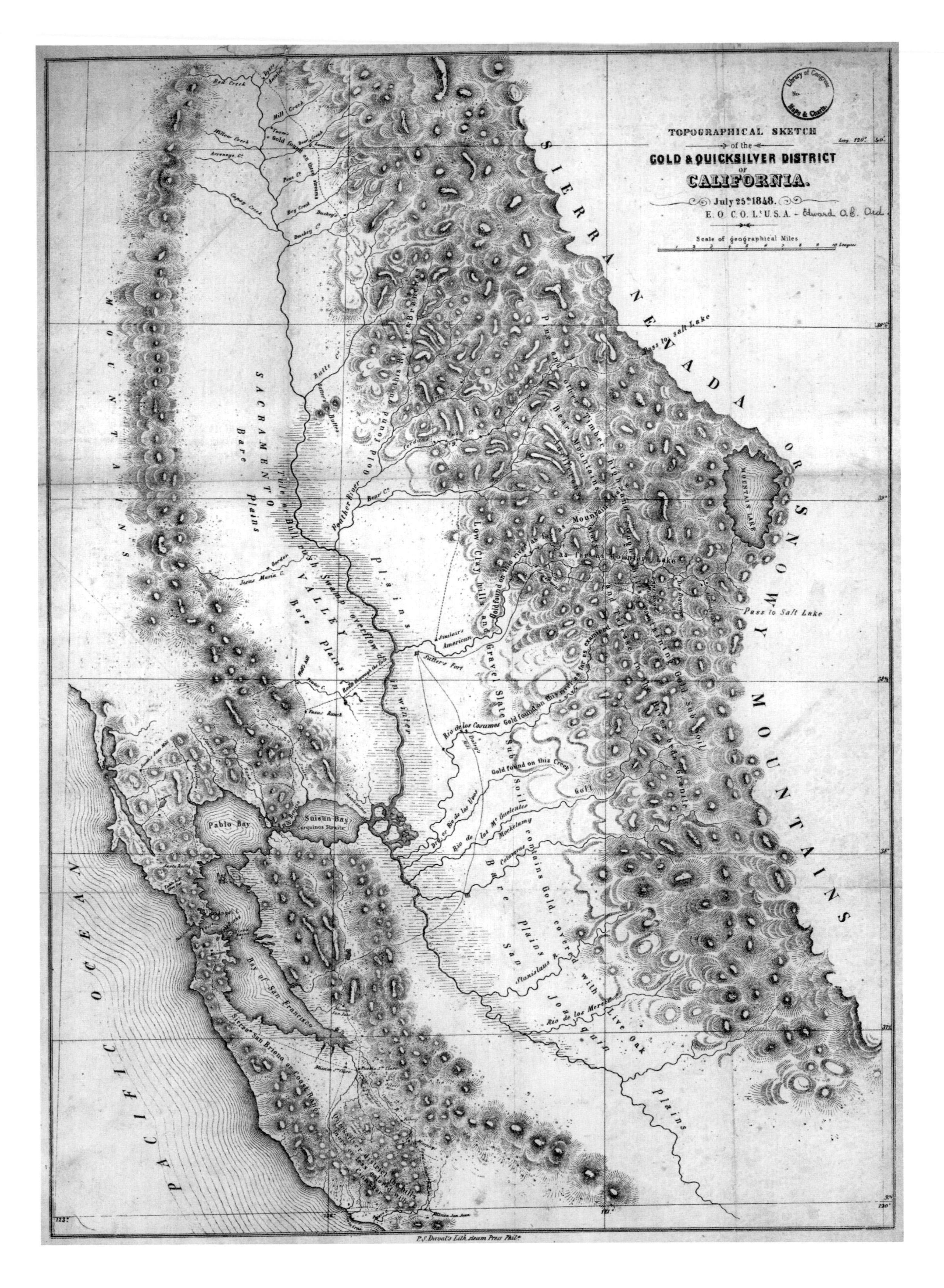
TOPOGRAPHICAL SKETCH
of the
GOLD & QUICKSILVER DISTRICT
OF
CALIFORNIA.
July 25th 1848.
E. O. C. O. Lt U.S.A. - Edward O. C. Ord.
Scale of geographical Miles
10 Leagues
Library of Congress
No.
Maps & Charts.
SIERRA NEVADA OR SNOWY MOUNTAINS
Pass to Salt Lake
Pass to Salt Lake
MOUNTAIN LAKE
SACRAMENTO
Bare
Plains
VALLEY
Bare
Plains
plains
Tule or Bull rush Swamp overflow in winter
Feather River
Bear Cr.
Butte Cr.
Buttes
Sutter's Fort
American
Red Creek
Mill Creek
Willow Creek
Pine Cr.
Dry Creek
Capay Creek
Jesus Maria Cr.
Suisun Bay
Carquinos Straits
Pablo Bay
Bay of San Francisco
Rio de los Cosumnes Gold found on this creek
Gold found on this Creek
Rio de las Mc. Guelenies
Mockelumy
Calaveras
Stanislaus R.
Rio de los Merced
San Joaquin
Plains
Live Oak
PACIFIC OCEAN
Sierra San Bruno or Coast Range
P. S. Duval's Lith. steam Press Phila.
123°
122°
121°
120°

THE GEOGRAPHY OF IMMIGRATION

Gotthelf Zimmerman, "Auswanderer-karte und Wegweiser nach Nordamerika" [Emigration Map and Guide to North America], 1853

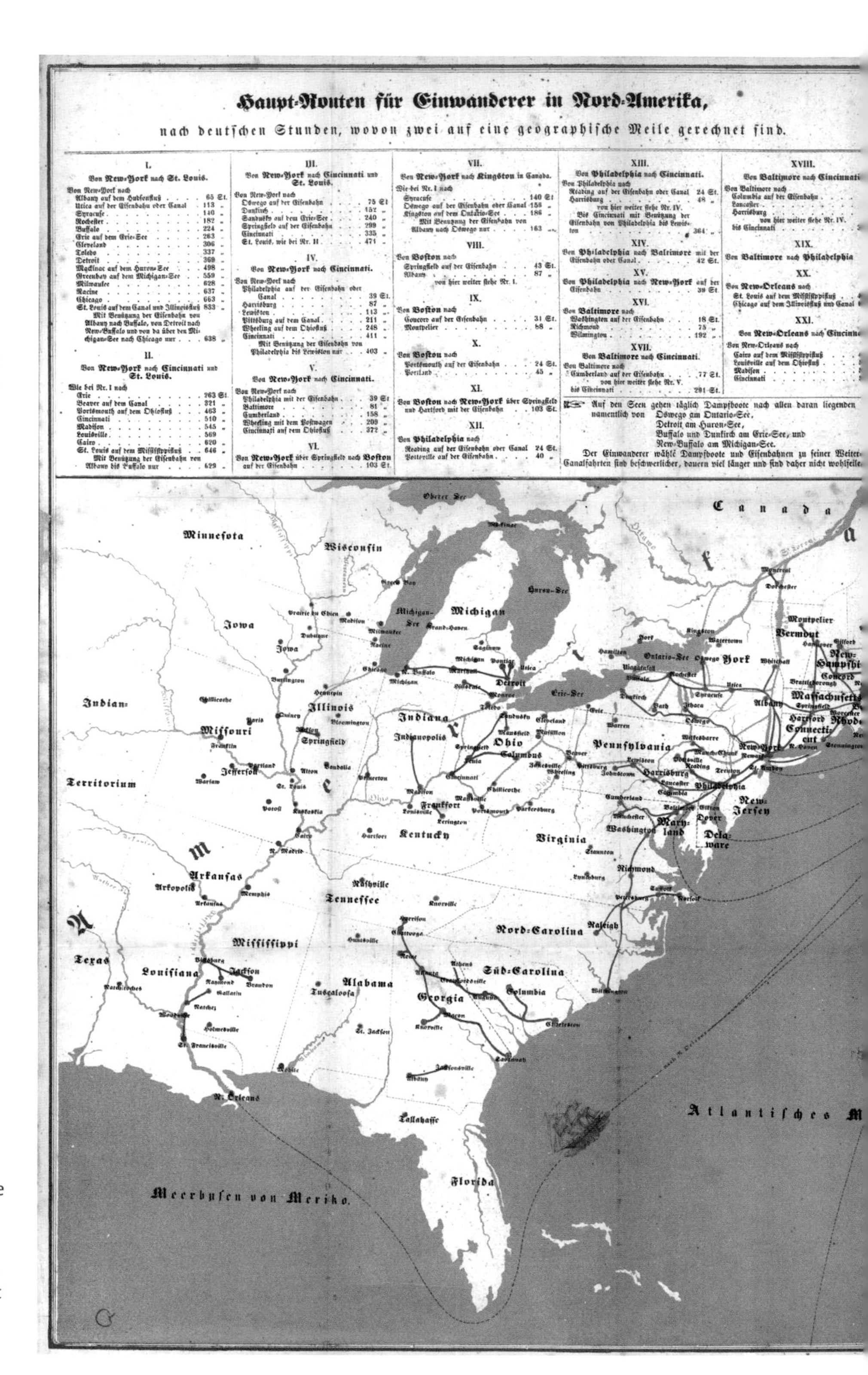

The national population expanded significantly in the first half of the nineteenth century, from 5.3 million in 1800 to 23 million by 1850. In the initial decades that growth was primarily a result of internal reproduction, but subsequently the population grew even faster owing to a robust wave of European immigration. From 1840 to 1860, over 4 million people entered the United States, chiefly from Germany and Ireland.

These two immigrant groups came from very different circumstances. The Irish potato famine killed over a million people and drove a million more to seek opportunities in the United States, Canada, and Australia. Without money or education, most took low-wage and unskilled manufacturing jobs in the emerging industrial centers of the Northeast, especially Boston, New York, and Philadelphia. By contrast, German immigrants to the United States were often refugees of the failed European Revolutions of 1848 seeking greater political and religious freedoms. Many of these "Forty-Eighters" were skilled and educated, and made significant contributions to American science and culture. Two maps in this chapter were the result of innovations introduced by German surveyors, engineers, and lithographers (pages 132 and 142).

German immigrants were also more likely than their Irish counterparts to seek out farming opportunities in the Midwest. This quest was stimulated by innumerable guidebooks in the 1850s advising immigrants on everything from cultural assimilation to promising occupations. The guides encouraged them to take up land further west if they hoped to recreate their home communities. Over time these chain migrations produced rural and urban German settlements throughout the Upper Midwest and in Cincinnati, St. Louis, Chicago, and Milwaukee.

This richly colored map was part of that immigrant literature. At first glance, it shows prospective emigrants a range of destinations that awaited them

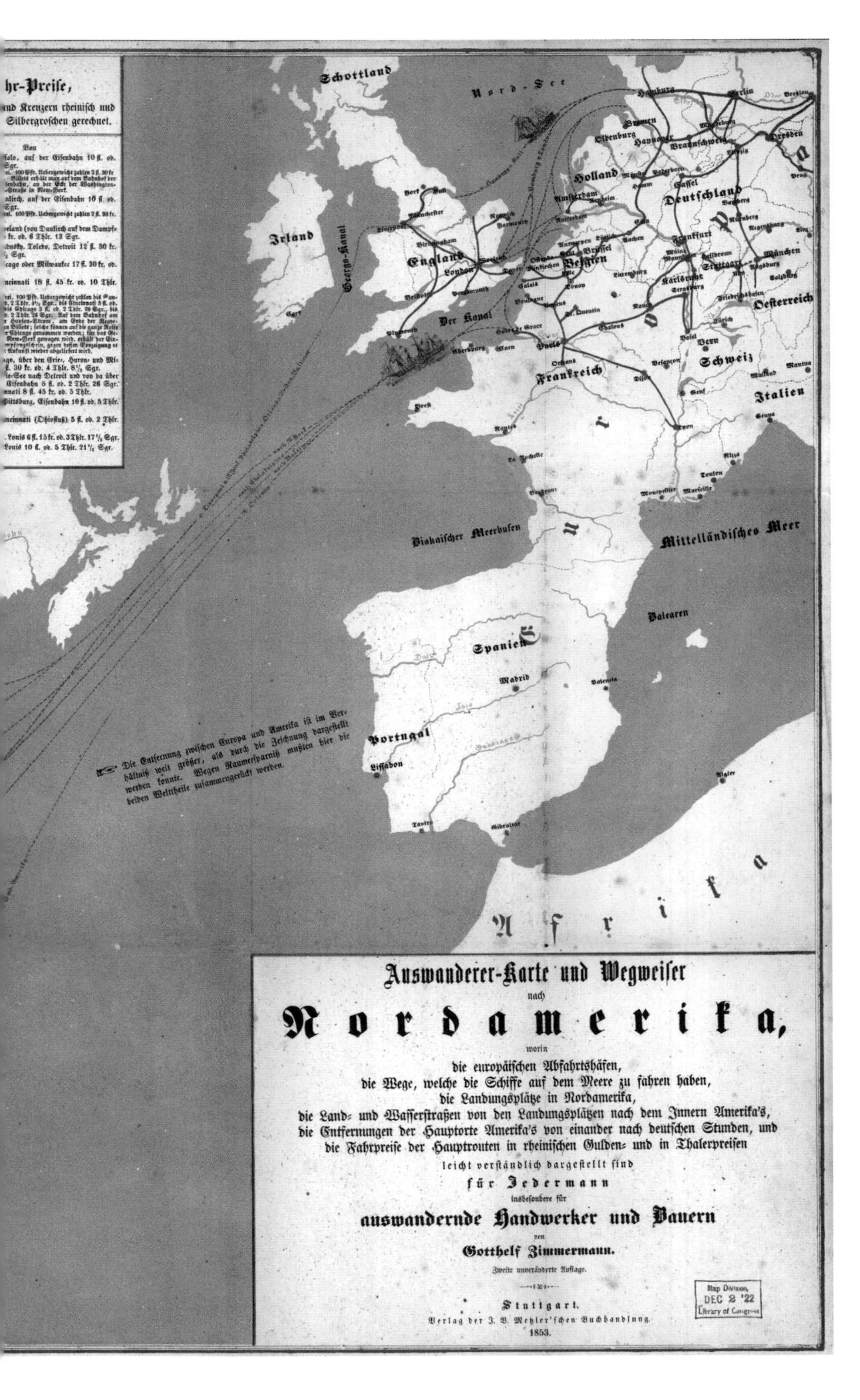

in America. But a closer look shows not only patterns of immigration but also the more fundamental dynamics that contributed to the Civil War. With traditional typography, which would have been familiar to its German readers, the map identifies cities and towns across the country and the distance from each to ports along the Atlantic coast. Railroads marked in red and canals in blue show modes of transportation, and together they conveyed greater regional networks. Canals, roads, and railroads closely integrated the Northeast and the Midwest, while transportation in the South was largely limited to rivers and a few railroads. This signaled not just the absence of industrialization in the South but also the relative isolation of that region within the nation.

While the Southern economy was also growing, it was not diversifying. Cotton production boomed in the 1850s, but the region invested little in infrastructure, industrialization, or transportation. Moreover, a system built on slavery left few opportunities for immigrant labor. In this sense, the map gives us not only a German perspective on migration but also a snapshot of the forces that were *differentiating* North from South. As the Northern states rushed headlong into urbanization and industrialization, Southern states did neither. Immigrants keenly understood the implications of these decisions. Germans settled small pockets in eastern Texas, but for the most part the South held little appeal or opportunity. And, while Germans may have gravitated more than the Irish toward rural areas and the Midwest, both groups decisively avoided the South.

German and Irish immigrants faced severe and ugly discrimination in the 1850s. Large concentrations of the Irish in urban areas sparked sharp resistance from native-born whites who feared competition for jobs as well as the influence of Catholicism in a largely Protestant culture. For a few years, anti-immigrant sentiment was as powerful a political force as opposition to slavery. "Nativism" burned hot in the 1850s, fueled by both ethnic prejudice and very real economic anxieties. But, despite virulent anti-foreign sentiment, immigrants continued to choose Northern and Midwestern destinations. In doing so, they further advanced the industrial trends that were separating the North from the South. In this regard immigration was both a cause and a consequence of the growing sectional divide within the United States.

THE GEOPOLITICS OF SLAVERY

John Jay, "Freedom and Slavery, and the Coveted Territories," 1856

On January 4, 1854, the Democratic Party's most powerful senator introduced a bill that prompted a national crisis and disrupted the entire political landscape. The Illinois senator, Stephen Douglas, designed the bill to unify the Democrats and bring him one step closer to the White House. Instead it tore his party apart and contributed to a violent reckoning over slavery in the West.

Douglas proposed the bill to organize the two large territories of Kansas and Nebraska for settlement, as shown on this map. They had been considered Indian territory until westward expansion led many to press for more land opportunities in the 1850s. This reconsideration of the territory was also driven by plans for a transcontinental railroad. Douglas preferred a northern route through his native Chicago, but knew that this would require support from the Southern wing of his party. For themselves, Southern Democrats insisted on the party's explicit commitment to defending and expanding slavery.

To curry favor with Southern Democrats, Douglas included a provision in the Kansas–Nebraska bill that granted future settlers of these two territories the right to determine the legality of slavery for themselves. He believed that this eminently "democratic" solution would appeal to everyone in his party, but he was utterly wrong. Thirty-four years earlier, Congress had made slavery illegal in the Louisiana Territory north of the 36° 30′ line, just south of Kansas. With his new bill, Douglas cavalierly overturned the Missouri Compromise, which had endured for over three decades. Slavery had historically been limited to the American South; now it had license to spread throughout the West.

This supreme political miscalculation inadvertently galvanized opposition to slavery across the political spectrum. Most Northerners cared little about the fate of African Americans, but the prospect of slavery expanding into the West both terrified and enraged them. The backlash was far worse than Douglas could have imagined, and it drove disaffected Whigs and Democrats together to form the new Republican Party. What united this diverse group was the conviction that Congress had not just the right but also the *obligation* to bar slavery from the western territories. Republicans believed that slavery was wrong, but even more pressing was their vision of the territorial west as protected for free whites.

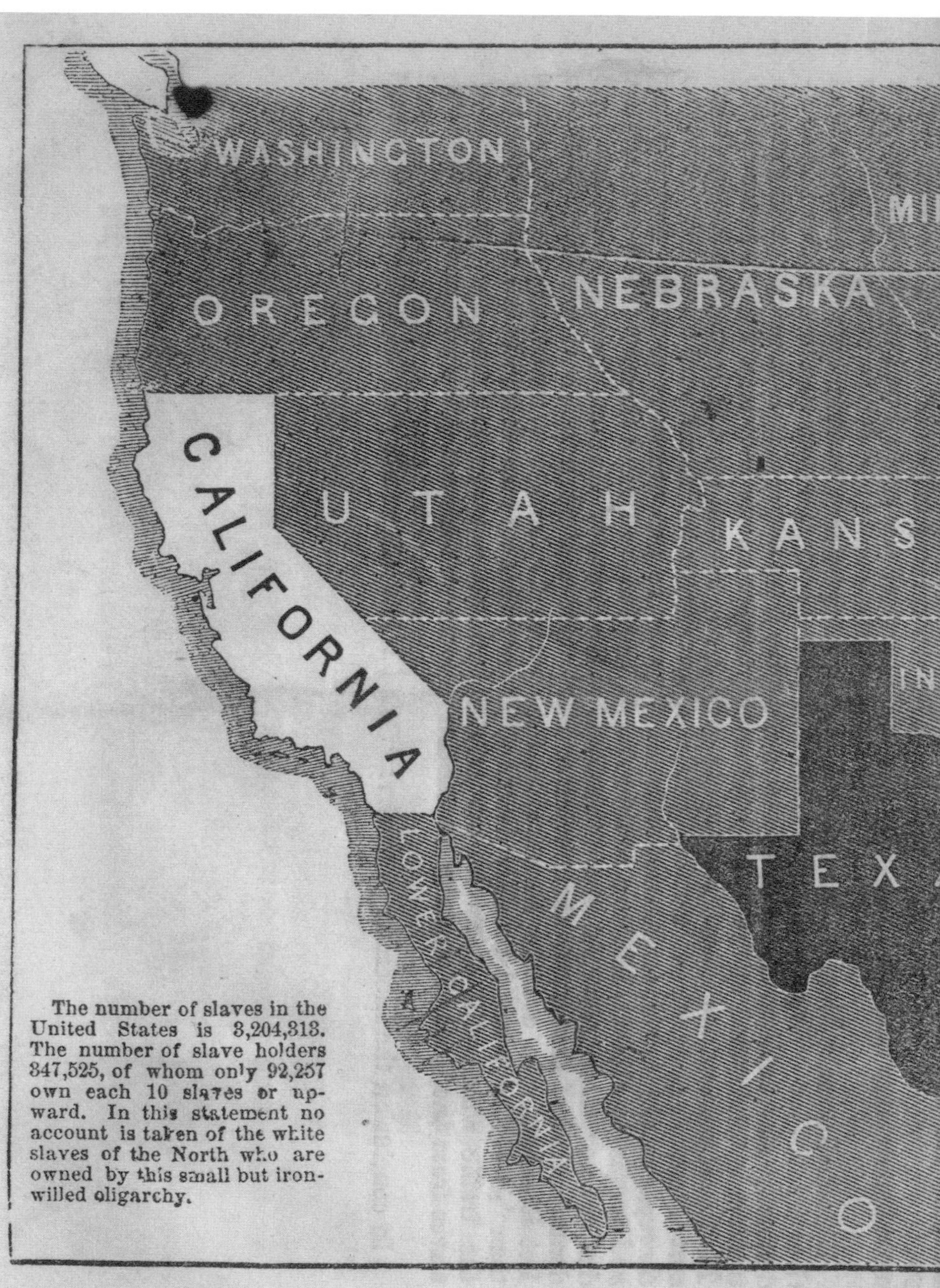
WASHINGTON
OREGON
NEBRASKA
CALIFORNIA
UTAH
KANS
NEW MEXICO
TEX
MEXICO
LOWER CALIFORNIA
The number of slaves in the United States is 3,204,313. The number of slave holders 347,525, of whom only 92,257 own each 10 slaves or upward. In this statement no account is taken of the white slaves of the North who are owned by this small but iron-willed oligarchy.
FREEDOM AND SLAVERY.

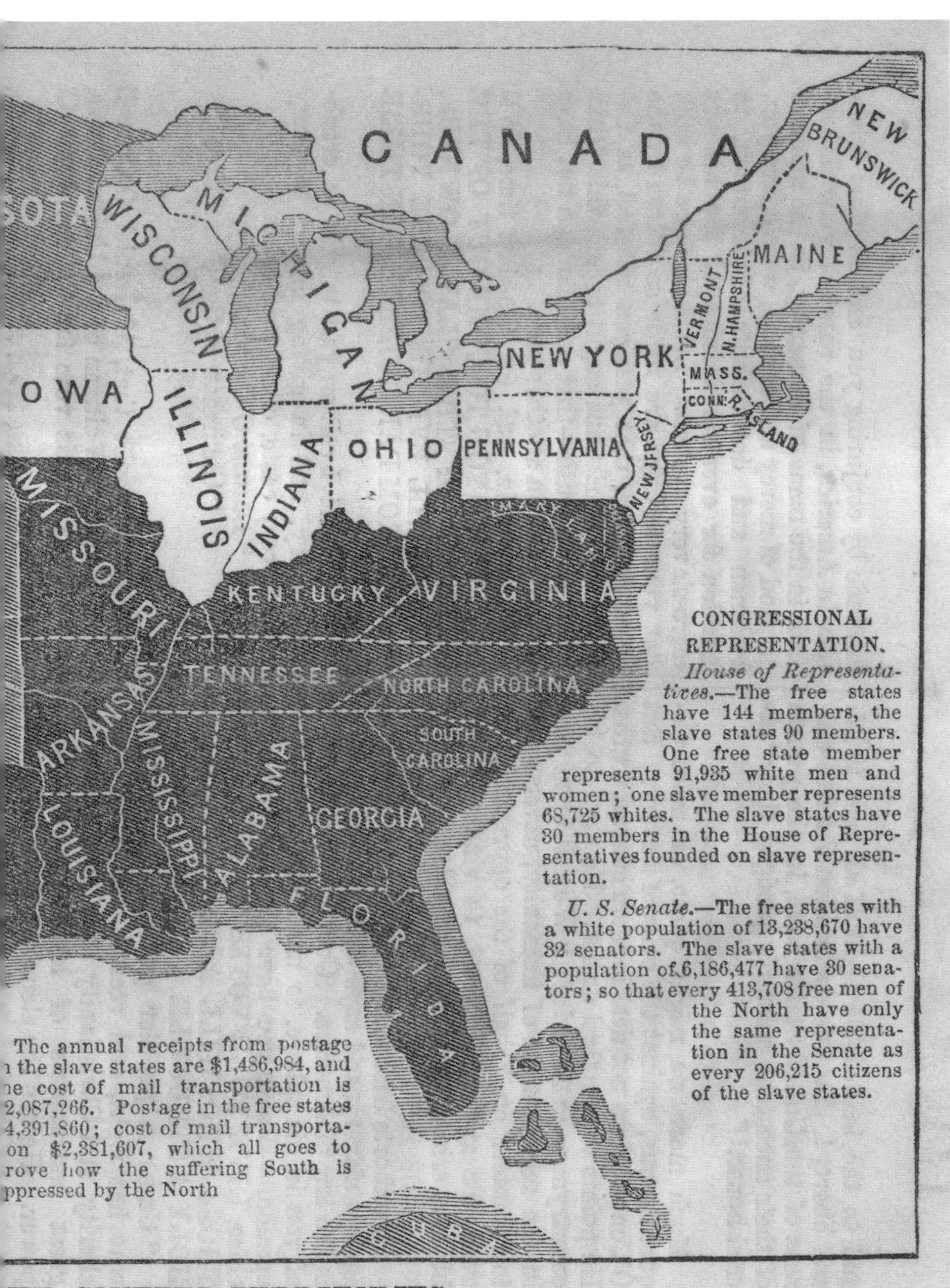

CANADA
SOTA
WISCONSIN
MICHIGAN
NEW BRUNSWICK
MAINE
VERMONT
N. HAMPSHIRE
NEW YORK
MASS.
CONN.
R. ISLAND
OWA
ILLINOIS
INDIANA
OHIO
PENNSYLVANIA
NEW JERSEY
MISSOURI
KENTUCKY
VIRGINIA
TENNESSEE
NORTH CAROLINA
ARKANSAS
MISSISSIPPI
ALABAMA
SOUTH CAROLINA
GEORGIA
LOUISIANA
FLORIDA
CONGRESSIONAL REPRESENTATION.
House of Representatives.—The free states have 144 members, the slave states 90 members. One free state member represents 91,935 white men and women; one slave member represents 68,725 whites. The slave states have 30 members in the House of Representatives founded on slave representation.
U. S. Senate.—The free states with a white population of 13,238,670 have 32 senators. The slave states with a population of 6,186,477 have 30 senators; so that every 413,708 free men of the North have only the same representation in the Senate as every 206,215 citizens of the slave states.
The annual receipts from postage
1 the slave states are $1,486,984, and
1e cost of mail transportation is
2,087,266. Postage in the free states
4,391,860; cost of mail transporta-
on $2,381,607, which all goes to
rove how the suffering South is
ppressed by the North
HE COVETED TERRITORIES.

When Kansas opened for settlement later that year, the worst fears of Northerners were realized. Armed and passionate opponents of slavery flooded into the territory to fight equally determined pro-slavery agitators. The violent skirmishes of "Bleeding Kansas" incensed Republicans, but they also provided a rallying cry for their party and enlarged its following. In 1856, Republicans mounted their first presidential campaign and nominated the heroic western explorer John Fremont as their candidate (see page 132). As the nation's first entirely sectional party, the Republicans faced an uphill battle. They worked to recruit votes by championing "Free Soil, Free Labor, Free Men, Fremont!"

This is just one of many maps issued by the Republicans during that campaign of 1856. All of these maps and broadsides excoriated the Democratic Party for betraying the nation's future. Here, John Jay—grandson of the revolutionary founder—warned of a future dominated by slavery. He annotated the map with census statistics to demonstrate that a conspiracy of slaveholders controlled the nation's institutions, resources, and political power.

Maps like Jay's were everywhere during the campaign, visual arguments that the fate of the West hinged on the election. To be sure, there were doubts about the viability of plantation agriculture in the arid West. Moreover, by this time Washington and Oregon had largely been established and settled as free territories. But the stark urgency of the map is the key to its power, for it reminded Americans that slavery continuously divided the nation and its parties along sectional lines. Similar maps were issued in German, aimed at drawing immigrants away from the Democrats and into the Republican fold.

Jay's map also reminds us that what heightened opposition to slavery in the 1850s was the prospect of its expansion into the West rather than its longstanding presence in the South. In 1857 the Supreme Court seemed to confirm Republican fears by expanding the rights of slaveholders in the territories through its *Dred Scott v. Sanford* decision. The majority opinion—like the Kansas–Nebraska Act—infuriated Republicans and swelled their ranks. Three years later, they elected Abraham Lincoln to the presidency.

SLAVERY, SECESSION, AND WAR

US Coast Survey, "Map Showing the Distribution of the Slave Population," 1861

In 1856 John Jay captured the urgency of the sectional crisis by starkly differentiating "America free, or America slave" (page 140). Four years later, those divisions were realized when Abraham Lincoln was elected president without any Southern support. Within weeks of Lincoln's victory, South Carolina had seceded from the Union; by February 1, seven states of the Deep South had followed, and together they formed the Confederacy. Lincoln took office in early March, and used his inaugural address to defuse the crisis by reminding Southerners that he had yet to take any action against slavery. For a time he kept the states of the upper South in the Union. In April, however, South Carolina attacked Fort Sumter, prompting Lincoln to call up a volunteer militia to suppress the rebellion. Many Virginians considered this a hostile act by the federal government, and voted to join the Confederacy. Within weeks Arkansas, North Carolina, and Tennessee had followed.

The secession crisis prompted the nation's military and civilian agencies to mobilize for war. The US Coast Survey harvested its extensive knowledge of Southern coastlines and harbors to prepare for a potential blockade. The Coast Survey also sent its men into the field to refine its understanding of both terrain and waterways, producing thousands of detailed charts to aid Union strategy. But even before the war began the agency had been a crucial source of experimental maps.

Among the most impressive of these were two groundbreaking maps of the distribution of slavery. The first profiled Virginia, and the second—shown here—covered the entire South. The map elegantly and innovatively used shading to indicate the density of the slave population in each county. While Jay's map on the previous page simply divides the country into two categories, the Coast Survey map reveals a far more complex and varied geography of slavery. Darkly shaded areas throughout the cotton belt and the Lower Mississippi River show high dependence on slavery, while lighter areas indicate its relative absence.

The Coast Survey's map was used to raise money for the Sanitary Commission, a civilian volunteer organization founded in 1861 to support the Union Army. The Superintendent of the Coast Survey—Alexander Dallas Bache—served as the Commission's treasurer, and no doubt provided the map for this fundraising effort. Its implicit message is that slavery caused the rebellion. At a glance, it is clear that the states of the Deep South that led secession were also those with the highest concentration of slaves. This message is reinforced by the table at the bottom of the map, which lists states according to their dependence upon slavery. This hierarchy almost precisely corresponded to the order in which the states left the Union. All of these cues embedded in the map reminded the public that the rebellion was driven by slavery.

The Coast Survey's decision to experiment with this type of statistical mapping shaped the uses of cartography during and after the Civil War. Among those who helped to create the map of slavery was Captain William Robert Palmer, an engineer with the Corps of Engineers who worked for the Coast Survey during the war. Palmer sent copies of the initial map of slavery in Virginia to military leaders as well as to members of President Lincoln's cabinet. Lincoln himself was captivated by the map of the South, and kept it close at hand.

In the president's mind, the map revealed the strengths and weaknesses of the Confederacy in a way that topographic maps could not. While many viewers were drawn to the dark spaces on the map, Lincoln used it to understand that the Confederacy might be most fragile in areas where slaves were absent. For instance, in the early months of the war the president held out hope that a railroad from Kentucky toward the Cumberland Gap or Knoxville might provide a lifeline to Southerners in eastern Tennessee who staunchly resisted the Confederacy. More generally, a map like this reinforced Lincoln's belief that secession had been imposed upon the South by a minority of slaveholders. With time and resources, he argued, the Confederacy might be overturned from within.

We know of Lincoln's close attention to this map through the diary of the painter Francis Bicknell Carpenter. In September 1862 Carpenter read about Lincoln's preliminary Emancipation Proclamation, which gave the military the power to liberate slaves as it moved through the rebel states. Carpenter was deeply moved by Lincoln's announcement, and believed it to be of national and moral importance second only to the Declaration of Independence. As an artist, he sought to capture the moment when the president revealed the plan of emancipation to his cabinet. Not only did Lincoln agree, but in early 1864 he invited Carpenter to set up a studio in the White House. The portrait that resulted is featured on the next page.

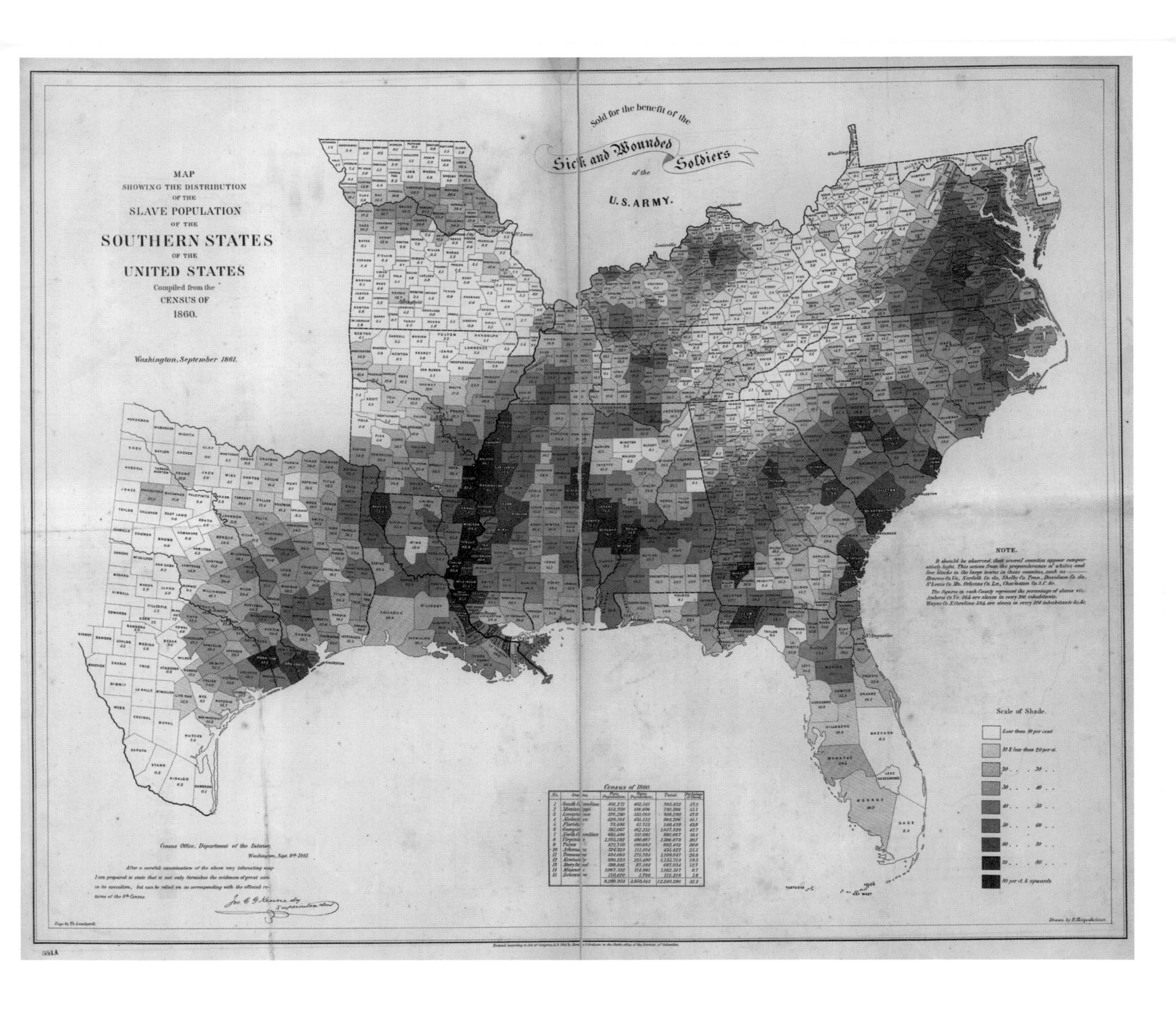
MAP
SHOWING THE DISTRIBUTION
OF THE
SLAVE POPULATION
OF THE
SOUTHERN STATES
OF THE
UNITED STATES
Compiled from the
CENSUS OF
1860.
Washington, September 1861.
Sold for the benefit of the
Sick and Wounded Soldiers
of the
U.S. ARMY.
NOTE.
Scale of Shade.
Census of 1860.
Census Office, Department of the Interior,
Washington, Sept. 9th 1861.
After a careful examination of the above very interesting map I am prepared to state that it not only furnishes the evidence of great care in its execution, but can be relied on as corresponding with the official returns of the 8th Census.

FANNIN 18.7
RED RIVER 35.6
BOWIE 52.5
YOUNG 15.6
JACK 5.0
WISE 4.1
DENTON 5.0
COLLIN 11.3
HUNT 8.7
HOPKINS 12.8
TITUS 25.3
CASS 41.3
LAFAYETTE 51.0
COLUMBIA 28.9
UNION 51.5
ASHLEY 43.6
CHICOT 81.3
WASHINGTON 92.3
SUN FLOWER 78.0
CARROLL 62.7
CHOCTAW 26.8
BUCHANAN 13.9
PALOPINTO 8.6
PARKER 5.8
TARRANT 14.1
DALLAS 12.4
KAUFMAN 13.5
VAN ZANDT 8.5
WOOD 20.1
UPSHUR 35.6
MARION 50.7
HARRISON 58.8
BOSSIER 73.3
CLAIBORNE 46.6
UNION 36.1
MOREHOUSE 63.5
CARROLL 77.1
ISSAQUENA 92.5
YAZOO 74.7
HOLMES 67.3
ATTALA 33.4
WINSTON 43.1
NOXUBEE 75.0
EAST LAND 0.0
ERATH 4.9
JOHNSON 11.9
HENDERSON 24.3
SMITH 37.2
RUSK 38.8
PANOLA 36.1
DE SOTO 64.0
BIENVILLE 45.9
JACKSON 43.8
FRANKLIN 55.2
MADISON 88.3
MADISON 77.5
HINDS 71.4
LEAKE 32.8
NESHOBA 26.5
KEMPER 49.1
COMANCHE 8.6
BROWN 0.0
HAMILTON 5.3
BOSQUE 14.6
NAVARRO 31.3
ANDERSON 33.3
FREESTONE 52.5
WINN 19.8
TENSAS 90.8
CLAIBORNE
JEFFERSON 81.0
SCOTT 36.4
NEWTON 35.0
LAUDER DALE 38.2
McLENNAN 38.7
LIMESTONE 23.6
SHELBY 27.5
CATAHOULA 52.7
COPIAH 51.7
SIMPSON 38.3
SMITH 28.7
JASPER 41.3
CLARKE 47.1
SAN SABA 9.7
LAMPASAS 14.9
CORYELLE 11.5
FALLS 47.5
LEON 38.6
HOUSTON 35.0
ANGELINA 16.1
SABINE 41.8
SABINE 29.4
CONCORDIA 91.0
ADAMS 71.7
FRANKLIN 57.6
LAWRENCE 40.1
COVINGTON 35.4
JONES 12.3
WAYNE 52.8
BELL 20.9
ROBERTSON 45.2
TRINITY 21.8
RAPIDES 61.2
AVOYELLES 54.9
WILKINSON 82.5
AMITE 64.1
PIKE 44.4
MARION 46.6
PERRY 28.4
GREENE 31.8
LLANO 4.9
BURNET 9.4
WILLIAMSON 19.7
MILAM 26.8
MADISON 30.2
TYLER 25.4
WALKER 50.5
POLK 50.6
WEST FELICIANA 82.5
EAST FELICIANA 72.2
ST HELENA 52.1
WASHINGTON 36.1
TRAVIS 38.3
Austin
BURLESON 35.2
GRIMES 53.1
St LANDRY
HARRISON 21.3
JACKSON 26.9
HAYS 37.5
WASHINGTON 52.2
MONTGOMERY 51.3
HARDIN 14.1
CALCASIEU 20.8
EAST BATON ROUGE
Baton Rouge
LIVINGSTON 29.6
St TAMMANY 36.9
HANCOCK 27.3
Biloxi
COMAL 4.8
CALDWELL 35.9
BASTROP 22.9
AUSTIN 38.6
LIBERTY 33.9
ORANGE 20.6
LA FAYETTE 50.9
HARRIS

LINCOLN'S MAP

Francis Bicknell Carpenter, *First Reading of the Emancipation Proclamation of President Lincoln*, 1864

Like many Americans, the artist Francis Bicknell Carpenter was heartened by President Abraham Lincoln's announcement of the Emancipation Proclamation in 1862. At the invitation of the president, Bicknell set up shop in the White House in 1864 in order to prepare a portrait to commemorate this policy, which fundamentally changed the meaning of the war. As the battle raged on, Carpenter set to work, studying each member of the cabinet in order to reproduce the scene as authentically as possible.

One day in the executive chamber, the artist was struck by a map "showing the slave population of the Southern States in graduated light and shade." It was in fact the Coast Survey's map of slavery (shown on the previous page), heavily used in the White House. Thereafter Carpenter noticed that President Lincoln frequently consulted the map, and so he decided to include it in the portrait. Carpenter took the map back to the studio in order to study it more closely.

The president often visited Carpenter's studio as a relief from the pressures of the war and to monitor the progress of the portrait; on his next visit he immediately exclaimed "*you* have appropriated my map, have you? I have been looking all around for it." The president then picked up the map and took it to the window. There he traced Hugh Judson Kilpatrick's recent raid around Richmond, and observed that, if it was successful, the Union Army would liberate quite a few slaves.

Carpenter took pains to include the map in his portrait because the president *used* it. Before emancipation it helped Lincoln assess the strength of Confederate sentiment, and thereafter it showed him the military's progress in destabilizing the enemy's greatest resource. The president had access to innumerable maps, but very few like this. It confirmed his belief that secession was not about state rights but about the defense of slavery. It revealed that slavery varied tremendously across the South, which in turn shaped his military strategy. Little wonder then that when President Lincoln saw Carpenter's finished portrait, he singled out the slave map in the lower right corner as one of its most satisfying details.

VIRGINIA

Map of
GEORGIA & ALABAMA
Representing RAILWAYS, POST-ROADS, POPULATION, AND AGRICULTURAL PRODUCTIONS.
Prepared at the Census Office.
Under direction of
Jos. C. G. Kennedy
Superintendent.
Statute Miles
References
4 Horse Mail Post Coach Roads
2 Do. Do. Stage Do.
1 Do. Do. or Sulkey Do.
Cross Do.
Rail Do.
Canals
W. No. of Whites.
F.C. Free Colored.
S. Slaves.
M. Military Males betw. 18 and 45
I.L. Acres of Improved Land.
H. and M. Horses and Mules.
N.C. Total Neat Cattle.
S. Swine.
W. Bushels of Wheat.
C. Corn.
O. Oats.
T. Tobacco.
H. Hay.
G.C. Ginned Cotton, Bales of 400 lbs.
R. Rice, lbs. of.
SOUTH CAROLINA
GEORGIA
OCEAN
HABERSHAM
WHITE
LUMPKIN
UNION
RABUN
FRANKLIN
HART
ELBERT
JACKSON
OGLETHORPE
WILKES
LINCOLN
COLUMBIA
TALLIAFERRO
GREENSBORO
WARREN
HANCOCK
GLASCOCK
JEFFERSON
BURKE
WAYNESBORO
WASHINGTON
SANDERSVILLE
MILLEDGEVILLE
JOHNSON
EMANUEL
SWAINSBORO
SCRIVEN
BULLOCH
STATESBORO
TATNALL
MONTGOMERY
LAURENS
PULASKI
HOUSTON
PERRY
DOOLY
IRWIN
WORTH
TELFAIR
COFFEE
APPLING
HOLMESVILLE
LIBERTY
WAYNE
PIERCE
WARE
LOWNDES
BERRIEN
COLQUITT
MITCHELL
DOUGHERTY
SUMTER
TAYLOR
SCHLEY
NEWTON
WALTON
GWINNETT
DE KALB
MORGAN
PUTNAM
JASPER
JONES
TWIGGS
WILKINSON
EDGEFIELD C.H.
BARNWELL C.H.
Aikens
Rail Road
PICKENS C.H.
Pendleton
Anderson
Abbeville
SAVANNAH
WASSAW SOUND
OSSABAW SOUND
ST. CATHARINES SOUND
SAPELO SOUND
Doboy Inlet
ALTAMAHA SOUND
ST. SIMONS SOUND
JEKYLL SOUND
ST. ANDREWS SOUND
JEFFERSONTON

GENERAL SHERMAN AND THE LOGIC OF DESTRUCTION

Map of Georgia (1839), annotated for military use, circa 1864

In early 1864, Union Commander Ulysses S. Grant instructed General William Tecumseh Sherman and his army to strike at the heart of the Confederacy. Sherman and his men were to pursue General Joseph Johnston's army into Georgia in order to destroy the enemy and its resources. Grant's directive became the basis for the Atlanta Campaign, in which Sherman's armies marched from northwest Georgia to Atlanta from May to September. Sherman then spread his men up to sixty miles wide on a march toward Savannah before heading north through the Carolinas. The timing of the campaign was crucial: it brought Georgia to its knees and gave the Union a crucial military victory. This in turn contributed to the re-election of President Lincoln in November.

Sherman's march has been judged brilliant and brutal, necessary and vindictive. Either way, it was the most ambitious campaign of the war, for it required him to take his armies far beyond the reach of Union lines. To execute the mission, Sherman relied on a crack team of mapmakers in the field who delivered detailed profiles of the terrain that were far superior to those of the enemy. But before Sherman even arrived in Georgia he immersed himself in census data to learn how and where his men could survive off the land after they were cut loose from their chain of supply. Early in the war he even asked the superintendent of the Census for strategic maps of Southern resources, seeking to harness this data for military purposes.

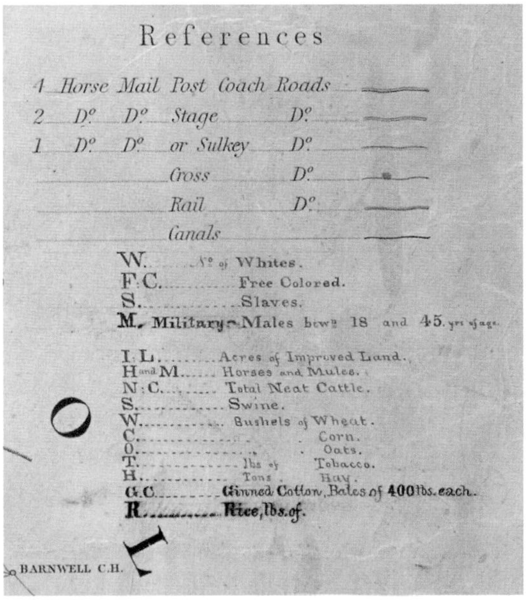

The Census Office had been experimenting with data maps since the beginning of the war in the hopes of aiding Union forces. Among the most relevant of these efforts for Sherman was this large 1839 postal map of Georgia, which marked roads and rivers. This limited information was ideal, for it enabled clerks to add new counties, then to annotate each with data from the most recent census. As shown in the updated legend, each county listed the population of whites, "free coloreds," slaves, and men of military age. That final figure enabled the army to calculate roughly how many men would be serving in the Confederate forces, and likely absent from the region.

The information on resources and livestock was just as valuable, for the war had turned routine census data into military intelligence. Corn and hogs could be eaten, while sugar and cotton could be burned. The information on slavery was no less useful. Sherman's army encountered approximately 90,000 slaves in the Georgia campaign: they constituted more than half of the population then present in those counties. A total of 14,000 emancipated men, women, and children attached themselves to his army. Those left behind struggled to survive, for Sherman attacked not just the Confederacy but also its food. By destroying corn as well as cotton, he subjected the weakest members of society to the greatest deprivation.

Sherman's actual march through Georgia was no doubt guided more by the sophisticated maps produced by his team of mapmakers and surveyors, as well as the terrain itself. But his quest to harvest this census information did prompt some of the nation's early data maps. These hastily compiled documents, in turn, helped Sherman to conceive and to undertake the operation in the first place. The general admitted as much at the end of the war, writing that the information supplied by the Census Office helped his armies to identify supply routes "which otherwise would have been subjected to blind chance, and it may be to utter failure." Armed with this information, he wrote, "I knew exactly where to look for food." Without it, "I would not have undertaken what was done and what seemed a puzzle to the wisest and most experienced soldiers of the world." In other words, it was the data that enabled Sherman to see what was possible as he prepared a march that shattered Confederate resolve.

THE DEFEAT OF RECONSTRUCTION

T. S. Hardee, "Battle of New Orleans for Freedom," 1874

Well before the end of the Civil War, congressional Republicans and President Lincoln began to anticipate reunification. Lincoln argued that the Confederate states ought to be swiftly *restored* to the Union once each abolished slavery by ratifying the Thirteenth Amendment. Many Republicans in Congress balked at this, believing that the severity of the rebellion and the human cost of the war gave the government latitude to *reconstruct* the defeated slave states before readmitting them to the Union.

At bottom, the debate over Reconstruction was a referendum on the war itself. Beyond the end of slavery, what did victory mean? Should the rebel states be brought back in as they were, or was it necessary to dismantle the power structure that led this insurrection? What punishment—if any—would Confederates face for taking up arms against the Union? Such questions were especially critical for the freedmen, since there were no guarantees that Southern states would respect their civil rights. These and other questions bedeviled Congress and the president long before General Lee conceded defeat in April 1865.

Congress passed a Republican plan of Reconstruction in 1867, though it was weakened by intraparty disagreement and Southern white intransigence. Former Confederates chafed at the federal government's continued occupation of the South, not to mention the imposition of political reforms. Many Southern Democrats also objected to the federal government's grant of citizenship privileges to the freedmen, such as the right to vote enshrined in the Fifteenth Amendment. In short, from 1867 to 1877 Reconstruction involved a series of overlapping and often violent power struggles between the national government and the states, North and South, Republicans and Democrats.

Confused and enraged by emancipation, and stung by military defeat, many Southern whites turned their opposition to Reconstruction into outright resistance. The "Battle of New Orleans" was one such act of Southern defiance that drew national attention. It began after Republican William Pitt Kellogg was elected governor of Louisiana in 1872. Kellogg had moved south from Vermont after the war, a "carpetbagger" sympathetic to Reconstruction. Democrats opposed to Kellogg coalesced around the White League, a paramilitary organization dedicated to preserving and restoring white rule in Louisiana. In rural parishes, the White League regularly threatened and assassinated Republicans in order to intimidate voters and undermine Reconstruction.

In the summer of 1874 the League quickly recruited 1,500 men to overturn Kellogg's administration, conducting armed drills around the city to demonstrate their strength. On September 12 the police sealed off the levee after learning that these men were expecting a large shipment of arms. Incensed White Leaguers converged on the Clay Statue on Canal Street to erect several barricades along Poydras Street. The police prepared for battle, while the newly elected governor retreated to the Custom House after commissioning the former Confederate General James Longstreet to defend the government and protect the city.

Thousands watched as the White League routed the police along Canal Street, driving them back along lines of "retreat" that are identified on the map. The League immediately declared victory, installed their own governor, and demanded that Kellogg resign. This map was made at that moment by the city surveyor and White League supporter T. S. Hardee, who framed the conflict as a battle for "freedom," an American "revolution" against Reconstruction. He described a revolt of "plain citizens" who heroically resisted a tyrannical and arrogant Republican government, martyrs for a noble cause rather than vigilantes. Hardee made no mention of the thirteen policemen killed by the White League on that day.

Within days President Grant restored Kellogg as governor, but the momentum behind Reconstruction had already begun to wane. Just as Kellogg lost the Battle of New Orleans, the Republicans lost the larger war over Reconstruction. After the highly contested 1876 election, the Republican president Rutherford Hayes agreed to withdraw the remaining federal troops from the South. White "Redeemer" Democrats swiftly returned to office and began to restrict black civil and voting rights. In state after state, armed resistance, intimidation, and violence ended Reconstruction. In Louisiana, the White League even served as the state militia.

This map of a single street fight captures the open, violent, and extralegal opposition to Reconstruction across the South in the 1870s. The Battle of New Orleans was celebrated for decades, and in 1891 the city even erected a statue to commemorate what locals considered a proper act of defiance. That monument to white supremacy remained for over a century before Governor Mitch Landrieu ordered its removal in the spring of 2017.

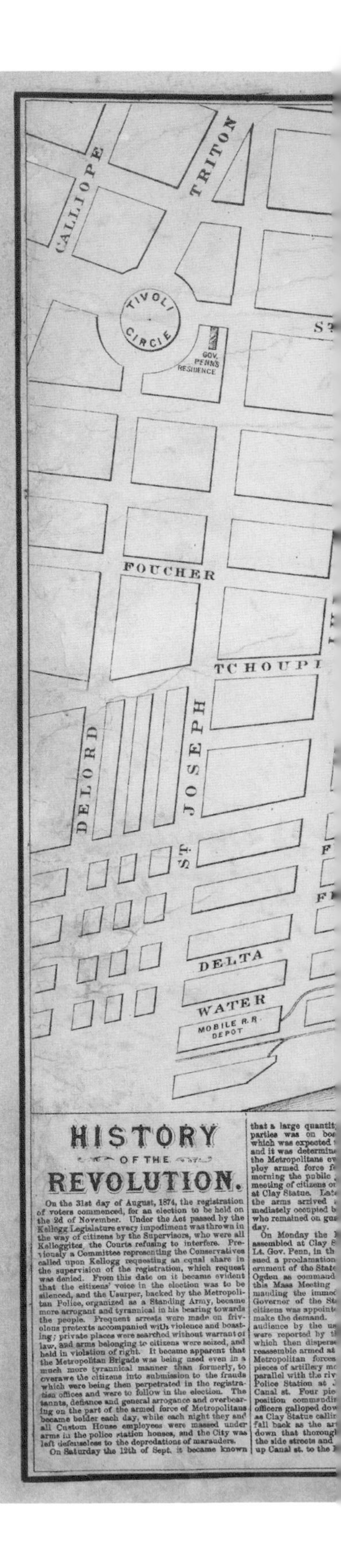

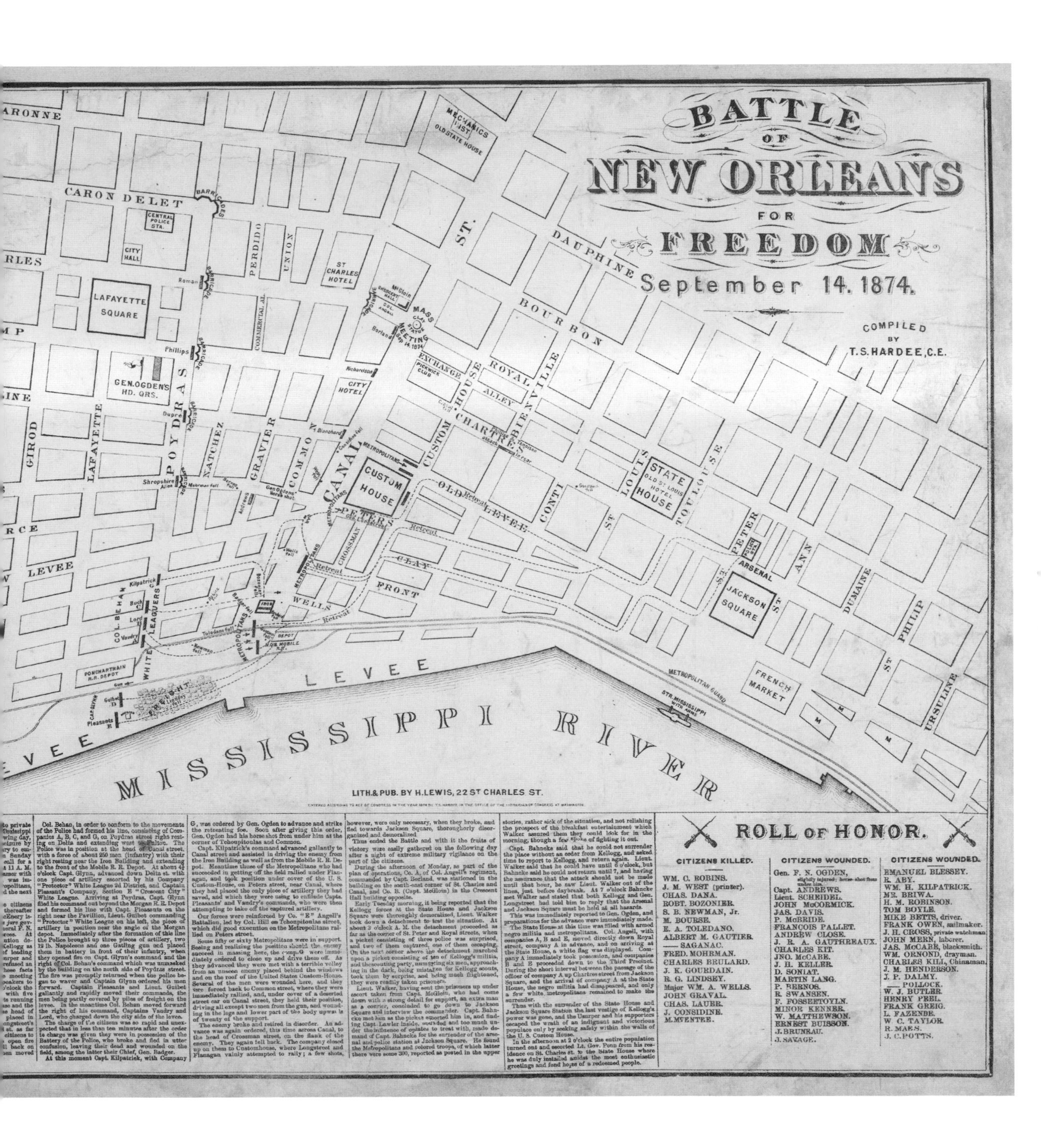

BATTLE OF NEW ORLEANS FOR FREEDOM
September 14. 1874.
COMPILED BY T.S. HARDEE, C.E.
LITH.& PUB. BY H.LEWIS, 22 ST CHARLES ST.
MISSISSIPPI RIVER
LEVEE
CUSTOM HOUSE
STATE HOUSE
JACKSON SQUARE
LAFAYETTE SQUARE
GEN. OGDEN'S HD. QRS.
CITY HALL
ST CHARLES HOTEL
CITY HOTEL
FRENCH MARKET
ARSENAL
MECHANICS INST. OLD STATE HOUSE
CENTRAL POLICE STA.
PONCHARTRAIN R.R. DEPOT
CANAL
POYDRAS
DAUPHINE
BOURBON
ROYAL
CHARTRES
OLD LEVEE
CLAY
FRONT
ROLL OF HONOR.
CITIZENS KILLED.
WM. C. ROBINS.
J. M. WEST (printer).
CHAS. DANA.
ROBT. BOZONIER.
S. B. NEWMAN, Jr.
M. BOURSE.
E. A. TOLEDANO.
ALBERT M. GAUTIER.
—— SAGANAC.
FRED. MOHRMAN.
CHARLES BRULARD.
J. K. GOURDAIN.
R. G. LINDSEY.
Major WM. A. WELLS.
JOHN GRAVAL.
CHAS. LAUER.
J. CONSIDINE.
M. McENTEE.
CITIZENS WOUNDED.
Gen. F. N. OGDEN, slightly injured; horse shot from under him.
Capt. ANDREWS.
Lieut. SCHEIDEL.
JOHN McCORMICK.
JAS. DAVIS.
P. McBRIDE.
FRANCOIS PALLET.
ANDREW CLOSE.
J. R. A. GAUTHREAUX.
CHARLES KIT.
JNO. McCABE.
J. H. KELLER.
D. SONIAT.
MARTIN LANG.
P. BERNOS.
R. SWANSEN.
F. FOSSEETOYLN.
MINOR KENNER.
W. MATTHEWSON.
ERNEST BUISSON.
J. BRUNEAU.
J. SAVAGE.
CITIZENS WOUNDED.
EMANUEL BLESSEY.
R. ABY.
WM. H. KILPATRICK.
MR. BRIWA.
H. M. ROBINSON.
TOM BOYLE.
MIKE BETTS, driver.
FRANK OWEN, sailmaker.
J. H. CROSS, private watchman.
JOHN MERN, laborer.
JAS. McCABE, blacksmith.
WM. ORNOND, drayman.
CHARLES KILL, Chinaman.
J. M. HENDERSON.
J. P. DALMY.
—— POLLOCK.
W. J. BUTLER.
HENRY PEEL.
FRANK GREIG.
L. FAZENDE.
W. C. TAYLOR.
R. MAES.
J. C. POTTS.

BATTLE
OF
NEW ORLEANS
FOR
FREEDOM
September 14, 1874.

COMPILED
BY
T.S. HARDEE, C.E.

LITH. & PUB. BY H. LEWIS, 22 ST CHARLES ST.

Col. Behan, in order to conform to the movements of the Police had formed his line, consisting of Companies A, B, C, and G, on Poydras street rights resting on Delta and extending west to Fulton. The Police was in position at the head of Canal street, with a force of about 250 men (infantry) with their right resting near the Iron Building and extending left to the front of the Mobile R. R. Depot. At about 9 o'clock Capt. Glynn, advanced down Delta st. with one piece of artillery escorted by his Company "Protector" White League 3d District, and Captain Pleasant's Company, Section E "Crescent City" White League. Arriving at Poydras, Capt. Glynn filed his command out beyond the Morgan R. R. Depot and formed his line with Capt. Pleasants on the right near the Pavillion, Lieut. Guibet commanding "Protector" White League on his left, the piece of artillery in position near the angle of the Morgan depot. Immediately after the formation of the line the Police brought up three pieces of artillery, two 12 lb. Napoleons and one Gatling gun and placed them in battery in front of their infantry, when they opened fire on Capt. Glynn's command and the right of Col. Behan's command which was unmasked by the building on the north side of Poydras street. The fire was promptly returned when the police began to waver and Captain Glynn ordered his men forward. Captain Pleasants and Lieut. Guibet gallantly and rapidly moved their commands, the men being partly covered by piles of freight on the levee. In the meantime Col. Behan moved forward the right of his command, Captains Vaudry and Lord, who charged down the city side of the levee.

The charge of these citizens was so rapid and unexpected that in less than ten minutes after the order to charge was given they were in possession of the Battery of the Police, who broke and fled in utter confusion, leaving their dead and wounded on the field, among the latter their Chief, Gen. Badger.

At this moment Capt. Kilpatrick, with Company

G, was ordered by Gen. Ogden to advance and strike the retreating foe. Soon after giving this order, Gen. Ogden had his horse shot from under him at the corner of Tchoupitoulas and Common.

Capt. Kilpatrick's command advanced gallantly to Canal street and assisted in driving the enemy from the Iron Building as well as from the Mobile R. R. Depot. Meantime those of the Metropolitans who had succeeded in getting off the field rallied under Flanagan, and took position under cover of the U. S. Custom-House, on Peters street, near Canal, where they had placed the only piece of artillery that had saved, and which they were using to enfilade Capt. Pleasants' and Vandry's commands, who were then attempting to take off the captured artillery.

Our forces were reinforced by Co. "E" Angell's Battalion, led by Col. Hill on Tchoupitoulas street, which did good execution on the Metropolitans rallied on Peters street.

Some fifty or sixty Metropolitans were in support. Seeing and realizing the position should the enemy succeed in massing here, the comp. G were immediately ordered to close up and drive them off. As they advanced they were met with a terrible volley from an unseen enemy placed behind the windows and on the roof of the United States Custom-House. Several of the men were wounded here, and they were forced back to Common street, where they were immediately rallied, and, under cover of a deserted street car on Canal street, they held their position, driving all except two men from the gun, and wounding in the legs and lower part of the body upwards of twenty of the support.

The enemy broke and retired in disorder. An advance was again ordered, this time across Canal, to the head of Crossman street, on the flank of the enemy. They again fell back. The company closed up on them to Customhouse, where Longstreet and Flanagan vainly attempted to rally; a few shots,

however, were only necessary, when they broke, and fled towards Jackson Square, thoroughly disorganized and demoralized.

Thus ended the battle and with it the fruits of victory were easily gathered on the following day after a night of extreme military vigilance on the part of the citizens.

During the afternoon of Monday, as part of the plan of operations, Co. A, of Col. Angell's regiment, commanded by Capt. Borland, was stationed in the building on the south-east corner of St. Charles and Canal, and Co. B. (Capt. McGloin) in the Crescent Hall building opposite.

Early Tuesday morning, it being reported that the Kellogg forces at the State House and Jackson Square were thoroughly demoralized, Lieut. Walker took down a detachment to test the situation. At about 3 o'clock A. M. the detachment proceeded as far as the corner of St. Peter and Royal streets, when a picket consisting of three police was surprised, and two of them captured, one of them escaping. On the next corner above the detachment marched upon a picket consisting of ten of Kellogg's militia, and the scouting party, numbering six men, approaching in the dark, being mistaken for Kellogg scouts, took them by surprise, and being much frightened, they were readily taken prisoners.

Lieut. Walker, having sent the prisoners up under escort taken from Capt. McGloin, who had come down with a strong detail for support, an extra man as a courier, continued to go down to Jackson Square and interview the commander. Capt. Bahncke met him as the picket escorted him in, and finding Capt. Lawler inside, wounded and too much under the influence of opiates to treat with, made demand of Capt. Bahncke for the surrender of the arsenal and police station at Jackson Square. He found the Metropolitans and colored troops, of which latter there were some 300, reported as posted in the upper

stories, rather sick of the situation, and not relishing the prospect of the breakfast entertainment which Walker assured them they could look for in the morning, though a few spoke of fighting it out.

Capt. Bahncke said that he could not surrender the place without an order from Kellogg, and asked time to report to Kellogg and return again. Lieut. Walker said that he could have until 8 o'clock, but Bahncke said he could not return until 7, and having the assurance that the attack should not be made until that hour, he saw Lieut. Walker out of the lines, just before daybreak. At 7 o'clock Bahncke met Walker and stated that both Kellogg and Gen. Longstreet had told him to reply that the Arsenal and Jackson Square must be held at all hazards.

This was immediately reported to Gen. Ogden, and preparations for the advance were immediately made.

The State House at this time was filled with armed negro militia and metropolitans. Col. Angell, with companies A, B and E, moved directly down Royal street, company A in advance, and on arriving at the State House, a white flag was displayed. Companies B and E proceeded down to the Third Precinct. During the short interval between the passage of the officer of company A up Chartres street from Jackson Square, and the arrival of company A at the State House, the negro militia had disappeared, and only a few white metropolitans remained to make the surrender.

Thus with the surrender of the State House and Jackson Square Station the last vestige of Kellogg's power was gone, and the Usurper and his supporters escaped the wrath of an indignant and victorious populace only by seeking safety within the walls of the U. S. Custom House.

In the afternoon at 3 o'clock the entire population turned out and escorted Lt. Gov. Penn from his residence on St. Charles st. to the State House where he was duly installed amidst the most enthusiastic greetings and fond hopes of a redeemed people.

ROLL of HONOR.

CITIZENS KILLED.

WM. C. ROBINS.
J. M. WEST (printer).
CHAS. DANA.
ROBT. BOZONIER.
S. B. NEWMAN, Jr.
M. BOURSE.
E. A. TOLEDANO.
——— SAGANAC.
FRED. MOHRMAN.
CHARLES BRULARD.
J. K. GOURDAIN.
R. G. LINDSEY.
Major WM. A. WELLS.
JOHN GRAVAL.
J. CONSIDINE.
M. MVENTE.

CITIZENS WOUNDED.

Gen. F. N. OGDEN, slightly injured; horse-shot from under him.
Capt. ANDREWS.
Lieut. SCHRIDEL.
JOHN McCORMICK.
JAS. DAVIS.
P. McBRIDE.
FRANÇOIS PALLET.
ANDREW CLOSE.
J. R. A. GAUTHERAUX.
CHARLES KIT.
JNO. McCABE.
J. H. KELLER.
D. SONIAT.
MARTIN LANG.
P. NEKNOS.
SWANSEN.
F. FOSSEETOYLN.
MINOR KENNER.
W. MATTHEWSON.
ERNEST BUISSON.
J. BRUNEAU.
J. SAVAGE.

CITIZENS WOUNDED.

EMANUEL BLESSEY.
R. ABY.
WM. H. KILPATRICK.
MR. BRIWA.
H. M. ROBINSON.
TOM BOYLE.
MIKE BETTS, driver.
FRANK OWEN, sailmaker.
J. H. CROSS, private watchman.
JOHN MERN, laborer.
JAS. McCABE, blacksmith.
WM. ORNOND, drayman.
CHARLES KILL, Chinaman.
J. M. HENDERSON.
J. P. DALMY.
——— POLLOCK.
W. J. BUTLER.
HENRY PEEL.
FRANK GREIG.
L. FAZENDE.
W. C. TAYLOR.
R. MAES.
J. C. POTTS.

6. 1874-1914

Industrialization and Its Discontents

The United States experienced one of the most rapid and thorough economic transformations in modern history after the Civil War. Industrialization was characterized by boom-and-bust cycles that affected urban and rural Americans alike. Though we commonly date this economic upheaval to the 1870s, in fact industrialization was accelerated by the exigencies of the war itself. Ironically, while the Civil War destroyed much of the economic capacity of the South, it catalyzed industrialization in the Northern states. The logistical demands of the Union war effort refined existing technologies such as the telegraph, which delivered intelligence to the military and news to the home front. Lucrative defense contracts expanded the nation's economic capacity, while the wartime mobilization of resources and capital ushered in a system of large-scale production that demanded a steady supply of inexpensive and relatively unskilled labor. By the end of the century, most Americans lived in urban areas and worked for wages.

At the center of this economic revolution were the railroads. Railroad track mileage tripled between 1860 and 1880, and then tripled again by 1920. This astounding growth brought railroad companies unprecedented power and influence. Initially hailed as technological wonders, the railroads provoked scorn and rage across the country by the 1870s, and for good reason. They benefited disproportionately from federal subsidies, especially the enormous land grants made to stimulate transcontinental rail construction. They fueled growth in some years, but sparked financial ruin in others. As feats of engineering, a mode of transportation, an economic force, and sources of corruption, railroads were both a cause and a consequence of industrialization.

More generally, this chapter explores the modernization of American society from the end of Reconstruction to the eve of World War I. We begin with maps designed to unearth one of the raw materials of industrialization: coal. The earliest geological map of Virginia on page 152 was part of a concerted effort by state leaders to rebuild the economy after the Civil War. Over the next two decades, coal mining drew Kentucky and the Virginias into an international web of economic relationships that fundamentally changed the lives of its people. All of this hinged on the construction of railroads, for without transportation—which required significant capital—the coal of Appalachia was of little value (page 154).

The outsized influence of the railroads was felt everywhere. When these corporations announced yet another round of wage cuts in the summer of 1877, angry workers in West Virginia organized a strike that quickly spread widely along the country's rail routes. In Pittsburgh, the strike culminated in a standoff between Pennsylvania Railroad workers and the state militia. Ultimately the militia fired on the large crowd of strikers, killing over twenty and driving the strikers to destroy train cars, tracks, and the roundhouse of the nation's largest private company. As strikes consumed industry, farmers similarly protested a system that left them at the mercy of a fluctuating world market. This discontent coalesced in a populist movement that advocated—among other things—public ownership of the railroads. In the last quarter of the nineteenth century, both farmers and industrial workers revolted against corporations that had not even existed a generation earlier.

The power of the railroads even extended to the experience of time. In 1883 the major railroads lobbied the federal government to create four standard time zones, which in turn facilitated consistent train schedules. Perhaps most contentious was the sway that the transcontinental railroads held over land and transportation in the West. The 1883 political broadside on page 158 captures the anger against the Northern Pacific Railroad, branded by the Democratic Party as a "soulless corporation" that controlled not just Washington Territory, but the entire Republican Party. Throughout the late nineteenth century, the two parties jockeyed for the upper hand by trading accusations of corruption in an era of high voter turnout and exceptionally competitive elections. That political landscape is captured by the first "red and blue" map of American electoral politics on page 156. This high turnout produced political stalemate and a series of one-term presidents, two of whom failed to win the popular vote.

Industrialization also fueled the growth of American cities, and maps were both tools and weapons in this era of unprecedented urbanization. The maps on pages 164, 166, and 168 reflect the contemporary enthusiasm for cartography as an instrument of urban reform. Political leaders in San Francisco designed sensationalistic maps to control the Chinese population. Florence Kelley and Agnes Holbrook made maps to investigate and publicize the condition of the immigrant poor on the south side of Chicago. And in Philadelphia W. E. B. DuBois mapped the living conditions of the black community to analyze patterns of segregation. All of these efforts were modeled on the work of Charles Booth, who famously attempted to map poverty in East London in the late nineteenth century. In turn, American suffragists creatively deployed maps to advertise and extend women's suffrage across the country (page 174).

The first states to grant women the right to vote were in the sparsely settled Far West, a region undergoing profound shifts in the late nineteenth century. The railroads and the federal government actively encouraged western settlement, which had a catastrophic effect on the tribes of the Great Plains. William Temple Hornaday mapped the systematic destruction of the bison at late century, a function of native hunting, fur trading, and especially the extension of the western railroads (page 160). His attention to the buffalo shocked the public and sowed the seeds of the modern conservation movement. While Hornaday warned against the extinction of the bison, John Wesley Powell urged Congress to address the unrestrained growth in the arid West. His solution was to reimagine western settlement not around the logic of the grid but around watersheds and local control of this limited resource (page 162). Powell failed to convince Congress, in part because the nation's industrialized economy was chiefly driven by the needs of producers rather than consumers.

Finally, industrialization at home led to a redefinition of American foreign policy. As overproduction saturated domestic markets, the US sought to expand trade with China and Latin America. These economic demands—compounded by a zealous sense of mission—brought the nation a host of new territories in the Pacific and the Caribbean. The nation's largest map producer, Rand McNally, redrew its map of the country in order to make room for the Philippines, Hawaii, Cuba, and Puerto Rico (page 170). Rand McNally itself was emblematic of the era: the company originally produced timetables and tickets for the Chicago railroads, then adopted inexpensive print techniques to produce maps, atlases, and school textbooks for a mass market. Its new map of the nation became a model for others to imitate, a visual assertion of American international power at the dawn of a new century.

Just a few years later, the nation was gripped by similar enthusiasm for the Panama Canal, one of the greatest engineering feats in American history (page 172). Capping an era of extraordinary growth and recession, the opening of the canal in 1914 was a source of tremendous national pride that also coincided with the outbreak of the world's most destructive war to date.

UNEARTHING COAL

Jedediah Hotchkiss, "Map of Virginia," in *Virginia: A Geographical and Political Summary*, 1874

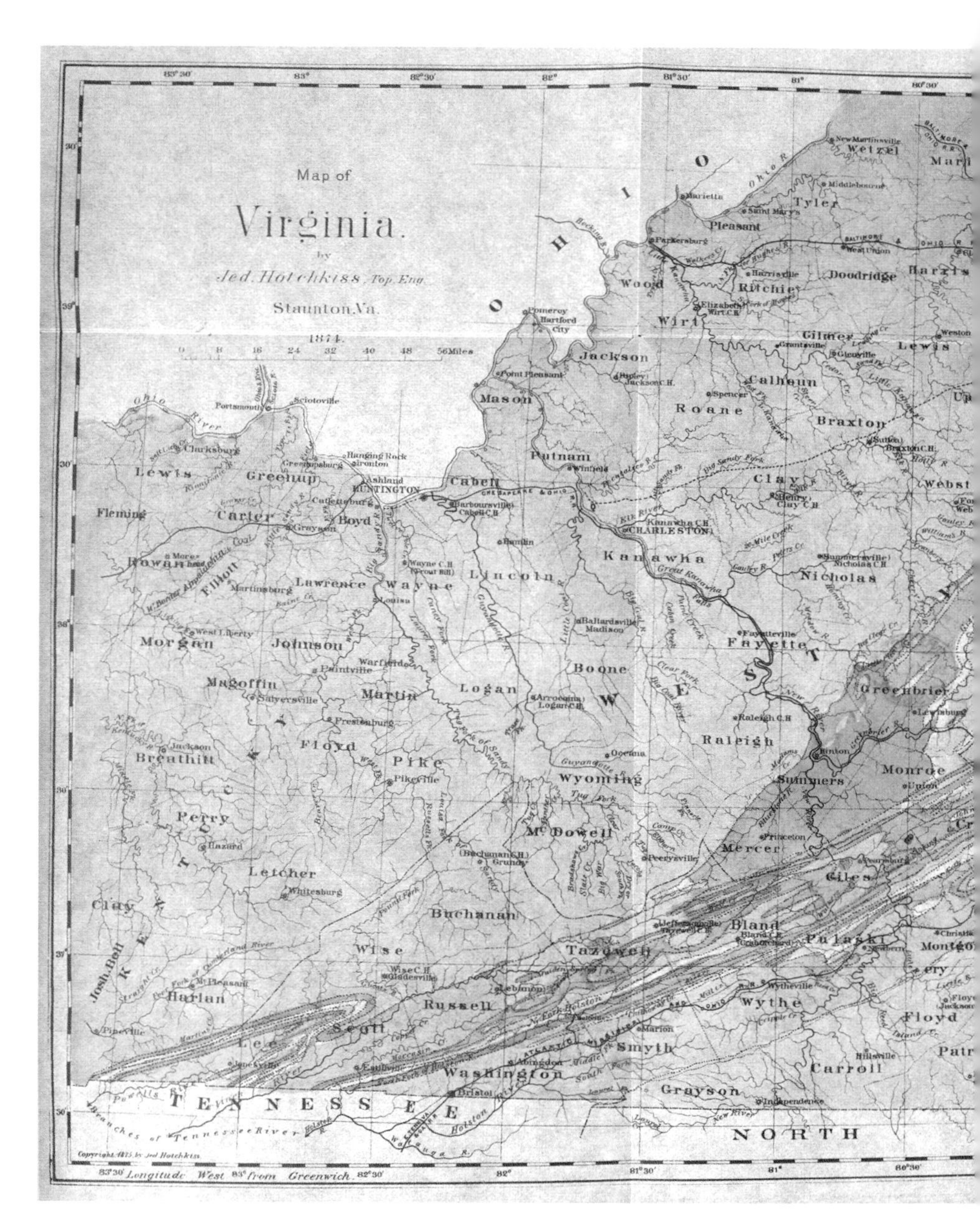

The United States modernized so quickly in the second half of the nineteenth century that by 1913 it had a greater industrial output than Germany, France, and Britain combined. Before the widespread use of oil and gas in the twentieth century, this new economy of factories, steamships, and railroads was fired largely by coal. Refined as coke, coal was also essential to the production of steel. And there was no more important source of coal in the nineteenth century than Virginia and West Virginia.

The exploitation of coal depended upon the relatively new science of geology. In the 1830s and 1840s Virginia geologist William Barton Rogers undertook the first geological survey of the state in order to advance prospects for mining in its western regions. Yet Barton found little support for his efforts in a state that was largely controlled by the slaveholders of the eastern Piedmont and Tidewater regions, who were thoroughly invested in agriculture. (The dependence upon slavery in the eastern part of Virginia is apparent on page 142. The virtual absence of slavery from the mountain regions also partly explains why those counties formed the new state of West Virginia during the Civil War and remained loyal to the Union.) In the early 1850s Rogers proposed a second geological survey of Virginia, which the legislature rejected. Discouraged, he returned to his position at the University of Virginia, and in 1853 left the state altogether to found the Massachusetts Institute of Technology.

Rogers may have left Virginia, but his geological research indelibly shaped the region's postwar development through Jedediah Hotchkiss. In fact, just as Rogers was heading north out of Virginia, Hotchkiss was heading south. Born in New York, Hotchkiss taught in the coal towns of Pennsylvania as a teenager, indulging his love of botany and geology in the surrounding area. He then settled in Virginia's Shenandoah Valley and took up surveying, a skill that made him invaluable to the Confederacy during the Civil War. As an aide to General Stonewall Jackson, Hotchkiss ranged widely through Virginia and produced some of its most detailed and accurate topographic maps.

After the war, Hotchkiss combined his surveying experience with his knowledge of geology to reinvent himself as a mining engineer and consultant. When state leaders asked for help promoting Virginia's economy in the 1870s, he responded by mapping the state's geology and mining prospects. In Hotchkiss' view, the state's future lay not in agriculture, but coal. Using Rogers' earlier research, Hotchkiss highlighted the enormous western coal beds in gray in order to attract both labor and capital. Though counties are named, they are secondary; more important to Hotchkiss is the relationship between geology and the progress of railroad and canal construction. This was, above all, a map outlining the state's economic future. Notice as well that Hotchkiss references the state survey undertaken by Rogers from 1835 to 1841, a reminder of how little support had been given to

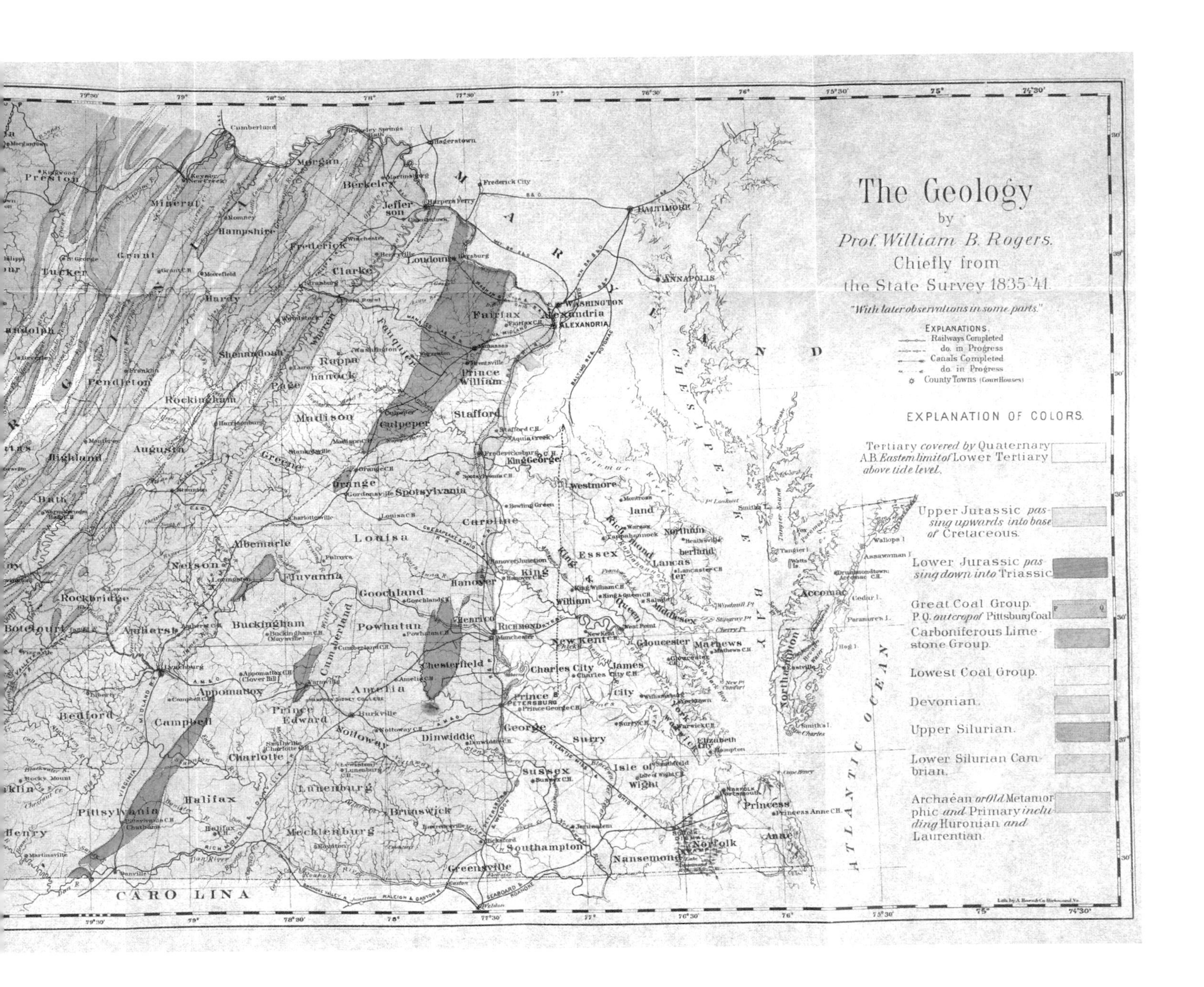

geological research since then.

While Hotchkiss was compiling the map in 1873 his task took on greater urgency, for Collis Huntington had just opened the first railroad line to the mines of West Virginia. Constructed along the Kanawha and New rivers, the railroad traversed gorges and tunneled through mountains to transport coal mined by the company to Richmond. By owning both the mines and the transportation, Huntington demonstrated the enormous profits to be made in Virginia, West Virginia, and eastern Kentucky. His success sparked a rush of railroad construction to the coal fields over the next several decades.

Hotchkiss directly advanced this boom through maps that guided both railroad investment and land sales. His mining journal of the 1880s, *The Virginias*, included dozens of original topographic and geological maps of this forbidding terrain. Hotchkiss understood that the biggest impediment to coal extraction in the region was the absence of reliable geological information. His efforts—such as the map on the next page—attracted the capital required to make the region one of the nation's most important sources of coal by the turn of the century.

MAPPING THE MOTHERLODE

Jedediah Hotchkiss, "Map of Part of the Great Flat-Top Coal-Field of Va. & W. Va.," 1886

The presence of coal in Virginia, West Virginia, and Kentucky was well established by the 1870s. Yet it was Jedediah Hotchkiss who mapped and profiled the economic potential of the region for investors. In his hands, maps were instruments of development, for the greatest obstacle to Appalachian mining was geography: without railroads, there was no way to transport coal to market, much less to make it profitable. The map at right highlights the role Hotchkiss played in attracting railroad investors from Philadelphia, New York, Boston, and London.

In 1873 Collis Huntington financed the first railroad across the mountains of West Virginia, designed to ship coal to Richmond. The Chesapeake & Ohio Railway was largely built by African Americans who had been freed during the Civil War. Many of these men lost their lives constructing this railroad through difficult and dangerous terrain. The completion of the C&O railroad launched a twenty-year boom in the region that was directly aided by maps such as this, which guided both land sales and future rail routes.

In 1881 Hotchkiss learned from his friend Frederick Kimball, vice president of the newly organized Norfolk & Western Railway, that the company would invest in the coal fields of southern Virginia. Kimball sought guidance from Hotchkiss, who in turn gave him maps and information detailing the unmatched potential of the Great Flat-Top Coalfield in Tazewell and Mercer Counties. Kimball's tip also led Hotchkiss to purchase 60,000 acres of rich coal fields further west in McDowell County.

Within two years Kimball had become president of the Norfolk & Western Railway, and took Hotchkiss' advice about building a route to the coal fields along the Bluestone River. In 1882 the railroad founded the town of Pocahontas as a hub for its nearby mining operations, and the following year it delivered its first load of coal to Norfolk. The company's headquarters in Roanoke transformed that sleepy town of under 700 in 1880 to over 16,000 within a decade.

The Norfolk & Western's decision to invest in the Great Flat-Top Coalfield was enormously profitable, and prompted Kimball to expand its railroad route further. Hotchkiss drew this map to encourage investment by others along that new route. Particularly audacious was Kimball's decision to extend the railroad 3,000 feet *through* Great Flat-Top Mountain (at the left edge of the map), to the motherlode in McDowell County. This became the most productive coal county in West Virginia. Kimball then pushed even further by forging the railroad north to the Ohio River to serve massive new coal markets in the Midwest and the Great Lakes. Hotchkiss was right: railroad companies could reap their greatest rewards by owning both the mines and the modes of transportation that brought coal to market.

Maps such as this drew capital to the Virginias and Kentucky, and Hotchkiss was uniquely equipped to produce them. The coal boom in this region meant that some of the first detailed maps of Appalachia profiled the region in economic terms for distant investors. These early maps emphasized railroads, coal outcroppings, and new mining prospects. Maps have long been instruments of economic decision making, but this region was initially mapped in detail *because* of its mining potential.

Before the Civil War, the people of southern Appalachia had been subsistence farmers. With the advent of mining they became enmeshed in an international market that would determine not just the price of coal but also the conditions of their work and life. Engaged in mining, they remained dependent upon the land, but now lived in isolated company towns that were often divided along racial and ethnic lines. At the mercy of outside forces, miners experienced the same lack of control that characterized the lives of their counterparts in factories and fields across the country. For those whose families had farmed independently for generations, the disruption made them more than a little suspicious of the outside world. Coal mining advanced American industrialization, but not without significant human and environmental costs.

Map of Part of
The Great Flat-top
Coal-field of Va. & W. Va.
Showing Location of
Pocahontas & Bluestone
Collieries
May, 1886.
McDowell County, W. Va.
Mercer County, W. Va.
Tazewell County, Va.
Great Flat-top Mtn.
"Reliance" Colliery
Wm. Booth & Co.
Stevenson, Mullin, & Co.
Caswell Creek Coal & Coke Co.
"Sterling" Colliery
John Cooper & Co.
Bluestone River.
Simmons
Bramwell
Cooper
Outcrop of "Pocahontas" Coal Bed.
East Mine
West Mine
Bluestone Junction
Pocahontas
Laurel Creek
New R. Br. o Norfolk & W. Rn.
Bluestone River.
Eng. Office of Jed. Hotchkiss, Staunton, Va.
0 1/4 1/2 1 Mile
Scale

RED AND BLUE AMERICA

"Popular Vote: Ratio of Predominant to Total Vote, by Counties ... 1880," 1883

Today Americans casually invoke "red and blue America" as a shorthand for the polarization that seems to have infected far more than national elections. Yet political partisanship has a long history in America, and reached one of its peaks in the late nineteenth century. Sectional and party divisions endured long after the end of Reconstruction, inflamed by debates over temperance, Catholicism, and the tariff. In the Gilded Age, these issues drove voters to the polls at rates that have never been matched since.

High voter turnout led to competitive elections, among the closest of which was the 1880 presidential race between Republican James Garfield and Democrat Winfield Scott Hancock. Garfield's slim majority was both visualized and analyzed in this innovative electoral map, issued in 1883 by Henry Gannett, superintendent of the census. As one of the first efforts to picture electoral returns at the local level, the map stunned viewers and challenged conventional political wisdom.

The map showed a nation organized according to prevailing Democratic and Republican majorities. (Note that while current conventions represent Democrats in blue and Republicans in red, here those colors are reversed. Moreover, in the 1880s the parties represented very different agendas than they do today.) Garfield's razor-thin margin of just 7,000 votes is graphed along the bottom of the page. Yet the electoral college was designed by the framers so that slim popular majorities such as Garfield's translate into wide electoral margins, even landslides. To highlight this incongruity, Gannett presented his readers with two maps shown here: an inset devoted to statewide electoral outcome and a much larger map profiling the results of individual voting districts.

On the inset map at lower right, the country appears deeply divided along a north–south axis, with Democrats dominating the South after defeating Republican Reconstruction governments. Though these statewide electoral returns determined the

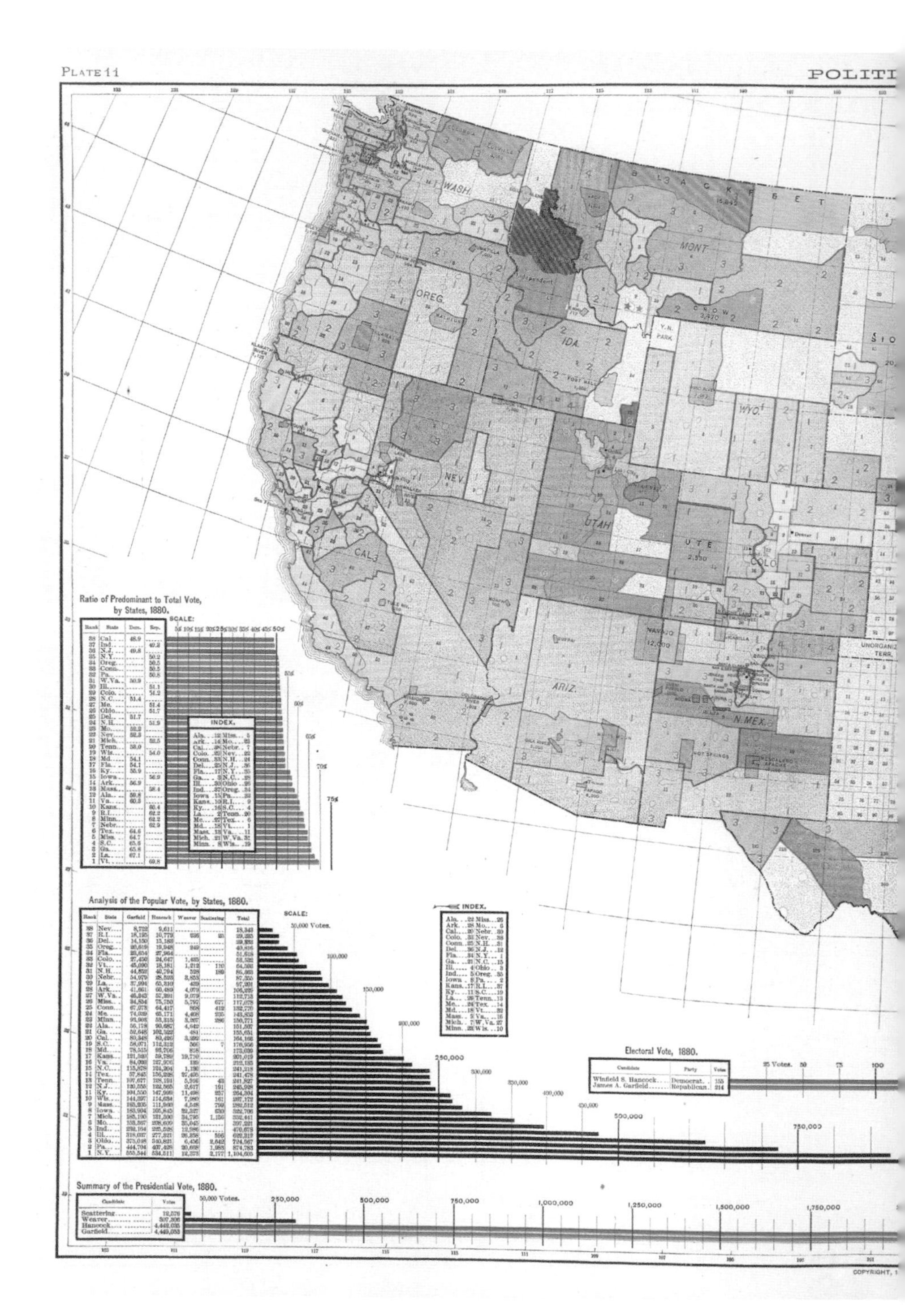

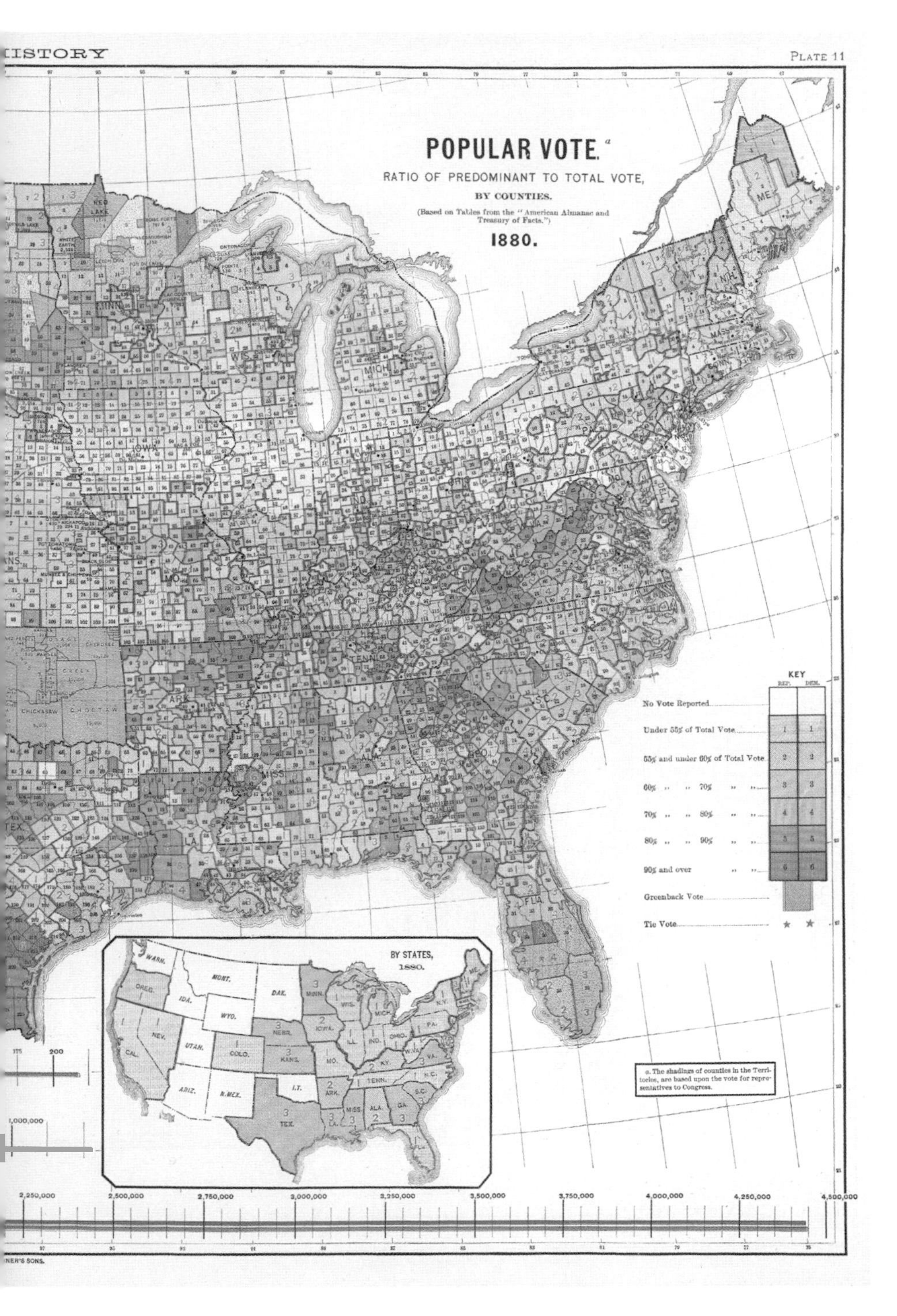

presidency, Gannett relegated the electoral map to an inset in order to feature a larger map of county-level returns that he knew would captivate his readers. Gannett's map was particularly innovative in going beyond the statewide outcome to assess local returns. He also used a scale of shade to assess the *strength* of parties at the county or parish level: the darker the color, the wider the margin of victory, with the lightest shade indicating the barest majority.

This shading revealed a political landscape that was far more complicated than that shown by the statewide results of the electoral college. While the former slave states leaned toward the Democrats, they also included significant pockets of Republican strength: for instance, the blue areas of eastern Tennessee hearkened back to the strong anti-Confederate sentiment during the Civil War that evolved into Republican loyalty thereafter. Similarly, broad swaths of red through Pennsylvania showed Republicans that their control over that state was weak, especially in light of recent economic volatility and labor unrest. Perhaps most revealing are the isolated Republican majorities throughout the Deep South, a sign of the freedmen's allegiance to the party of Lincoln. Tragically, those blue pockets would disappear once white Southern Democrats disenfranchised African American voters by the turn of the century.

Gannett's detailed profile of election returns upended assumptions about party dominance in several states. By the early twentieth century, tools such as this had become ordinary instruments of political strategy alongside disciplined parties and research-driven campaigns. With his visual innovations, Gannett would have been entirely at home in our own world of data-driven cartography and analysis. Significantly, his map revealed dynamics that are concealed—and sometimes suppressed—by the electoral college.

STRANGLED BY THE RAILROADS

"Under a Black Cloud!," 1883

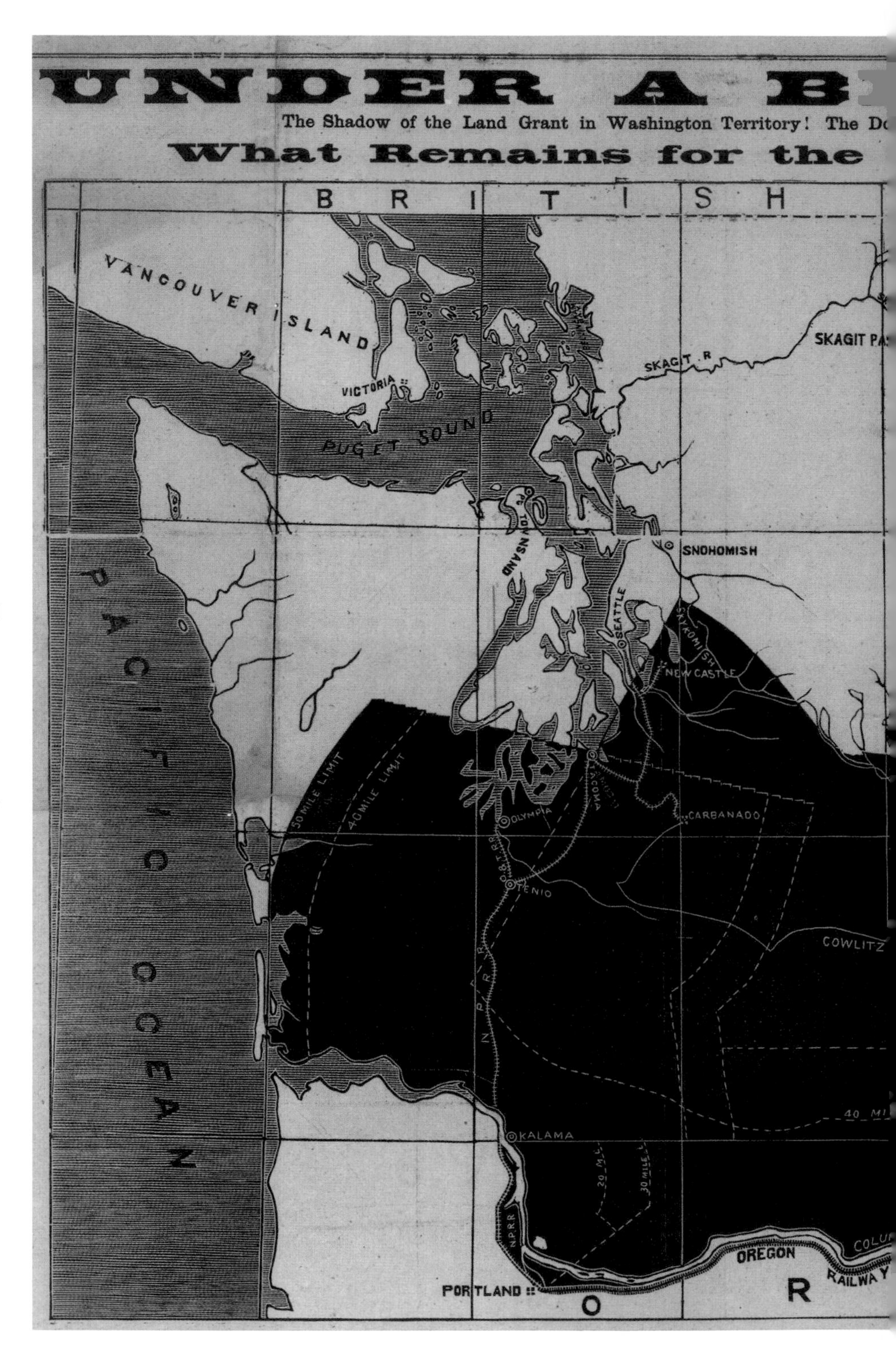

In the 1840s and 1850s the railroads were regarded as an astonishing technological feat that had the potential to integrate the nation and extend its commercial power. They were, in short, a sign of wonder, progress, and promise. That perception shifted dramatically in subsequent decades. By the 1860s the federal government had granted the railroad companies 130 million acres of western land to encourage the construction of transcontinental routes.

No company benefited more than the Northern Pacific Railway, which received 40 million acres to subsidize construction of a route from Lake Superior to the Puget Sound—much of it across Native American land. This was the largest single land subsidy to any railroad, and included grants in the western territories that were twice the size of those given to the Union Pacific and the Central Pacific. Overnight, the NP became the nation's most powerful private landlord.

This federal largesse enabled the railroads to influence not just patterns of settlement but the government itself. Members of Congress who sat on the board of the NP ensured that the company survived despite consistent losses, corruption, and ineptitude. Several western railroads manipulated the terms of their land grants to secure their economic position, a strategy mastered by Henry Villard.

In 1881 Villard quietly began to amass shares to control the Northern Pacific out of concern that it might compete with the Columbia River route operated by his own Oregon Railroad & Navigation Company. By the next year, Villard had secured control of both railroads. This allowed the NP to claim that it had "completed" the transcontinental route, for which it received those extensive land grants. That enraged the public and rival railroads, both of whom saw the Northern Pacific as the owner of "unearned lands" in Washington Territory. The company failed to construct the promised final leg of the route to the Pacific, yet it remained the largest landholder in the territory.

The partial completion of the Northern Pacific only demonstrated how little demand there was for that route. In the headlong rush to construct the

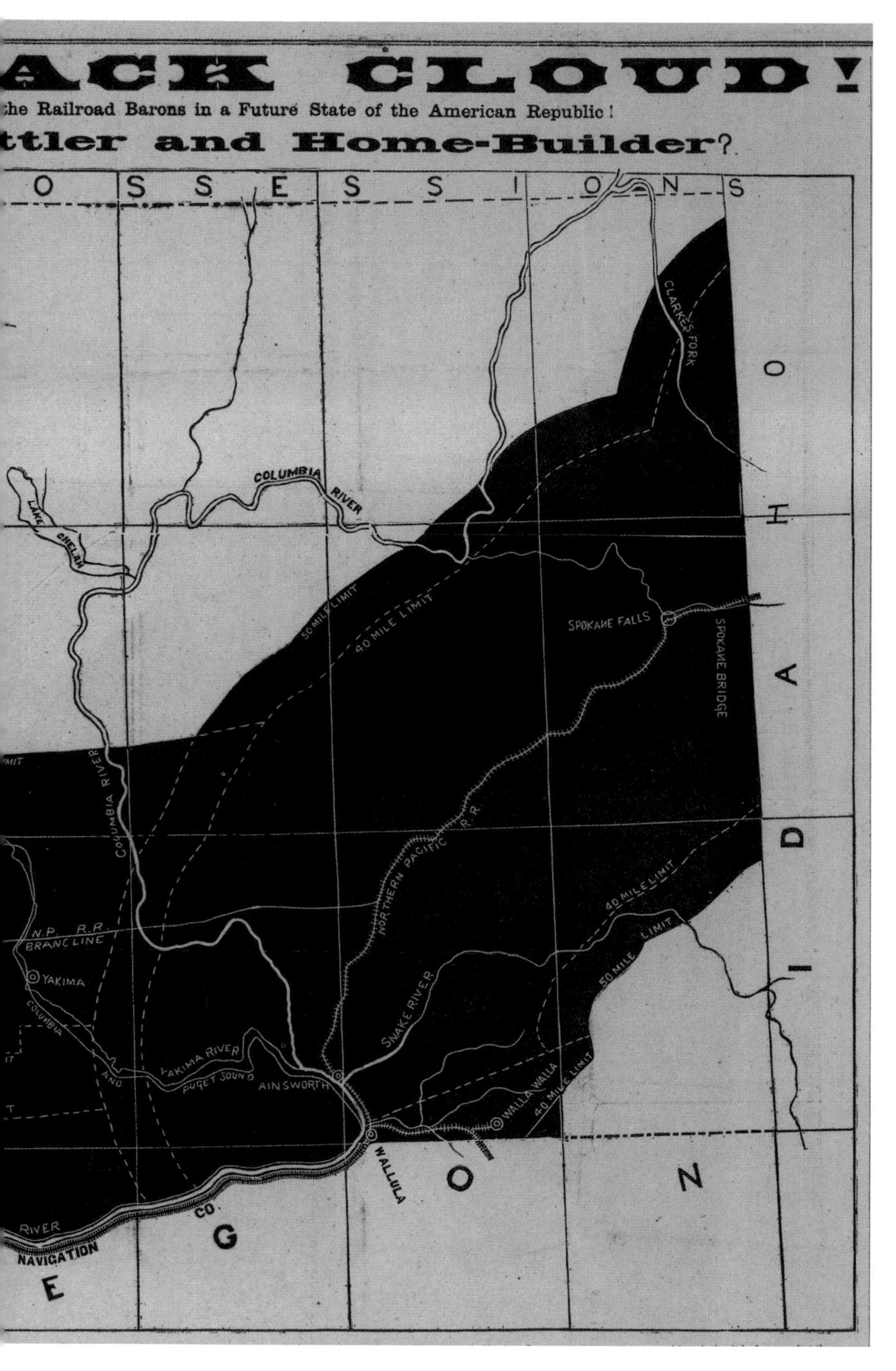

transcontinental railroads, it seems, few had asked whether they were actually necessary. To add insult to injury, the railroads had become so powerful that they were too big to fail: their economic troubles and misdeeds fueled nationwide financial convulsions in 1873 and 1893. By 1884 the Northern Pacific's increasing debt and falling stock price forced Villard out, yet the company continued to grow while the public bore the costs of its losses.

All of this monopolistic behavior generated a backlash. In 1877 a series of wage cuts on the Baltimore & Ohio Railroad prompted a strike by workers that spread along its own route and then to other lines. In Washington Territory, anger focused on the Northern Pacific's ongoing control of 7.7 million acres. Congress had the authority to confiscate the land, which made this a powerful political issue for Democrats given that Republicans had largely been responsible for authorizing the original grants. In broadsides such as this, Democrats portrayed a cozy relationship between Republicans and these "soulless corporations." The map itself is designed to stoke the anger of voters by dramatically highlighting the vast areas held in reserve by the NP, even as its promised route remained unbuilt.

The Democrats used the broadside to charge the Republicans with recklessly squandering the public domain. Nowhere in the country, they argued, did "the ingenious, crooked, and devious railroad lines" control so much as in Washington Territory. In its 1884 platform the Democratic Party called for all lands that had been "improvidently granted to railroad corporations" by the Republican Party to be restored to the public domain. Though both parties courted the railroads, the political tactic paid off for the Democrats: for the first time in nearly thirty years, they won the White House, and by 1890 Congress had reclaimed 28 million acres in the West.

TO THE BRINK OF EXTINCTION

W. T. Hornaday and Henry Gannett, "Map Illustrating the Extermination of the American Bison," 1889

In 1886 the Smithsonian Institution sent its chief taxidermist to Montana to bring back specimens of the American bison. William Temple Hornaday was well suited to the task, for he had traveled to the ends of the earth to find exotic animals for his popular dioramas. But this assignment left him outraged and saddened, since in Montana he discovered a species that was nearly extinct. In his ensuing report, Hornaday documented this decline, but it was the small map tucked into its endpapers that packed the greatest punch. With the help of skilled geographer Henry Gannett and zoologist Joel Allen, Hornaday visualized the devastation of an animal that had once ranged across North America. His map shocked a public that had come to see the bison as the quintessential creature of the West and the embodiment of the frontier.

On the map, red numbers indicate the date by which the bison could no longer be found in any given geography. This starkly reminded the public that these herds once ranged from Mexico to the Mississippi Delta, and from east of the Allegheny Mountains to the northern reaches of Idaho. The contraction of bison herds can be traced to the Spanish introduction of horses to the Great Plains around 1700. Tribes that adopted the horse were able to roam across a much wider geographical area, but ironically that mobility made them more dependent upon the buffalo for survival. By the early nineteenth century, a growing demand for fur and hides led to more aggressive hunting practices. At the same time, settlers moving into the trans-Mississippi West further encroached upon bison rangelands. Even as buffalo became fixtures in the American vision of the West, more advanced firearms enabled scouts and hunters to kill entire herds at once.

Immediately after the Civil War, smaller herds of buffalo could still be seen roaming across the Great Plains. But the advent of the transcontinental lines divided the bison into a northern and a southern herd and sharply curtailed the grasslands on which they fed. Two years later, the construction of the Kansas branch of the Union Pacific Railroad reduced the southern herd even further. The climax of that slaughter occurred between 1870 and 1873, and Hornaday marked the pathetic remains of that southern herd with a blue caterpillar line.

To the north, the bison ranged across a much wider expanse, but there too were devasted after the Civil War. At mid-century, the Sioux were the largest and most powerful tribal culture of the northern plains, with a domain that extended from the Rocky Mountains in the West to Minnesota in the East, and from the Platte River north to the Yellowstone River. Within that vast region, they were sustained by millions of buffalo. But the contraction of bison herds correspondingly weakened the Sioux. To be sure, Sioux Indians seeking robes and hides were responsible for some of the decline of the buffalo. But far more consequential was the building of the Northern Pacific Railway in the 1880s. With green circles, Hornaday highlighted the few small bands that remained by 1889. The largest of these was a herd of 200, which was protected within the boundaries of the country's first national park at Yellowstone.

Hornaday's report exposed Americans to the dark side of western development, and sparked a discussion around conservation that flourished under Theodore Roosevelt and other influential leaders. The decline of the bison also had a far more immediate and profound effect upon the equestrian tribes of the plains. Some federal Indian agents even saw this destruction of the bison as a means to force natives onto the new reservation system. Just as Hornaday published his report, the federal government seized the remains of the Sioux territory and divided it into five smaller units. This was the final stage in a transformation of Sioux life, punctuated by violence on the Pine Ridge Reservation in South Dakota at the end of 1890. The Wounded Knee Massacre was the last major fight between the tribe and federal troops, concluding decades of armed conflict and forced relocation to reservations throughout the West. In 1910, Congress opened large areas of the Pine Ridge Reservation to non-Indian homesteaders. In this respect, Hornaday's map documents not only the decline of the bison but also the parallel confinement and destruction of the Plains tribes.

MAP
ILLUSTRATING
THE EXTERMINATION OF
THE AMERICAN BISON
PREPARED BY
W. T. HORNADAY.
EXPLANATION.
Boundary of the area once inhabited by the American Bison (Mainly after J. A. Allen.)
Approximate boundary between the area of desultory extirpation and that of systematic destruction for robes and hides
A Area of gradual extermination by desultory methods
B Area of wholesale slaughter by systematic methods
Range of two great herds in 1870
Range of the herds in 1880
Range of the scattered survivors of the southern herd in 1875, after the great slaughter of 1870-1873
Range of the northern herd in 1884, after the great slaughter of 1880-1883.
Dates in red figures represent the year of the Bison's extermination in the localities over which the figures are placed.
Figures in green represent the locality and number of wild Bison in existence January 1st 1889.
HUDSON BAY
CANADA
GULF OF MEXICO
BAFFIN BAY
DAVIS STRAIT
HUDSON STRAIT
BAHAMA ISLANDS
550
10
200
20
25
1880
1878
1850
1840
1883
1886
1870
1869
1832
1825
1867
1876
1830
1820
1837
1828
1795
1800
1730
1810
1760
1720

ARID REGION
OF THE
UNITED STATES
Showing Drainage Districts.

Scale:
100 50 25 0 50 100 200 300 STAT. MILES.

AN ALTERNATIVE VISION FOR THE AMERICAN WEST

John Wesley Powell, "Arid Region of the United States Showing Drainage Districts," 1890

With the Homestead Act as an incentive, Americans flooded into the Great Plains and the interior West after the Civil War. The 1890 Census counted over a million residents in Nebraska, and close to 1.5 million in Kansas, many of whom had been attracted by the persistent but false hope that rain follows the plow. One of the few to voice skepticism about the capacity of the West to support large-scale farming was John Wesley Powell, director of the United States Geological Survey. Powell had lost an arm fighting rebels at Shiloh, and thereafter led several western surveys and a heroic expedition down the Colorado River.

In the 1870s he began to articulate what generations of explorers, settlers, and Native Americans had long known: that the regular rainfall and humidity that characterized the eastern United States evaporated west of the one-hundredth meridian. In the West, precipitation concentrated in the high country and fell irregularly elsewhere. Outside the Pacific Northwest and a few other spots, the western half of the country could not be farmed unless it was irrigated. Without water, the land had no agricultural value.

Powell was no enemy of western settlement, and in fact optimistically believed that this vast arid region could be redeemed through the systematic management of water. To that end, in 1888 he undertook an ambitious survey of irrigation practices throughout the West, and then testified at length before Congress to warn of the dangerous trends that were already in place. Migrants assumed that streams flowed year-round, only to find themselves facing dry creeks and little rain. Competition over water led settlers to continuously move upstream into zones of elevation with limited potential for cultivation. Conflicts over water led to endless litigation between individuals, counties, and states. Unpredictable rainfall and recurrent drought led to the abandonment of homesteads, and sometimes entire communities.

In contrast, Powell detailed successful examples of irrigation throughout the arid West that could be scaled up to expand the region's potential. Without reservoirs, canals, and other techniques of redistributing water, he argued, the West could sustain neither settlement nor agriculture.

In this respect, Powell's observations were hardly controversial. Since the 1820s American schoolbooks and maps had labeled much of the West the "Great American Desert." If Powell's diagnosis of the problem was correct, however, his solution was far more difficult for Congress to accept. Through maps such as this, Powell proposed a West organized not around the logic of the grid, but around watersheds that he distinguished with brilliant color. He then divided those watersheds into regional communities, such as the Platte River basin. In Powell's view, only with the community control of water would individuals be forced to collectively determine the best use of this precious resource. This principle had long been practiced in the Mormon settlements of Utah and older Hispano communities of southern Colorado and northern New Mexico. The premise was that water was not a private right but a public good, and that local control and investment ensured the best outcome. In Powell's mind, this reorganization of water rights was a necessary remedy for a system where corporations and speculators had historically cornered the best land, resources, and water rights.

Yet Powell was testifying before a Congress that had created the very problem he sought to address. In the 1860s the federal government had made extraordinary land grants to the railroads in an effort to encourage the construction of transcontinental routes that might invite settlement, despite the aridity of these regions. By 1890 the newly organized Bureau of Indian Affairs had forcibly placed Native Americans on reservations shown here in order to liberate land for homesteaders. These and other federal policies were directly at odds with Powell's vision of planned growth. His own outsized confidence—some might say arrogance—did little to help his cause, and after his testimony Congress terminated his irrigation survey altogether.

Powell's vision proved too disruptive to existing patterns and practices. Western cities continued to grow and homesteaders continued to settle on marginal land. In fact, by promoting irrigation Powell inadvertently influenced the rise of California's agricultural empire as well as its subsequent water woes. Yet his map remains a challenging reminder of decisions made and paths not taken, and a very different vision for the American West.

MAPPING VICE IN SAN FRANCISCO

Willard Farwell et al., "Official Map of 'Chinatown' in San Francisco," in *The Chinese at Home and Abroad*, 1885

The completion of the transcontinental railroad in 1869 was regarded as a national triumph, and commemorated with a ceremonial golden spike that linked the Union Pacific with the Central Pacific at Promontory Point in Utah. None of it would have been possible without abundant low-wage labor: the Union Pacific was largely built by Irish immigrants and Union veterans, while Chinese immigrants were chiefly responsible for the dangerous construction of the Central Pacific through the Sierra Nevada Mountains.

When the railroad work ended, many Chinese immigrants settled throughout the West while others moved to San Francisco. In the severe economic depression of the 1870s, these immigrants became easy scapegoats and targets. Chinese workers—so desperately needed just a few years earlier—were accused of undercutting wages just as the Irish immigrants had been in the 1850s. The earliest anti-Chinese group formed in 1867, and by the 1870s a new political party was advocating immigration restriction as a way to protect white labor. Anti-Chinese riots throughout the west, alongside this political party, led to the nation's first attempt to prevent a particular ethnic group from immigrating. The Chinese Exclusion Act of 1882 halted the immigration of Chinese laborers and prohibited those already in the country from seeking naturalized citizenship.

In response to this widespread antagonism, more Chinese sought refuge in San Francisco. By the mid-1880s, they constituted one-tenth of the city's population, largely concentrated in a neighborhood of fifteen square blocks known as "Chinatown." There, too, they were greeted with harassment, largely in the form of ordinances designed to control everything from their business practices to their physical appearance. The city supervisors also launched an investigation into the living and working conditions in Chinatown, and their ensuing report captures the extraordinary racism that gripped San Francisco in the 1880s.

Billed as a public health effort to uncover the "unvarnished truth," the report described the Chinese "race" as "the rankest outgrowth of human degradation that can be found upon this continent." The supervisors presented an extensive catalogue of sanitary code violations, most of which amounted to inadequate plumbing and drainage. They described residents of Chinatown "living scarcely one degree" above waterfront rats, in unimaginably crowded conditions. Instead of nuclear families, the investigators found men sleeping in shifts in filthy lodging houses while women were widely enslaved as prostitutes. Such living conditions, the authors speculated, surely bred leprosy and drug addiction.

To drive publicity for the report, the authors commissioned an elaborate map that relentlessly focused on vice. Originally folded into the report, the map was then reissued in a much larger format that measured five feet wide, a portion of which is shown here. Chinese lodgings and businesses are marked in tan, gambling houses in pink, and opium dens in yellow. A special effort was made to distinguish Chinese prostitution (green) from white prostitution (blue), in order to highlight the growing demand that drew white women into the sex trade. Red marks Chinese "joss houses," or places of worship. The authors also noted that the map identified only the street level activity in each establishment, while the report went further in describing the sins committed in the labyrinthine world below.

The map is presented as an authoritative urban plan, quite similar to the contemporary Sanborn insurance maps that undergirded the growth of modern cities. Its elegance and precision hide its racially charged and sensationalistic profile of the Chinese population. By mapping vice—and the absence of nuclear families—the authors argued that there was something fundamentally alien about the Chinese. The overall message was clear: if the Chinese were unable to convert to Christianity, and if they continued to sow "immorality, vice, and disease," they must be expelled from the city. The map was just one example of the growing use of cartography to address social problems in the Gilded Age. In identifying the human geography of the city, the map anticipated the next two examples on pages 166 and 168, yet here the goal was segregation rather than assimilation.

The story of the Chinese in California has echoes throughout American history, where a market economy alternately demands, then rejects, low-cost immigrant labor. The limits on Chinese immigration served as a precedent for the more comprehensive 1924 restrictions. Yet Chinatown continued to thrive, especially after the liberalization of immigration restrictions in 1965: the neighborhood once derided as the vortex of filth in San Francisco remains one of the largest Chinatowns outside of Asia.

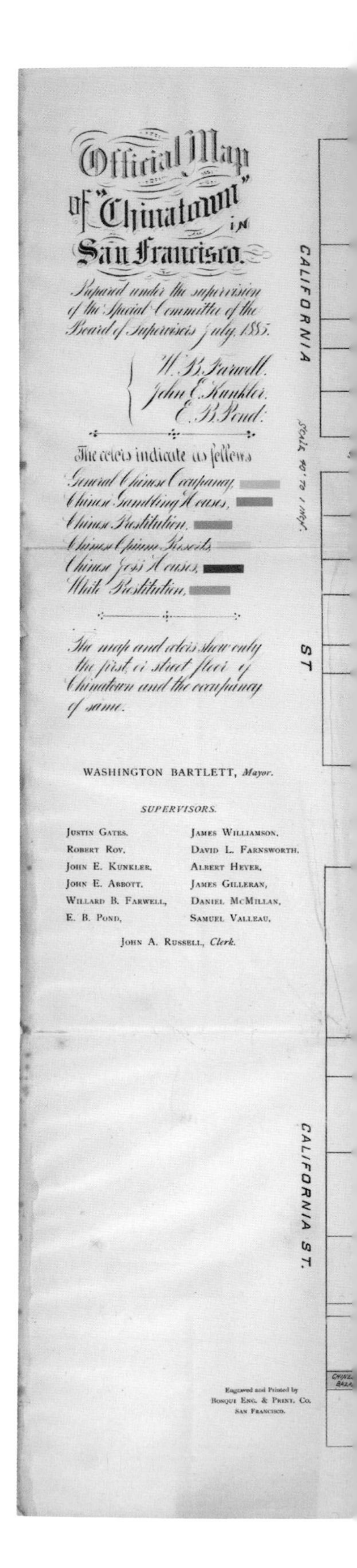

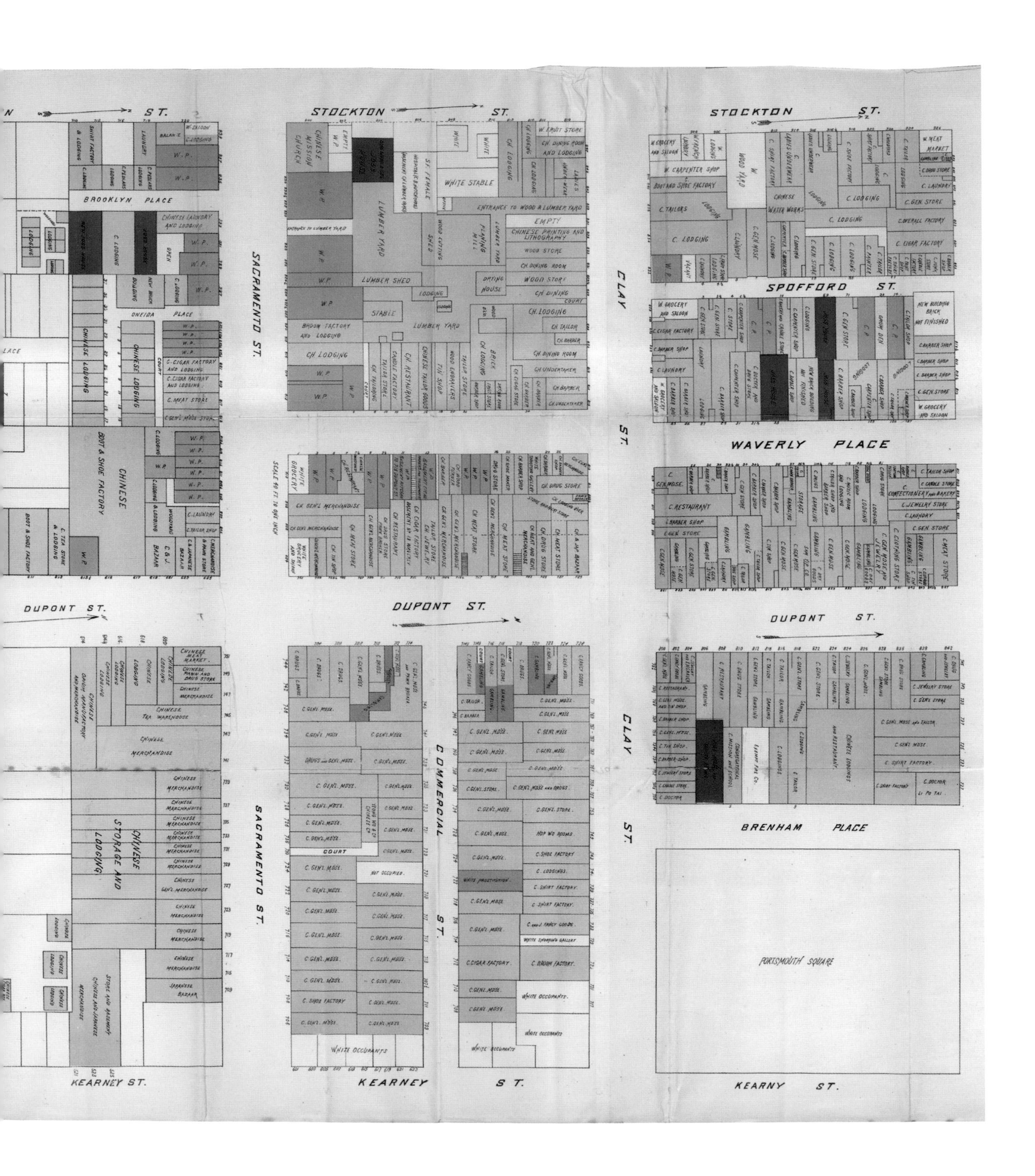
STOCKTON ST.
STOCKTON ST.
BROOKLYN PLACE
ONEIDA PLACE
SACRAMENTO ST.
CLAY ST.
SPOFFORD ST.
WAVERLY PLACE
DUPONT ST.
DUPONT ST.
DUPONT ST.
COMMERCIAL ST.
SACRAMENTO ST.
CLAY ST.
BRENHAM PLACE
PORTSMOUTH SQUARE
KEARNEY ST.
KEARNEY ST.
KEARNY ST.
SCALE 40 FT TO ONE INCH
CHINESE MISSION CHURCH
LUMBER YARD
WHITE STABLE
CHINESE BOOT & SHOE FACTORY
CHINESE STORAGE AND LODGING
CHINESE LODGING

THE HUMAN LANDSCAPE OF CHICAGO

Agnes Sinclair Holbrook and Florence Kelley, "Nationalities Map No.1," 1895

The intense antagonism toward the Chinese in San Francisco was one response to the massive urbanization of the late nineteenth century. Across the nation, cities struggled to assimilate immigrants, newly emancipated slaves, and desperate farmers seeking work. In Chicago, the social reformer Jane Addams found her calling by building a settlement house in the Nineteenth Ward, the city's poorest and most densely populated neighborhood. Working with Ellen Gates Starr, Addams attracted several talented women to Hull House, including Florence Kelley, a highly educated single mother of three from New York. Addams helped Kelley find work with the Illinois Bureau of Labor Statistics, where the latter quickly honed skills of data collection and analysis that would serve her well on the south side of Chicago.

Along with her fellow settlement house workers, Kelley focused on the conditions of the Nineteenth Ward in order to understand urban poverty more generally. She turned to maps in order to publicize the conditions of a slum that more fortunate Chicagoans would never see firsthand. Armed with extensive data gathered by the women of Hull House, she adapted existing survey maps provided by Samuel Greeley to profile the wages and ethnic backgrounds of its residents. Through a kaleidoscope of color, Kelley presented an inventory of the conditions in this diverse and struggling neighborhood.

The report that accompanied these maps reads like a trenchant indictment of late nineteenth-century capitalism. From their study of wages, the reformers concluded that poverty was not an individual failing but rather the outgrowth of a system where pay was persistently lowered by new waves of labor. Immigrants, blacks, women, and children were all used to drive down wages. In an era when union representation remained limited to a few white male artisans, the poor had little choice but to accept available jobs. This fueled the rise of sweatshops and child labor, both of which Kelley sought to regulate through the Illinois legislature.

But beyond basic workplace safety measures, the authors of the Hull House maps did not advocate specific solutions. Immersed in the neighborhood, they focused more on publicizing poverty than on engineering social policy. At the time, the country as a whole was engrossed in a debate over inequality and its remedies. In the 1870s the Yale scholar William Graham Sumner forcefully argued that any effort to ameliorate the conditions of the poor would only perpetuate harm. His brand of social Darwinism raised the question of the moment: What—if anything—did the nation owe to its most vulnerable citizens? Hull House reformers answered the question in a very different way through their maps. While they did not endorse a specific remedy, their decision to map wages and ethnicity reflected their belief that one could, through systematic investigation, determine both the sources of poverty and its remedies.

The female researchers at Hull House were committed, educated, and determined to make a difference long before they were permitted to vote, run for office, or occupy positions in universities. But their work in the settlement house movement also reflected contemporary attitudes. Along the right edge of the map they categorize Irish immigrants apart from "English-residents," reasoning that the former were "so distinct" in character as to necessitate a separation. African Americans were similarly separated out as distinct from "English-speaking" residents, underscoring the more pervasive discrimination that they faced. Moreover, the maps conceal as much as they reveal. Though they identify the ethnic identity of individual residences, little indication is given as to the density of the population, a crucial point in light of the grave concerns about slum congestion.

In both San Francisco and Chicago, city leaders and reformers mapped immigrant neighborhoods as a way of exposing their plight. But while the map on the previous page was designed to blame the Chinese, this map frames immigrants as victims of a much larger capitalist system.

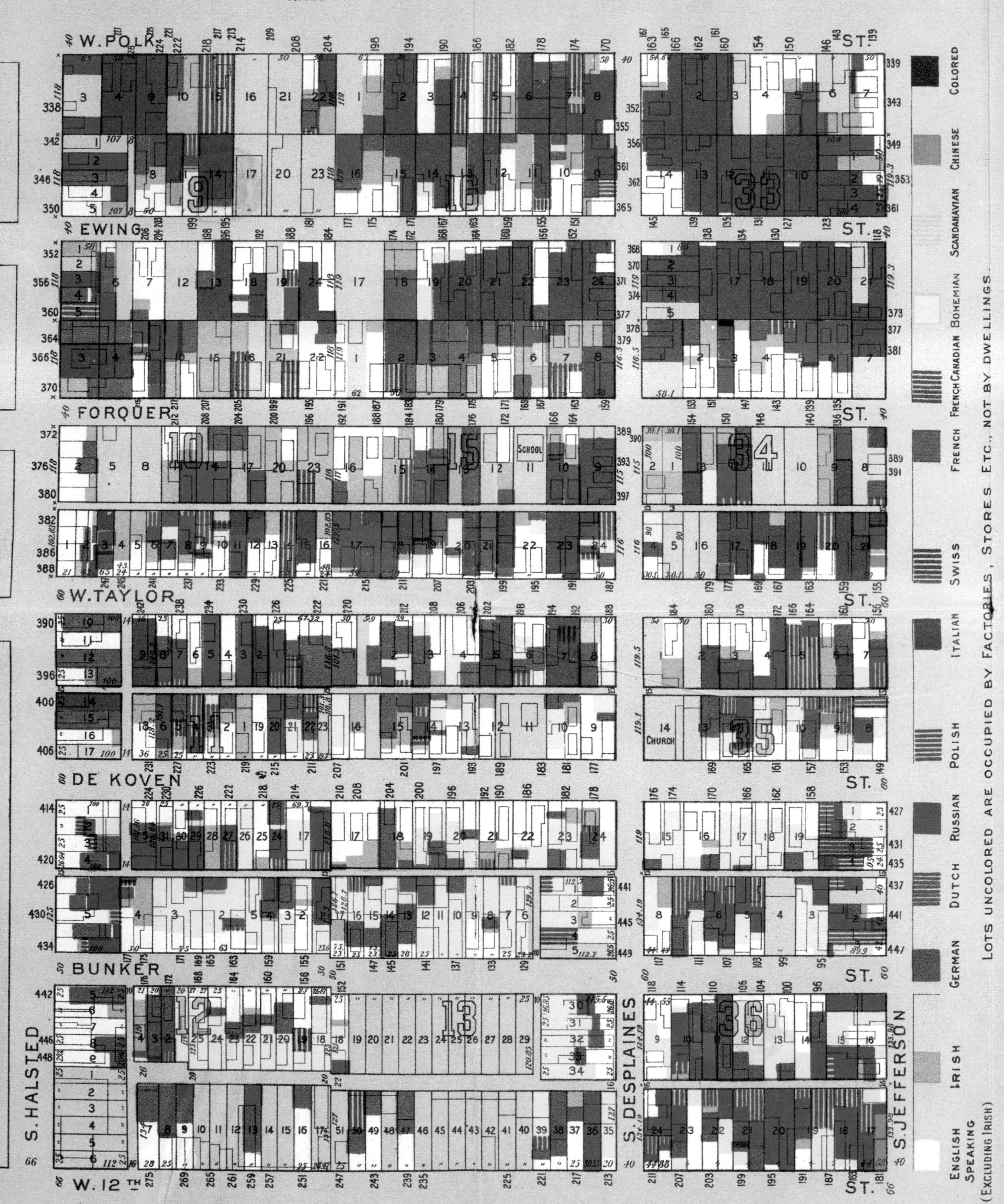

(NORTH)
NATIONALITIES MAP No.1.- POLK STREET TO TWELFTH,
HALSTED STREET TO JEFFERSON, CHICAGO.
W. POLK ST.
EWING ST.
FORQUER ST.
W. TAYLOR ST.
DE KOVEN ST.
BUNKER ST.
W. 12TH ST.
S. HALSTED
S. DESPLAINES
S. JEFFERSON
SCHOOL
CHURCH
COLORED
CHINESE
SCANDANAVIAN
BOHEMIAN
FRENCH CANADIAN
FRENCH
SWISS
ITALIAN
POLISH
RUSSIAN
DUTCH
GERMAN
IRISH
ENGLISH SPEAKING (EXCLUDING IRISH)
LOTS UNCOLORED ARE OCCUPIED BY FACTORIES, STORES ETC., NOT BY DWELLINGS.
(SOUTH)

RACE AND THE LIMITS OF MOBILITY

W. E. B. DuBois, "The Seventh Ward of Philadelphia," in *The Philadelphia Negro*, 1899

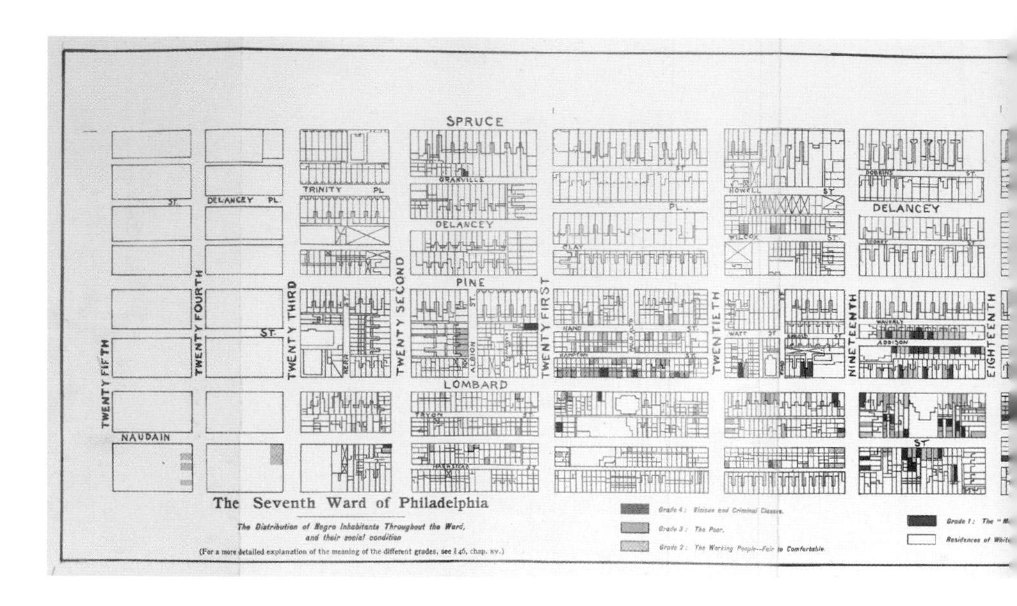

At the turn of the twentieth century, Philadelphia had the largest and oldest African American urban population in the North. Most of the 40,000 blacks residing in the city were concentrated in a long narrow strip that stretched from Seventh Street to the Schuylkill River in the narrow band from Spruce to South streets. Despite this substantial presence, blacks remained even more segregated than the waves of immigrants who arrived throughout the late nineteenth century. One of the leading intellectuals of the era, W. E. B. DuBois, sought to investigate this curious pattern.

DuBois was raised in Vermont, and graduated from Fisk University before earning his graduate degree at Harvard. While in Cambridge he spent time with the intellectual luminaries William James and Charles Sanders Peirce, who exposed him to new ideas in philosophy, sociology, and history. They also insisted that it was culture, and not heredity or race, that defined individuals. And if one's culture could be changed, so too could one's behavior and identity.

Despite his extraordinary education, DuBois faced tenacious discrimination and struggled to gain stable academic work in his field of sociology. Upon landing a position at the University of Pennsylvania, he immediately launched an ambitious sociological study of Philadelphia's black population, which remained mired in poverty, crime, and unemployment. For the next two years, his research team conducted one of the most thorough inventories of an African American community to date, investigating everything from work and education to the presence of churches and family structure.

Social reformers considered themselves to be neutral arbiters whose education equipped them to see problems with a reasoned, detached, and morally enlightened eye. As DuBois himself wrote, "we must study, we must investigate, we must attempt to solve." In this framework, maps were exciting tools of observation that could replace misinformation with a more objective picture. Yet the map reveals DuBois' own assumptions. He mapped African Americans according to their social and economic condition, classifying the "vicious and criminal" classes in black, the poor and working classes in blue and green respectively, and the middle and upper classes in red. DuBois' profile of the black neighborhood showed considerable variation, with some rough blocks very near prosperous ones. But even economically comfortable blacks, he found, were invisible in a segregated city.

Like all progressive reformers, DuBois had values of his own, and his study is full of admonitions to both the white and the black communities. He held African Americans responsible for a significant amount of the city's crime, which constituted "a menace to a civilized people." He stressed an ethic of thrift and industry, and commented extensively on what he saw as the profligate and immoral ways of lower-class blacks, who shunned traditional family life and middle-class values. It was incumbent on the white community, he wrote, to examine its own racism and bigotry. But it was the "talented tenth" whom he addressed most pointedly, advocating an intellectual and moral aristocracy of black leaders that would challenge white perceptions and raise up all African Americans.

The larger importance of DuBois study—and his map—was the emphasis placed on the malleability of society and the importance of circumstance. Black poverty, he insisted, had less to do with innate inferiority than the persistent segregation and discrimination that inhibited mobility. If such arguments seem patently obvious to us today, it is because DuBois brought a more detached lens to the study of race. Despite his own assumptions, he insisted that both racial attitudes and behaviors had been learned, and could therefore be changed.

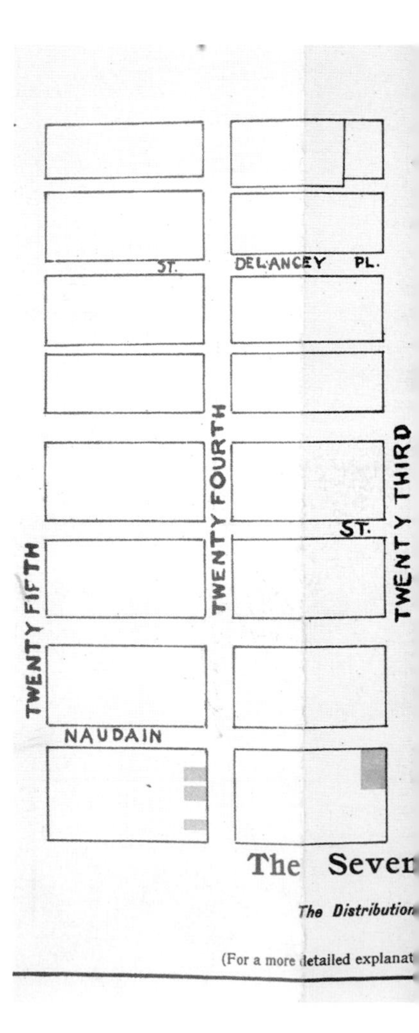

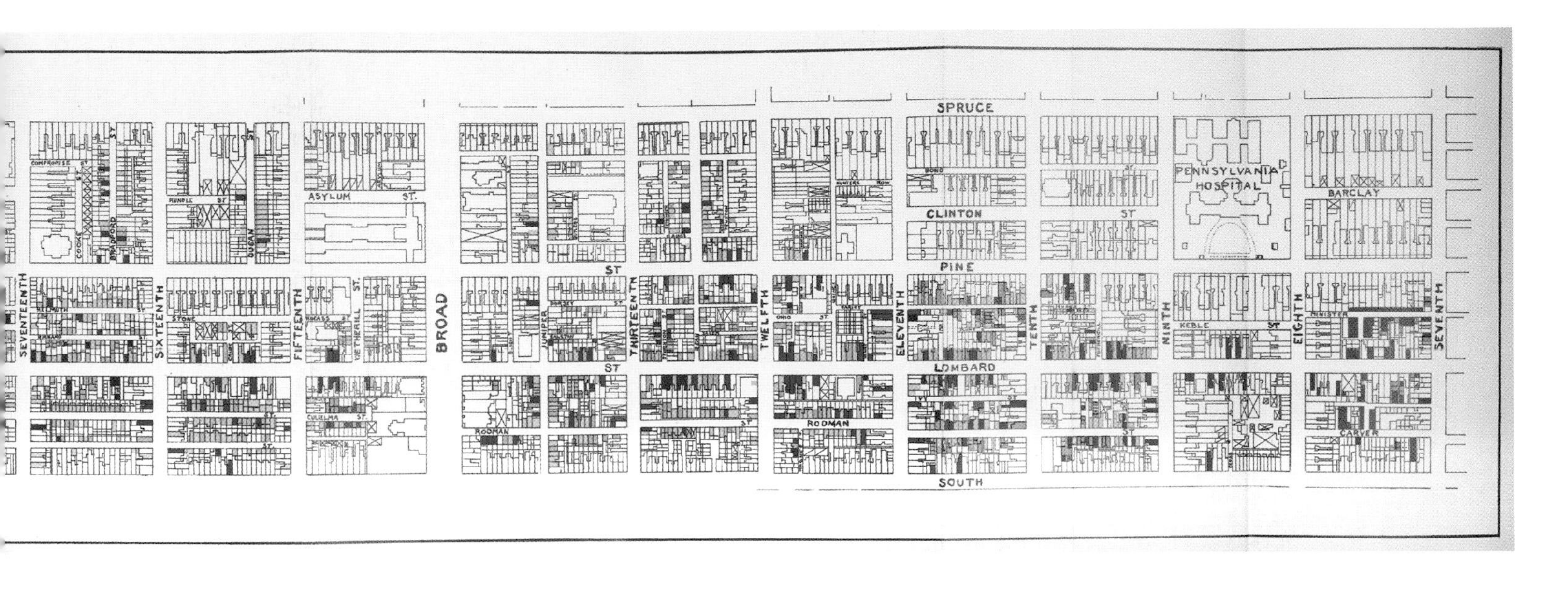
SPRUCE
PENNSYLVANIA HOSPITAL
BARCLAY
ASYLUM ST.
CLINTON ST
PINE
LOMBARD
RODMAN
SOUTH
SEVENTEENTH
SIXTEENTH
FIFTEENTH
BROAD
THIRTEENTH
TWELFTH
ELEVENTH
TENTH
NINTH
EIGHTH
SEVENTH
KEBLE ST
CARVER

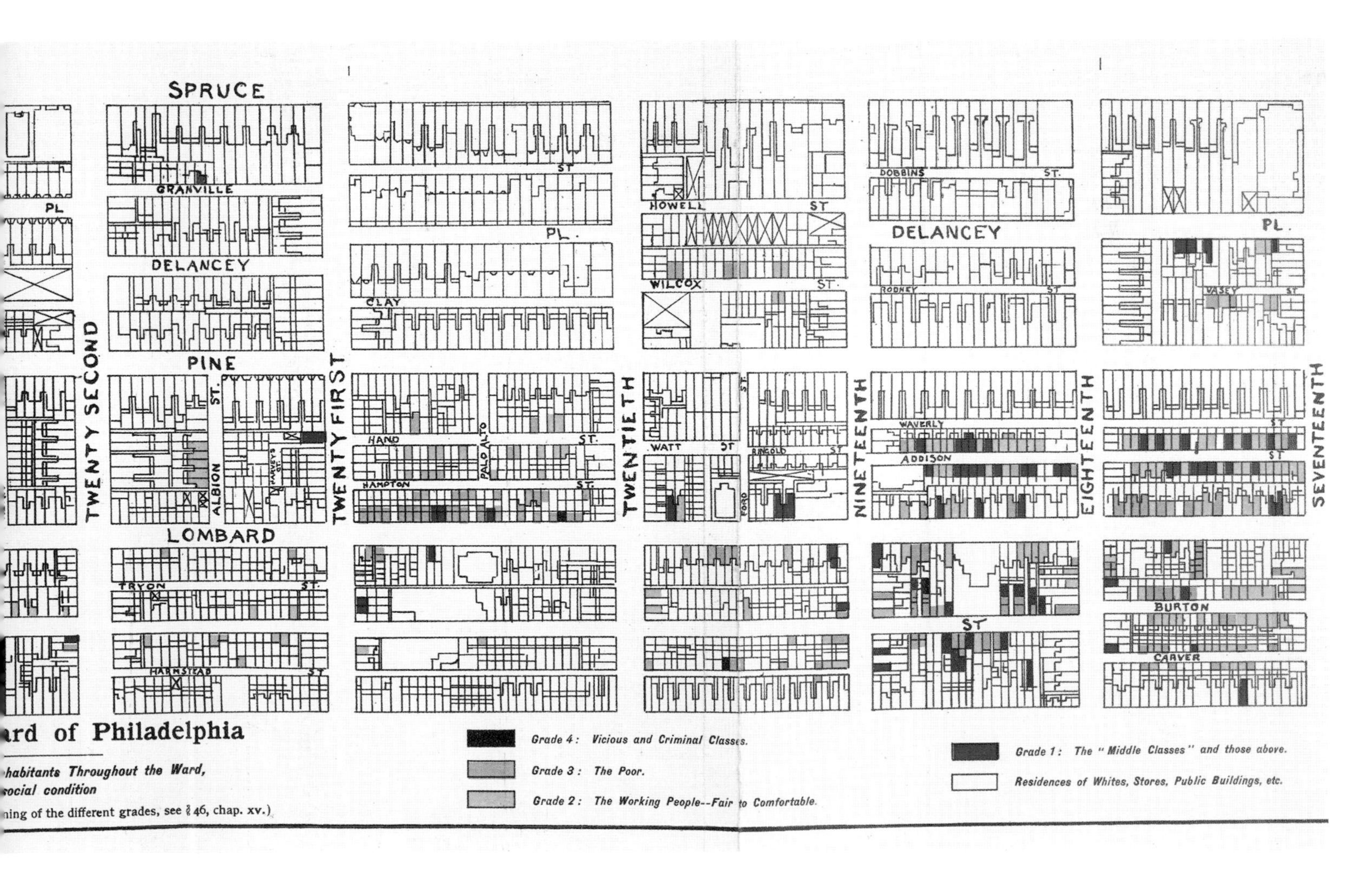
SPRUCE
GRANVILLE
PL
DELANCEY
PINE
LOMBARD
TRYON ST.
TWENTY SECOND
ALBION
TWENTY FIRST
PL.
CLAY
HAND ST.
HAMPTON ST.
PALO ALTO
TWENTIETH
HOWELL ST
WILCOX ST.
WATT ST
RINGOLD
NINETEENTH
DOBBINS ST.
DELANCEY
RODNEY ST
WAVERLY
ADDISON
ST
EIGHTEENTH
PL.
VASEY ST
ST
ST
BURTON
CARVER
SEVENTEENTH
ard of Philadelphia
habitants Throughout the Ward,
social condition
ning of the different grades, see § 46, chap. xv.)
Grade 4 : Vicious and Criminal Classes.
Grade 3 : The Poor.
Grade 2 : The Working People--Fair to Comfortable.
Grade 1 : The "Middle Classes" and those above.
Residences of Whites, Stores, Public Buildings, etc.

AN AMERICAN EMPIRE?

Rand McNally, map of US acquisitions, 1904

The expansion of American trade and military activity abroad in the 1890s prompted a sustained debate about the nation's foreign commitments. To be sure, the nation had been entangled with affairs beyond its borders since its birth. With the Louisiana Purchase, the war against Mexico, and the nineteenth-century Indian wars, the United States had engaged with foreign powers and extended its territorial domain. But in each of these cases, this growth had been limited to North America. By contrast, in the 1890s the nation began to set its sights on more distant arenas.

Much of this reorientation was brought by the Spanish–American War. The name Americans gave to the conflict is itself revealing, for it began with Cuba's fight for independence from Spain. When hostilities broke out between Cubans and Spanish colonizers in 1898, the US Navy sent the battleship USS *Maine* to Havana harbor to protect Americans. An accidental explosion destroyed the ship and killed hundreds of servicemen aboard. American tabloids relentlessly framed this as a national insult, prompting public outrage and the cry "Remember the Maine, to hell with Spain!" In April, President William McKinley requested that Congress declare war. The "splendid little war" that ensued was just long enough for Theodore Roosevelt to raise a company of volunteers and—very publicly and dramatically—charge up San Juan Hill in Cuba. When Spain surrendered that summer, the United States inherited a far-flung collection of colonies from the Pacific to the Caribbean.

The US officially granted Cuba independence, though it retained a right to intervene in its affairs until 1934 (the US still leases a portion of Guantánamo Bay). The situation in the Philippines was far more complicated, for the departure of the Spanish led to a brutal and protracted war with the United States. This ugly conflict prompted many Americans—including Jane Addams, Andrew Carnegie, Mark Twain, and former presidents Grover Cleveland and Benjamin Harrison—to declare themselves "anti-imperialists." Though they had diverse motives and agendas, all agreed that it was unacceptable for the US to intervene in the Philippines, much less to annex the islands. By the time the war ended in 1902, 4,000 Americans and 20,000 Filipinos had been killed in combat, and many times that number had died from disease and starvation.

Though the anti-imperialists were vocal and articulate, most Americans celebrated and supported the nation's activist foreign policy. Before his presidency, Roosevelt pushed for an enlarged navy to protect these new overseas territories. The nation's largest mapmaker, Rand McNally, even redrew its national map to incorporate these new holdings. Like contemporary maps of the British empire, this one asserted overseas acquisitions as part of the nation's long history of expansion that began with the thirteen colonies and continued westward. Rather than convincing its readers to support the Spanish–American War, Rand McNally simply framed the territories as part of a national evolution. Through the map, a divisive war was presented as the most recent stage in the country's history.

The enthusiasm for territorial expansion extended to the inclusion of Alaska on the redesigned national map. Though the purchase of that territory from Russia had been negotiated by Secretary of State William Seward in 1867, for decades skeptics considered it an empty frozen wasteland. Only after the discovery of gold in the Yukon did Alaska—formerly derided as "Seward's Folly"—regularly appear on the map of the United States.

Rand McNally's reconfigured map of the nation proved immensely popular, and was reproduced in contemporary atlases and textbooks as an announcement of the nation's arrival on the world stage. The larger historical question is whether these commitments abroad were a break or a continuation with the past. Ten years before he was elected president, Woodrow Wilson urged his country to take its civilizing mission seriously. He considered the Spanish–American War a turning point not for Cuba and the Philippines but for the United States, a country poised to share its hard-won wisdom with peoples emerging from the yoke of imperialism at the dawn of a new century. Like so many of his contemporaries, Wilson saw his country as a source of benevolence and uplift, motivated by idealism rather than by commercial or political gain. This combination of sincerity and arrogance would shape American foreign policy for the next century.

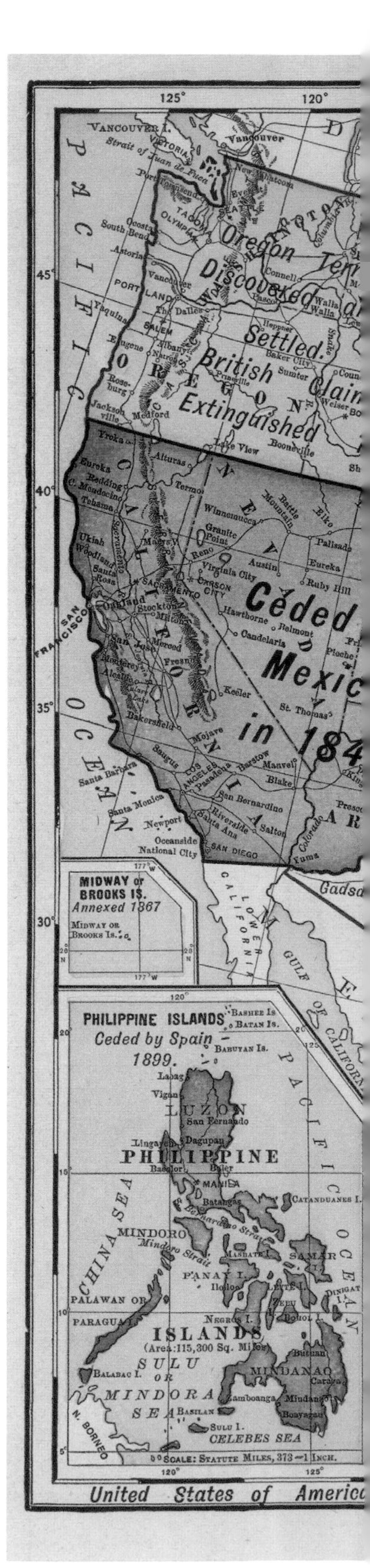

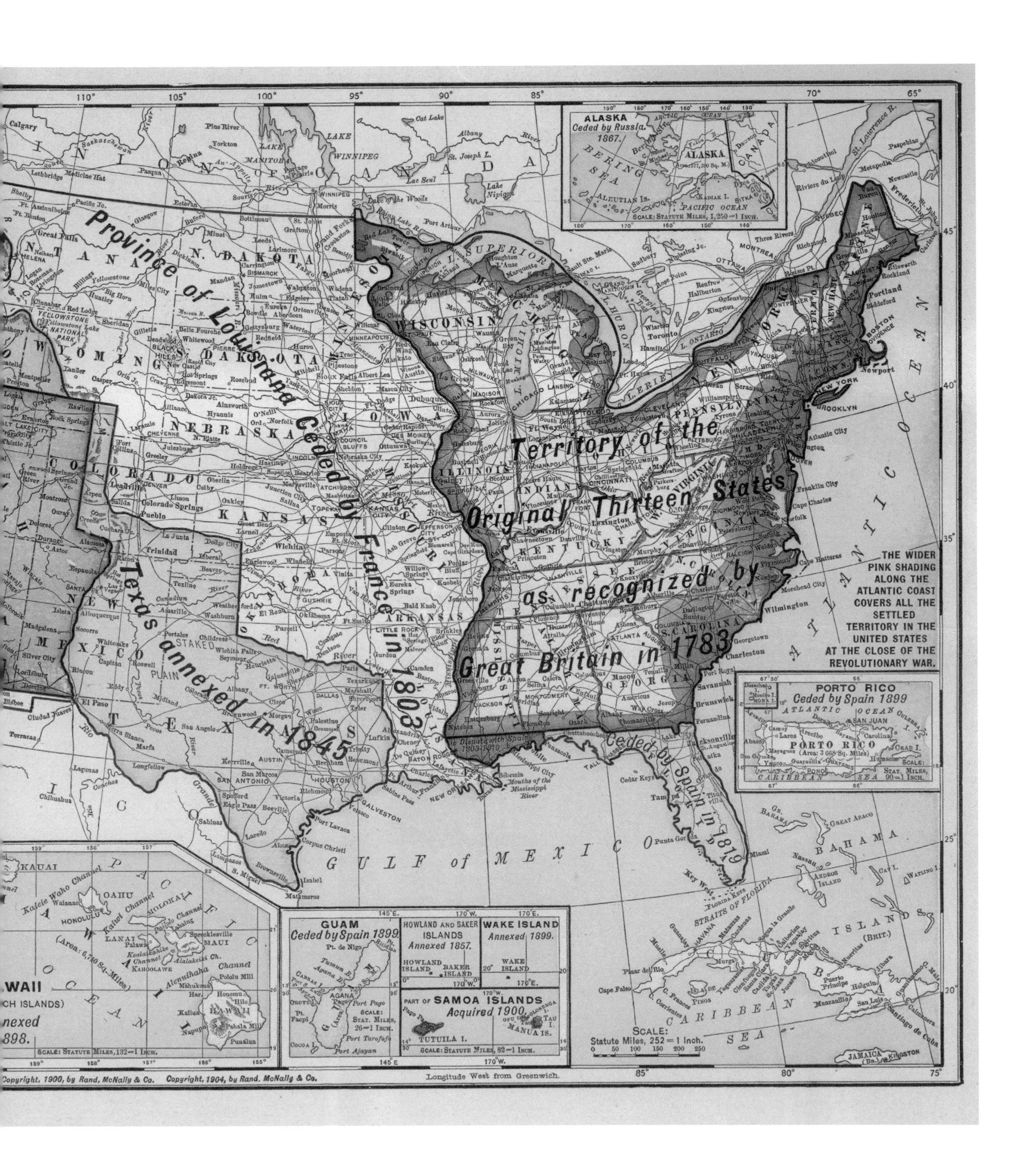
Province of Louisiana Ceded by France in 1803
Texas annexed in 1845
Territory of the Original Thirteen States as recognized by Great Britain in 1783
Ceded by Spain in 1819
ALASKA Ceded by Russia. 1867.
PORTO RICO Ceded by Spain 1899
GUAM Ceded by Spain 1899.
HOWLAND AND BAKER ISLANDS Annexed 1857.
WAKE ISLAND Annexed 1899.
PART OF SAMOA ISLANDS Acquired 1900.
THE WIDER PINK SHADING ALONG THE ATLANTIC COAST COVERS ALL THE SETTLED TERRITORY IN THE UNITED STATES AT THE CLOSE OF THE REVOLUTIONARY WAR.
GULF of MEXICO
ATLANTIC OCEAN
Copyright, 1900, by Rand, McNally & Co. Copyright, 1904, by Rand, McNally & Co.
Longitude West from Greenwich.

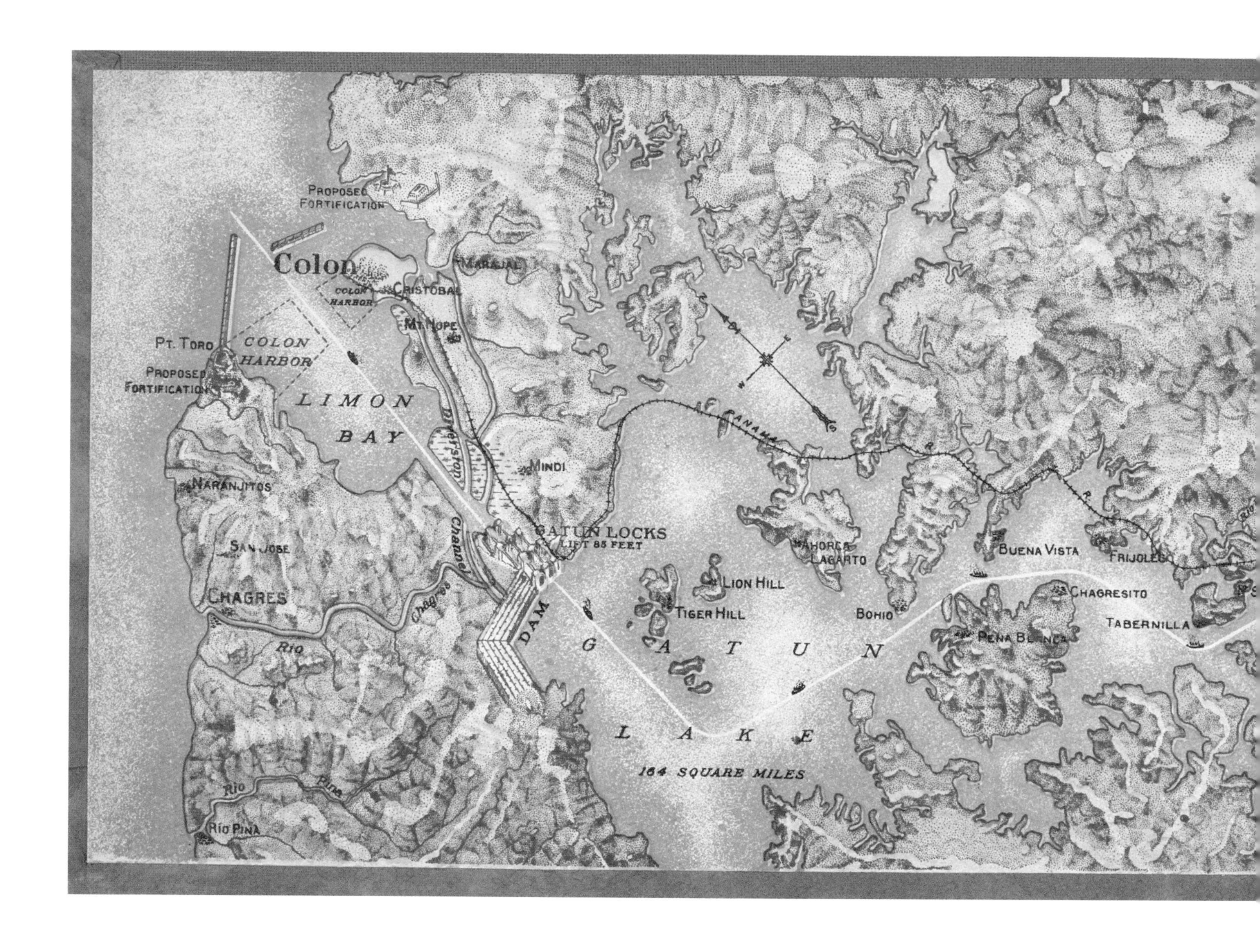

BALBOA'S DREAM REALIZED

C. P. Gray, "Aeronautical View of the Panama Canal," 1911

The Panama Canal remains one of the most remarkable engineering feats in history, the realization of a dream that stretched back to the European discovery of America itself. In 1492 Christopher Columbus sailed west to find a route to Asia, and subsequent explorers sought a similar passage across North America. Vasco Núñez de Balboa conclusively disproved that any such strait existed, but in 1513 came across a narrow isthmus just forty miles wide that separated the Atlantic and the Pacific oceans.

The discovery of gold in California in 1848 unleashed a global wave of migration to San Francisco, once again demonstrating the need for a passage between the two oceans. The French began to build the canal at Panama in 1881, but seemed doomed from the start. Ignorance of the climate led to yellow fever and malaria, while heavy rains foiled the effort to excavate a sea-level canal. After thirteen years of frustration and thousands of deaths, the French abandoned their effort at Panama in 1894.

The timing was crucial, for in the 1890s the United States actively extended its military and commercial

reach, as shown on the previous page. This new posture led McKinley—and the Senate—to endorse a canal to advance commerce and expand the power of the US Navy. In 1903, Panamanian nationalists declared independence from Colombia, and then immediately signed a treaty that gave the United States extensive control of a proposed canal zone in exchange for protection of their new country. This agreement gave Americans the access they needed to undertake such a major project, though it created problems throughout the twentieth century.

After the route was determined, engineers set to work. The Panama Railway removed mountains of dirt that had contributed to earlier landslides. Gargantuan quantities of steel, cement, and machinery were shipped over vast distances to the site. Thousands of laborers came to work under difficult, dangerous, and segregated conditions. Ten years later, the canal was complete. In August 1914, just days after the outbreak of war in Europe, the Panama Canal opened for business. The distance from New York to San Francisco was reduced by 8,000 miles, and travel time was cut in half. It was the canal, rather than the transcontinental railroad, that truly integrated US coasts and ports. Americans celebrated its completion with souvenirs such as this view made by the mapmaker and illustrator C. P. Gray for the Central Novelty toy company.

Votes for Women a Success

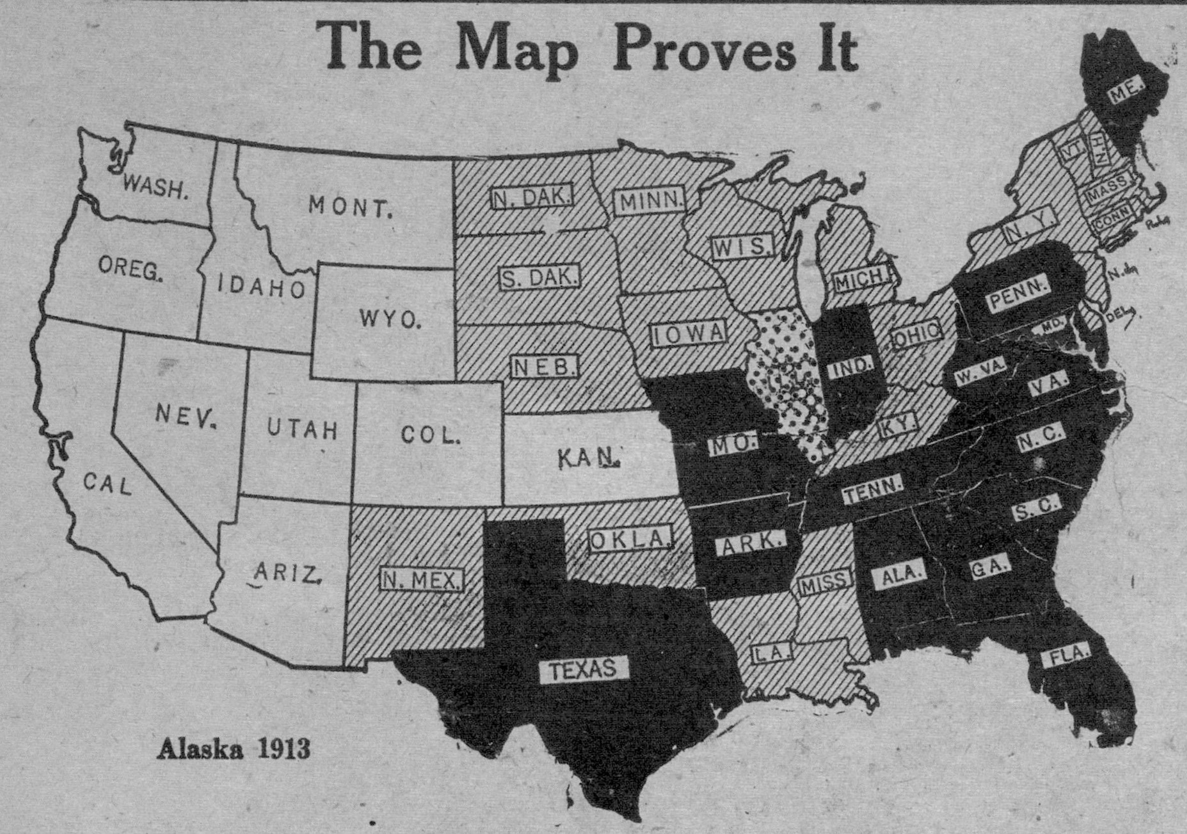

WHITE STATES; Full Suffrage SHADED STATES: Taxation, Bond or School Suffrage
DOTTED STATE: Presidential, Partial County and State, Municipal Suffrage BLACK STATES: No Suffrage

SUFFRAGE GRANTED

1869—WYOMING
1893—COLORADO
1896—UTAH
1896—IDAHO
1910—WASHINGTON
1911—CALIFORNIA
1912—OREGON
1912—ARIZONA
1912—KANSAS
1913—ILLINOIS
1913—ALASKA
1914—MONTANA
1914—NEVADA

Would any of these States have adopted EQUAL SUFFRAGE if it had been a failure just across the Border?

NATIONAL WOMAN SUFFRAGE PUBLISHING COMPANY, Inc
PUBLISHERS FOR THE
NATIONAL AMERICAN WOMAN SUFFRAGE ASSOCIATION
505 Fifth Avenue New York City

BEFORE THE NINETEENTH AMENDMENT

National American Woman Suffrage Association, "Votes for Women a Success: The Map Proves It," circa 1914

Among the most commonly reproduced maps in American history is this depiction of the progress toward universal women's suffrage. First designed a decade before the Nineteenth Amendment, it appeared in multiple forms and styles, and was constantly updated to mark the expansion of voting rights in individual states. Its widespread use mirrors the ongoing local, state, and national activism that was necessary for women to win the right to vote.

The American movement for woman suffrage debuted at the Seneca Falls Convention of 1848. Among the supporters was Frederick Douglass, who also pressed for the guarantee of black male suffrage through the Fifteenth Amendment after the Civil War. That victory proved ephemeral, however, when the end of Reconstruction brought sustained and violent segregation. Through poll taxes, literacy clauses, voter intimidation, and other means, black men lost the right to vote in every Southern state by the turn of the century, and would not regain it until the Voting Rights Act of 1965.

As black men lost the vote in the South, white women began to win it out West. Wyoming, Colorado, Utah, and Idaho all passed female suffrage by 1896, but thereafter the movement stalled. In response, suffragists began to adopt new techniques to amplify their message. In 1907 *Appleton's Magazine* published a simple black-and-white map reporting the status of suffrage laws around the country. Suffrage groups continuously adapted, updated, and reprinted the map to reflect changing state laws. Through its incremental changes, the map chronicled the rising tide of support for suffrage. This 1914 edition used white to denote the states where women had full voting rights, and shaded those with limited rights. Recent victories in Illinois, New York, and Pennsylvania brought the issue to a tipping point, leaving a minority of states still resisting the change.

The key to the map's enduring power was its simple and straightforward design, which enabled activists around the country to adapt it to local conditions and purposes. From 1907 until the ratification of the Nineteenth Amendment in 1919, various iterations of the map appeared across the country on billboards, and in marches and parades. Suffragists also took a cue from the emerging profession of advertising by printing the map in baseball programs and college yearbooks, and on paper fans, drinking glasses, and calendars. Through their efforts, the map became ubiquitous and familiar, a constant reminder of the movement's progress. Here was a message of reform tailored to the advertising age: bold, engaging, and with a clear message.

With its stark geopolitical message, the suffrage map recalls those produced in the 1850s to warn against the westward expansion of slavery (page 140). In both, the map wields a political message by showing the potential for change. In the earlier map, geography is a source of dread that marks the growing threat of slavery, while here geography is a source of optimism. As this broadside boasted, "Would any of these States have adopted equal suffrage if it had been a failure just across the Border?" The progress toward equality seemed to be marching from west to east, inverting assumptions about the westward march of civilization.

Ironically, however, opponents of women's suffrage used the same map to point out that the vast majority of Americans lived in states that had yet to extend the vote to women. Furthermore, they argued that the map showed that states with equal suffrage laws were geographical outliers, located not in the densely settled and established East but in the wild (and perhaps less civilized) West. A map adjusted for population, they pointed out, would have actually demonstrated how unpopular female suffrage really was. Both opponents and supporters of women's suffrage used persuasive maps to advance their cause, and in the process demonstrated the power of cartography itself.

7. 1914–1940

Prosperity, Depression, and Reform

In light of the byzantine alliances and swollen arsenals that had been building across Europe for years, the outbreak of war in the summer of 1914 seemed almost inevitable. President Woodrow Wilson spent the next two years urging Americans to remain impartial in both thought and action, and for good reason. While the United States had longstanding ties with England and France, millions of immigrants at the turn of the century held loyalties of their own that complicated any immediate support for the Allies.

The German sinking of the *Lusitania* in 1915 horrified Americans, and prompted at least one to join the British Army on the Western Front (page 178). Yet Wilson steadfastly maintained an official policy of neutrality. The German decision to resume submarine warfare in early 1917 was partly a response to Wilson's hypocrisy in claiming neutrality while the United States deepened its trade links with the Allies. Once the United States declared war on Germany in April, the president authorized a bold and unprecedented drive to generate support for the Allies through its Committee on Public Information (CPI). In countless pamphlets, maps, newsreels, posters, lectures, and press releases, the CPI valorized the Allied cause and demonized the German enemy. The map on page 180 is one of the most widely circulated examples of that elaborate propaganda campaign.

Americans made up less than 1 percent of the 18 million killed in the war, yet World War I indelibly shaped the nation's future. Wartime mobilization demonstrated the power of the state to coordinate production, regulate industry and labor, control transportation, and expand taxation. This mobilization extended to all areas of American life: just as the CPI fostered patriotism and support for the Allies, Congress outlawed any criticism of the war or the draft. This crackdown on dissent created a climate of fear that was compounded by the harrowing Spanish flu epidemic of 1918 (page 182). The training and deployment of troops actually accelerated the spread of the disease, which killed far more people than died in the entire war.

To frame the United States' mission in the war, Wilson proposed a new world order grounded in democracy, open trade, freedom of the seas, self-determination, and an end to secret alliances. European leaders greeted this vision with skepticism, noting more than a little arrogance in Wilson's determination to instruct the world in freedom and democracy. Yet Wilson's ambitious attempt in the "Fourteen Points" to align these ideals with national interests became an abiding theme in American foreign policy.

More immediately, World War I ushered in a nationwide ban on the production and sale of alcohol, and stimulated a tremendous demand for labor in Northern industry. With wartime curbs on immigration, that demand was largely met by African Americans: from 1910 to 1920, 500,000 blacks left the South for opportunities in New York, Philadelphia, Chicago, St. Louis, Detroit, Pittsburgh, and Cleveland. In the 1920s that number reached 1 million. The mass migration of blacks into Harlem generated a rich culture that attracted some of the leading musicians, artists, and writers of the interwar period (page 190). But urban density also brought new forms of crime and delinquency, which social scientists energetically investigated through maps (page 188).

The economic activity jump-started by the war was further accelerated by aviation, electronic, and communication technologies. Refrigeration, for example, brought about an era of competition and consolidation in the American meatpacking industry (page 184). Radio and telephone inventions integrated the population in entirely unexpected ways. But the most consequential of these innovations was the automobile, which proved to be far more than a means of transportation. The automobile changed the organization of work with the rise of Henry Ford's assembly line. It changed the relationship of the individual to the landscape by liberating travelers from rail routes and schedules. And it facilitated a new kind of "city" that was organized less around a dense urban core than its periphery.

The automobile in turn generated new products and services. With increased highway travel came a market not just for glass, rubber, and metal but also for billboards, the motor lodge, and diners. Even during the Great Depression, Americans hit the road in startling numbers. They sought out spots touted as cultural destinations and even obligations of citizenship, such as national parks, state capitals, and the "real" landscapes that lay beyond the main highways (page 196). The road trip quickly became an essential feature of American life, even a rite of passage. Yet if automobile travel was a hallmark of American culture, it was also one experienced in radically different ways by African Americans in the era of segregation (page 198).

The roads that knit this sprawling nation closer together also brought more homogeneous forms of consumption, such as chain restaurants, stores, service stations, and hotels. This emerging mass culture was exemplified by motion pictures. By the 1920s the American film industry was dominated by Hollywood, where studios had migrated to take advantage of California's congenial climate and diverse geography (page 194). With an ample supply of water imported (if not stolen) from the Sierra Nevada mountains, Los Angeles was poised for unlimited growth (page 186). Little surprise, then, that after World War II Southern California became home not just to Disneyland but also to the defense industry.

The prosperity of American cities in the 1920s also highlighted a growing cultural divide between an urban sensibility and its rural counterpart that was on full display in the Scopes Trial of 1925. The entire nation, it seemed, followed the prosecution of a young instructor who defied the law against teaching evolution in Dayton, Tennessee. Daniel Wallingford slyly caricatured these contemporary regional stereotypes, taking aim at the parochialism not of the country but rather of the city (page 192). While Wallingford pointedly skewered the self-centered attitude of New Yorkers, his map underscores the more general human tendency to see the world through our individual and geographically determined experience.

The economic prosperity of the 1920s was highly dependent upon the automobile. That uneven—and unstable—growth became clear once the evaporation of consumer demand, the failure of banks, and soaring unemployment brought the economy to a standstill in the Great Depression. President Franklin D. Roosevelt had little sense of how to end the crisis when he was inaugurated in early 1933, but he quickly convened Congress to consider a flurry of legislation proposing everything from bank regulation to direct relief. At the center of this legislation was the Tennessee Valley Authority (TVA), an ambitious project designed to bring electricity to the poor and rural South (page 202). The TVA generated both widespread excitement and lasting opposition to Roosevelt's New Deal principles of planning and limited government spending. It was also a sign of things to come, for the largest expenditure of the New Deal was on public works: roads, dams, airports, bridges, and housing projects which put Americans to work and built national infrastructure.

From public works and the insurance of bank deposits to the expansion of labor protections and industry regulations, the New Deal affected every American. Among the many programs of this reform era was the Federal Housing Administration, which sought to stabilize the housing market by buttressing the mortgage and credit business. As shown on page 200, that program enlisted local housing authorities to assess the credit worthiness of different neighborhoods through "security" maps that had lasting, and sometimes pernicious, effects. There is no doubt that the New Deal failed to end the depression, but by erecting a limited welfare state it fundamentally realigned the relationship between individuals and their government.

OVER THE TOP

Arthur Guy Empey, "Diagram Illustrating Typical Fire Trench, Second Line and Communication Trenches," 1917

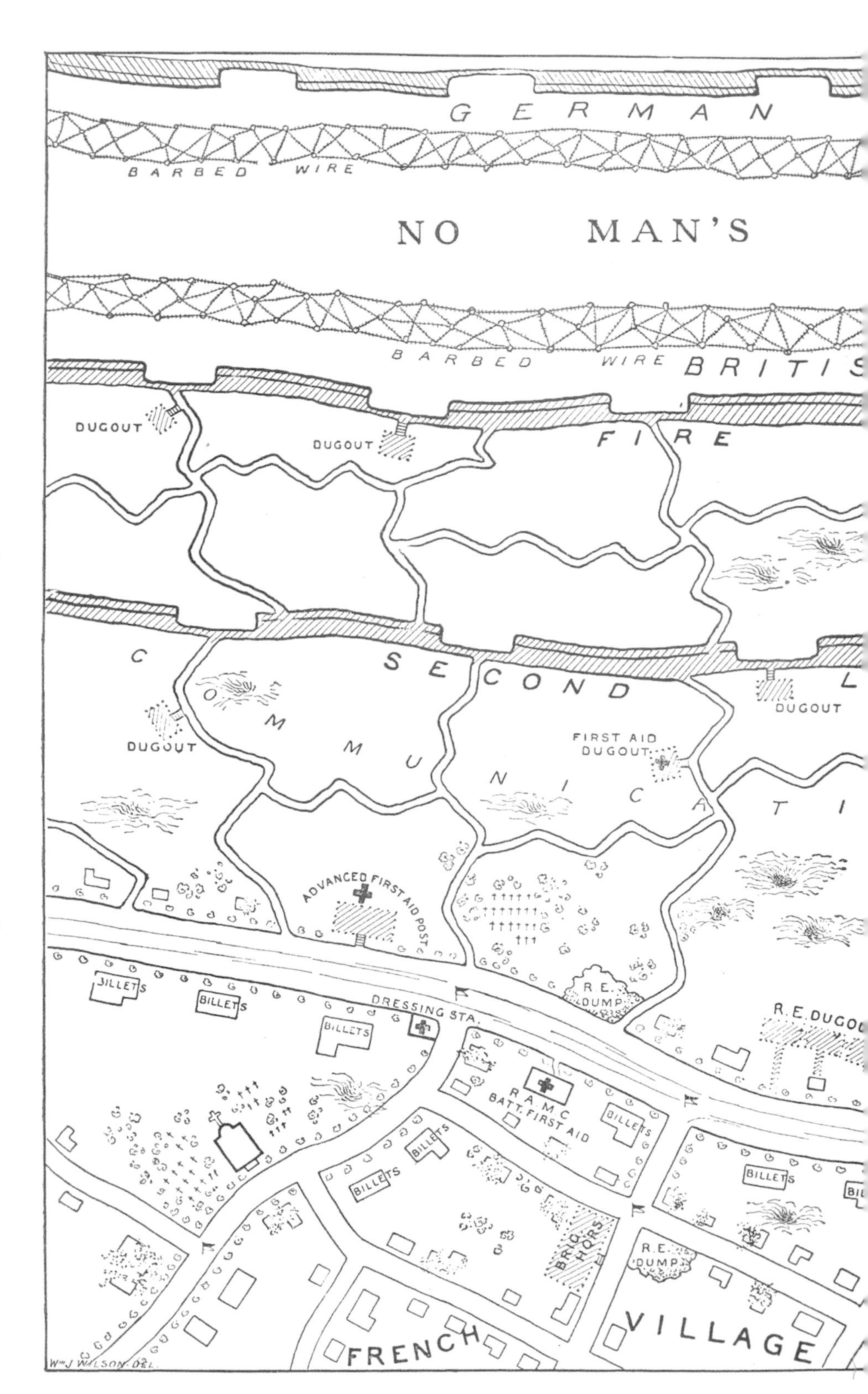

On the morning of May 7, 1915, a German submarine torpedoed the British ocean liner RMS *Lusitania* off the coast of Ireland. The next day newspapers across the United States highlighted the loss of American lives in the attack, stoking outrage and calls for a declaration of war. In a New Jersey military recruiting office Guy Empey read the news, then watched his superior solemnly drape an American flag over a map of Europe on the wall. The United States, they assumed, would now join the Allied cause to avenge this German atrocity. They quickly prepared the office for an onrush of volunteers.

Yet in the coming weeks and months President Wilson issued no call for war, and instead doggedly reiterated a policy of neutrality. Back at the recruiting office, the flag came down from the wall, and Empey took matters into his own hands. At the end of 1915 he sailed to London to join the British Army. Initially recruiters refused his offer of service out of deference to American neutrality, but a few weeks later he was sent to the Western Front in France as a machine gunner. By this point the front stretched from the North Sea through Belgium, northeastern France, and Alsace-Lorraine all the way to the Swiss border. Sergeant Empey's first task was to dig miles and miles of trenches in preparation for the "big push" against Germany in the summer of 1916.

From July to November, over 1 million men were killed in the Battle of the Somme, the most destructive of the entire war. Empey himself was injured at the Somme, and upon his return home wrote a memoir of the war entitled *Over the Top* that captured both the horror and humor of the war. Empey portrayed his fellow soldiers—the British "Tommies"—as both stoically and heroically facing the constant threat of death. To illustrate his account, Empey included a map that profiled not a specific location in the conflict, but rather the general dynamics of trench warfare. Between 1915 and 1917 the front was nearly immovable, so Empey's map might apply to a spot anywhere along the line.

It was this uniformity that made the front so destructive. From the rigid front lines to the fortified rear defenses, the armies tore at each other. The

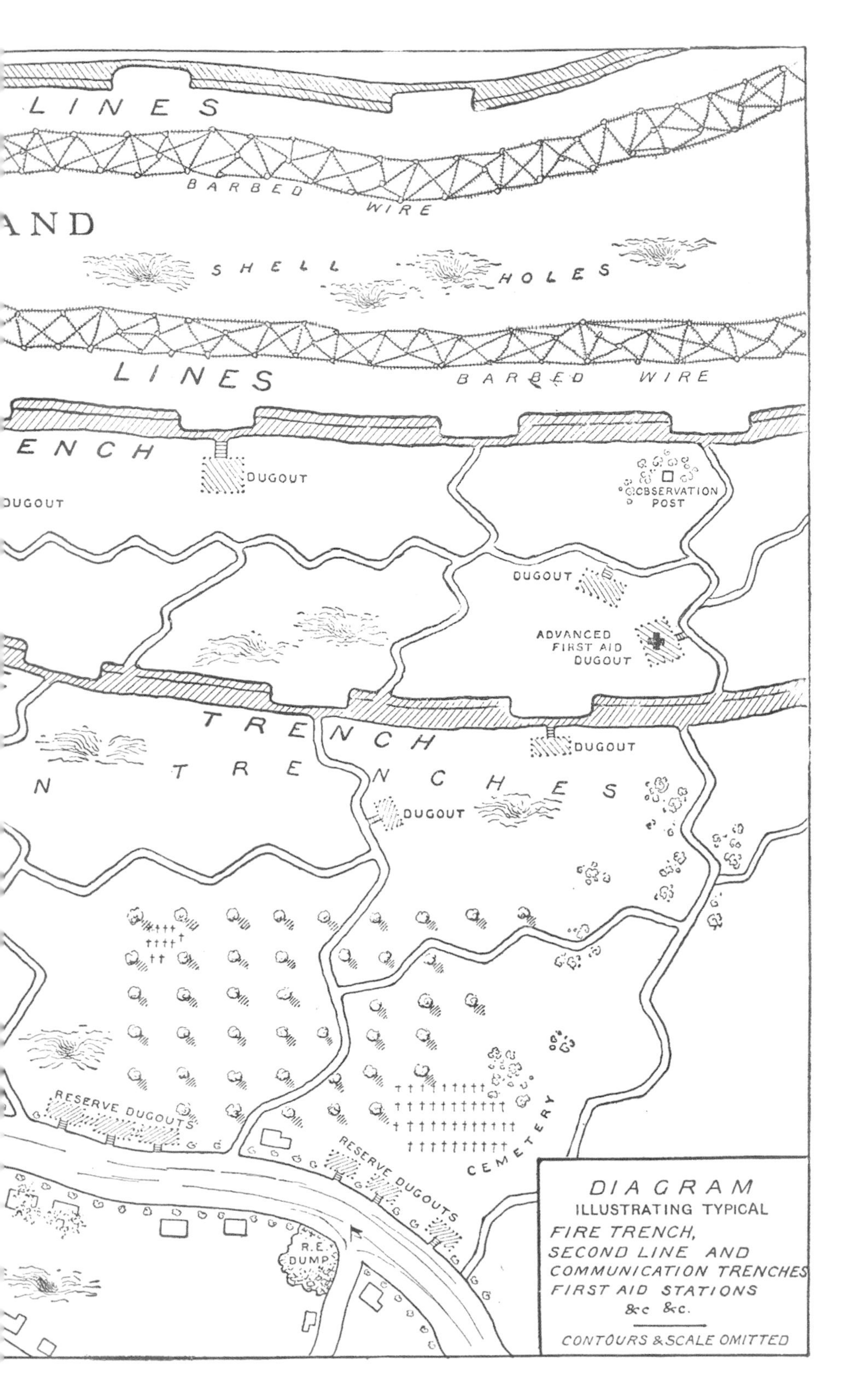

phrase "no-man's-land" was used to designate the exposed and immovable zone between the two belligerents. The barbed wire marked along the front was an innovation of the American West, originally designed to rein in cattle but here used to prevent soldiers from rushing the line. Behind the wire were two lines to keep watch and fire upon the enemy. The zigzag lines of communication and movement enabled men to sustain the line for years, rotating soldiers and sending supplies in as needed from the rear. Empey even includes shell holes to remind the viewer of the dangers not just across the front, but above the line as well.

Empey's stylized and sanitized sketch may not overtly express the terror of the front, yet therein lies the importance of his map: by "diagramming" these general geographical patterns across Europe he inadvertently revealed the futility of the years of trench warfare that were to consume a generation. By World War II, the advent of air power as well as tanks would render trenches a thing of the past.

Empey's memoir was published just as the nation declared war on Germany, and that timing made it an instant bestseller. Newspapers across the nation excerpted passages, billing it as an authoritative account from an American who was "two years ahead of his country." For months Empey traveled the US to aid the war effort and the draft, testifying to its importance at a time when many Americans remained skeptical of the Allies. Wilson had promoted neutrality for over two years, and so the administration treated *Over the Top* as an essential text for the Allied cause and Empey as a model of national service.

In interviews, Empey promoted enlistment, and gave plenty of advice. He warned the Americans not to present themselves as the saviors of Europe, and to acknowledge the years that the French and British had already been at war. He downplayed the dangers of service, advising his fellow Americans that the newspapers had painted a lopsided view of combat. Trench warfare was grim enough, he remarked, but the experience also brought a sense of camaraderie. Above all, he wrote, the war made him a man, and could do the same for any other American brave enough to serve.

In a twist that says much about the contemporary moment, once Empey finished rallying support for the war he went to California to star in the motion picture version of his memoir. It was one of the first feature films produced in Hollywood. Thereafter he made a career in silent films, part of a growing cohort in Hollywood that is detailed on page 194.

THE CREATION OF A GERMAN ENEMY

Committee on Public Information, "Why Germany Wants Peace Now," 1917

While the carnage wrought by World War I indelibly shaped Europe, the United States suffered comparatively few casualties and no territorial losses. Yet the conflict deeply shaped American society by forging a modern state and establishing the relationship between government and the public in wartime. Nowhere was this more apparent than in the realm of official propaganda.

For the first three years of the war, President Wilson relentlessly urged American neutrality. The United States had historical ties to Britain and France, but ten million Americans had direct loyalties to Germany and Austria. Even more Irish Americans had a longstanding hatred of the English, and a sizable population of Russian Jews opposed any aid to the czar. Moreover, from the American perspective the war presented no clear mission worth defending; many considered Britain's high-handed control of the Atlantic as unwarranted as German military tactics.

In practice, however, the position of the United States from 1914 to 1917 was hardly neutral. American manufacturers traded heavily with the Allies, and this activity directly stimulated the economy. Most revealing is the balance sheet of credit: by April 1917 American banks had loaned $2.3 billion to the Allies during the war, and only $27 million to Germany. Trade similarly favored Britain and France over Germany, all but ensuring that if the United States were to enter the war it would do so on the side of the Allies.

That moment came when Germany resumed submarine warfare in 1917, prompting Wilson to request a declaration of war from Congress. To generate support for this reversal from neutrality to belligerency, the president created the nation's first state-sponsored propaganda agency. The Committee on Public Information was overseen by journalist George Creel, who used every conceivable form of mass communication—posters, newsreels, public lectures, books, newspapers, and pamphlets—to recruit soldiers, stimulate patriotism, and raise money for the war.

The CPI's best-known materials are the bold and vibrant posters enlisting public support, such as Uncle Sam's iconic "I Want *You* for the U.S. Army." But its most effective messages were not those illustrating what Americans were fighting *for*, but rather what they were fighting *against*. Posters characterized the Germans as rapacious and subhuman enemies of civilization, while an avalanche of print warned of a longstanding German plan to expand from "Hamburg to the Persian Gulf." This concept of "Pan-Germanism" was directly imported from French and British propaganda to convince a skeptical public that American participation in the war was imperative.

The CPI's campaign to publicize Pan-Germanism spread quickly. With maps such as this, the agency scorned German peace overtures as an opportunistic plan to preserve the country's territorial conquests. The map is stripped of topography or any other information not directly relevant to the argument at hand. Instead, it foregrounds a German geopolitical threat that stretched from the North Sea to the Middle East. The map was reprinted over a million times in the anti-German tract *Conquest and Kultur*, and the CPI distributed an additional 122,000 copies to the public. Military training camps posted large copies of the map in mess halls to expose recruits to the geopolitics of the war.

The CPI's relentless demonization of the enemy fostered terrible anti-German hysteria in 1917 and 1918. In California, public schools discontinued German-language instruction, while across the country orchestras ceased to perform the music of German composers. It was a short step from portraying German culture as "poisonous" to vilifying German Americans themselves, and they were viciously singled out during the war.

Even more pernicious was the more general crackdown on dissent. The nation's leading papers immediately began to report on "Pan-Germanism," demonstrating how successful the CPI had become in disseminating a coordinated message during the war. Radicals and pacifists who criticized the war and challenged the expansion of state power were swiftly prosecuted through the Espionage and Sedition acts. Among these was Eugene V. Debs, a union leader and four-time Socialist Party candidate for president. Debs spoke publicly against the war, particularly highlighting the irony of a nation that had gone to war for democracy abroad while refusing to tolerate dissent at home. The prosecution and conviction of Debs left little doubt that the first casualty of war was freedom itself.

WHY GERMANY WANTS PEACE NOW
THE PANGERMAN PLAN
as realised by War
IN EUROPE AND IN ASIA
"Central Europe" and its Annexe in the Near East (Germany, Austria, Hungary, Bulgaria, Turkey)
The Entente Powers
Territory occupied by Central Powers
Territory occupied by Entente Powers
GERMANY'S MAIN ROUTE TO THE EAST
(Berlin-Bagdad, Berlin-Hodeida, Berlin-Cairo-Cape)
Supplementary Routes
(Berlin-Trieste, Berlin-Salonica-Athens, Berlin-Constantza-Constantinople)
Uncompleted sectors
Stockholm
Petrograd
Moscow
RUSSIAN EMPIRE
Poland
Berlin
Breslau
Prague
Cracow
Lemberg
Budapest
HUNGARY
Odessa
Trieste
Belgrade
Bucharest
Constantza
BLACK SEA
Nish
Sofia
Varna
BULGARIA
Constantinople
Salonica
TURKISH EMPIRE
Athens
Konia
Aleppo
Mesopotamia
R. Tigris
R. Euphrates
Bagdad
Cyprus
Damascus
Malta
CASPIAN SEA
PERSIA
AFGHANISTAN
CHINA
INDIA
Cairo
(Italian)
ARABIA
Persian Gulf
OMAN
Assuan
Medina
Mecca
Wadi Halfa
ANGLO-EGYPTIAN SUDAN
RED SEA
Khartum
ERITREA (Italian)
Hodeida
ARABIAN SEA
Bombay
Aden
Gulf of Aden
Socotra (British)
ABYSSINIA
FR. SOMALILAND
BR. SOMALILAND
ITALIAN SOMALILAND
C.S.Hammond & Co., N.Y.
PEACE OFFER
opulation (in round figures)
68,000,000
52,000,000
5,500,000
21,000,000
146,500,000
6,500,000
6,000,000
18,000,000
5,000,000
5,000,000
40,500,000
187,000,000 People
NTS PEACE

DEADLIER THAN WAR

US Public Health Service, "Chronological Map of the Influenza Epidemic of 1918," 1919

The outbreak of Spanish flu in 1918 was the worst pandemic in modern history: it infected nearly a third of the world's population and killed far more than the fighting in World War I. In the United States, at least half a million people died from the flu, a number exceeding American casualties of *all* twentieth-century wars combined. The most terrifying aspect of the disease was the speed of infection, which defied any effort either to contain it or to locate its source. This unassuming map was an attempt by the US Public Health Service to track the disease as it ravaged the country.

Years later, medical researchers traced the earliest cases of infection to Camp Funston in Kansas, where soldiers reported to the camp hospital with fever, headache, and backache. Within a few days most of them had returned to duty. Though a few developed pneumonia, the entire outbreak seemed to subside within two weeks. Other bases reported similar episodes soon after, but military and public health officials took little notice of this. After Congress declared war on Germany in April, the military created thirty-two camps to house and train up to 45,000 men each. These "cities" brought together large numbers of men from around the country who transmitted the contagion and then carried it to Europe as they were deployed. By May, American and French troops on the front had been infected.

A second and far deadlier wave of the influenza virus returned to the United States that summer, when a few soldiers fell ill upon arriving at Boston's Commonwealth Pier. Eight new cases were reported within a day, and sixty-eight the day after that. On September 20, the army counted 9,313 cases of influenza in its ranks; three days later there were 20,000. By September 26, the disease had spread to coastal bases at Puget Sound, San Francisco, Louisiana, and even the Great Lakes. A few days later it had reached twenty military camps in the interior of the country. Most of the isolated spots of early outbreak marked on the map (September 14–28)

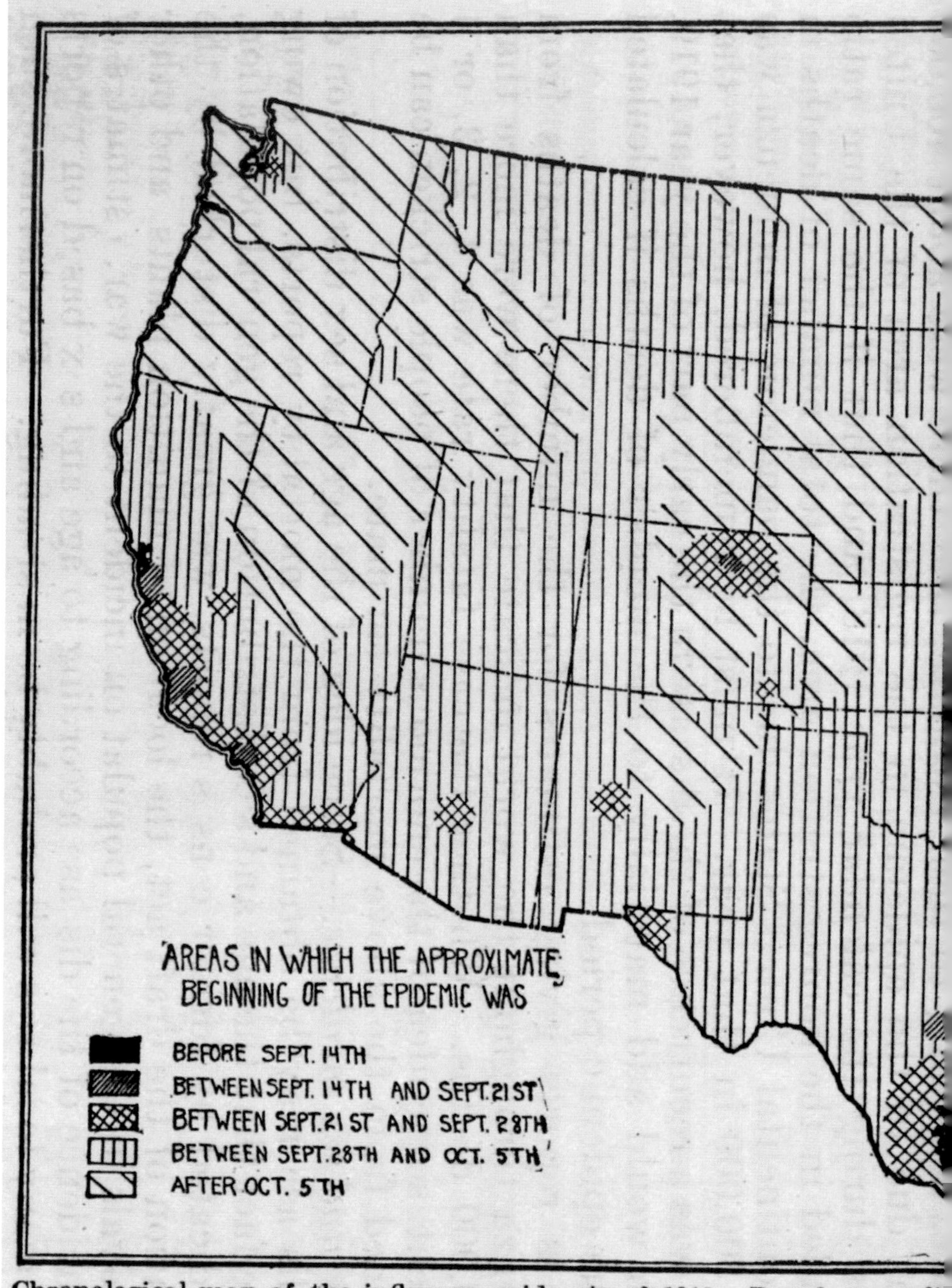

Chronological map of the influenza epidemic of 1918. Based on preli
which the disease re

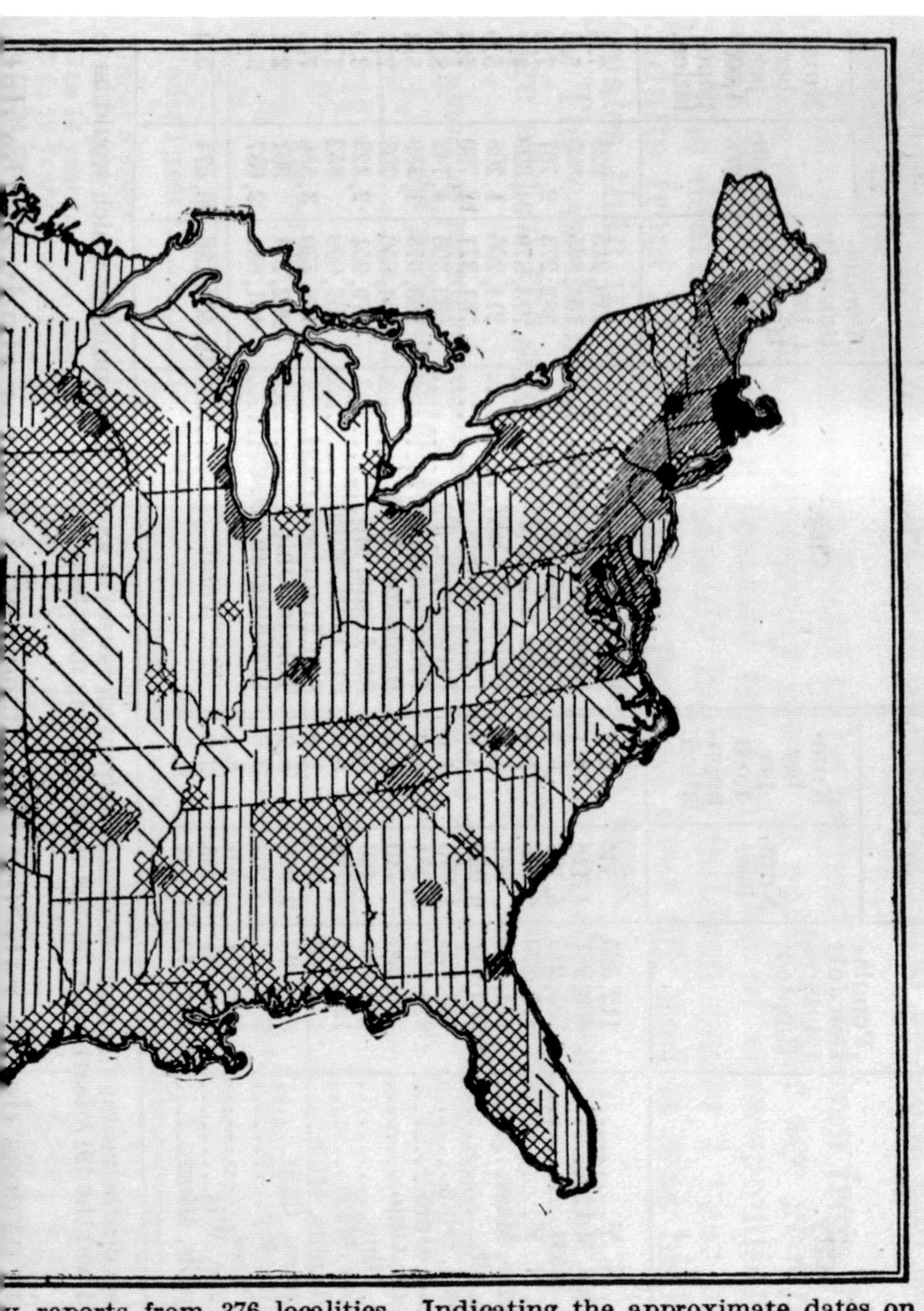
y reports from 376 localities. Indicating the approximate dates on
an epidemic stage.

correspond to the largest training and cantonment camps, for military mobilization facilitated the epidemic.

The virus soon spread from the military bases to the general population. In October 1918 over 20,000 stateside soldiers perished from the disease; in Philadelphia 700 died in a single day. In New York City, the public health commissioner staggered the hours of schools, shops, and workplaces in order to lower human interaction. Military officials briefly halted the draft and quarantined soldiers, but to little avail. Oddly, those between the ages of 20 and 40 were the most likely to die from infection, while higher proportions of children and the elderly managed to survive.

The rapid spread of the disease in the fall of 1918 sparked rumors that the German enemy had intentionally contaminated American soldiers in order to infect the home front. But the conditions of war worsened the pandemic. Soldiers were crowded together in military camps, then went abroad as carriers of the virus. Equally important were the conditions of modern life: the concentration of Americans in urban environments—often with poor sanitation—made it easy for the virus to spread. Moreover, modern transportation—steamships and railroads—gave the pathogen ongoing access to new hosts that it needed to survive.

The pandemic took the emerging profession of public health by surprise. Just a few years earlier, US Army physician Walter Reed had identified the cause of yellow fever, while advances in vaccination, sanitation, and bacteriology had begun to bring typhus, cholera, and typhoid under control. These medical advances made the flu pandemic all the more shocking. As the war came to a close, the disease inflicted its greatest damage. Influenza deaths peaked in October 1918, and continued through that winter before subsiding in early 1919.

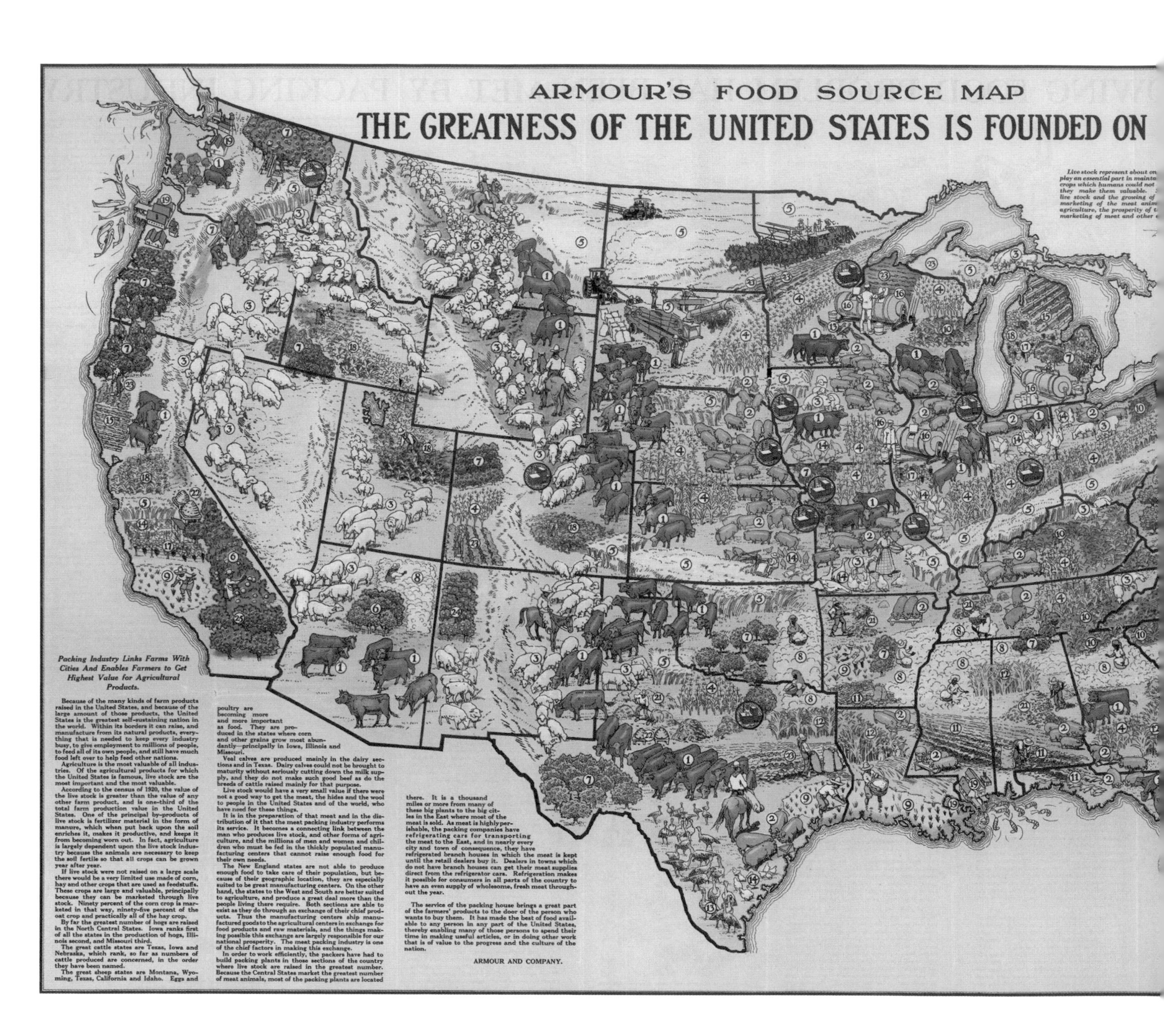
ARMOUR'S FOOD SOURCE MAP
THE GREATNESS OF THE UNITED STATES IS FOUNDED ON
Live stock represent about
play an essential part in
crops which humans could not
they make them valuable.
live stock and the growing
marketing of the meat
agriculture, the prosperity
marketing of meat and other
Packing Industry Links Farms With Cities And Enables Farmers to Get Highest Value for Agricultural Products.
Because of the many kinds of farm products raised in the United States, and because of the large amount of those products, the United States is the greatest self-sustaining nation in the world. Within its borders it can raise, and manufacture from its natural products, everything that is needed to keep every industry busy, to give employment to millions of people, to feed all of its own people, and still have much food left over to help feed other nations.
Agriculture is the most valuable of all industries. Of the agricultural products for which the United States is famous, live stock are the most important and the most valuable.
According to the census of 1920, the value of the live stock is greater than the value of any other farm product, and is one-third of the total farm production value in the United States. One of the principal by-products of live stock is fertilizer material in the form of manure, which when put back upon the soil enriches it, makes it productive, and keeps it from becoming worn out. In fact, agriculture is largely dependent upon the live stock industry because the animals are necessary to keep the soil fertile so that all crops can be grown year after year.
If live stock were not raised on a large scale there would be a very limited use made of corn, hay and other crops that are used as feedstuffs. These crops are large and valuable, principally because they can be marketed through live stock. Ninety percent of the corn crop is marketed in that way, ninety-five percent of the oat crop and practically all of the hay crop.
By far the greatest number of hogs are raised in the North Central States. Iowa ranks first of all the states in the production of hogs, Illinois second, and Missouri third.
The great cattle states are Texas, Iowa and Nebraska, which rank, so far as numbers of cattle produced are concerned, in the order they have been named.
The great sheep states are Montana, Wyoming, Texas, California and Idaho. Eggs and poultry are becoming more and more important as food. They are produced in the states where corn and other grains grow most abundantly—principally in Iowa, Illinois and Missouri.
Veal calves are produced mainly in the dairy sections and in Texas. Dairy calves could not be brought to maturity without seriously cutting down the milk supply, and they do not make such good beef as do the breeds of cattle raised mainly for that purpose.
Live stock would have a very small value if there were not a good way to get the meat, the hides and the wool to people in the United States and of the world, who have need for these things.
It is in the preparation of that meat and in the distribution of it that the meat packing industry performs its service. It becomes a connecting link between the man who produces live stock, and other forms of agriculture, and the millions of men and women and children who must be fed in the thickly populated manufacturing centers that cannot raise enough food for their own needs.
The New England states are not able to produce enough food to take care of their population, but because of their geographic location, they are especially suited to be great manufacturing centers. On the other hand, the states to the West and South are better suited to agriculture, and produce a great deal more than the people living there require. Both sections are able to exist as they do through an exchange of their chief products. Thus the manufacturing centers ship manufactured goods to the agricultural centers in exchange for food products and raw materials, and the things making possible this exchange are largely responsible for our national prosperity. The meat packing industry is one of the chief factors in making this exchange.
In order to work efficiently, the packers have had to build packing plants in those sections of the country where live stock are raised in the greatest number. Because the Central States market the greatest number of meat animals, most of the packing plants are located there. It is a thousand miles or more from many of these big plants to the big cities in the East where most of the meat is sold. As meat is highly perishable, the packing companies have refrigerating cars for transporting the meat to the East, and in nearly every city and town of consequence, they have refrigerated branch houses in which the meat is kept until the retail dealers buy it. Dealers in towns which do not have branch houses can get their meat supplies direct from the refrigerator cars. Refrigeration makes it possible for consumers in all parts of the country to have an even supply of wholesome, fresh meat throughout the year.
The service of the packing house brings a great part of the farmers' products to the door of the person who wants to buy them. It has made the best of food available to any person in any part of the United States, thereby enabling many of those persons to spend their time in making useful articles, or in doing other work that is of value to the progress and the culture of the nation.
ARMOUR AND COMPANY.

THE MASS PRODUCTION OF FOOD

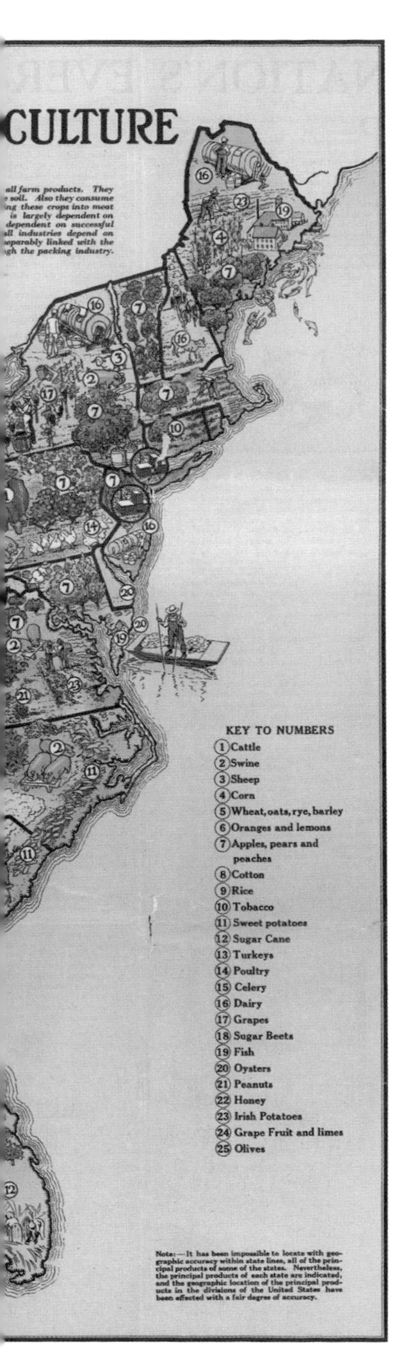

Joseph Pennell and Armour & Company, "Armour's Food Source Map," 1922

In 1906 Upton Sinclair published *The Jungle*, a searing account of working conditions in Chicago's meatpacking industry. He hoped to generate sympathy for labor, but found his readers far more disturbed by the factories that processed their meat. As he later commented, "I aimed at the public's heart, and by accident I hit in the stomach." His investigation caught the attention of President Roosevelt and other leaders, and led to one of the nation's first attempts to regulate food safety. The largest companies initially balked at the Federal Meat Inspection Act, but soon these regulations enabled them to consolidate their control over the industry by putting smaller operators out of business.

Ten years later, producers such as Armour and Swift had begun to resemble monopolies, drawing criticism and regulatory scrutiny. Armour responded by organizing one of the earliest corporate public relations campaigns in history. In 1918 the company established a "Bureau of Agricultural Research and Economics" to address what it termed "misleading headlines" about its practices and its size. Through pamphlets and books, maps, and even lavishly illustrated children's literature, Armour framed itself as less a meatpacking company than a selfless organization that endured razor-thin margins in order to feed a hungry world. Armour's public relations campaign continued after the war with this bright pictorial map designed by the noted artist Joseph Pennell. In the accompanying narrative, Armour described its successful efforts to solve the "nation's ever-growing food problem."

Armour's national distribution capacity was just one example of a much larger and more fundamental shift in food production that would transform American agriculture, ranching, dairying, and ultimately eating habits. In the 1860s Americans could not have imagined a future where beef raised on the prairie would be consumed on the East Coast. That process began soon after the Civil War, when bison ranging from Texas to the High Plains were replaced by cattle (page 160). With refrigeration and a growing rail network, cured, smoked, and pickled pork gave way to fresh beef. Philip Armour led the Midwestern effort to compete with eastern producers by reorganizing meat processing and making the most of his geographical location. Soon Chicago was the hub of this new meatpacking industry.

Armour industrialized meat production in several ways: he halved the time it took to fatten cows by feeding them corn rather than grass; he built a "disassembly line" to render the cattle and hogs, thereby obviating the need for skilled and expensive butchers; he used refrigerated railroad cars to transport meat around the country. The company even sold kosher beef by having a rabbi sever the jugular of cattle. Armour also made use of every part of the animal, not just for food but also for wool, glue, glycerine, leather, and oils.

This revolution in meat processing paralleled the rapid expansion of dairy farming in the Upper Midwest, the spread of grain belts across the country, and the cultivation of produce in California and Florida. By the early twentieth century Americans could eat cornflakes at breakfast, fresh beef at dinner, and even oranges at Christmas. All of this would have been unimaginable to their grandparents.

"Armour's Food Source Map" was originally designed as an advertisement. Yet it nicely captures the degree to which the country was now linked by commodities. By extension, individual producers and consumers were enmeshed in this web of food production. Armour advertised itself as the benevolent architect of this new system. Its public relations department published pamphlets and juvenile literature to portray the company's myriad operations in a positive and even patriotic light. It turned accusations of monopoly into an asset, framing itself as the industry leader in progress, productivity, efficiency, sanitation, and convenience.

As Armour explained, the millions it spent to meet federal regulations supported the welfare of thousands. By extension, the company's distribution system created "a cash market as broad as the world is wide," encompassing farmers in Iowa, ranchers in Texas, dairymen in Wisconsin, and even miners in Wales. It was the meatpacking industry, Armour argued, that fed the world. True enough: Armour continued to publish this map through the 1960s, and the practices developed by the company still shape the way the nation eats a century later.

WATER FOR LOS ANGELES

Los Angeles Bureau of Power and Light, utility bill, 1922

In 1890 John Wesley Powell tried to convince Congress that the future of the West hinged on the management of its limited water (page 162). Though Congress largely ignored Powell's warnings, Southern Californians paid attention, for they knew firsthand that Los Angeles could not grow without an imported water supply. The history of California is, to a great extent, the history of its water. Among the first to realize this was Frederick Eaton.

Eaton became the superintendent of the Los Angeles City Water Company in 1875 at the age of nineteen. He served alongside William Mulholland, who shared his conviction that Los Angeles was destined to grow despite its aridity. And grow it did: in 1900 Los Angeles had 100,000 residents, and within four years that figure had doubled. Eaton convinced Mulholland that the only way to sustain the city was to locate a reliable source of water. They settled on the Owens River, 250 miles due north of Los Angeles.

The Owens River rises out of the Eastern Sierra near Yosemite, and flows south through a valley bounded on the east by the White Mountains. The area was initially settled and farmed by the Paiute Indians, who were displaced by white farmers in the latter half of the nineteenth century. These settlers relied on the river to build the towns of Bishop, Big Pine, Independence, and Lone Pine. The federal Bureau of Reclamation believed that the agriculture of the Owens Valley could expand even further with improved irrigation.

Eaton, however, saw things differently. He believed that the water of the Owens Valley was wasted on local farming, and would be better used to slake the growing thirst in Los Angeles. In 1904 he and Mulholland formed the Los Angeles Department of Water and Power and began to surreptitiously secure rights to the Owens River. In 1905, with the support of the Board of Water Commissioners, they rapidly purchased land and water rights in the valley. Their ultimate goal was both ambitious and outlandish: first take control of the water, and then pass a bond measure to build an aqueduct to transport the water to Los Angeles. The voters approved the measure, and construction began in 1907.

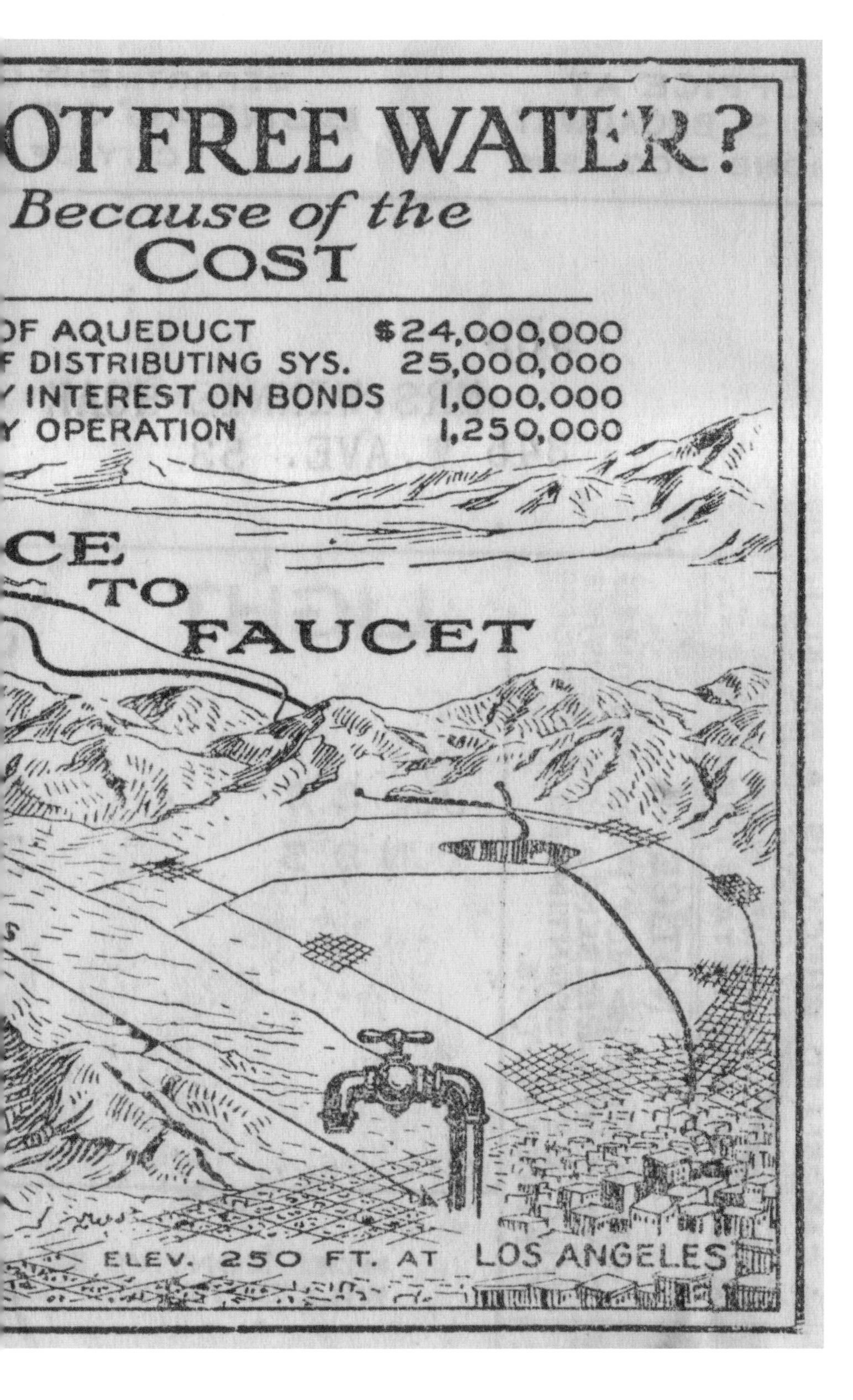

OT FREE WATER?
Because of the
COST
OF AQUEDUCT $24,000,000
F DISTRIBUTING SYS. 25,000,000
Y INTEREST ON BONDS 1,000,000
Y OPERATION 1,250,000
CE
TO
FAUCET
ELEV. 250 FT. AT LOS ANGELES

At the time it was built, the aqueduct was the longest in the western hemisphere. Through tunnels, conduits, and reservoirs, it carried water by gravity across 223 miles to Los Angeles. Upon opening in 1913 it delivered eight times the amount of water the city consumed, which inadvertently created pressure for Los Angeles to expand. From 1913 to 1927, the city swiftly annexed surrounding communities to become the nation's largest urban territory governed by a single entity. The annexation of the San Fernando Valley alone doubled the size of Los Angeles, and enriched those who had bought land knowing that the aqueduct would ensure its agricultural future.

By the 1920s, farmers in the San Fernando Valley used three times the amount of water that was going to the 1.2 million residents of Los Angeles. This diversion of water for large-scale agriculture enraged farmers in the Owens Valley who had lost their water rights. Some took matters into their own hands and dynamited the aqueduct in May 1924. Ultimately, however, they could do little but watch Los Angeles expand while their own fortunes stagnated. The absence of water in the Owens Valley not only ended farming, but made the region far more prone to dust storms.

The Los Angeles aqueduct also helped to generate electricity for the city. Here the Bureau of Power and Light used the back of its electric bill to explain rate increases to its customers. Perhaps Californians had already begun to take water and electricity for granted, though both were the work of enormous engineering feats. Within a few years, the continued expansion of Los Angeles had led the city to secure even more water, this time from the Colorado River.

Through a combination of political will, legal maneuvering, and engineering, water was drained from the Owens Valley, making it possible for Los Angeles to become home to millions of people over the next few decades. In this sense, Eaton and Mulholland transformed not just that city, but the dynamics of twentieth-century migration. The maps of Hollywood and Disneyland on pages 194 and 226 attest to the growth of Southern California and its abiding influence over American life.

UNDERSTANDING THE UNDERWORLD

Frederic M. Thrasher, "Chicago's Gangland," 1927

In the 1920s and 1930s newspapers relentlessly portrayed Chicago as a lawless town. Breathless headlines covered the St. Valentine's Day Massacre in 1929, when Al Capone and his men set about murdering rival gang members in a bid to control the city's organized crime. This coverage continued through the 1930s, though almost all American cities were reckoning with similar challenges that were only exacerbated by Prohibition.

The explosion of urban crime in the interwar years was closely studied by sociologists at the University of Chicago. Robert Park, a product of the Progressive Era, taught his students to think of cities in ecological terms, as dynamic systems. The city of Chicago became their laboratory. One of Park's leading students was Frederic Thrasher, who believed the key to controlling crime was a more thorough study of gang behavior. In his seven-year investigation, Thrasher counted 1,000 gangs, composed of over 25,000 members. These ranged from "embryonic" groups of young boys to sophisticated networks that controlled organized crime. Thrasher explored gangs as social units with their own internal logic and social structure. If they were to be overcome, gangs must first be understood.

Thrasher's study is full of extraordinary detail based on census data, institutions, and his own surveys. He joined gang members in their haunts and hangouts to observe their rituals, codes, and language. He interviewed them at length about their values and experiences. And, though he did not use the term "rape," he discussed shocking patterns of sexual behavior reported by his subjects. Later sociologists regarded Thrasher's work as unsystematic and dated, but it remained the most comprehensive investigation of the subject for decades.

Trained by Park to think in spatial terms, Thrasher mapped Chicago's criminal landscape. In the portion shown here, he identified the location of ethnic enclaves across the city. Poles, Italians, Lithuanians, Germans, and the Dutch had settled in Chicago at the turn of the century while African Americans had migrated during and after World War I. Thrasher also identified large populations of native born Americans who were equally involved in the gang culture of the interwar era.

This ethnic landscape formed the backdrop of Thrasher's map. Then he used red ink to identify the presence and influence of various gangs across the city landscape. Clear red lines denote the stronger and more territorial gangs, while smaller icons mark their less powerful counterparts. A close look at the map indicates that Thrasher aimed to do justice to the local vernacular, from Dukies to Mickies, Lake Front Jungles to No Man's Land. "Back of the Yards," "Wop Park," and "Bum Park" all signal particular neighborhoods, if not specific locations. Similarly, across the city he marked sites where individual gangs came up against one another.

Like many interwar sociologists, Thrasher was influenced by the community of Hull House reformers who had dedicated themselves to Chicago's immigrant communities on the South Side at the turn of the century (page 166). That earlier generation used maps to isolate particular classes of information—such as ethnicity or wages—in order to investigate urban structure. Here Thrasher similarly mapped the distribution of gangs and their realm of influence in order to discern their relationship to the urban landscape. He found that gangs clustered around canals, rivers, railroad tracks, and industrial areas, what he termed "interstitial" spaces. Stepping back further, he noticed that the gangs formed a kind of semicircle around the central loop, hemmed in by the more settled and safer residential areas that formed the outer ring of Chicagoland.

Thrasher's study was part of the emerging field of urban sociology. It bore directly on the new realm of juvenile justice, which aimed to treat young offenders not in terms of their personal moral failings but as part of a larger social order. Thrasher observed that Chicago's gangs were produced by the colossal influx of new residents—mostly immigrants—who had endured massive dislocation. For young men who were bored at school and too young to work, gangs offered structure, hierarchy, and a sense of belonging and purpose. In other words, gangs had a role to play, and until this was acknowledged and addressed they would continue to thrive.

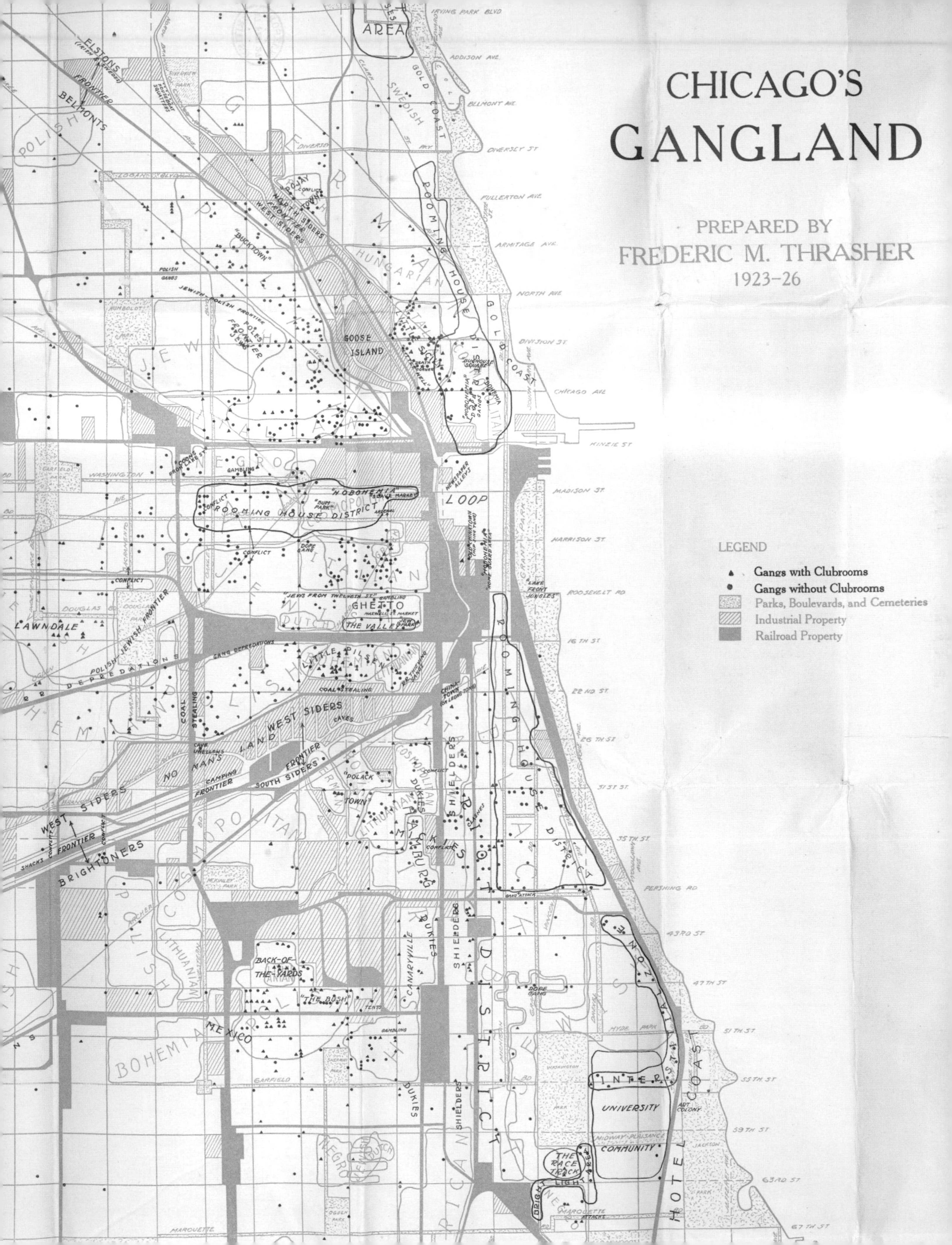

CHICAGO'S
GANGLAND
PREPARED BY
FREDERIC M. THRASHER
1923–26
LEGEND
Gangs with Clubrooms
Gangs without Clubrooms
Parks, Boulevards, and Cemeteries
Industrial Property
Railroad Property
IRVING PARK BLVD.
ADDISON AVE.
BELMONT AVE.
DIVERSEY ST.
FULLERTON AVE.
ARMITAGE AVE.
NORTH AVE.
DIVISION ST.
CHICAGO AVE.
KINZIE ST.
MADISON ST.
HARRISON ST.
ROOSEVELT RD.
16TH ST.
22ND ST.
26TH ST.
31ST ST.
35TH ST.
PERSHING RD.
43RD ST.
47TH ST.
51ST ST.
55TH ST.
59TH ST.
63RD ST.
67TH ST.
GOLD COAST
SWEDISH
ROOMING HOUSE DISTRICT
HUNGARIAN
GOOSE ISLAND
LITTLE SICILY
BOHEMIA
LOOP
NEWSPAPER ALLEYS
HOBOHEMIA
CONFLICT
GHETTO
THE VALLEY
LITTLE PILSEN
CHINATOWN
WEST SIDERS
NO MAN'S LAND
SOUTH SIDERS
FRONTIER
BRIGHTONERS
LAWNDALE
POLISH
JEWISH
ITALIAN
NEGRO
GERMAN
COSMOPOLITAN
LITHUANIAN
BOHEMIAN
MEXICO
BACK-OF-THE-YARDS
"THE BUSH"
CANARYVILLE
DUKIES
SHIELDERS
RIOT DISTRICT
BLACK BELT
UNIVERSITY COMMUNITY
INTERSTITIAL ZONE
HOTEL COAST
ART COLONY
THE RACE TRACK
BRIGHT LIGHT AREA
MIDWAY PLAISANCE
WASHINGTON PARK
JACKSON PARK
GARFIELD
MARQUETTE
ELSTONS
BELMONTS
NORTH SIDERS
"BUCKTOWN"
"POLACK TOWN"

MUSIC AND MAYHEM IN NEW YORK

E. Simms Campbell, "A Night-Club Map of Harlem," 1932

Cities across the Northeast and Midwest were transformed by the waves of African American migration after the turn of the century. The neighborhood of Harlem at the northern end of Manhattan became known as the cultural capital of black America, one with a rich and vibrant community of writers, artists, and musicians. Prohibition produced speakeasies across the country, but in Harlem it also fostered a rich musical culture that drew whites and blacks alike. In the era of Prohibition, Harlem became famous for jazz. This map—drawn by one of the century's most successful African American illustrators—captures the energy of that moment.

Elmer Simms Campbell studied at the Art Institute of Chicago before moving to Manhattan in 1929 in search of work as a cartoonist. As a black man, he faced a string of rejections before catching a break with the founding of *Esquire* magazine in 1933. For the next thirty-eight years he supplied the magazine with smart, appealing cartoons featuring the predicaments of beautiful young white women. In fact, it was Campbell who established *Esquire*'s knowing visual style: playful, often risqué, and thoroughly urban. Throughout the 1930s he also drew covers for the *New Yorker* and cartoons for other national magazines. Campbell's sheer artistic talent helped him cross the color line, but no doubt his success also had much to do with his shrewd decision to focus on white culture rather than black life.

As a young man, Campbell met band leader Cab Calloway, who, along with Duke Ellington, presided over the legendary performances at the Cotton Club. Calloway and Campbell became fast friends, drinking buddies, and regulars at many of Harlem's famed speakeasies and jazz clubs. Before he was discovered by *Esquire*, Campbell drew this whimsical and insightful roadmap to Harlem's hotspots for *Manhattan* magazine. Part tourist guide, part spoof, and part loving tribute, the map captures the boundless vitality of Harlem at the height of its popularity.

Campbell's map pulses with energy, but also hints at Harlem's complex racial dynamics. Many of these nightclubs catered to middle-class whites in search of slightly dangerous urban thrills. Black residents of the neighborhood had little say in the permissiveness that characterized their neighborhood. The Cotton Club—like most establishments in Harlem—was segregated and owned by whites. Other clubs such as Small's Paradise, at the right edge of the map, were owned by blacks and catered to the same. Smaller clubs were often open to both races, and all of these establishments collectively widened the audience for American jazz.

Prohibition also inadvertently made it more acceptable for women to drink in public, or at least in these semi-public clubs and speakeasies. The map shows blacks and whites drinking and dancing together, at a moment when both legal and customary segregation dominated much of the nation. Campbell references Calloway's smash hit "Minnie the Moocher" on the map, with the bandleader belting out "HO-DE-HI-DE-HO." No doubt this explains why Calloway kept a copy of the map in his office. Just outside the Savoy Ballroom—across Lenox Avenue—"Marahuana cigarettes" are sold at two for twenty-five cents. Campbell's exuberant rendering of Harlem's nightlife practically leaps off the page. And, while he celebrates the social and racial diversity in some of these clubs, he also acknowledges the general mayhem in the streets. At right, two officers play cards inside a shiny new police station, while drunks sow chaos outside.

All of this nightlife and culture was fueled by twelve years of Prohibition. While it curtailed the incidence of alcoholism, Prohibition also fostered an illegal market that worsened crime. In turn, this led to the rise of the penal state, anticipating the war on drugs at the turn of the twentieth century. In 1928 Al Smith ran for president on a platform that openly called for repeal of the Eighteenth Amendment. Though Smith lost to the Republican Herbert Hoover, his position attracted many new voters to the Democratic Party who would go on to support Franklin Roosevelt—and the repeal of Prohibition—in 1932.

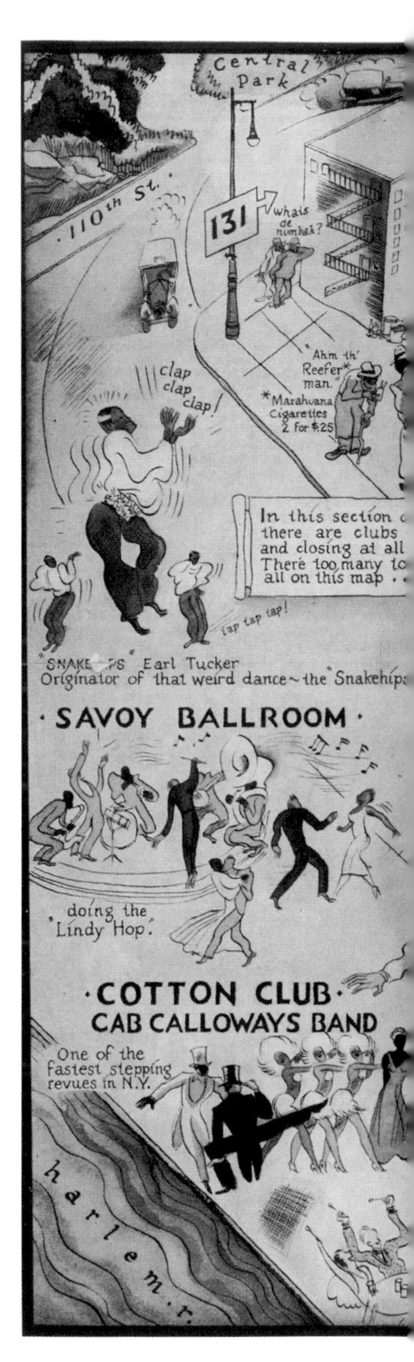

A NIGHT-CLUB MAP OF HARLEM
The stars indicate the places that are open all night ..... The only important omission is the location of the various speakeasies but since there are about 500 of them you won't have much trouble .....
CONNIE'S INN
LAFAYETTE THEATRE
Crooner - Harlan Lattimore
Friday nite is the midnight show — Most Negro Revues begin and end here.
"Bill" Bojangles" Robinson The Worlds Greatest tapdancer
his gold gun.
Harlem Moon
CLUB HOT-CHA
Nothing happens before 2 a.m. ask for Clarence.
The nice new POLICE STATION
SMALL'S PARADISE
Café au lait girls and dancing waiters.
LOG CABIN
TILLIES
Pods and Jerry ~ Proprietors Ask for Geo Woods An intimate little spot.
Specializes in fried chicken ~ and its really good!
BLUESSS
"taxi!"
THEATRICAL GRILL
Dickie Wells and Gene ~ prop. Jimmie Mordecai ~ manager.
Youve never heard a piano really played until you hear Garland Wilson
HI-DE-HO
SEVENTH avenue or heaven
A Sheik
GLADYS' CLAM HOUSE
Gladys Bentley wears a tuxedo and high hat and tickles the ivories
The YEAH MAN
"Are we going in here?" "Yeah Man!"
Go late!
LENOX avenue
RADIUM CLUB
Big breakfast dance every Sunday morning 4 or 5 a.m.
Jeff Blount ~ "headman"
"Crab Man!"
"whats de numbah?"
"its $.35
Actual size of Harlems national drink a shorty of Gin.
Engraved and copyrighted 1932 by . . . E. Simms Campbell
light wines and beer.
133
134
135
136
138
142

THE MAPS IN OUR HEADS

Daniel K. Wallingford, "A New Yorker's Idea of the United States of America," 1939

Just as Harlem was known as the capital of black America in the interwar era, New York assumed itself to be the nation's cultural and financial center. By any measure, this assertion was hard to dispute, but it also fostered a rather notorious insularity that is gently mocked by Daniel K. Wallingford's popular comic map of the 1930s.

Wallingford was from Indianapolis, and followed his father's path by pursuing a degree in architecture before serving in World War I. Thereafter he spent time in New England, and drew a map to poke fun at the elitism and self-importance among Bostonians. After moving to New York, he drew an elaborate pictorial view of "Architectural Manhattan," which demonstrated his talent for artistic design and execution. From there he returned to satirical mapping, this time imagining the country as seen by New Yorkers. Yet in mapping New York's geographical provincialism, Wallingford pointed out a much more general—and rather profound—human tendency to see the world in self-referential terms.

With a bird's-eye view and in a pictorial style, Wallingford drew the country according to perception rather than geographical scale. Most egregious are the locational errors: Yellowstone is placed in Colorado, Denver in Utah, and Nebraska in Illinois, while Montana abuts Alaska. The Great Lakes are all mixed up—and even include the Great Salt Lake—while St. Paul lies just south of Chicago and Milwaukee. To a New Yorker, Wallingford suggests, it matters not: everything is somewhere "out west."

Small vignettes serve as pointed remarks on these geographical stereotypes: oil and cowboys in Texas, movies in Southern California, and the auto industry in Detroit. But perhaps most revealing of all is the treatment of urban life more generally. However misplaced, cities appear on the map, while absolute geographical space is obliterated. The Great Plains are completely absent, an entire region that failed to register in the minds of New Yorkers. Instead, while the area east of the Mississippi River bears some passing resemblance to actual geography, west of the river we find a foreshortened land that simply skips over the vast interior.

Further, note that the five boroughs dwarf their New England and mid-Atlantic neighbors. The only other place on the map given the same attention is Florida, which figured prominently in the New York imagination. Miami and (presumably) Miami Beach are acknowledged as the vacation playground for New Yorkers, but beyond that confusion reigns regarding Nashville and New Orleans. To the west, three "Swanee Rivers" flow into the Gulf of Mexico.

The map has a deceptively simple—even innocent—appearance, but it reveals crucial dynamics that are by no means unique to New Yorkers. Regardless of our origins, all of us construct mental maps that govern our sense of value and distance, and situate us in the larger scheme of things. To his lasting credit, Wallingford was able to translate these slippery yet meaningful impressions expressed by New Yorkers into visual form. Perhaps his Midwestern roots gave him the distance to appreciate this northeastern geographical myopia.

Wallingford first published the map in the early 1930s, and was soon producing several different sizes to meet public demand. The map was widely reissued as a souvenir of the 1939 New York World's Fair, which drew 44 million visitors to witness "the world of tomorrow" at Flushing Meadows. That Wallingford's map remains relevant and funny indicates how deeply entrenched these regional attitudes and geographical perceptions can be. How else to explain the enduring popularity of Saul Steinberg's similar "View of the World from 9th Avenue," the iconic cover of the *New Yorker* from 1976? Both Wallingford and Steinberg may owe their ideas to an even earlier model: John McCutcheon's similar "New Yorker's Idea of the Map of the United States," which appeared in the *Chicago Tribune* in 1922.

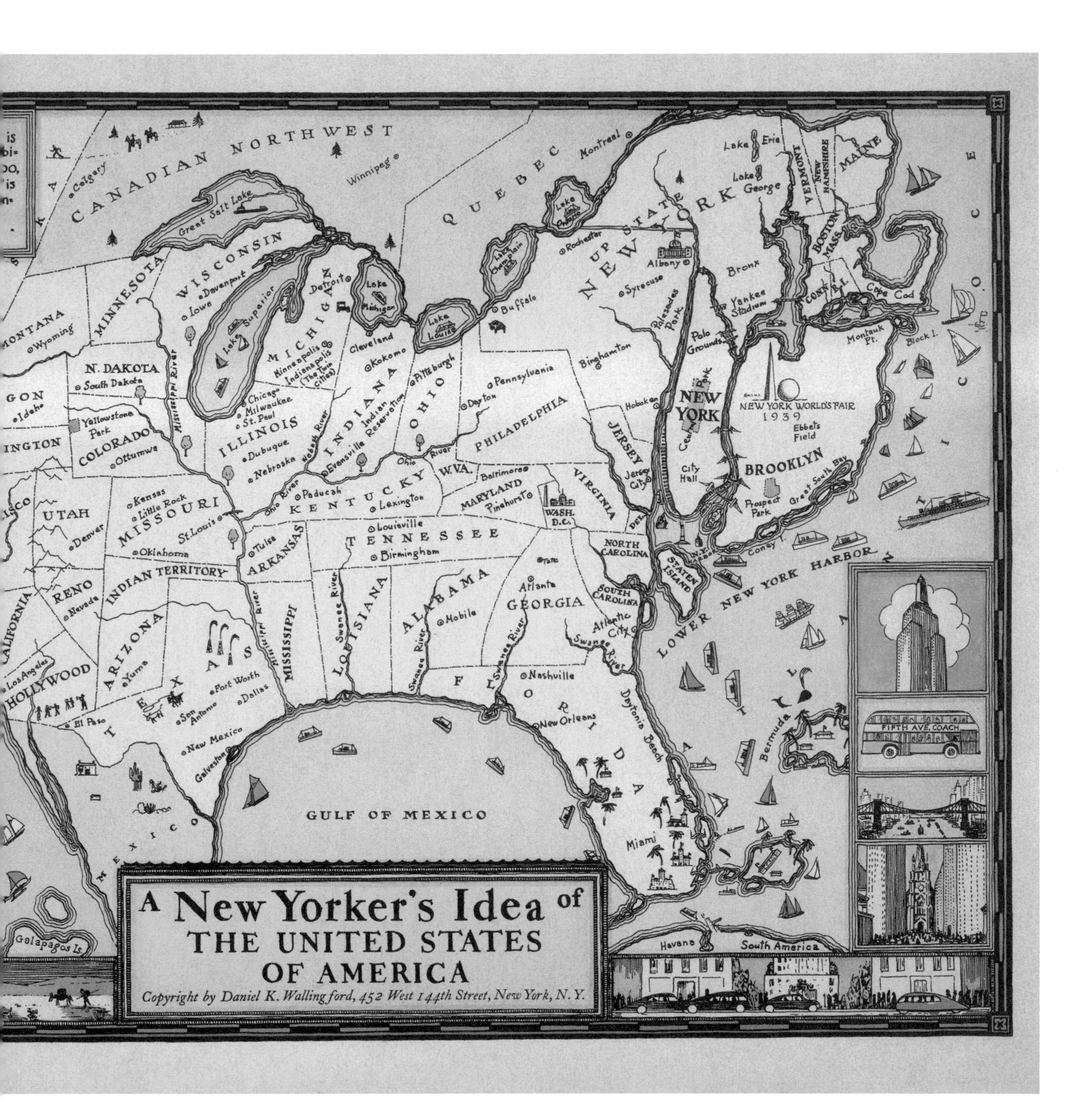
A New Yorker's Idea of
THE UNITED STATES
OF AMERICA
Copyright by Daniel K. Wallingford, 452 West 144th Street, New York, N.Y.
CANADIAN NORTHWEST
QUEBEC
Montreal
Winnipeg
Calgary
Great Salt Lake
WISCONSIN
MINNESOTA
MONTANA
Wyoming
N. DAKOTA
South Dakota
Idaho
Yellowstone Park
COLORADO
Ottumwa
UTAH
Denver
RENO
Nevada
CALIFORNIA
Los Angeles
HOLLYWOOD
ARIZONA
Yuma
TEXAS
El Paso
San Antonio
Fort Worth
Dallas
New Mexico
Galveston
INDIAN TERRITORY
Oklahoma
MISSOURI
Kansas
Little Rock
St. Louis
Tulsa
ARKANSAS
MISSISSIPPI
LOUISIANA
Swanee River
ALABAMA
Mobile
TENNESSEE
Louisville
Birmingham
KENTUCKY
Paducah
Lexington
ILLINOIS
Chicago
Milwaukee
St. Paul
Dubuque
Nebraska
Mississippi River
Davenport
Lowe
Lake Superior
MICHIGAN
Minneapolis
Indianapolis (The Twin Cities)
Detroit
Lake Michigan
Lake Louise
Cleveland
Kokomo
INDIANA
Indian Reservation
Evansville
Wabash River
OHIO
Pittsburgh
Dayton
Ohio River
W. VA.
Pennsylvania
PHILADELPHIA
Baltimore
MARYLAND
Pinehurst
WASH. D.C.
VIRGINIA
DEL.
NORTH CAROLINA
SOUTH CAROLINA
Atlantic City
GEORGIA
Atlanta
FLORIDA
Nashville
New Orleans
Daytona Beach
Miami
GULF OF MEXICO
MEXICO
Galapagos Is.
Havana
South America
Bermuda
Lake Ontario
Lake Champlain
Buffalo
Rochester
Syracuse
Albany
Binghamton
UP STATE NEW YORK
Lake Erie
Lake George
VERMONT
NEW HAMPSHIRE
MAINE
BOSTON MASS.
CONN.
R.I.
Cape Cod
Bronx
Yankee Stadium
Polo Grounds
Palisades Park
Hoboken
NEW YORK
Central Park
City Hall
JERSEY
Jersey City
NEW YORK WORLD'S FAIR 1939
Ebbets Field
BROOKLYN
Prospect Park
Great South Bay
Coney I.
STATEN ISLAND
LOWER NEW YORK HARBOR
Montauk Pt.
Block I.
ATLANTIC OCEAN
FIFTH AVE. COACH

THE GEOGRAPHY OF HOLLYWOOD

"Around the World in California in 4 Days," *Los Angeles Sunday Times*, March 4, 1934

If the last map captured the geographical imagination of New Yorkers, this one reveals a view of the world from Hollywood. This Sunday insert in the *Los Angeles Sunday Times* promoted the film industry by showing readers how California's landscapes enabled producers to simulate any conceivable geography without leaving the state.

The film industry is so inextricably associated with Hollywood that we might be forgiven for assuming that it began there. In fact, the earliest American motion pictures were produced and financed in New York in the 1890s before studios began to migrate to Southern California after 1900. By the 1920s, Hollywood dominated the film industry in the same way that Pittsburgh cornered steel and Chicago dominated meatpacking. Motion pictures had become the most popular form of American entertainment, with box office revenues of $750 million by 1927. Hollywood accounted for 88 percent of the films created each year in the 1920s. By 1927 weekly attendance had reached 100 million, nearly the population of the entire country.

Hollywood gained control over motion pictures in several ways. In the early years, studios were lured to Southern California by the capacity for growth, as shown on page 186. Los Angeles was also a non-union city, which was highly appealing for a labor-intensive industry. Just as compelling was the region's climate. Perennial sunshine meant there was more time available for filming, which in turn accelerated production schedules just as the industry was becoming more competitive.

California's physical landscape also gave Hollywood a tremendous advantage. The state's geographical diversity meant that almost any type of film could be produced there, which minimized travel costs. Even geographically specific and exotic locales could be approximated in its wide variety of topography and landscapes. As the *Los Angeles Sunday Times* explained here, everything imaginable was right in California's backyard. The Gobi Desert could be found in Barstow, Siberia in Truckee, the Swiss Alps near Lake Tahoe, and the Dead Sea in the inland Salton Sea. From the Kentucky backcountry to the deserts of Arabia, California could conjure up any possible geographical illusion.

In the late 1920s Hollywood released its first talking pictures, which raised hopes even further for the new medium. Boosters argued that films could be used not just for entertainment, but also as a tool of education and commerce. To some degree this prediction was realized in the newsreels that opened feature films from the 1930s through World War II. Air-conditioning technology also boosted the industry: a traditional summer slump caused by overheated theaters was transformed into a season of profits as patrons sought relief and escape in these cool, darkened spaces.

The seven major studios constituted a virtual monopoly by the interwar period. They exerted enough power to force theaters to buy blocks of their films together and to charge fixed rates of admission around the country. But this golden age of film came to an end with the advent of television in the 1950s. Between 1946 and 1956 theater attendance fell by half, and it never recovered. Moreover, the changing economics of the film industry meant that movie stars were no longer chained to a single studio, and independent filmmakers began to draw more attention after mid-century. But this map captures Hollywood's enduring influence, and the way that it shaped popular understandings of geography for generations.

day Times   March 4, 1934   Los Angeles Sunday Times

orld
ia in 4 days

ld in four days
eling and with-
you don't have
River country
a little closer,
d time consider-
om which much
ere taken, show
he contiguous
ly all the scenes
p of the entire
en photographed
o add rice fields
going time.

NAPLES—Catalina.

Left—SAHARA DESERT—Death Valley.

ALPINE or Scandinavian falls—Vernal Falls, Yosemite.

MAMMOTH CAVE—Cave in Providence Mountains, twenty miles north of Fenner.
—Auto Club of Southern California.

EUROPEAN ESTATE—Busch gardens, Pasadena.

ALASKA—Russian River country.

ALPS—Lake Dorothy, Sierra Nevada.

"LOCATION" MAP OF CALIFORNIA, a key to photos shown on this page. Every world spot is in your Golden State.

LAKES OF KILLARNEY—Sherwood Lake.

NEW ENGLAND COAST—Laguna Beach.
—Loyd Cooper.

IN SEARCH OF FREEDOM ON THE OPEN ROAD

"The Great American Roadside," *Fortune* (September 1934)

Before World War I, the average American worker would have had to spend two years' worth of wages to purchase an automobile. This changed dramatically with Henry Ford's mass production techniques: by 1925 a Model T was rolling off the assembly line every ten seconds, and by the end of the decade the same worker needed only three months' salary to buy one of the 26 million cars that were now on the road.

As early as 1920 the automobile business employed 4 million workers and accounted for 10 percent of the nation's wealth. It stimulated the market for glass, rubber, gasoline, and construction, and generated a new form of advertising: the roadside billboard. The automobile extended the distance between home and work and created Sunday drives and auto camping. Even dating rituals were transformed, for young people were no longer confined to the front porch or parlor. Entire cities—most importantly Los Angeles—were structured around the automobile, with growth irrespective of an urban core. As Donald Meinig once observed, Detroit may have invented the car, but Los Angeles taught us how to use it.

Even in the Great Depression, Americans rarely sold their cars. The automobile had become a symbol of freedom and possibility, evidenced by the ever-growing number of road trips taken during the 1930s. Well before the advent of interstate highways, auto vacations spawned roadside attractions and amenities across the country. *Fortune* magazine predicted that roadside businesses would gross $3 billion in 1934. The American road trip had become not just a rite of passage but an emerging and profitable market.

Among the earliest transcontinental roads was the Lincoln Highway, first billed as a national heritage tour from New York to San Francisco. Though the route was determined more by commerce than by history, it held out the promise of rediscovering America through the auto. This ideal even influenced the public works projects of the New Deal, for among its best-known legacies are the American Guide Series produced by the Federal Writers' Project. Each of these state guides is organized around auto tours and itineraries designed to showcase local color, regional identity, and history. Ironically, as the automobile and the highway homogenized the country, the road trip was idealized as a way to access the "real" America beyond the realm of mass culture.

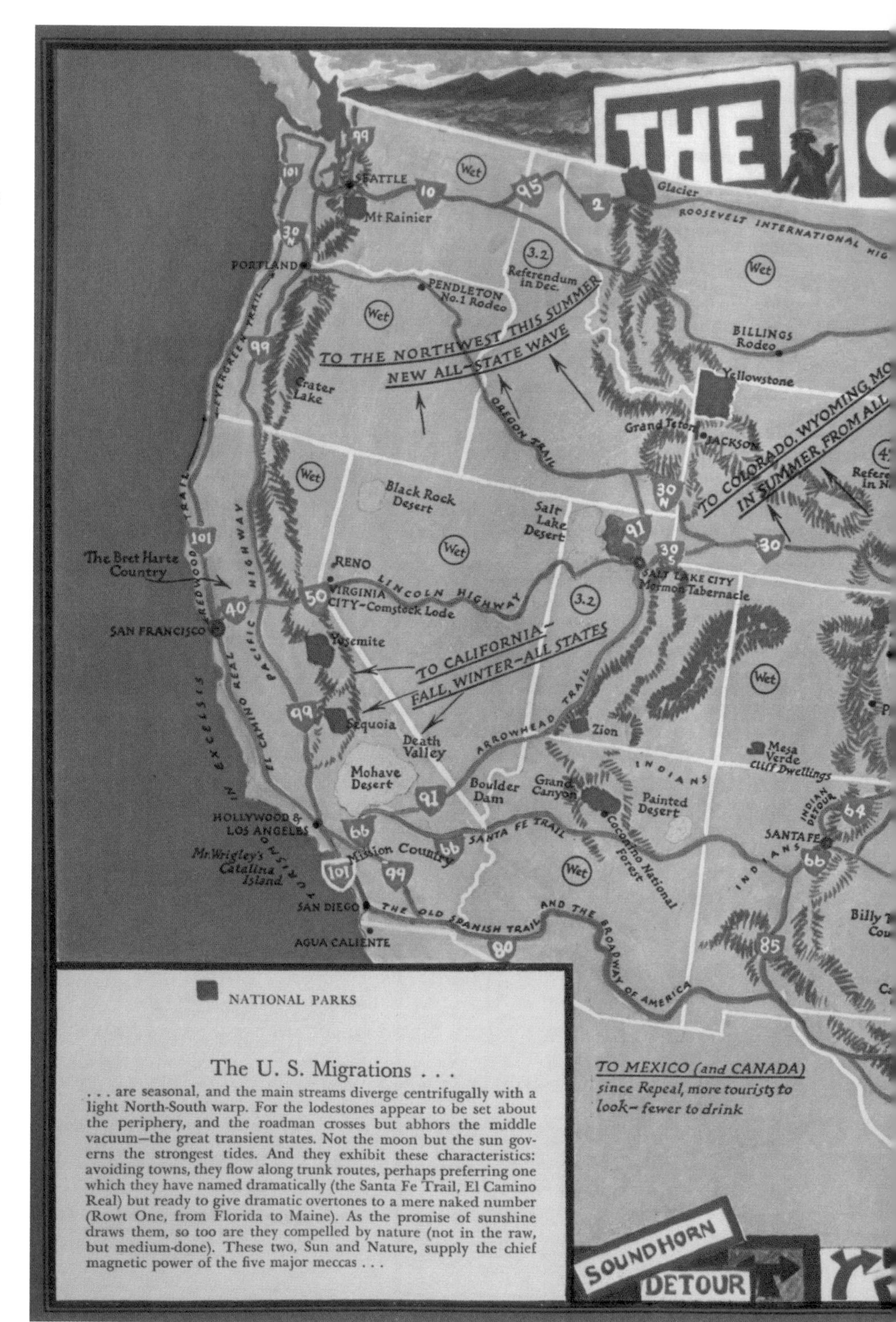

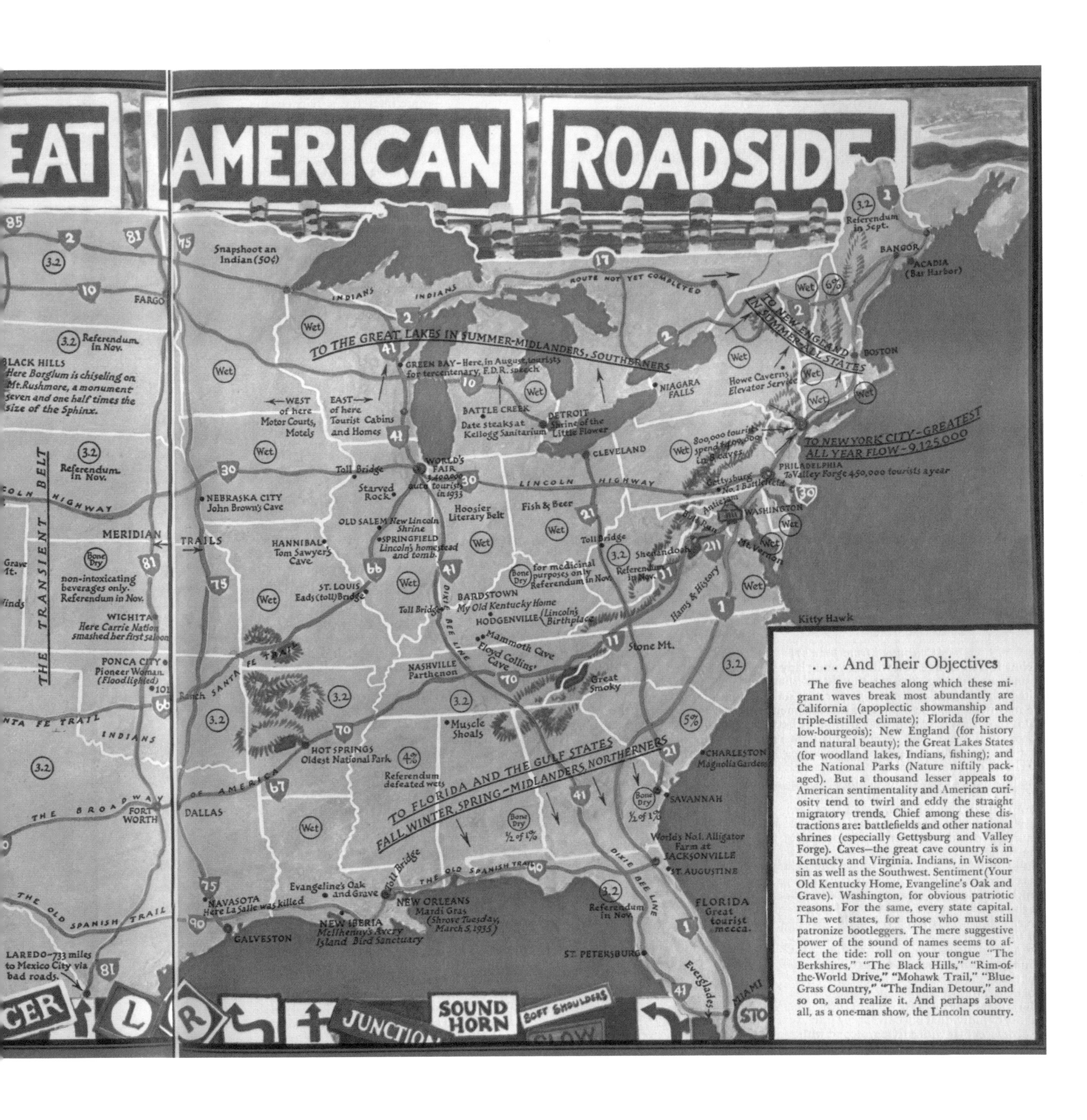

. . . And Their Objectives

The five beaches along which these migrant waves break most abundantly are California (apoplectic showmanship and triple-distilled climate); Florida (for the low-bourgeois); New England (for history and natural beauty); the Great Lakes States (for woodland lakes, Indians, fishing); and the National Parks (Nature niftily packaged). But a thousand lesser appeals to American sentimentality and American curiosity tend to twirl and eddy the straight migratory trends. Chief among these distractions are: battlefields and other national shrines (especially Gettysburg and Valley Forge). Caves—the great cave country is in Kentucky and Virginia. Indians, in Wisconsin as well as the Southwest. Sentiment (Your Old Kentucky Home, Evangeline's Oak and Grave). Washington, for obvious patriotic reasons. For the same, every state capital. The wet states, for those who must still patronize bootleggers. The mere suggestive power of the sound of names seems to affect the tide: roll on your tongue "The Berkshires," "The Black Hills," "Rim-of-the-World Drive," "Mohawk Trail," "Blue-Grass Country," "The Indian Detour," and so on, and realize it. And perhaps above all, as a one-man show, the Lincoln country.

AFRO AMERICAN TRAVEL MAP
DOMINION of CANADA
LAKE SUPERIOR
LAKE MICHIGAN
LAKE HURON
LAKE ERIE
L. ONTARIO
MAINE
AUGUSTA
OLD ORCHARD
PORTLAND
MOOSEHEAD LAKE
BANGOR
SWAMPSCOTT
BURLINGTON
CONCORD
DOVER
RUTLAND
N. H.
SARATOGA SPRS.
SYRACUSE
GREAT BARRINGTON
BOSTON
SPRINGFIELD
MASS.
OAK BLUFFS
HARTFORT
WATERBURY
R. I.
PROVIDENCE
NEW HAVEN
CONN.
NIAGARA FALLS
NEW YORK
BUFFALO
WISCONSIN
ELLSWORTH
MICHIGAN
BITELY
MILWAUKEE
MADISON
ANN ARBOR
IDLEWILD
DETROIT
CHICAGO
ANGOLA
TOLEDO
CLEVELAND
AKRON
YOUNGSTOWN
ERIE
WILKES-BARRE
PENNSYLVANIA
PITTSBURGH
HARRISBURG
PHILADELPHIA
WEST CHESTER
TRENTON
JERSEY
ASBURY PARK
ATLANTIC CITY
N. J.
WILDWOOD
OCEAN CITY
CAPE MAY
DEL.
DOVER
RIVER
ILLINOIS
INDIANA
OHIO
SPRINGFIELD
WHEELING
COLUMBUS
SPRINGFIELD
INDIANAPOLIS
CINCINNATI
ZANESVILLE
FAIRMONT
HAGERSTOWN
BALTIMORE
MD.
HARPERS FERRY
WASHINGTON
DANVILLE
TERRE HAUTE
FRENCH LICK
FRANKFORT
W. VA.
HUNTINGTON
CHARLESTON
STANTON
CHARLOTTESVILLE
FREDRICKSBURG
ST. LOUIS
E. ST. LOUIS
LOUISVILLE
NATURAL BRIDGE
LEXINGTON
RICHMOND
WELCH
ROANOKE
LYNCHBURG
PETERSBURG
BLUEFIELD
GLASGOW JT.
KENTUCKY
VIRGINIA
NORFOLK
PADUCAH
MT. STERLING
BRISTOL
GREENSBORO
WINSTON-SALEM
DURHAM
RALEIGH
WILSON
ALBEMARLE SOUND
SANFORD
GOLDSBORO
MISSISSIPPI
CLARKSVILLE
FISK
NASHVILLE
KNOXVILLE
TENNESSEE
NORTH CAROLINA
ASHEVILLE
CHARLOTTE
FAYETTEVILLE
NEW BERN
CHATTANOOGA
MEMPHIS
CORINTH
SHEFFIELD
SPARTANBURG
SOUTH CAROLINA
COLUMBIA
MISSISSIPPI
COLUMBUS
BIRMINGHAM
ATLANTA
CHARLESTON
AUGUSTA
BEAUFORT
ALABAMA
GEORGIA
JACKSON
MACON
YAZOO CITY
MERIDIAN
AUBURN
TUSKEGEE
MACON
MONTGOMERY
ALABAMA STATE
VICKSBURG
COLUMBUS
SAVANNAH
MCCOMB
LAUREL
BRUNSWICK
LA.
MOBILE
BREWTON
CAIRO
PENSACOLA
TALLAHASSEE
BATON ROUGE
NEW ORLEANS
MOBILE BAY
MISSISSIPPI DELTA
JACKSONVILLE
DAYTONA BEACH
FLORIDA
LAKE LAND
TAMPA
MIAMI
ATLANTIC OCEAN
GULF of MEXICO
BAY SHORE BEACH
BUCKROE, VA.
★ Hotel and Dining Room
★ Dance Pavilion
★ Leading Dance Bands
Picnics, Excursions, Conventions Given Special Consideration
Address
NEW BAY SHORE CORP.
BUCKROE, VA.
CHARLES WILLIAMS, President
RUFUS McHERN, Manager

"Afro American Travel Map," 1942

The success of the Lincoln Highway, Route 66, and other roads in the interwar era increased the pressure for a more consistent and coordinated national highway system. In 1938 President Roosevelt asked the Bureau of Public Roads to report on a possible network of north–south and east–west transcontinental highways. The result was a comprehensive study of road use, which laid the foundation for the Federal-Aid Highway Act of 1956 (page 228).

"The Great American Roadside" map on the previous page—and the "Afro American Travel Map" here—show the country before those interstate highways. The first was part of a *Fortune* magazine study of auto tourism, and presents the country and the economy in terms of highways and destinations, travel and leisure. In this regard, the map also unwittingly highlights a particular view of American history. From Valley Forge and Mount Vernon in the east to Buffalo Bill's grave out west, the map marks spots that would become fixtures of domestic tourism. Indians—when they appear at all—are little more than curiosities to be photographed (as noted in northern Minnesota).

As the map describes, the most popular destinations were the picturesque landscapes in California, Florida, the Rockies, New England, the Great Lakes, and the national parks. Civic tourism ranked high as well: Gettysburg and Valley Forge had become national shrines, while the construction of Mount Rushmore was well underway. All of these spots beckoned Americans to take to the road and "See America First."

Yet the ideal of the open road and the promise of the automobile was severely limited for African Americans in the era of Jim Crow. Though auto travel liberated American blacks from the degradation of segregated rail cars, it also exposed them to the risk of being refused service at filling stations, barred from restrooms, and excluded from hotels and restaurants. Such possibilities forced African Americans to plan differently for road trips, packing food, gasoline, and even portable toilets in order to make it safely to their destination.

While segregation throughout the Southern states was written into the law, African Americans had come to expect exclusion everywhere. Throughout the country, the proliferation of "sundown towns"—where blacks were unwelcome after dark—indicated the extent of racial hostility. In response to this pervasive discrimination, segregation, and physical danger, blacks began to publish guides and maps of their own. Families on holiday, salesmen traveling for work, and even entertainers on tour relied on these for safety.

This map was part of a "Travel Guide of Negro Hotels and Guest Houses," which listed places where weary motorists would not be refused service because of their race. The map and the guide were issued by the Afro Travel Bureau, which was a division of the *Afro-American* newspaper chain founded in 1892. The map marks national landmarks such as the Smoky Mountains and Natural Bridge, but it also identifies spots of special interest to the black community, such as the Tuskegee Institute and Fisk University. Subsequent pages list accommodations that did not discriminate, and even included private homes in towns where blacks might otherwise be unwelcome.

Similar guides, such as *The Negro Motorist Green Book*, carried far more extensive listings, including hotels as well as barber shops, beauty parlors, night clubs, restaurants, and service stations. The guide advertises, quite seriously, "Carry your Green Book with you—you may need it!" As the 1962 map on page 230 illustrates, blacks faced extreme hostility as they challenged Southern segregation. Once the 1964 Civil Rights Act outlawed discrimination in public accommodations—and the rise of interstates and chain hotels gave black travelers new options—the guides ceased publication. But they remain poignant reminders that black drivers faced circumstances and dangers that were scarcely visible to most whites, regardless of class.

The very existence of such guides demonstrates that the freedom of the open road was circumscribed by race. Though mid-century prosperity extended to blacks as well as whites, segregation governed all areas of life, even travel and leisure. Whites could assume that they would be served without incident, while blacks were forced to travel with care and more than a little anxiety. Like these guides, the rise of black resorts at mid-century was another response to this discrimination, an attempt to create spaces where African Americans could enjoy themselves without risk of humiliation, or worse. In 1925, for instance, two African American businessmen founded Lincoln Hills, a resort in the foothills west of Denver where blacks could purchase cottages or rent rooms in the spacious clubhouse. While the resort struggled in the Great Depression, thereafter it drew middle-class families from around the country who were eager to enjoy the splendor of the Rocky Mountains with their children at a time when they were barred from owning property in neighboring counties.

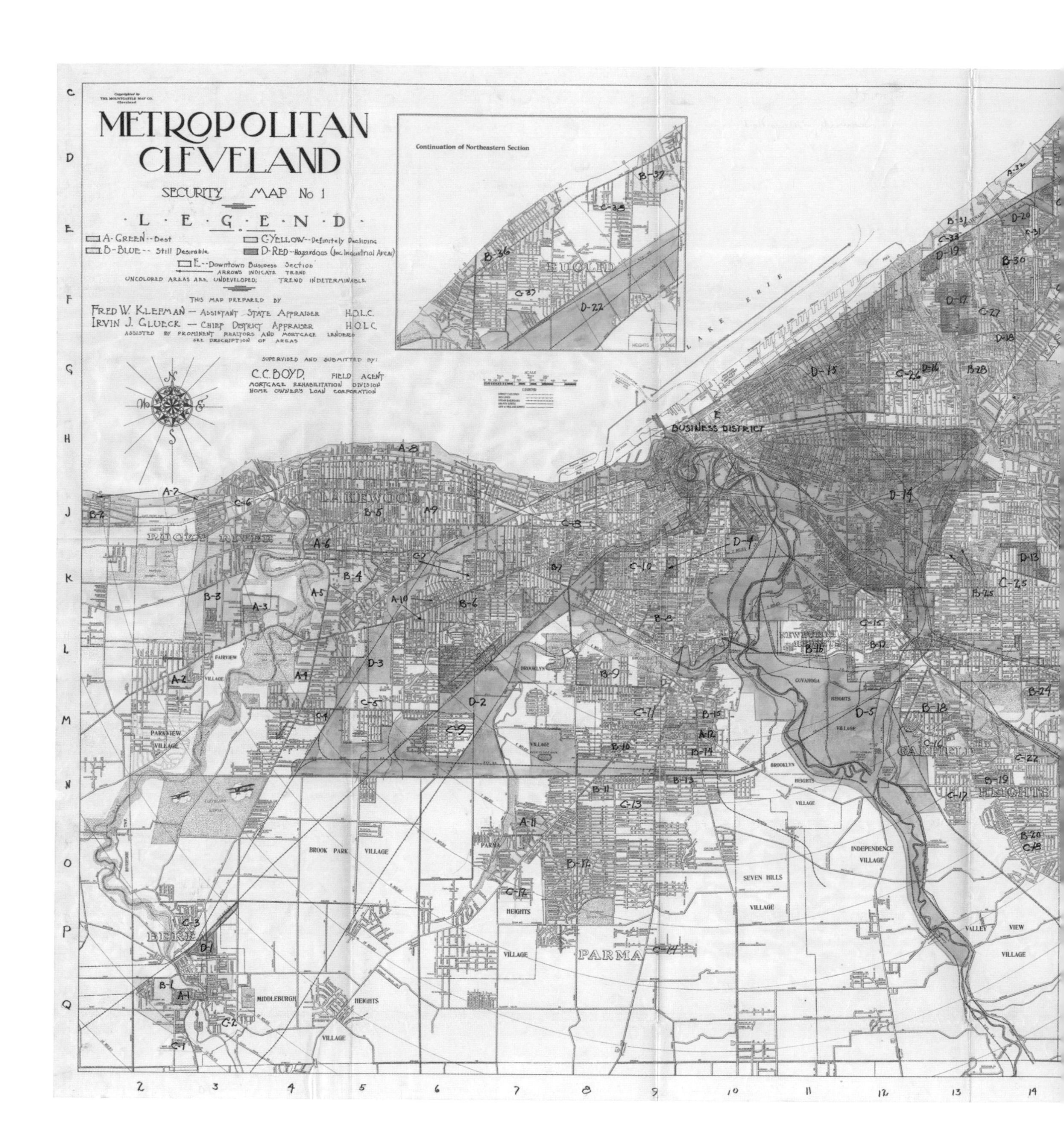
Copyrighted by
THE MOUNTCASTLE MAP CO.
Cleveland
METROPOLITAN CLEVELAND
SECURITY MAP No 1
·L·E·G·E·N·D·
A-GREEN--Best
B-BLUE-- Still Desirable
C-YELLOW--Definitely Declining
D-RED--Hazardous (Inc. Industrial Areas)
E--Downtown Business Section
ARROWS INDICATE TREND
UNCOLORED AREAS ARE UNDEVELOPED; TREND INDETERMINABLE
THIS MAP PREPARED BY
FRED W. KLEFMAN — ASSISTANT STATE APPRAISER H.O.L.C.
IRVIN J. GLUECK — CHIEF DISTRICT APPRAISER H.O.L.C.
ASSISTED BY PROMINENT REALTORS AND MORTGAGE LENDERS
SEE DESCRIPTION OF AREAS
SUPERVISED AND SUBMITTED BY:
C.C. BOYD, FIELD AGENT
MORTGAGE REHABILITATION DIVISION
HOME OWNER'S LOAN CORPORATION
N
W
E
S
Continuation of Northeastern Section
EUCLID
LAKE ERIE
BUSINESS DISTRICT
LAKEWOOD
ROCKY RIVER
FAIRVIEW
VILLAGE
PARKVIEW
VILLAGE
BROOKLYN
CUYAHOGA
HEIGHTS
VILLAGE
NEWBURGH
HEIGHTS
GARFIELD
HEIGHTS
BROOK PARK VILLAGE
PARMA
HEIGHTS
PARMA
SEVEN HILLS
VILLAGE
INDEPENDENCE
VILLAGE
VALLEY VIEW
VILLAGE
BEREA
MIDDLEBURGH
HEIGHTS
VILLAGE
C
D
E
F
G
H
J
K
L
M
N
O
P
Q
2
3
4
5
6
7
8
9
10
11
12
13
14

REDLINING, HOME OWNERSHIP, AND CIVIL RIGHTS

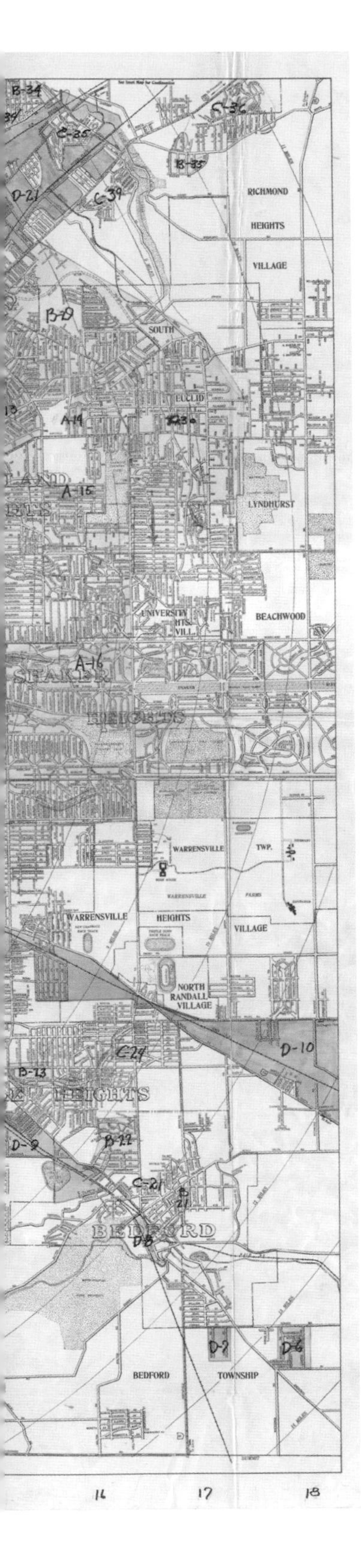

Home Owners' Loan Corporation, "Metropolitan Cleveland Security Map," 1936

The Great Depression ravaged the nation's housing market. The widespread failure of banks dramatically limited the availability of loans, while millions of Americans faced foreclosure. The Roosevelt administration responded by regulating lending and banking practices through the creation of agencies such as the Federal Housing Administration (FHA) and the Home Owners' Loan Corporation (HOLC).

Through the introduction of longer-term mortgages and mortgage insurance, the FHA sought to stabilize the market and to extend homeownership to new segments of the population. At the same time, the HOLC helped to buy and refinance mortgages on more favorable terms to help homeowners avoid foreclosure. To systematize the process of property appraisals, the HOLC also sponsored detailed assessments of lending risks in 239 urban areas across the country. From Buffalo to Seattle, from Dallas to Atlanta, the HOLC relied on local real-estate professionals, lenders, and appraisers to evaluate and map the credit risks of America's cities.

These assessments were based not just on housing stock and economic stability, but also on the "social status of the population." That category became the source of lasting controversy, for the agency judged this status according to the presence of "Negro," "foreign born," or other "undesirable" groups. The HOLC asked that neighborhoods be rated from A to D, with A (green) being the most desirable and stable, B (blue) as "still desirable," C (yellow) as "definitely declining," and D (red) as "hazardous." In one region after another, areas with heavy populations of blacks, Mexican Americans, or poor ethnic whites were deemed hazardous and least desirable, marked with red (or "redlined"). Yellow areas were often described as having been "infiltrated" by blacks. As the HOLC officials explained, this did not mean that good mortgages did not exist in these lower areas, but instead that these loans and mortgages ought to be extended and serviced in different ways.

This HOLC map of Cleveland was typical, assessing the neighborhoods in terms of their racial and economic cast. It shows wealthy neighborhoods concentrated in the eastern and western suburbs, with the "declining" and "hazardous" areas—inhabited mostly by working-class whites and African Americans—clustered around the urban center. Shaker Heights was marked in green on the eastern edge of the city as "A-16," classified as one of the area's best neighborhoods. This suburb was founded as a planned residential community in the early twentieth century, and became an incorporated city in 1931. The community originally exercised racial and ethnic restrictions on homeownership, though it successfully integrated in the 1970s and 1980s.

These maps, and the practices behind them, influenced the nation's housing market. They defined norms that were then institutionalized by the FHA's regulation, and exposed a generation of appraisers, lenders, and real-estate experts to policies that segregated by class and race. Some scholars argue that the HOLC maps reveal a federal agency actively codifying discriminatory lending practices. The maps institutionalized segregation by characterizing black neighborhoods as unworthy of credit in ways that only hardened after World War II. And by preventing lower-income blacks from accessing credit and homeownership, they contributed to the wealth gap between blacks and whites.

Other scholars disagree, pointing out that in many instances the maps were confidential and did not circulate, and that "redlining" practices of the FHA and private lenders alike long predated the HOLC. In some cities, lower-rated areas actually received loans at higher rates than more desirable neighborhoods. No doubt the maps were used differently in different cities. But even if these maps only reflected practices that were already in place, they are a sober reminder that discrimination is compounded across generations to widen the gulf between those who have access to credit and those who do not. As one study of these maps observed, the lower-rated areas only became *more* segregated over time, an indication that local lenders and brokers were paying attention to the boundaries. With these maps, the Roosevelt administration expanded homeownership for some but not for others.

LET THERE BE LIGHT

Stephen Voorhies for Rand McNally, "Raw Materials for a U.S. 'Ruhr,'" *Fortune* (October 1933)

Franklin Roosevelt became president at a moment of acute national crisis. Over 4,000 banks failed in the first few months of 1933, and the unemployment rate reached nearly 25 percent. Within days of his inauguration in March, Roosevelt sent a series of bills to Congress to alleviate the Great Depression and to stimulate economic growth. The most ambitious of these was the Tennessee Valley Authority, a massive public works project designed to address what Roosevelt considered the nation's most pressing problem: the poverty of the South.

The Tennessee Valley is almost the size of England, stretching 44,000 square miles across the heart of the Southeast. Its watershed reaches north into Virginia and Kentucky, east into North Carolina, and south into Georgia, Alabama, and Mississippi. Much of the region is mountainous, and large areas are subject to heavy rainfall and recurrent flooding. By 1930, decades of intensive cotton, tobacco, and corn farming had impoverished the soil and left it vulnerable to erosion, while deforestation further denuded the landscape. The residents of the valley had little or no access to electricity or other basic modern conveniences. The TVA was Roosevelt's effort to address this poverty through flood control, electrification, reforestation, and agricultural reform.

The project began in the lower left corner of the full map above, at Muscle Shoals, where Wilson Dam generated some of the first electricity for the valley. To the north is Norris Dam, designed to control the erratic flow of the Tennessee River in order to revive farming throughout the valley. (As shown on page 198, the Norris Dam even became a tourist destination after its completion in 1936.) These twin goals of flood control and rural electrification drove the construction of six dams along the river and its tributaries by 1940, with dozens more planned. The map marks the waterways to be improved by the TVA, as well as the existing roads and proposed air routes out of Nashville Airport (a New Deal public works project which opened in 1937).

The map conveys the ambitions of the project, which captured widespread news coverage throughout the 1930s. But the TVA also provoked intense criticism. Some noted that, though agricultural reform increased production, in the long run small farmers still could not compete against their larger counterparts. But far more consequential was the political resistance to the TVA as a form of regional planning. This criticism first appeared in the pages of *Fortune*, the monthly magazine devoted to commerce and industry founded by Henry Luce, an ardent opponent of Roosevelt's New Deal. Soon after Congress authorized the TVA, *Fortune* published this map by illustrator Stephen Voorhies to investigate the program. Voorhies used a bird's-eye view to underscore the sheer geographical scope of the project. The editors revealed their skepticism in the accompanying article by labeling the TVA as a governmental plan to remake the region into a "social-industrial-agrarian machine," an American counterpart to Germany's industrial Ruhr Valley.

American history is replete with large public works projects, such as the Panama Canal, so why did the TVA provoke such resistance? The editors of *Fortune* grudgingly acknowledged the importance of electrification, flood control, and economic development. But critics scoffed at the power and utopian aims of this new federal entity, and worried that regional planning posed a threat to state power. Power companies feared that the TVA would compete with their services, and unsuccessfully challenged its constitutionality in the courts. The president of the largest electric utility holding company in the country, Wendell Willkie, then ran against Roosevelt in the 1940 presidential election.

Willkie lost, but criticism of the TVA seeded conservative opposition to Roosevelt and the New Deal that would flourish later in the century. These critics rejected the principle that the federal government should undertake large-scale programs, planning, and regulation on behalf of the people. Resistance to the New Deal became the foundation of the modern conservative movement, even as many of these programs—chiefly social security—won bipartisan support. The TVA brought electricity to much of the rural South, provided work for thousands, and regulated flooding within the valley. But it also engendered an enduring resistance to national planning in American political culture.

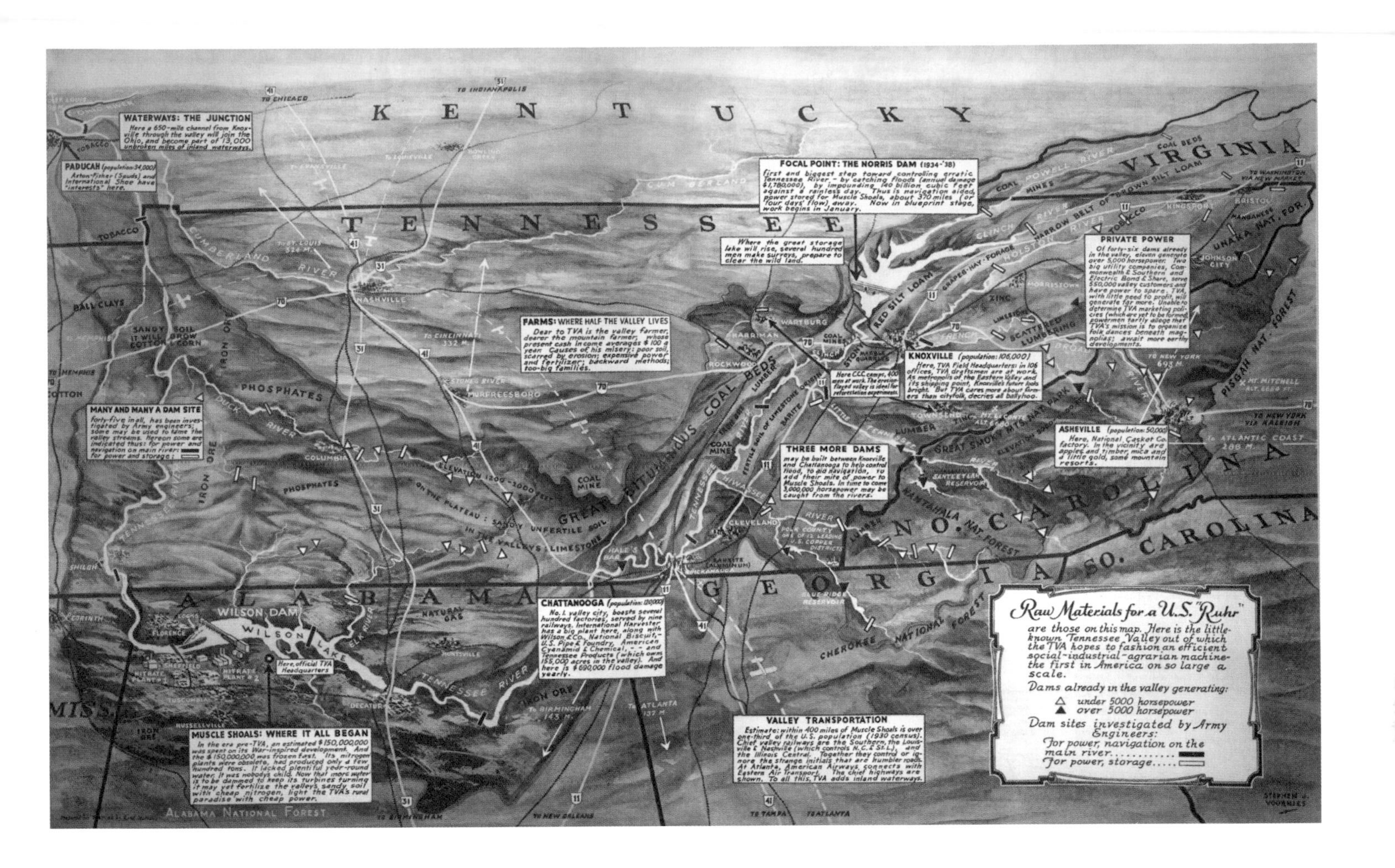
KENTUCKY
TENNESSEE
VIRGINIA
ALABAMA
GEORGIA
NO. CAROLINA
SO. CAROLINA
MISS.
TO CHICAGO
TO INDIANAPOLIS
WATERWAYS: THE JUNCTION
Here a 650-mile channel from Knoxville through the valley will join the Ohio, and become part of 13,000 unbroken miles of inland waterways.
PADUCAH (population: 34,000)
Aston-Fisher (Spuds) and International Shoe have "interests" here.
TOBACCO
CUMBERLAND RIVER
NASHVILLE
BALL CLAYS
SANDY SOIL IT WILL GROW COTTON & CORN
TO MEMPHIS
COTTON
IRON ORE
PHOSPHATES
DUCK RIVER
COLUMBIA
MANY AND MANY A DAM SITE
forty-five in all, has been investigated by Army engineers; some may be used to tame the valley streams. Hereon some are indicated thus: for power and navigation on main river: for power and storage:
FARMS: WHERE HALF THE VALLEY LIVES
Dear to TVA is the valley farmer, dearer the mountain farmer, whose present cash income averages $ 100 a year. Causes of his misery: poor soil, scarred by erosion; expensive power and fertilizer; backward methods; too-big families.
TO CINCINNATI 332 M.
STONES RIVER
MURFREESBORO
ELEVATION 1200-2000 FEET
ON THE PLATEAU: SANDY UNFERTILE SOIL
IN THE VALLEYS: LIMESTONE
COAL MINE
GREAT BITUMINOUS COAL BEDS
FOCAL POINT: THE NORRIS DAM (1934-'38)
first and biggest step toward controlling erratic Tennessee River - by catching floods (annual damage $1,780,000), by impounding 140 billion cubic feet against a rainless day. Thus is navigation aided, power stored for Muscle Shoals, about 370 miles (or four days' flow) away. Now in blueprint stage, work begins in January.
Where the great storage lake will rise, several hundred men make surveys, prepare to clear the wild land.
KNOXVILLE (population: 106,000)
PRIVATE POWER
THREE MORE DAMS
ASHEVILLE (population: 50,000)
SHILOH
CORINTH
WILSON DAM
WILSON LAKE
FLORENCE
SHEFFIELD
NITRATE PLANT # 2
TUSCUMBIA
RUSSELLVILLE
Here, official TVA Headquarters
NATURAL GAS
ELK RIVER
HUNTSVILLE
TENNESSEE RIVER
DECATUR
IRON ORE
TO BIRMINGHAM 143 M.
TO ATLANTA 137 M.
MUSCLE SHOALS: WHERE IT ALL BEGAN
In the era pre-TVA, an estimated $150,000,000 was spent on its War-inspired development. And the $150,000,000 was frozen fast. Its nitrogen plants were obsolete, had produced only a few hundred tons. It lacked plentiful year-round water. It was nobody's child. Now that more water is to be dammed to keep its turbines turning it may yet fertilize the valley's sandy soil with cheap nitrogen, light the TVA's rural paradise with cheap power.
CHATTANOOGA (population: 120,000)
No. 1 valley city, boasts several hundred factories, served by nine railways. International Harvester has a big plant here, along with Wilson & Co., National Biscuit, U.S. Pipe & Foundry, American Cyanamid & Chemical, - - and Tennessee Products (which owns 155,000 acres in the valley). And here is $690,000 flood damage yearly.
HALE'S BAR
BAUXITE (ALUMINUM)
CLEVELAND
BLUE RIDGE RESERVOIR
CHEROKEE NATIONAL FOREST
VALLEY TRANSPORTATION
Estimate: within 400 miles of Muscle Shoals is over one-third of the U.S. population (1930 census). Chief valley railways are the Southern, the Louisville & Nashville (which controls N.C. & St.L.), and the Illinois Central. Together they control or ignore the strange initials that are humbler roads. At Atlanta, American Airways connects with Eastern Air Transport. The chief highways are shown. To all this, TVA adds inland waterways.
ALABAMA NATIONAL FOREST
TO BIRMINGHAM
TO NEW ORLEANS
TO TAMPA
TO ATLANTA
Raw Materials for a U.S. "Ruhr"
are those on this map. Here is the little-known Tennessee Valley out of which the TVA hopes to fashion an efficient social-industrial-agrarian machine - the first in America on so large a scale.
Dams already in the valley generating:
under 5000 horsepower
over 5000 horsepower
Dam sites investigated by Army Engineers:
For power, navigation on the main river
For power, storage
STEPHEN J. VOORHIES

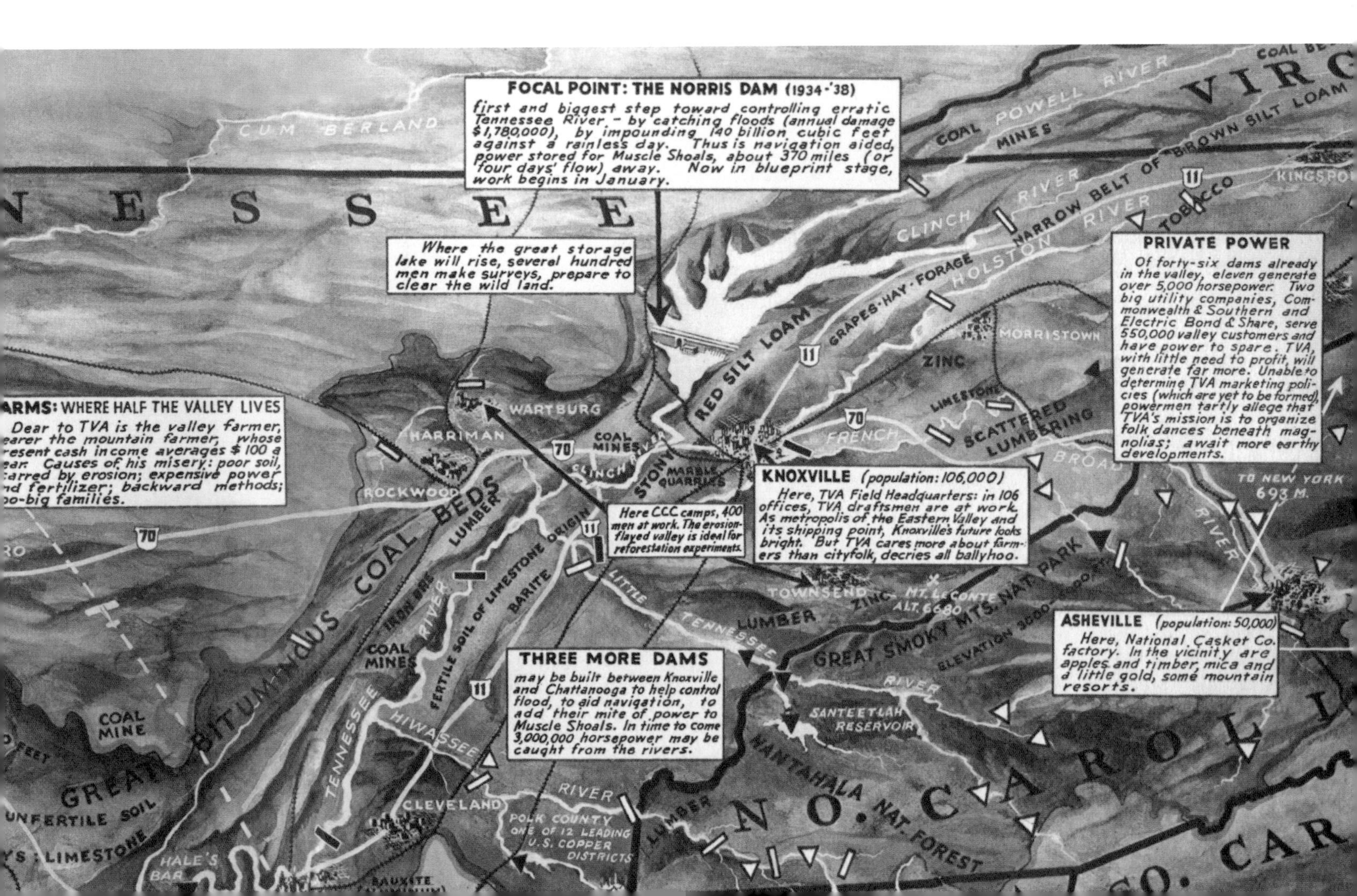
FOCAL POINT: THE NORRIS DAM (1934-'38)
first and biggest step toward controlling erratic Tennessee River - by catching floods (annual damage $1,780,000), by impounding 140 billion cubic feet against a rainless day. Thus is navigation aided, power stored for Muscle Shoals, about 370 miles (or four days' flow) away. Now in blueprint stage, work begins in January.
CUMBERLAND
COAL MINES
POWELL RIVER
COAL BEDS
NARROW BELT OF BROWN SILT LOAM
KINGSPORT
TOBACCO
CLINCH RIVER
HOLSTON RIVER
GRAPES · HAY · FORAGE
RED SILT LOAM
Where the great storage lake will rise, several hundred men make surveys, prepare to clear the wild land.
PRIVATE POWER
Of forty-six dams already in the valley, eleven generate over 5,000 horsepower. Two big utility companies, Commonwealth & Southern and Electric Bond & Share, serve 550,000 valley customers and have power to spare. TVA, with little need to profit, will generate far more. Unable to determine TVA marketing policies (which are yet to be formed), powermen tartly allege that TVA's mission is to organize folk dances beneath magnolias; await more earthy developments.
MORRISTOWN
ZINC
LIMESTONE
SCATTERED LUMBERING
FRENCH BROAD RIVER
ARMS: WHERE HALF THE VALLEY LIVES
Dear to TVA is the valley farmer, earer the mountain farmer, whose resent cash income averages $ 100 a ear. Causes of his misery: poor soil, carred by erosion; expensive power nd fertilizer; backward methods; oo-big families.
WARTBURG
HARRIMAN
ROCKWOOD
COAL MINES
CLINCH
STONY
MARBLE QUARRIES
KNOXVILLE (population: 106,000)
Here, TVA Field Headquarters: in 106 offices, TVA draftsmen are at work. As metropolis of the Eastern Valley and its shipping point, Knoxville's future looks bright. But TVA cares more about farmers than cityfolk, decries all ballyhoo.
TO NEW YORK 693 M.
Here C.C.C. camps, 400 men at work. The erosion-flayed valley is ideal for reforestation experiments.
GREAT BITUMINOUS COAL BEDS
LUMBER
IRON ORE
FERTILE SOIL OF LIMESTONE ORIGIN
BARITE
LITTLE TENNESSEE
TOWNSEND
ZINC
MT. LECONTE ALT. 6680
GREAT SMOKY MTS. NAT. PARK
ELEVATION 3000-6000 FT.
LUMBER
ASHEVILLE (population: 50,000)
Here, National Casket Co. factory. In the vicinity are apples and timber, mica and a little gold, some mountain resorts.
COAL MINES
COAL MINE
THREE MORE DAMS
may be built between Knoxville and Chattanooga to help control flood, to aid navigation, to add their mite of power to Muscle Shoals. In time to come 3,000,000 horsepower may be caught from the rivers.
TENNESSEE
HIWASSEE
SANTEETLAH RESERVOIR
NANTAHALA NAT. FOREST
RIVER
CLEVELAND
POLK COUNTY ONE OF 12 LEADING U.S. COPPER DISTRICTS
LUMBER
UNFERTILE SOIL
S: LIMESTONE
HALE'S BAR
BAUXITE
NO. CAROLI
SO. CAR

8. 1940–1962

Between War and Abundance

In 1934 Senator Gerald Nye launched a high-profile investigation of American entry into World War I. Nye was responding to longstanding accusations that munitions manufacturers had influenced the decision to go to war against Germany in 1917. After eighteen months, his committee found little direct evidence for these claims, but the investigation drew wide attention and revived the debate about whether the nation ought to have entered such a costly war in the first place. The Nye Committee's work also signaled the deeply isolationist mood of the country. Americans watched apprehensively as fascist governments took power in Europe and Asia during the 1930s, but expressed little desire to intervene.

President Franklin D. Roosevelt publicly voiced concern about the growth of fascism in 1937, but he faced a legislature that was determined to prevent the entanglements that had pulled the country into war two decades earlier. Congress first attempted to legislate neutrality by limiting American overseas trade and travel. After Germany invaded Poland in 1939, Congress required France and Great Britain to "cash and carry" any arms purchased from the United States. Even after the swift Nazi invasion against France and the Low Countries in 1940, many Americans wondered whether a rearmed and expansionist Germany necessarily posed a threat to national security.

American isolationism was reinforced by the geographical perception that the Atlantic and Pacific oceans separated the United States from Europe and Asia. The advent of aviation challenged that assumption, but only with the entry into World War II were Americans forced to reckon with these new spatial realities. Among the most forceful apostles of this new "air age" geography was Richard Edes Harrison, whose creative and unconventional maps simultaneously shocked and dazzled the public during the war. His 1940 effort to map the nation from unfamiliar angles (page 206) was just one of many attempts to reassess world geography in the age of aviation.

The following summer, Harrison issued an equally disruptive map centered on the North Pole to demonstrate that this "European" war had everything to do with the United States (page 208). Just as American readers were poring over that map, Roosevelt and the British prime minister Winston Churchill met secretly off the coast of Newfoundland to develop principles that would govern the postwar world. Roosevelt was formulating such a vision even before the United States had entered the war, and the principles of the Atlantic Charter echo President Wilson's "Fourteen Points" in 1918. In both cases, the US prioritized open access to the foreign markets and raw materials. The British artist MacDonald Gill rendered an ebullient and optimistic profile of that new world, one ordered and integrated through commerce (page 210).

Franklin Roosevelt's own experience as Assistant Secretary of the Navy in the 1910s gave him a keen appreciation of geography. On several occasions during the war, he asked Americans to consult their maps and globes as he explained various aspects of strategy over the radio. On the title page of this book we see the president consulting his 1942 Christmas gift from the Army, a 750-pound globe that was designed to rotate freely—without an axis—in order to facilitate strategic thinking in a world governed by aviation. Like Harrison, Roosevelt was attuned to a global sense of geography.

Through 1942 and much of 1943, most American troops were sent to the Pacific, which led the Soviet leader Joseph Stalin to charge that his country was bearing the brunt of the fight against Germany. With the invasion of Italy in 1943, but more importantly France in June 1944, the Americans forcefully joined the Allied battle to destroy the Third Reich in Europe. One soldier on the front lines—Henry MacMillan—mapped the military and logistical challenges of that brutal fight (pages 212–215). His two maps reveal not just the Allied defeat of Germany, but the delicate relationship between American and Soviet forces as they both marched toward Berlin.

In that final year of war, American troops battling the enemy in Germany and Austria also confronted the unimaginable horror of the Final Solution. The immensity of the Holocaust is impossible to comprehend, but the maps on pages 216 and 217 record one young man's survival of the war and the concentration camps. Michal Kraus chronicled his imprisonment and ultimate liberation through a lengthy illustrated memoir written immediately after the war. Like MacMillan, Kraus chose to *map* his experience. His first map documents the brutal network of camps that tortured him and killed his family. Once the Americans had liberated Austria, he began the long and harrowing journey home to Czechoslovakia, passing through lands that were now controlled not by the Germans but by the Soviets. In those closing days of World War II, the beginnings of the Cold War were already apparent.

The war made the United States the most powerful nation on earth. Its territory was unscathed, and wartime mobilization more than doubled the nation's economic capacity. That power—combined with a sense of international stewardship—launched a new era of foreign policy that was accelerated by the swift deterioration of relations with the Soviet Union. The world of free trade outlined by the Atlantic Charter—a logical extension of American capitalism—was treated as a direct threat by the Soviet Union. While the United States helped to rebuild western Europe after the war through the Marshall Plan, it also began to expand its military presence in Europe and around the world. The Soviets similarly asserted their sphere of influence, raising the stakes in a cold war where each considered the other the aggressor (pages 222 and 224).

World War II and the Cold War profoundly shaped domestic life as well. Wartime spending quickly ended the Great Depression, and postwar federal priorities both altered the nation's economic structure and fueled widespread prosperity. In 1944 the Army Map Service printed an astonishing series of maps tracing the evolution of the lower Mississippi River over centuries. The research undergirding these maps was designed to control flooding on the river, and reflected a heightened commitment to domestic infrastructure (page 218). Similarly, the Serviceman's Readjustment Act—known as the G.I. Bill—created a generation of homeowners and college students, while the Federal-Aid Highway Act of 1956 integrated the country and advanced suburban development (page 228). These large-scale federal investments—as well as ongoing defense spending—accelerated the westward migration of the population, and by 1963 California had eclipsed New York as the nation's largest state.

While highways facilitated mobility on the ground, the rise of commercial aviation expanded mobility in the air (page 220). Here too the war mattered, for thousands of surplus aircraft stimulated the explosion of postwar air travel. In the 1950s, the jet engine revolutionized aviation, shortening travel between the coasts from nine hours to five. Just as the railroads had transformed geography in the late nineteenth century, the advent of airline travel transformed it in the twentieth, integrating different industries and regions and stimulating economic growth.

Postwar prosperity and technological innovation also brought new forms of leisure and entertainment, such as television. Among the early adopters of this new medium was Walt Disney, who began to broadcast his animated characters into American living rooms by the early 1950s. At the same time, he began to envision a large theme park where these characters would come to life. The preliminary sketch of Disneyland on page 226 captures his vision for this modern form of leisure, one that would further enhance the allure of Southern California.

But this era of abundance was severely limited by racial discrimination. Among the most egregious examples was the wartime executive order that evacuated Japanese Americans to remote internment camps throughout the West. The armed forces were also segregated by race during the war, prompting black leaders to decry the hypocrisy of fighting tyranny abroad while racial injustice remained unchallenged at home. After the war, the Supreme Court's *Brown v. Board of Education* decision determined that segregation indeed violated the Fourteenth Amendment's equal-protection clause. But the limited gains of the Civil Rights Movement left many activists frustrated by the early 1960s. The modest newspaper map of the Freedom Rides on page 230 illustrates the crucial role of the media in exposing Southern resistance to federal desegregation orders. The fight against Soviet repression abroad also nudged civil rights forward at mid-century, demonstrating the deep connection between foreign and domestic policy in these decades.

WORLD WAR II AND THE REINVENTION OF CARTOGRAPHY

Richard Edes Harrison, "Three Approaches to the U.S.," *Fortune* (September 1940) and "The World Divided," *Fortune* (August 1941)

More Americans encountered maps during World War II than in any previous era in American history. From elaborate inserts in *National Geographic* to schematic views in the daily news, maps were everywhere. On September 1, 1939, the Nazis invaded Poland, and by the end of the day maps of Europe had sold out across the United States. Two years later, the attack on Pearl Harbor sparked another rush to buy maps. The country's leading mapmakers reported their largest sales to date in 1941, and in early 1942 *Newsweek* dubbed Washington, D.C. "a city of maps," where "it is now considered a *faux pas* to be caught without your Pacific arena."

If the war drove popular interest in geography, it was the advent of aviation that transformed its meaning. Aviation collapsed distance and realigned geographical relationships, and nowhere was this more apparent than in the changing look of maps. In the first half of the twentieth century, most Americans had been reared on world maps using the Mercator projection (page 210), which dates back to the sixteenth century. This cylindrical projection was invaluable in the age of ocean navigation, for directions are true even though geography is distorted as one moves away from the equator. But though the Mercator projection had become ubiquitous, it could not adequately depict geographical relationships in the age of aviation. Oceans, be they vast seas or the frozen Arctic to the north, were no longer insurmountable physical barriers or buffers. Air routes could not be accurately charted on the Mercator projection, nor did it adequately illustrate the importance of the North Pole. For all these reasons, by mid-century Americans needed to relearn geography, and to reacquaint themselves with the spherical nature of the earth.

One of the most effective teachers of this new geography was Richard Edes Harrison, an artist and designer who drew a series of maps for *Fortune* magazine during the war. *Fortune* was founded by Henry Luce in the 1930s with an intent to convey an internationalist message to the business community. Harrison's maps helped to visualize this new posture of American stewardship. With his creative use of color and perspective, Harrison redrew the world map in a way that engaged and challenged readers to consider the new realities of geography.

Shown here are Harrison's wildly popular global views, which he used to highlight particular geographical relationships. Consider his well-executed "Three Approaches to the U.S.," first published in September 1940 as part of a series designed to challenge American isolationism. With a stylized approach, Harrison brought the war home to Detroit (from Berlin), to the West (from Tokyo), and to the "soft belly" of the South (from South America). No longer were the Atlantic and Pacific oceans the protective buffers they had been on a Mercator projection; instead, viewers were startled to see how unfamiliar their own nation appeared with just a minor shift of perspective. In Harrison's schematic view, the interior was as vulnerable as the coasts—a direct challenge to the comfortable isolationism voiced by groups such as America First.

Notice also that, unlike most reference maps, Harrison's included *only* the information needed to make his point: by depicting terrain in an exaggerated manner, and only presenting selected cities, he conveyed particular relationships without distracting the reader with unnecessary detail. This visual style electrified *Fortune*'s readers by pulling them into the landscape as if they were pilots hovering over the land. No such aerial view exists, where the curvature of the earth is visible alongside topographic details, but this imaginative rendering of the globe enabled Harrison to visualize direction in a way that resonated with readers. His oblique perspective—colorfully and confidently executed and with a keen eye for design—helped viewers rediscover relationships that lay obscured in traditional maps. Geography was made new.

Harrison was equally known for rejecting the Mercator projection. Among his most striking efforts was "The World Divided" (shown on the next page). Here, the world is organized around the North Pole to highlight the proximity brought about by aviation. Though the map grossly distorts the southern latitudes, Harrison used a polar projection to demonstrate the relationship of the belligerents over the North Pole, where the great circle routes of aviation created new geographical truths.

THREE APPROACHES TO THE U.S.

Map 9

...FROM BERLIN

A great-circle route from Berlin here passes through Detroit. The fanciful can see, if they wish, a pincers movement extending from Newfoundland down the New England coast and down the St. Lawrence to the continent's heart; a third arm reaching to the south shore of Hudson Bay, where the terrain permits quick construction of landing fields. And there is no east-west highway north of the Great Lakes.

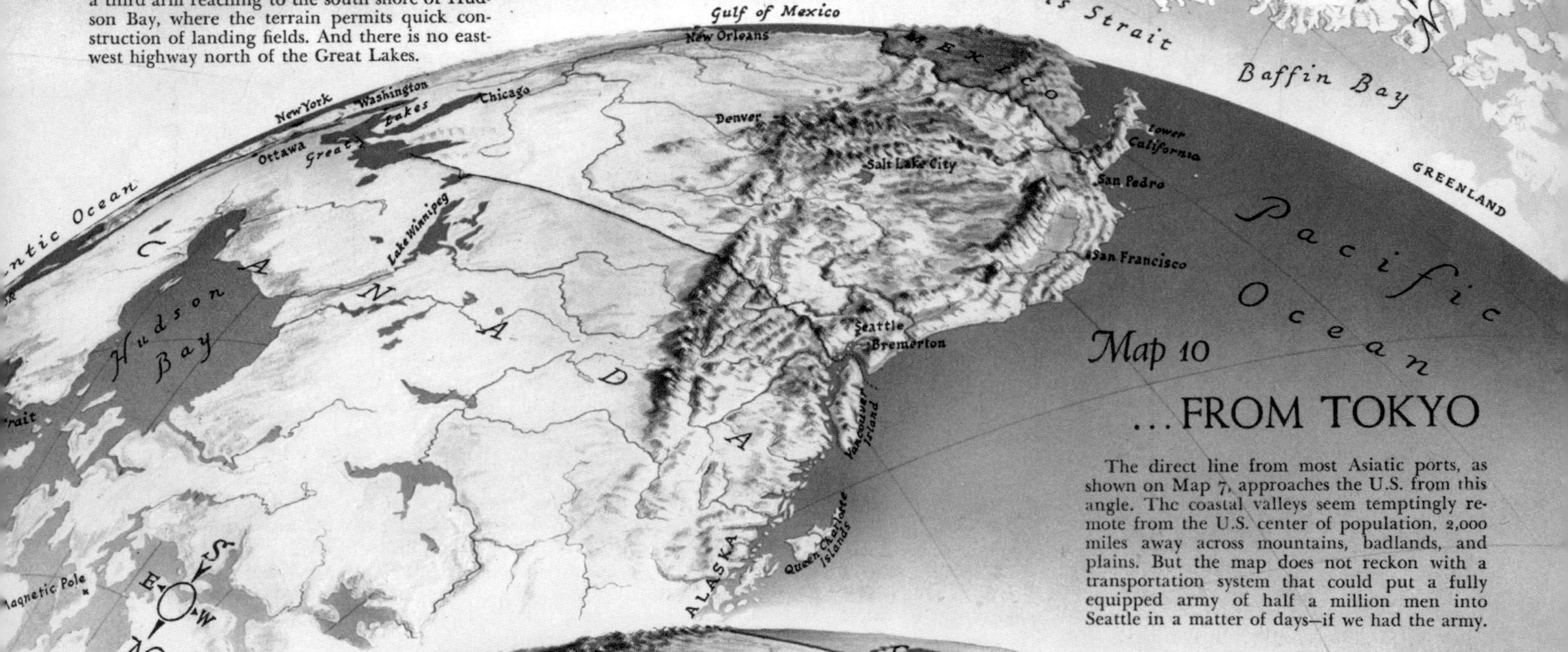

Map 10

...FROM TOKYO

The direct line from most Asiatic ports, as shown on Map 7, approaches the U.S. from this angle. The coastal valleys seem temptingly remote from the U.S. center of population, 2,000 miles away across mountains, badlands, and plains. But the map does not reckon with a transportation system that could put a fully equipped army of half a million men into Seattle in a matter of days—if we had the army.

Map 11

...FROM CARACAS

If an enemy should ever establish himself on the northern shore of South America or in the mazes of the West Indies, he would cut first at the U.S. G-string, the Canal, then rip at its soft belly here displayed. For the Gulf Coast—with oil, salt, sulfur, coal, and gas—is becoming a great chemical stewpot nourishing, shaping, and extending industry, a modern analogue to the earlier iron-ore economy of the Great Lakes.

Drawn in July 1941, the map captures the harrowing moment when the German invasion of the Soviet Union left the British alone to fight the Axis. In August, Roosevelt met with Churchill to forge an alliance, but the United States was not yet at war, and many Americans continued to believe that the oceans protected them from these foreign crises. With this map, Harrison showed Americans how *close* these conflicts were, and in the process buttressed support for the recent Lend-Lease program of aid to Britain. With orange lines underscoring the web of American involvement, Harrison made it impossible to dismiss the war as a purely European affair. Moreover, his accessible explanation in the upper left corner helped to demystify the concept of map projections. In this realigned world, the United States lay at the heart of the conflict, not at its periphery. A perennial bestseller, the map was updated and reissued throughout the war.

Once Congress had declared war on Japan and Germany, Harrison's maps seemed prescient, and their popularity grew even further. They imaginatively and persuasively captured a world disrupted by both war and aviation. His *Look at the World* atlas—specifically designed to explain wartime strategy—sold out before it even reached bookstores. The Army and Navy reprinted 250,000 copies of Harrison's maps for servicemen abroad. His perspective maps were used to expose bomber pilots to aerial views of European and Asian terrain. Dozens of corporations, newspapers, civic organizations, citizens, and towns wrote to Harrison in praise of his iconoclastic and visually provocative style, which was widely imitated during and after the war.

But in each of Harrison's riveting maps, accuracy in one area came at the expense of inaccuracy elsewhere. Most professional geographers welcomed Harrison's ability to capture the public's attention and to foster a more global sensibility. However, some bristled at a pictorial approach that seemed to prize design over mathematical precision. Such criticisms did not faze Harrison. He proudly identified as an artist—not a mapmaker—and believed that his training in architecture and design gave him the critical distance to disrupt static views of geography. The widespread imitation of his maps confirms that Americans were willing to reconsider their understanding of world geography and their place within it.

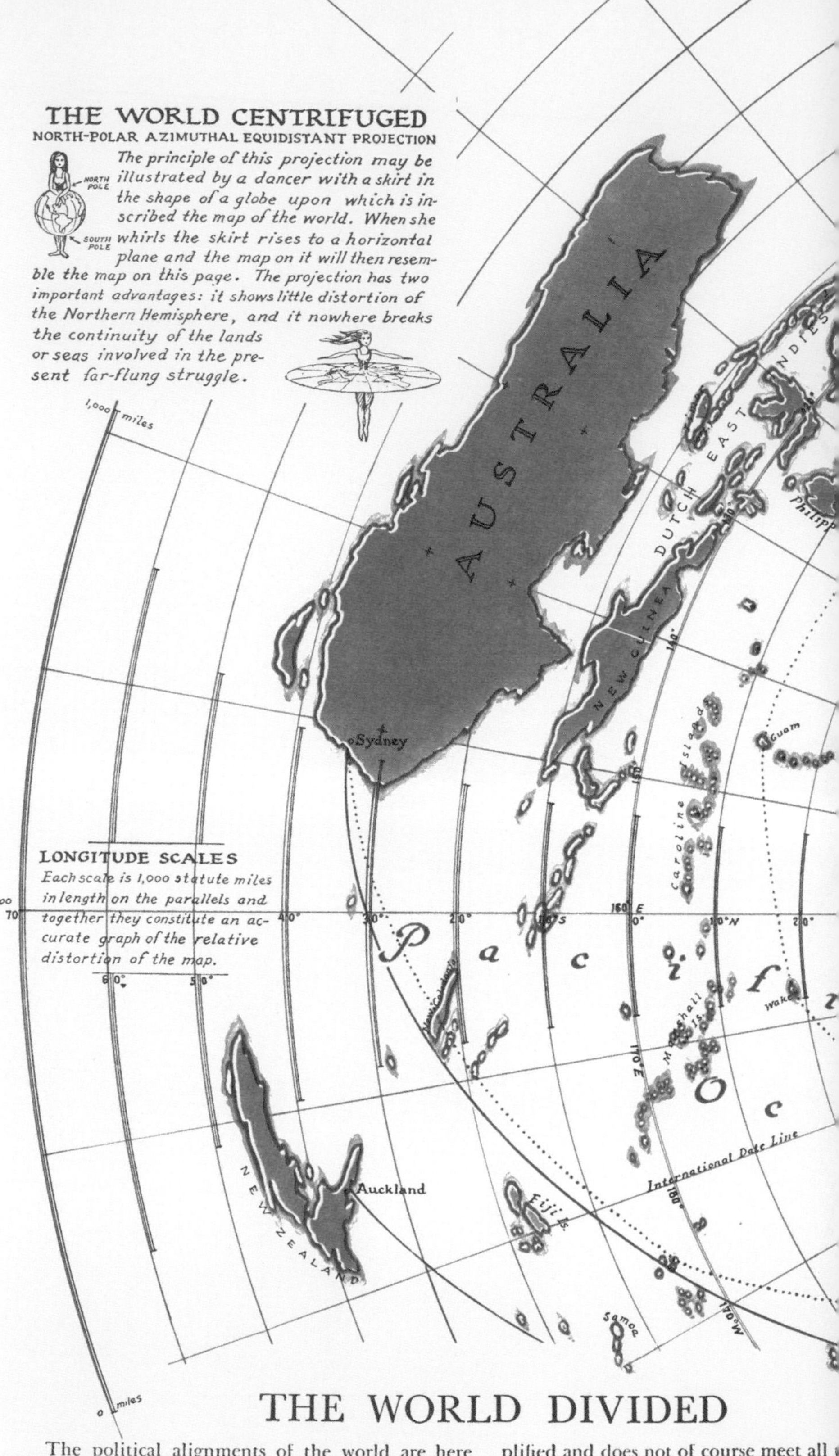

THE WORLD CENTRIFUGED
NORTH-POLAR AZIMUTHAL EQUIDISTANT PROJECTION
The principle of this projection may be illustrated by a dancer with a skirt in the shape of a globe upon which is inscribed the map of the world. When she whirls the skirt rises to a horizontal plane and the map on it will then resemble the map on this page. The projection has two important advantages: it shows little distortion of the Northern Hemisphere, and it nowhere breaks the continuity of the lands or seas involved in the present far-flung struggle.
NORTH POLE
SOUTH POLE
1,000 miles
AUSTRALIA
DUTCH EAST INDIES
NEW GUINEA
Sydney
Guam
LONGITUDE SCALES
Each scale is 1,000 statute miles in length on the parallels and together they constitute an accurate graph of the relative distortion of the map.
Pacific Oc
International Date Line
Auckland
NEW ZEALAND
Fiji Is.
Samoa
0 miles
THE WORLD DIVIDED
The political alignments of the world are here shown, centering geographically around the North Pole but ideologically and economically around the U.S. The classification keyed below is necessarily simplified and does not of course meet all
of politics and diplomacy—for instance,
position of Vichy and its colonies. The s
the U.S.S.R. is acknowledged by a speci

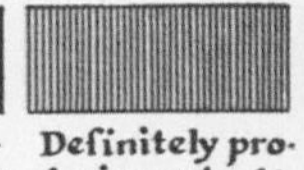

Indian Ocean
MADAGASCAR (VICHY)
EQUATOR
MOZAMBIQUE
UNION OF SOUTH AFRICA
RHODESIA
BECHUANALAND
Cape of Good Hope
TANGANYIKA
KENYA
ITALIAN EAST AFRICA
UGANDA
BELGIAN CONGO
ANGOLA (Port.)
S.W. AFRICA
SUMATRA
Singapore
Rangoon
Bombay
INDIA
THAILAND
BURMA
VICHY INDO-CHINA
Burma Road
NEPAL
TIBET World's purest isolationism
AFGHANISTAN
IRAN
Basra
SAUDI ARABIA
ANGLO-EGYPTIAN SUDAN
DE GAULLE AFRICA
AFRICA
Hong Kong
Manila
Chungking
CHINA
SINKIANG
MONGOLIA
Controlled by Moscow
MANCHUKUO
ASIA
U. S. S. R.
Siberia
Count this block if Nazis win a quick and complete victory.
Black Sea
TURKEY
Suez
EGYPT
LIBYA
NIGERIA
VICHY AFRICA
Mediterranean Sea
TUNISIA
ALGERIA
MOROCCO
Meridian of Greenwich
GOLD COAST
St. Helena
Tokyo
SWEDEN
London
VICHY
SPAIN
PORT.
OCCUPIED FRANCE
Str. of Gibraltar
SIERRA LEONE
Bathurst
Dakar
Ascension
Jan Mayen I.
NORTH POLE
Iceland U.S. outpost
Greenland
Icecap
Bering Str.
LATITUDE SCALE
STATUTE MILES
Correct anywhere on meridians (radii).
Azores (Port.)
Cape Verde Is. Port.
Atlantic Ocean
Midway I.
ALASKA
possible summer route
NORTH AMERICA
CANADA
U. S. A.
Glasgow 3,404
Bathurst 3,956
Suez 13,844
Basra 13,793
Bombay 13,059
New York
Archangel 6,983
Vladivostok 5,256
Singapore 8,486
Rangoon 8,732
Honolulu
San Francisco
Sydney 7,415
Auckland 6,532
MEXICO
CUBA
HAITI
Jamaica
Martinique (Vichy)
GUIANAS
Br.
Neth.
Vichy
VENEZUELA
COLOMBIA
ECUADOR
PERU
BOLIVIA
PARAGUAY
BRAZIL
SOUTH AMERICA
ARGENTINA
CHILE
German colonists
Falkland Is. (Br.)
GUAT.
HOND.
EL SAL.
COSTA RICA
PAN.
Panama Canal
Galapagos Is.
Marquesas
All distances are given in statute miles (in nautical miles they are 1/7 less).
Richard Edes Harrison July 1941

THE WARTIME ROOTS OF THE AMERICAN CENTURY

MacDonald Gill, "The 'Time & Tide' Map of the Atlantic Charter," 1942

After Congress approved Roosevelt's Lend-Lease program of massive military aid to Britain in 1941, Roosevelt and Churchill secretly met off the coast of Newfoundland to advance their alliance and establish a set of postwar principles that would be dubbed the Atlantic Charter. While the charter may appear to be a general statement of universal principles, embedded within these principles were very specific American goals. Echoing President Wilson's Fourteen Points of 1918, the charter promised that neither country would seek territorial gains in the war, and that self-determination would guide postwar settlements. Equally essential were Roosevelt's priorities of lowered trade barriers and freedom of the seas, which sent a message that the US was eager to access the goods that became available as the colonial era waned.

Four months after the charter was signed, the attack on Pearl Harbor led the US to declare war on Germany and Japan. In 1942, *Time & Tide* magazine commissioned the noted and admired commercial artist MacDonald Gill to boost British morale with this buoyant map promising better days ahead. Powerful steamships busily transport passengers, mail, and goods across the seas. In the foreground a man takes a sledgehammer to military weapons, while horses pull ploughshares just beyond. This is a world at peace, and above all one defined through trade.

American entry into World War II was necessary to destroy fascism. At the same time, the United States was extraordinarily fortunate to emerge with its national economy, territory, and population largely intact relative to the European belligerents. This structural integrity—buttressed by the 1944 authorization of the World Bank and the International Monetary Fund—directly fueled the postwar economic boom that made the US the world's most powerful country. For a generation reared on depression and war, such prosperity was both welcome and deserved. While the war hobbled Europe and Asia, it created sustained postwar growth for the US.

Just before the Atlantic Charter was issued, the publisher Henry Luce urged Americans to accept their position of international stewardship. Luce predicted an "American Century," where the United States—not Britain—would use its economic and moral power to promote democracy and freedom around the world. This posture would profoundly shape American foreign policy for the rest of the century.

Map of THE ATLANTIC CHARTER

HARTER

ly expressed wishes of the peoples concerned.
which they will live; and they wish to see
ibly deprived of them.
, to further the enjoyment by all States,
trade and to the raw materials of the

tions in the economic field, with the
ncement and social security.

SIXTH, after the final destruction of Nazi tyranny, they hope to see established a peace which will afford to all nations the means of dwelling in safety within their own boundaries, and which will afford assurance that all the men in all the lands may live out their lives in freedom from fear and want.

SEVENTH, such a peace should enable all men to traverse the high seas and oceans without hindrance.

EIGHTH, they believe all of the nations of the world, for realistic as well as spiritual reasons, must come to the abandonment of the use of force. Since no future peace can be maintained if land, sea or air armaments continue to be employed by nations which threaten, or may threaten, aggression outside of their frontiers, they believe, pending the establishment of a wider and permanent system of general security, that the disarmament of such nations is essential. They will likewise aid and encourage all other practicable measures which will lighten for peace-loving peoples the crushing burden of armaments.

Franklin D Roosevelt Winston S. Churchill 1941

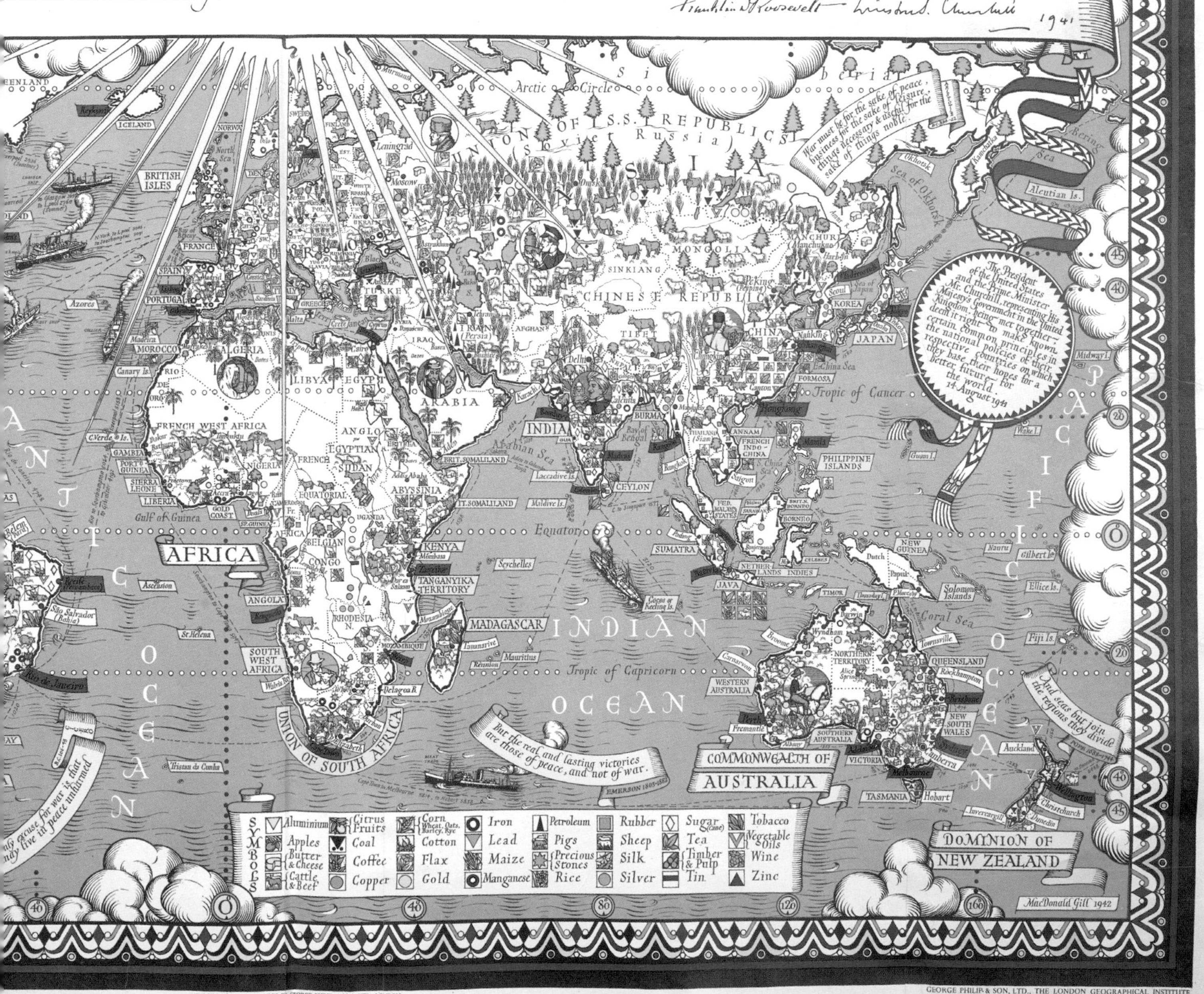

PRINTED IN GREAT BRITAIN BY GEORGE PHILIP & SON, LTD., LONDON
GEORGE PHILIP & SON, LTD., THE LONDON GEOGRAPHICAL INSTITUTE

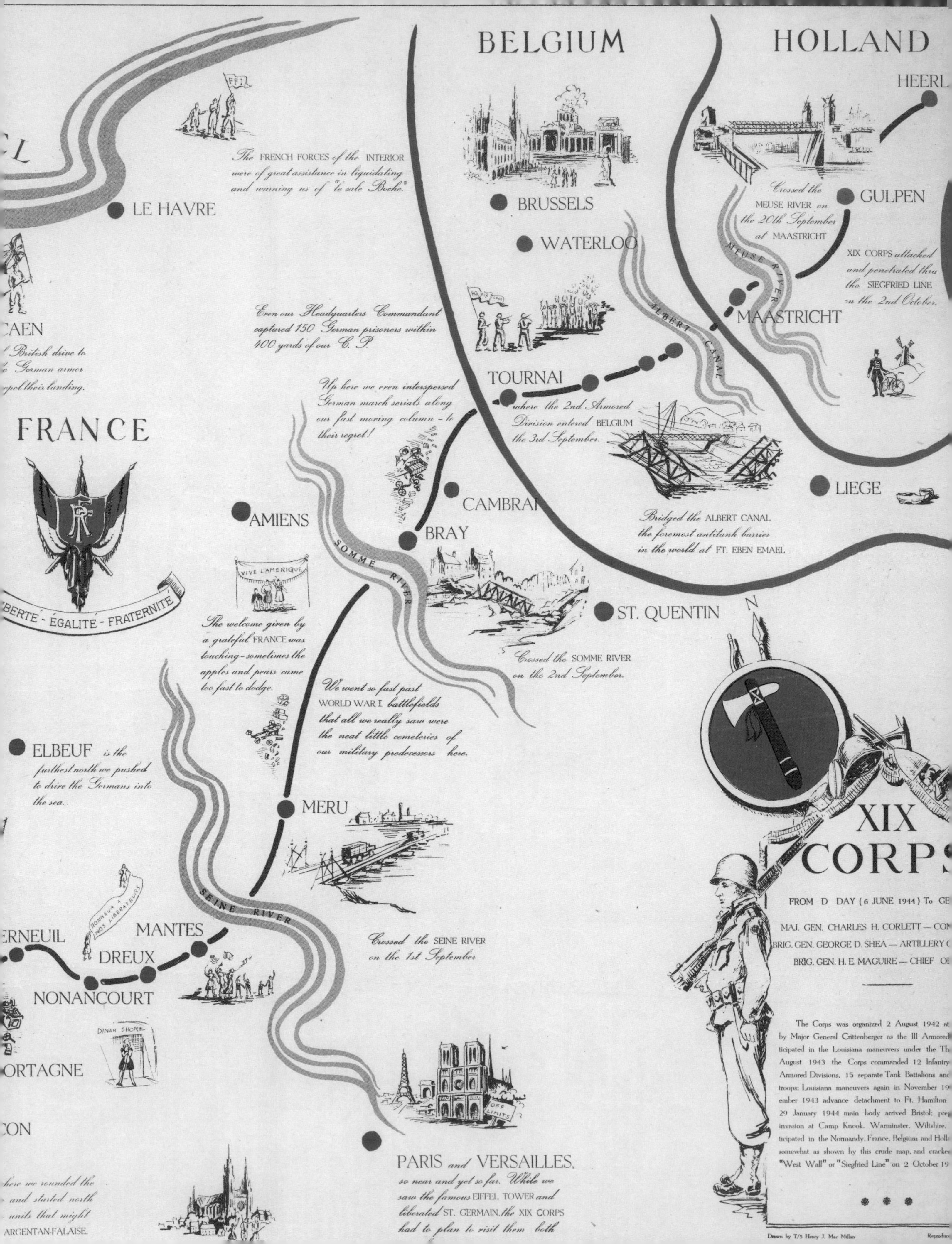

BELGIUM
HOLLAND
HEERL
FFI
L
The FRENCH FORCES of the INTERIOR were of great assistance in liquidating and warning us of "le sale Boche."
LE HAVRE
BRUSSELS
WATERLOO
Crossed the MEUSE RIVER on the 20th September at MAASTRICHT
GULPEN
MEUSE RIVER
XIX CORPS attacked and penetrated thru the SIEGFRIED LINE on the 2nd October.
CAEN
British drive to
German armor
pel their landing.
Even our Headquarters Commandant captured 150 German prisoners within 400 yards of our C. P.
ALBERT CANAL
MAASTRICHT
TOURNAI
Up here we even interspersed German march serials along our fast moving column – to their regret!
where the 2nd Armored Division entered BELGIUM the 3rd September.
FRANCE
LIEGE
CAMBRAI
AMIENS
BRAY
Bridged the ALBERT CANAL the foremost antitank barrier in the world at FT. EBEN EMAEL
VIVE L'AMERIQUE
SOMME RIVER
BERTE - ÉGALITÉ - FRATERNITÉ
ST. QUENTIN
N
The welcome given by a grateful FRANCE was touching – sometimes the apples and pears came too fast to dodge.
Crossed the SOMME RIVER on the 2nd September.
We went so fast past WORLD WAR I battlefields that all we really saw were the neat little cemeteries of our military predecessors here.
ELBEUF is the furthest north we pushed to drive the Germans into the sea.
MERU
XIX
CORPS
FROM D DAY (6 JUNE 1944) To GE
HONNEUR A NOS LIBERATEURS
SEINE RIVER
MAJ. GEN. CHARLES H. CORLETT — COM
RNEUIL
MANTES
BRIG. GEN. GEORGE D. SHEA — ARTILLERY
DREUX
Crossed the SEINE RIVER on the 1st September
BRIG. GEN. H. E. MAGUIRE — CHIEF O
NONANCOURT
DINAH SHORE
The Corps was organized 2 August 1942 at by Major General Crittenberger as the III Armored ticipated in the Louisiana maneuvers under the Th August 1943 the Corps commanded 12 Infantry Armored Divisions, 15 separate Tank Battalions and troops; Louisiana maneuvers again in November 19 ember 1943 advance detachment to Ft. Hamilton 29 January 1944 main body arrived Bristol; pre invasion at Camp Knook, Warminster, Wiltshire, ticipated in the Normandy, France, Belgium and Holla somewhat as shown by this crude map, and cracke "West Wall" or "Siegfried Line" on 2 October 19
ORTAGNE
OFF LIMITS
CON
PARIS and VERSAILLES, so near and yet so far. While we saw the famous EIFFEL TOWER and liberated ST. GERMAIN, the XIX CORPS had to plan to visit them both
here we rounded the and started north units that might ARGENTAN-FALAISE.
Drawn by T/5 Henry J. Mac Millan

GERMANY
LAST ACTE
" DEUTSCHLAND
UNTER
ALLIES"
will be the 1944 edition
of the German national
anthem.
ARMEE BLANCHE
We were helped in
numerous ways by
the Belgian resistance.

THE DEFEAT OF GERMANY

Henry MacMillan, US Army, "XIX Corps in Action" and "XIX Corps in Action: From Siegfried Line to Victory," 1945

More than 16 million men and women served in the American armed forces in World War II. Staggering as that may seem, the war killed 60 million people worldwide. About 320,000 Americans died fighting the war, and 800,000 were injured; their commitment helped to defeat the brutal regimes in Germany and Japan. At the front lines of the European theater was the Nineteenth Corps, which landed on the beaches of Normandy in June 1944. Through summer and fall the men pushed through western Europe, where they awaited instructions from Supreme Commander Dwight Eisenhower before driving into Germany.

Henry MacMillan, an Army engineer, mapped the maneuvers of the Corps in that last year of the war. At left is a portion from his first map, which detailed the battle to liberate France. On both maps, MacMillan depicted the individual encounters that so profoundly affected the millions of Americans serving in Europe. Alongside depictions of tactical victories and enemy fire we find careful renderings of the local people as well as landmarks that MacMillan and so many others were seeing for the first time as they fought through Europe: the Eiffel Tower, Chartres, and—poignantly—the battlefield cemeteries of World War I. Shown here is the upper corner of that map, where the Corps pushed through France into Belgium and Holland toward Germany.

On the following page is the entire second map, which documented the Corps as it fought through Germany toward victory. As an engineer with the 62d Division, MacMillan took care to note the logistical support and tactical ingenuity that made it possible to cross five rivers, capture German soldiers, and liberate thousands of prisoners.

The push began at the lower left of the next map, where the First and Ninth armies held the northern shoulder of the Battle of the Bulge through the brutal winter of 1945. From there they turned toward the Roer River, but found that the enemy had destroyed a dam in order to flood the region. The river had widened from 20 to 300 yards, making any Allied advance impossible. For two weeks, engineers worked around the clock to build fifteen bridges while under enemy fire, withstanding a powerful current that brought dangerous debris and broken boats down the river.

Once they had finally crossed the Roer, the infantry moved quickly past Jülich, grateful for earlier American air strikes that had pacified the city. The capture of München-Gladbach, a manufacturing center, led them toward the Rhine; they were prepared to cross by March 1, but were ordered to delay until preparations could be made for the final push into the heart of Germany. The crossing of the Rhine on March 28 set up what MacMillan considered the most important moment in the campaign: the conquest of the Ruhr Pocket by the First and Ninth armies. The two armies stretched across 150 miles to corner and conquer Germany's most critical industrial area, further weakening the Reich and capturing over 300,000 enemy soldiers.

The men then raced toward the Elbe River at right, the last major obstacle to Berlin. On one day, they logged a record fifty-seven miles. The urgency of the march was heightened by the corresponding progress of the Soviet Army from the east. While the Americans welcomed Soviet support in defeating the German enemy, this dual advance also required some diplomatic maneuvering in light of the mutual suspicion between the two powers. In mid-April, Commander Eisenhower ordered the armies to stand on the Elbe River rather than cross it and continue on toward Berlin. At the banks of the river, the Corps established a preliminary military government to evacuate thousands of prisoners and displaced persons.

On April 11, the 83rd Infantry arrived at Langenstein, a sub-camp of Buchenwald that had been built just a year earlier to provide labor for the German war effort. The Germans had forced prisoners to labor sixteen-hour days to construct underground factories at nearby Halberstadt; those who were too weak to work were executed. Just before the Americans arrived, Langenstein reached the peak of its operation, with 5,000 prisoners at work. After the liberation, only 1,000 were still alive, and almost all died soon thereafter. On April 30, just seven days before the Allies declared victory in Europe, the corps met their Soviet counterparts at Wittenberg.

The Nineteenth Corps landed on D-Day, sent the first men into Belgium and Holland, and led the assault on the Lower Rhine. MacMillan's maps capture the pride and anguish of soldiers in the final year of war, a narrative of a corps in action.

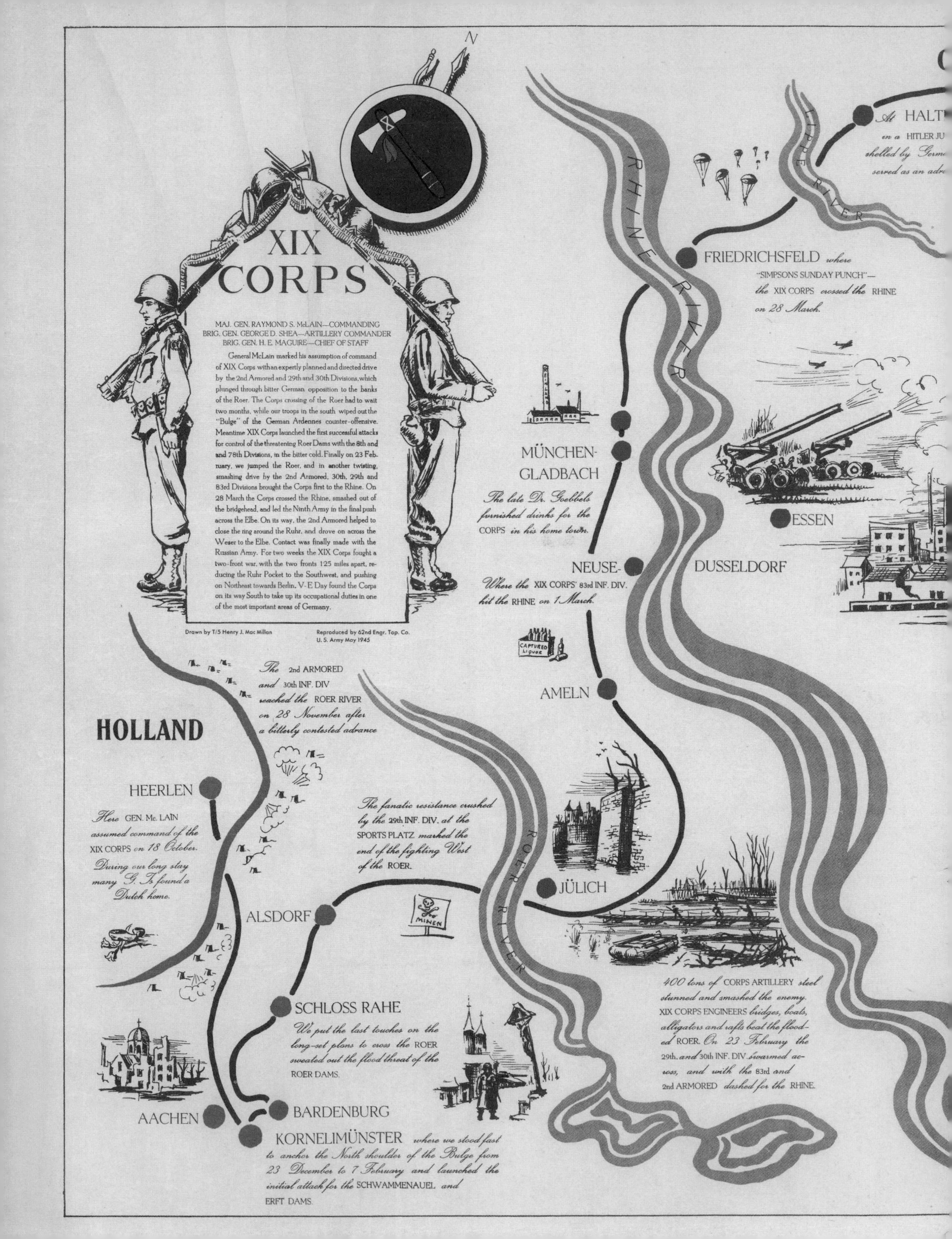
N
XIX
CORPS
MAJ. GEN. RAYMOND S. McLAIN—COMMANDING
BRIG. GEN. GEORGE D. SHEA—ARTILLERY COMMANDER
BRIG. GEN. H. E. MAGUIRE—CHIEF OF STAFF
General McLain marked his assumption of command of XIX Corps with an expertly planned and directed drive by the 2nd Armored and 29th and 30th Divisions, which plunged through bitter German opposition to the banks of the Roer. The Corps crossing of the Roer had to wait two months, while our troops in the south wiped out the "Bulge" of the German Ardennes counter-offensive. Meantime XIX Corps launched the first successful attacks for control of the threatening Roer Dams with the 8th and and 78th Divisions, in the bitter cold. Finally on 23 February, we jumped the Roer, and in another twisting, smashing drive by the 2nd Armored, 30th, 29th and 83rd Divisions brought the Corps first to the Rhine. On 28 March the Corps crossed the Rhine, smashed out of the bridgehead, and led the Ninth Army in the final push across the Elbe. On its way, the 2nd Armored helped to close the ring around the Ruhr, and drove on across the Weser to the Elbe. Contact was finally made with the Russian Army. For two weeks the XIX Corps fought a two-front war, with the two fronts 125 miles apart, reducing the Ruhr Pocket to the Southwest, and pushing on Northeast towards Berlin. V-E Day found the Corps on its way South to take up its occupational duties in one of the most important areas of Germany.
Drawn by T/5 Henry J. Mac Millan
Reproduced by 62nd Engr. Top. Co. U. S. Army May 1945
RHINE RIVER
LIPPE RIVER
FRIEDRICHSFELD where "SIMPSONS SUNDAY PUNCH"— the XIX CORPS crossed the RHINE on 28 March.
MÜNCHEN-GLADBACH
The late Dr. Goebbels furnished drinks for the CORPS in his home town.
ESSEN
NEUSE-
DUSSELDORF
Where the XIX CORPS' 83rd INF. DIV. hit the RHINE on 1 March.
CAPTURED LIQUOR
AMELN
The 2nd ARMORED and 30th INF. DIV reached the ROER RIVER on 28 November after a bitterly contested advance
HOLLAND
HEERLEN
Here GEN. Mc. LAIN assumed command of the XIX CORPS on 18 October. During our long stay many G. I's found a Dutch home.
The fanatic resistance crushed by the 29th INF. DIV. at the SPORTS PLATZ marked the end of the fighting West of the ROER.
ROER RIVER
JÜLICH
ALSDORF
MINEN
400 tons of CORPS ARTILLERY steel stunned and smashed the enemy. XIX CORPS ENGINEERS bridges, boats, allegators and rafts beat the flooded ROER. On 23 February the 29th. and 30th INF. DIV. swarmed across, and with the 83rd and 2nd ARMORED dashed for the RHINE.
SCHLOSS RAHE
We put the last touches on the long-set plans to cross the ROER sweated out the flood threat of the ROER DAMS.
AACHEN
BARDENBURG
KORNELIMÜNSTER where we stood fast to anchor the North shoulder of the Bulge from 23 December to 7 February and launched the initial attack for the SCHWAMMENAUEL and ERFT DAMS.

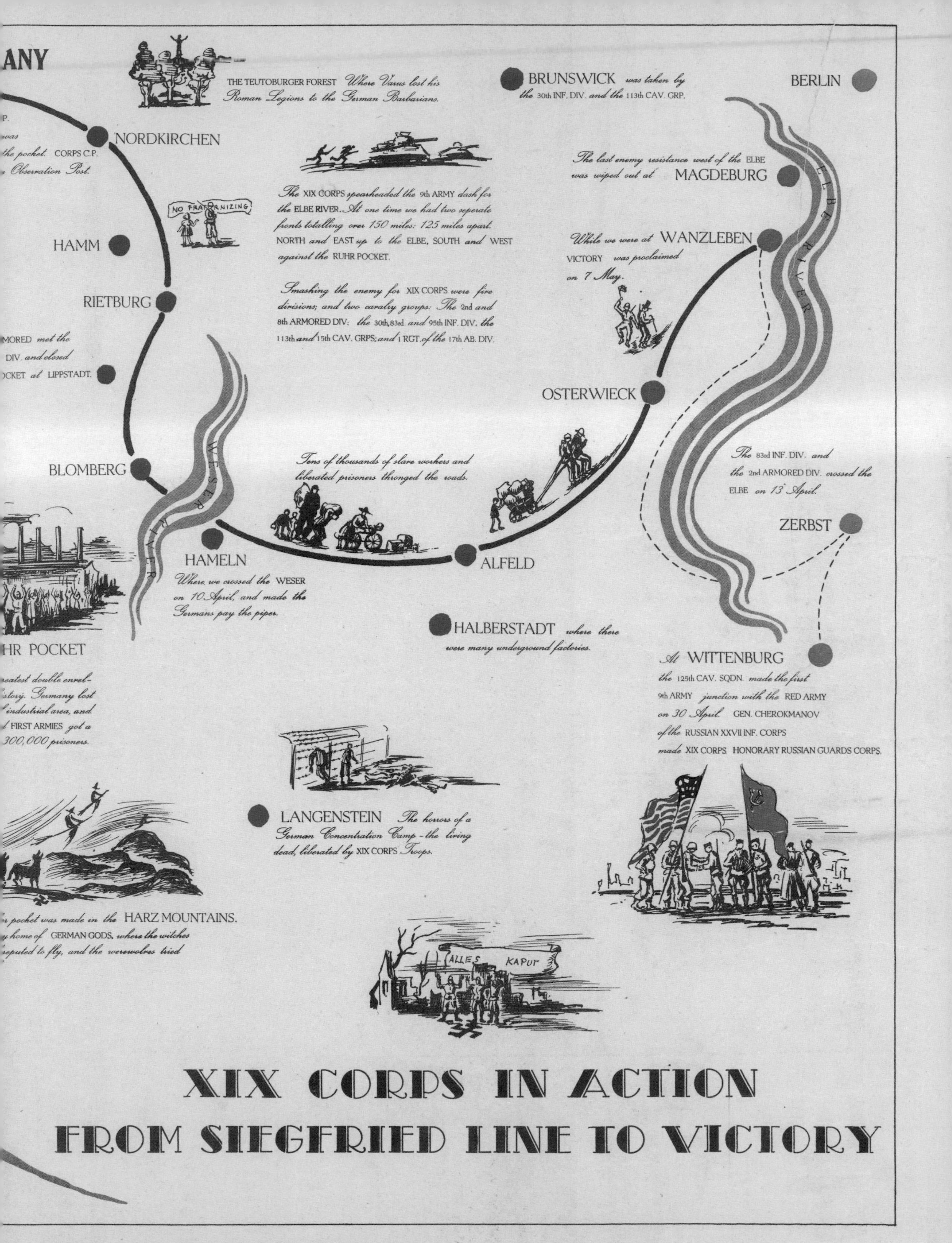
ANY
THE TEUTOBURGER FOREST Where Varus lost his Roman Legions to the German Barbarians.
BRUNSWICK was taken by the 30th INF. DIV. and the 113th CAV. GRP.
BERLIN
P.
was
the pocket. CORPS C.P.
Observation Post.
NORDKIRCHEN
The last enemy resistance west of the ELBE was wiped out at MAGDEBURG
ELBE RIVER
The XIX CORPS spearheaded the 9th ARMY dash for the ELBE RIVER. At one time we had two seperate fronts totalling over 150 miles: 125 miles apart. NORTH and EAST up to the ELBE, SOUTH and WEST against the RUHR POCKET.
NO FRATERNIZING
HAMM
While we were at WANZLEBEN VICTORY was proclaimed on 7 May.
RIETBURG
Smashing the enemy for XIX CORPS were five divisions; and two cavalry groups: The 2nd and 8th ARMORED DIV: the 30th, 83rd and 95th INF. DIV. the 113th and 15th CAV. GRPS; and 1 RGT. of the 17th AB. DIV.
MORED met the
DIV. and closed
CKET at LIPPSTADT.
OSTERWIECK
WESER RIVER
Tens of thousands of slave workers and liberated prisoners thronged the roads.
BLOMBERG
The 83rd INF. DIV. and the 2nd ARMORED DIV. crossed the ELBE on 13 April.
ZERBST
HAMELN
Where we crossed the WESER on 10 April, and made the Germans pay the piper.
ALFELD
HALBERSTADT where there were many underground factories.
HR POCKET
reatest double envel-
story. Germany lost
industrial area, and
FIRST ARMIES got a
300,000 prisoners.
At WITTENBURG the 125th CAV. SQDN. made the first 9th ARMY junction with the RED ARMY on 30 April. GEN. CHEROKMANOV of the RUSSIAN XXVII INF. CORPS made XIX CORPS HONORARY RUSSIAN GUARDS CORPS.
LANGENSTEIN The horrors of a German Concentration Camp - the living dead, liberated by XIX CORPS Troops.
n pocket was made in the HARZ MOUNTAINS.
y home of GERMAN GODS, where the witches
reputed to fly, and the werewolves tried
ALLES KAPUT
XIX CORPS IN ACTION
FROM SIEGFRIED LINE TO VICTORY

HOLOCAUST

Michal Kraus, map of Austrian concentration camps, and "From Linz to Nachoda," both circa 1945–7

Between 1940 and 1945, the Nazi regime systematically murdered 6 million Jews. No artifacts can convey the scope of this atrocity, but we rely on material evidence to record, remember, and understand the Holocaust. The two maps here were drawn by a young Jewish boy at the end of the war: the first documents his life in the concentration camps, while the second traces his perilous journey home after the defeat of Germany.

Michal Kraus was born in Czechoslovakia in 1930, and at the age of eight he watched the Germans invade his hometown of Náchod. The Nazis quickly began to isolate and oppress Jews, expelling Michal from school and forcing the family into a ghetto in 1941. A year later the family was deported to Terezín, a camp where 33,000 prisoners perished from the brutal conditions alone; the Germans sent 90,000 more to extermination camps throughout Nazi-controlled territories further east. Despite the desperation at Terezín, the large population of prisoners fostered a culture and sense of community. Michal himself drew comics and portraits for the young boys' newspaper *Kamarad*.

In December 1943 the Nazis sent Michal and his family to Auschwitz-Birkenau in Poland; six months later Michal's mother was removed yet again, to a camp at Stutthof. Michal's father had fallen ill and his condition worsened throughout the spring and early summer of 1944. In July he was tortured and killed by the Germans. For the next six months the fourteen-year-old Michal remained at the camp; every day he watched as new prisoners were sent directly to the gas chambers. Over a million Jews were murdered at Auschwitz between 1940 and 1945.

In the fall of 1944 Michal began to hear the sound of Soviet and American planes flying overhead, and by the following January the Soviets had pressed the Eastern Front toward Auschwitz. The Nazis responded by abandoning the camp and forcing prisoners to march for three days in the depths of winter to a train station sixty-five kilometers away. Most of the prisoners froze to death or were killed by German guards. Michal and others who survived were crammed into an open railway car for four excruciating days as it lumbered toward Austria.

The map below marks the next five months of Michal's life, spent in concentration camps along the Danube River. He remembered this as the worst time of his life.

Michal's first destination was the camp at Mauthausen, which he knew as the site where many of his fellow Czech Jews had been murdered. He arrived there after a full week without food. From Mauthausen he was deported to Melk, where prisoners worked twelve-hour shifts to construct an underground factory for the German war effort. The labor alone killed some, while others were executed once they proved too weak to work. Michal's sole source of hope was the occasional American air raid above; he recalled how "gorgeous" it was "to behold those avenging Allies, the large bombers glittering in the rays of the sun."

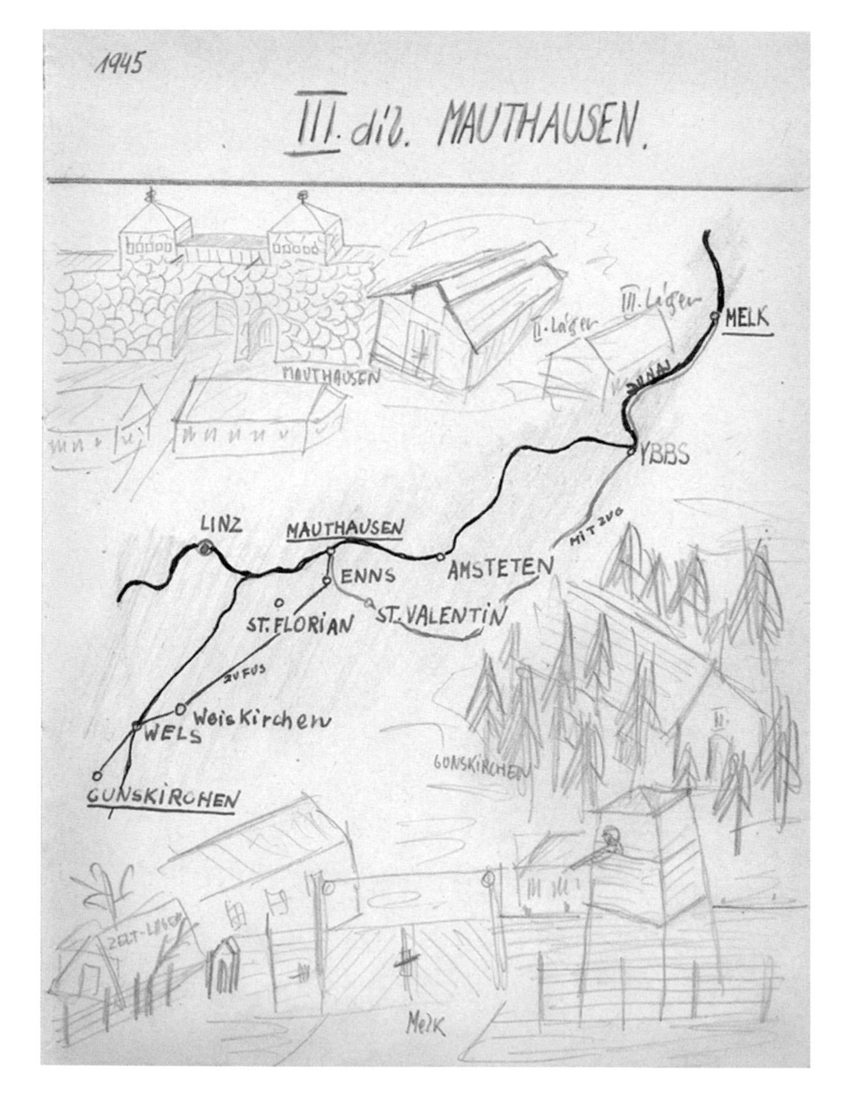

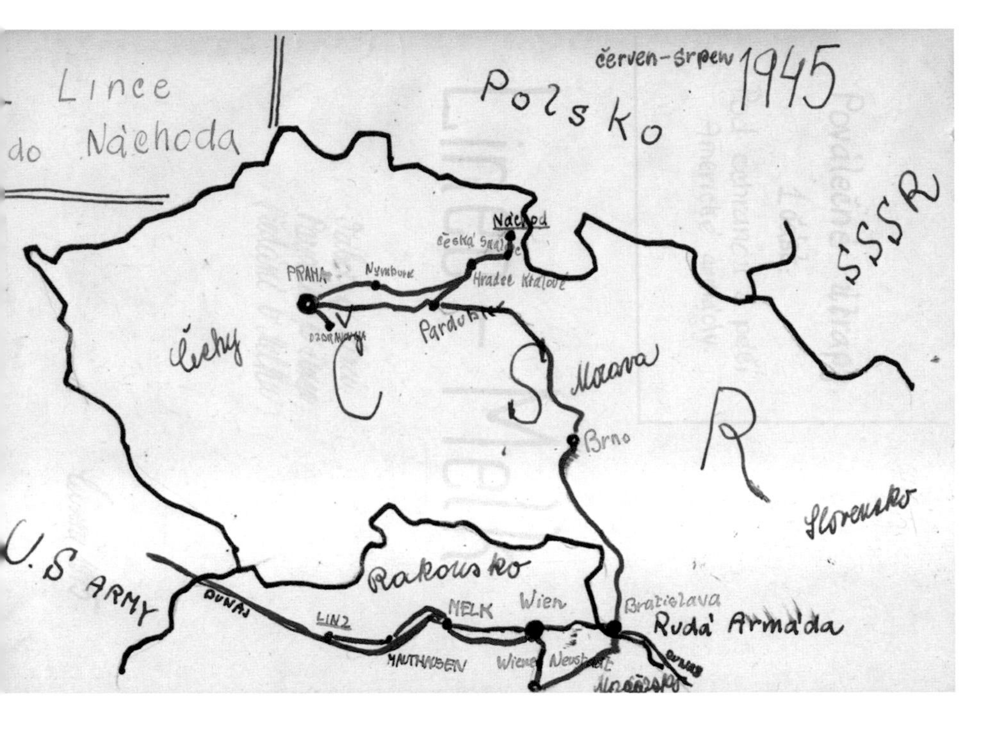

As American forces pressed from the west and the Soviets advanced from the east, Nazi brutality continued. The Germans sent Michal and others from Melk back to Mauthausen (along the green line on the map at left), where prisoners began to hear rumors of a German surrender. The rumble of American bombers could now be heard day and night. The Nazis relentlessly forced the prisoners on another march (along the red line) of over sixty kilometers to a dense forest at Gunskirchen. Those who fell behind were executed.

Upon arrival at Gunskirchen, exhausted and near death, Michal and his fellow prisoners were sequestered for ten days without food or blankets. On May 5 the sound of gunfire signaled the approach of American troops, prompting the German guards to abandon the area. Michal and other survivors emerged from the forest to find themselves among thousands of other prisoners in a uniformly desperate condition. After four years of imprisonment, just shy of his fifteenth birthday, he was free.

Now liberated, Michal thrilled at the sight of American soldiers—both black and white—driving tanks and convoys along roads that had imprisoned him for so long. Everything he saw confirmed the destruction of the Wehrmacht. But he was utterly alone, nowhere near home, and immediately became so sick with typhus that he was forced to remain in Austria to recover. As soon as he regained his health, he quite literally jumped at the opportunity to join 500 Czech and Slovak Jews heading home aboard a convoy of trucks.

Michal drew the map to the left to record the route of his long and perilous journey home in June. At lower middle he marks the American military presence in Austria ("Rakousko"), and the Danube River in blue. Red ink marks his route home to Czechoslovakia, bordered on the north by Poland and on the east by the Soviet Union ("SSSR"). His trek began on a steamboat up the Danube from Linz through his two former concentration camps, Mauthausen and Melk. In a terrible irony, he was once again held at Melk while awaiting passage home, this time guarded by the Russian rather than the German army. While at Melk he was horrified to see the ovens and other devices that the Germans had used to murder prisoners just a few weeks earlier.

After the Russians released Michal from Melk, he traveled by train through Vienna to Wiener Neustadt, where he met other Czech refugees. Rail lines were so badly damaged by bombing raids that Michal traveled on foot over one hundred kilometers to the Czech border at Bratislava. As he walked in a procession that stretched for miles, he carefully avoided the Russian soldiers who were plundering their way west through Czechoslovakia.

In Bratislava Michal boarded a train that arrived in Prague on June 28, 1945. It was only then that he learned his mother had been killed in the camps. When he finally returned to his hometown of Náchod, at the upper center of the map, he began to compose a diary to honor his parents and document the horror he had survived. Over three volumes, he detailed and illustrated the camps he had endured in Poland, Czechoslovakia, and Austria. His maps are a first draft of history, a vivid and specific record of the geography of genocide.

On his long walk home, Michal passed through regions that would soon be contested in the emerging Cold War. Just a few months after he finished the diary in 1947, communists staged a coup in Czechoslovakia. Soon afterwards Michal went to Canada to complete high school, and from there he studied architecture at Columbia University. After traveling and working in Europe, in 1951 he settled permanently in the United States with his wife. Michal lost every member of his extended family in the Holocaust except for an aunt and a cousin.

MAPPING THE MIGHTY MISSISSIPPI

Harold Norman Fisk, "Ancient Courses [of the] Mississippi River Meander Belt" (sheet 9), 1944

The Mississippi River is arguably the most important waterway in the US, and certainly among the most consequential in its history. The river drains 1.2 million square miles—40 percent of the continental United States—into the Mississippi Delta. The rich soils on either side of the river became the heart of the slave plantation system in the 1850s, and created enormous wealth at the height of the cotton trade. Thereafter control of the river was central to Union strategy in the Civil War, and it remained vital to commerce well into the twentieth century. But the river's power was matched by its unpredictability, which prompted the creation of the Mississippi River Commission in 1879 both to improve navigation and to control the recurrence of flooding.

Such efforts were repeatedly thwarted, most importantly with the Great Flood of 1927. New Orleans was largely spared when levees downriver were destroyed to alleviate pressure on the city. The hardest-hit areas of the delta were further north near Greenville, Mississippi, where hundreds of thousands of African Americans were displaced. The long-term consequences were significant: within one year of the flood, 50 percent of the black population of the delta had moved out of the South into cities such as Chicago and Detroit.

With the advent of the New Deal, more ambitious public works—such as the Tennessee Valley Authority (page 202)—encouraged engineers and scientists to stabilize the banks and the flow of the Mississippi River. Among the most creative and determined of these men was Harold Norman Fisk, an irascible geologist with the Louisiana Geological Survey who convinced the commission to fund his comprehensive study of the lower river in 1941.

Fisk argued that the river ought to be understood as an evolving, dynamic entity given that it had shifted its course over time. For two years he pushed his colleagues relentlessly to reconstruct the geological history of the river through fieldwork, aerial photography, and archival research. The photographs allowed him to search for long-abandoned meanders and channels, and to see layered patterns of soil, vegetation, and drainage. Historical maps found in archives showed him how the river had changed its course over the prior century and a half.

The result was a landmark study that transformed the way water engineers understood the river itself. The report featured a series of elaborate and beautifully executed maps that charted the wild behavior of the river from the prehistoric era to the present. With imaginative—even psychedelic—use of color and shading, Fisk recreated the river's history in three dimensions: fifteen maps—stretching from Cape Girardeau in Missouri down to Donaldsonville, just above New Orleans—graphically illustrated the patterns of flow and sedimentation that ignored fixed riverbanks. Through these maps, scientists could identify the formation of the meander belt of the river about 6,000 years ago, and then trace its subsequent evolution. With these images Fisk made a contribution not just to geology and science, but also to map design. His innovative approach to capturing change over time has caught the attention of mapmakers and graphic designers ever since.

It is worth noting that when Fisk completed the maps during World War II they were printed by the Army Map Service. The war revived transportation on the river and intensified the federal government's attention to geological research. This section of the map includes some of the area worst hit by the 1927 flood, illustrating its many diversions and cutoffs. By examining the deep history of sedimentation, Fisk was able to explain the rich alluvial plain that extended into seven Southern states. And by studying hydrographic surveys, he was able to document the shifting banks and erosion that plagued the river. Putting all these studies together, Fisk created a crucial foundation for engineers to understand how the river behaved, and how to address both the incremental erosion and the more radical shifts that led to the endemic flooding of the Lower Mississippi Valley.

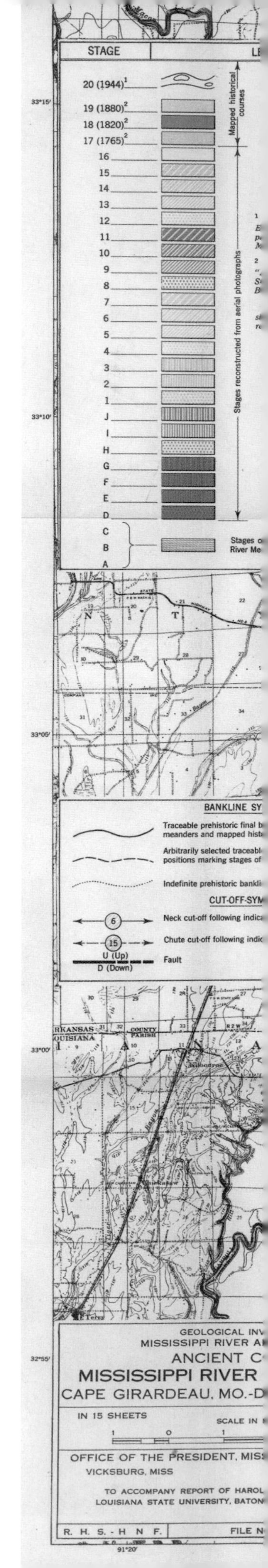

n the 12th
e Mississip-
the Gulf of
uary 1944.
n maps of
ver, Early
ro, Ill., to
lished Mis-
sion Quad-
LEY
ER BELT
ONVILLE, LA.
SHEET 9
ER COMMISSION
1944
CONSULTANT
ED 1 DEC. 1944
SH. 33-1
ARKANSAS
MISSISSIPPI RIVER
LAKE LEE
WASHINGTON
LEROY PERCY
STATE PARK
ISLAND NO. 87
HOWARD
TOWHEAD
WORTHINGTON POINT
ISLAND NO. 88
BEND
GRAND LAKE
WASHINGTON COUNTY
LAKE WASHINGTON
MISSISSIPPI
ISLAND NO. 89
ARKANSAS
LOUISIANA
CHICOT
GASSOWAY LAKE
SARAH ISLAND
SARAH CUT-OFF 1936
OPOSSUM CHUTE
PITTMAN ISLAND
ISSAQUENA
Longwood
Marathon
Mayersville
91°15'
91°10'
91°05'
91°00'
90°55'
33°00'
32°55'

THE ADVENT OF AIR TRAVEL

Hal Shelton, "Denver–Chicago," for United Air Lines, 1949

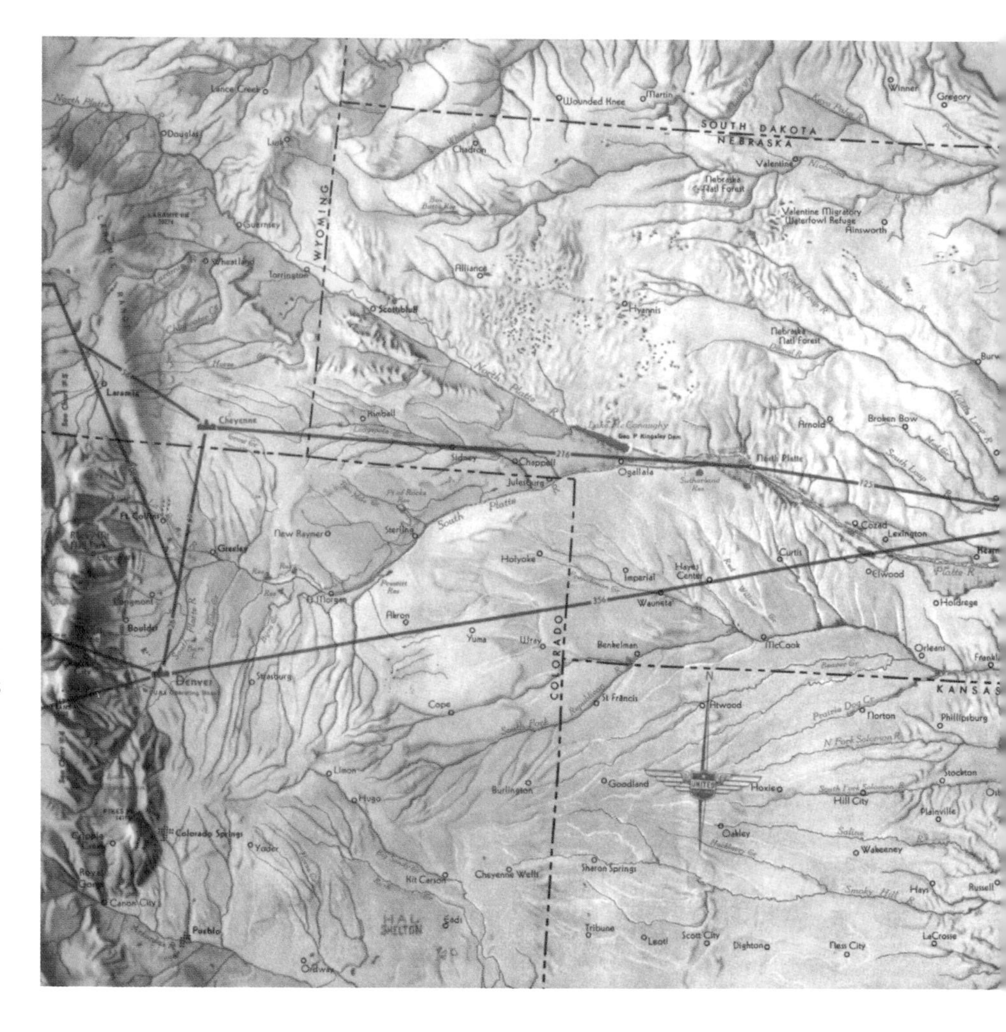

Given the sheer size of the United States, the advent of commercial aviation had a profound effect on American life. Charles Lindbergh tapped the public's excitement around air travel when he flew solo from New York to Paris in 1927. Three years later United Air Lines offered the first transcontinental route, from New York to San Francisco, with an overnight stay in Chicago.

The introduction of the DC-3 airplane in the mid-1930s made it possible to fly across the country non-stop, and the surplus of aircraft after World War II substantially expanded the number of cities that could be served. The earliest passengers tended to be bankers and businessmen flying from New York to Chicago. The subsequent advent of transcontinental routes helped to integrate geographically distant industries, for instance the financial resources of New York with the broadcasting and film industry in Los Angeles.

In the early years of commercial aviation, the in-flight map was not just a schematic diagram but an active way for passengers to follow the journey. These planes flew under the clouds at relatively low elevations, so the maps included landmarks such as radio towers to guide pilots as well as passengers. Airline maps gave travelers a way to pass the time while demystifying what for many was a new and perhaps unsettling experience.

By 1950, the number of miles traveled by plane exceeded that of rail, and the introduction of jet aircraft in 1958 cut the length of a transcontinental flight from nine to five hours. As flight speeds increased, airline maps began to encompass far greater geographical scope at a much higher elevation. This United Air Lines route map from 1949 was part of a series drawn by Hal Shelton, one that had implications far beyond air travel. Shelton had trained as an artist, and his visual education enabled him to think differently about map design. Upon finishing a degree in scientific illustration in 1938, he did field survey work with the United States Geological Survey (USGS). While mapping a remote region of the Sierra Nevada mountains, he discovered something: even though local residents could name all the peaks of the nearby ranges, they could not read the contour maps produced by the USGS. This experience taught Shelton that to be successful a map must communicate as much as possible without the use of abstract symbols such as contour lines.

Shelton was also spending a good deal of time flying with his brother, a pilot, and this led him to reconsider aeronautical charts. Rather than using traditional conventions, he sought to present the landscape more intuitively, as the viewer might see it. His experimental charts caught the eye of Elrey Jeppesen, a pilot who found Shelton's technique ideally suited for airline passengers. Together they collaborated on a series of innovative charts for United Air Lines.

Shelton's creativity is at work throughout the chart. By avoiding the use of symbols he removed the need for a legend. He also deemphasized boundaries, cities, and roads, minimizing the human presence on the land in order to present the earth below as it might be apprehended—or even imagined—from

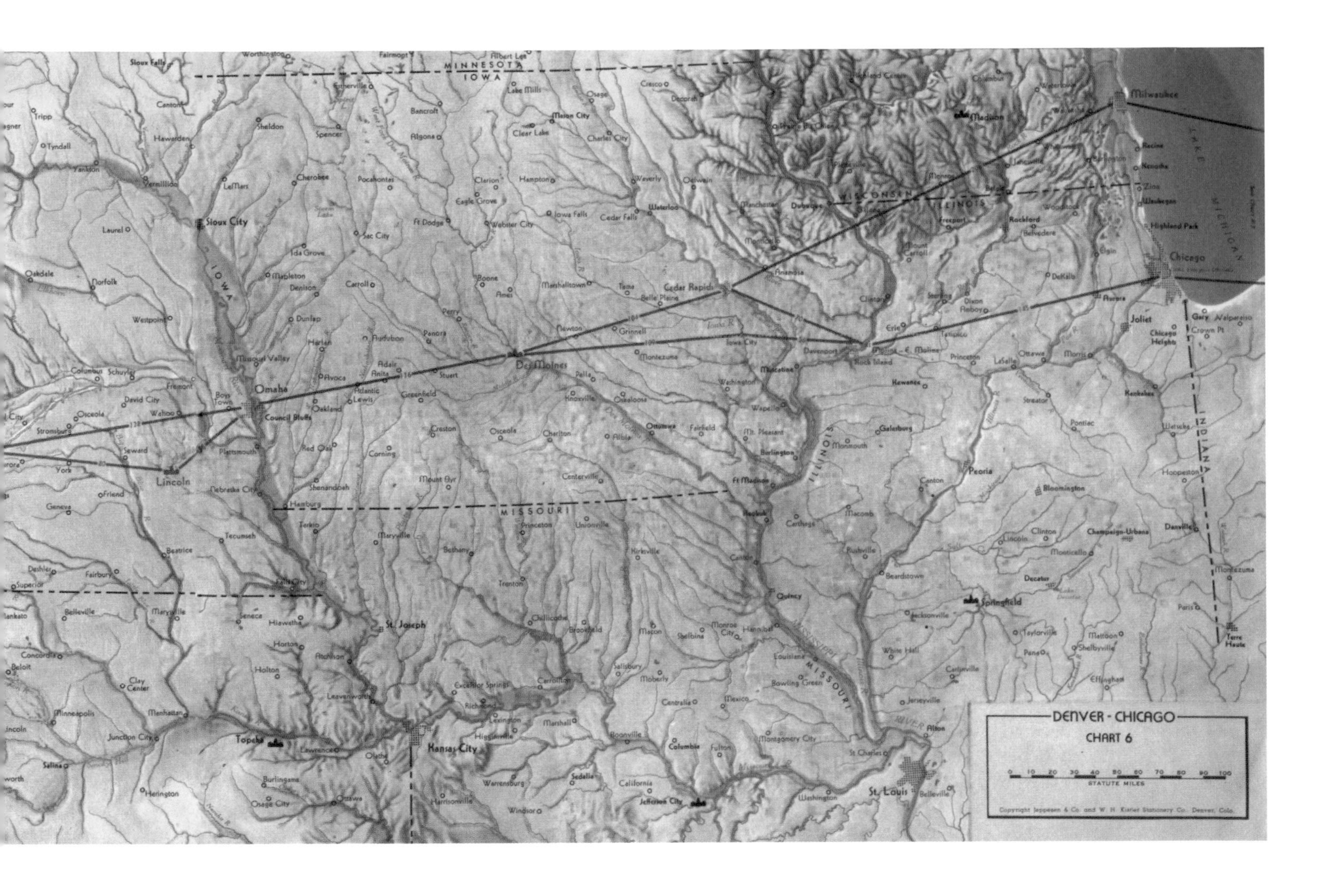

higher elevations. The color scheme mimics the landscape: green forests, beige deserts, blue water, and white snowcaps. Within that range, Shelton subtly adjusted tone to depict depth or elevation. Though the effect is understated and restrained, there is tremendous nuance embedded in the depiction of valleys, river drainage systems, and mountain chains. The overall effect is a physiographic landscape that appears—above all—natural and coherent. The map approximates the land itself.

Yet tremendous effort was required to make these "natural" maps. Dozens of geographers and mapmakers generated the information on climate, rainfall, landforms, and drainage. Shelton used hundreds of aerial color photographs to help organize this data, then began etching and painting the charts—inch by inch—through a secret process. The result was a visual masterpiece, where "the mountains jump up" and the valleys appear to descend.

Long before satellite imagery, Shelton's maps were so realistic that the National Aeronautics and Space Administration (NASA) used them to identify photographs of the earth taken on early space missions. By capturing the vantage point of a pilot above the terrain, Shelton brought maps ever closer to photographs. This combination of an oblique perspective with a detailed rendering of the terrain below was used to great effect in his iconic maps of the Colorado ski country. Shelton, like Richard Edes Harrison (pages 206 and 208), used artistry to render the terrain, and in the process changed the conventions that had governed mapmaking for decades.

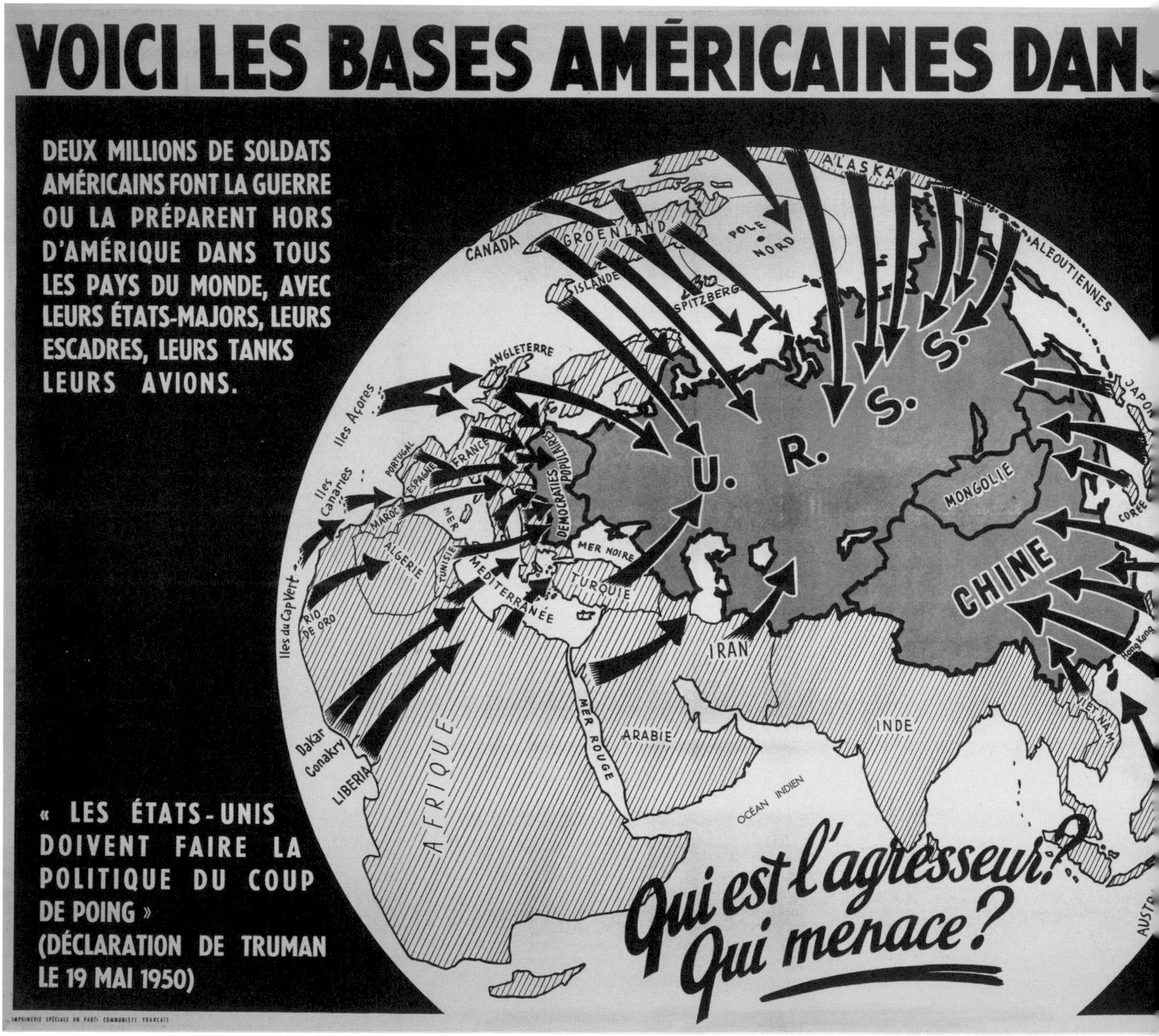

VOICI LES BASES AMÉRICAINES DAN
DEUX MILLIONS DE SOLDATS AMÉRICAINS FONT LA GUERRE OU LA PRÉPARENT HORS D'AMÉRIQUE DANS TOUS LES PAYS DU MONDE, AVEC LEURS ÉTATS-MAJORS, LEURS ESCADRES, LEURS TANKS LEURS AVIONS.
« LES ÉTATS-UNIS DOIVENT FAIRE LA POLITIQUE DU COUP DE POING » (DÉCLARATION DE TRUMAN LE 19 MAI 1950)
ALASKA
POLE NORD
GROENLAND
CANADA
ALEOUTIENNES
ISLANDE
SPITZBERG
ANGLETERRE
Iles Açores
PORTUGAL
FRANCE
ESPAGNE
Iles Canaries
MAROC
DÉMOCRATIES POPULAIRES
U. R. S. S.
MONGOLIE
CORÉE
JAPON
CHINE
MER NOIRE
ALGÉRIE
TUNISIE
MER MÉDITERRANÉE
TURQUIE
Iles du Cap Vert
RIO DE ORO
IRAN
Hong Kong
VIET NAM
Dakar
Conakry
LIBERIA
AFRIQUE
MER ROUGE
ARABIE
INDE
OCÉAN INDIEN
Qui est l'agresseur? Qui menace?
IMPRIMERIE SPÉCIALE DU PARTI COMMUNISTE FRANÇAIS

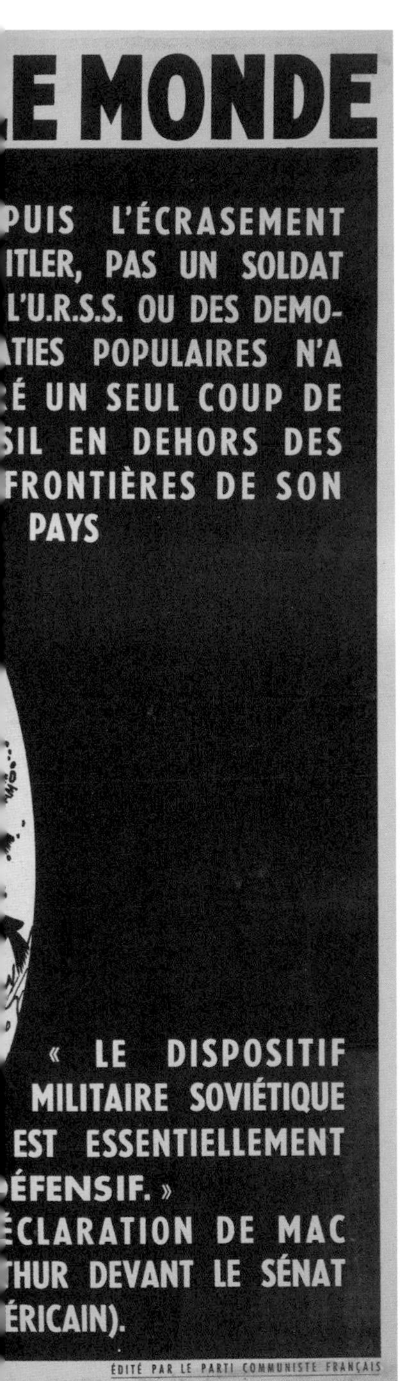

French Communist Party, "Voici les bases Americaines dans le monde" [Here are the American bases throughout the world], 1951

The French Communist Party designed this propaganda map at the height of the early Cold War, urgently asking "Who is the aggressor? Who is the menace?" If there was any doubt, the party insisted that it was the American military that drove the epic struggle between communism and the West.

The Cold War began even before World War II ended. Though formal allies, the Soviet Union and the United States had a history of mutual suspicion. Joseph Stalin saw the United States as capitalist and expansive, a nation that deliberately delayed the opening of a second front in Europe so that the Soviets would shoulder the fight against Germany. During the war, President Roosevelt believed—perhaps naively—that he could "handle" Stalin, but later generations wondered if more could have been done to resist Soviet demands at the Yalta meeting in February 1945. Within two months Roosevelt was dead, leaving his inexperienced vice president, Harry Truman, to lead the country through the final stages of the Pacific War.

That summer, the United States detonated an atomic bomb, which Truman believed would give him an uncontested advantage over the Soviets. The U.S. used the bomb on Hiroshima and Nagasaki, which simultaneously defeated the Japanese and demonstrated American military prowess. Relations with the USSR deteriorated rapidly from there. In 1946 Winston Churchill described an "iron curtain" that divided free Europe from the Soviet sphere. The next year, a young American diplomat stationed in Moscow sent a lengthy telegram to his superiors describing Russian behavior as governed chiefly by a sense of international insecurity. Such a mindset, leavened with a zealous commitment to communism, meant that, while the Soviets might not pose a direct threat to the United States, they must be met with patient, firm, and vigilant containment.

In perhaps the first example of containment, President Truman committed military support to Greece and Turkey to stave off communist influence. He then proposed a massive infusion of economic aid to those western European countries willing to support democratic institutions and free trade. While the Soviets and their eastern European allies initially considered accepting the aid, their ultimate rejection of the Marshall Plan heightened Cold War divisions further. These tensions worsened in 1948, when a communist coup in Czechoslovakia was followed by a blockade of Berlin. Western Europeans responded with the North Atlantic Treaty Organization (NATO), which obliged the United States to extend its military commitments in Europe. Before the end of the year, the Soviets detonated an atomic bomb, and the communists mounted a successful revolution in China. Fears of a communist axis mounted when North Korea invaded its southern neighbor in 1950.

This rapid sequence of threatening events drove Truman toward his more hawkish advisers. The result was an increase in American commitments abroad, and by 1950 more than a million American military personnel were stationed across thirty-five foreign countries. That military presence prompted French communists to produce this broadside, most likely part of the party's campaign in the parliamentary election of June 1951. The French Communist Party won a quarter of the vote that summer.

While the broadside portrays the United States as the aggressor, the country does not appear on the map save for a sliver of Alaska over the North Pole. Instead, it is the growth of American military bases that encircle the Soviets and their communist allies, who, as the text explains, have yet to fire a shot beyond their borders. The map presents a communist heartland that is encroached upon on all sides by American militarism. By this time Eastern Europe was largely controlled by the Soviet Union, though in an Orwellian twist the map identifies these as "Democraties Populaires." With its global perspective and suggestive symbols, the French Communist Party urgently points to a need to join the worldwide network of support for the Soviets and resistance to American capitalism. The oblique perspective used by Harrison on page 206 to highlight American vulnerability in World War II is here used to highlight Soviet vulnerability in the Cold War.

Though an exaggerated example, the map reflects the ambivalence felt by many in western Europe after the war: grateful for American aid and military protection but concerned by their dependence upon the same. Even as propaganda, the map captures the global nature of the emerging Cold War. The conflict took hold not just in Europe, but also in the growth of communist parties in Asia, Africa, and Latin America. The next map reflects the American response, which frames the Soviets as the aggressors and the Americans as the defenders of freedom.

THE VIGILANT FIGHT AGAINST COMMUNISM

Research Institute of America, "How Communists Menace Vital Materials," 1956

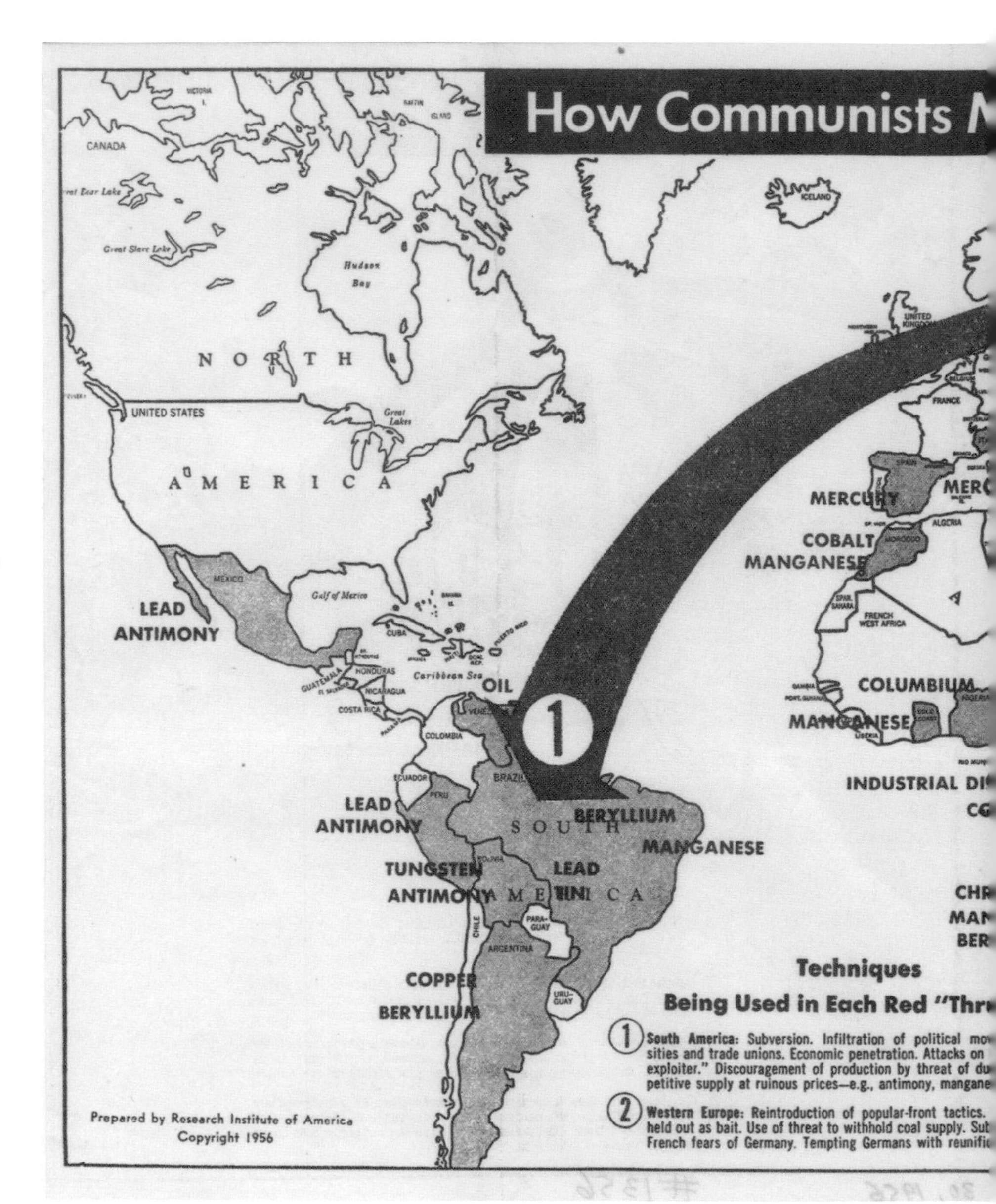

Ten years after the end of World War II, Democrats and Republicans remained convinced that Soviet communism posed the greatest threat to national security. The country's growing defense budgets bore out that political consensus. Domestically, however, the influence of the American Communist Party was negligible by the mid-1950s. Vigorous prosecutions of the party under the Smith Act had crushed its leadership, while purges of communists from universities and the entertainment industry had effectively silenced any communist voice in education and popular culture.

The domestic fight against communism began at the end of World War I, when a young J. Edgar Hoover used the new Federal Bureau of Investigation to monitor those whom he considered politically dangerous or radical. But the prosecution of internal communism was complicated when the United States joined with the Soviet Union to defeat fascism during World War II. This era of the "Popular Front" gave rise to an attitude of tolerance toward the American Communist Party, and a few civil servants in the Roosevelt and Truman administrations were even convicted of spying for the Soviet Union. Such convictions prompted widespread suspicions of an extended espionage network in America. These suspicions also justified ongoing investigations conducted by the House Committee on Un-American Activities (HUAC). Truman further heightened fears of treason by requiring in 1947 that civil servants take oaths of loyalty.

That fear of communism turned into outright panic once Senators Joseph McCarthy and Pat McCarran escalated accusations of domestic subversion. While Republicans might have had an upper hand in these attacks, anti-communism was a thoroughly bipartisan issue. In the 1950s Pennsylvania Democrat Francis Walter continued the work of the HUAC. Walter had successfully collaborated with McCarran to prevent current and former communists from immigrating to the United States. After both McCarthy and McCarran died in 1956, Walter worried that the country had begun to accommodate itself to the Soviet Union, settling for "peaceful coexistence" rather than defiant

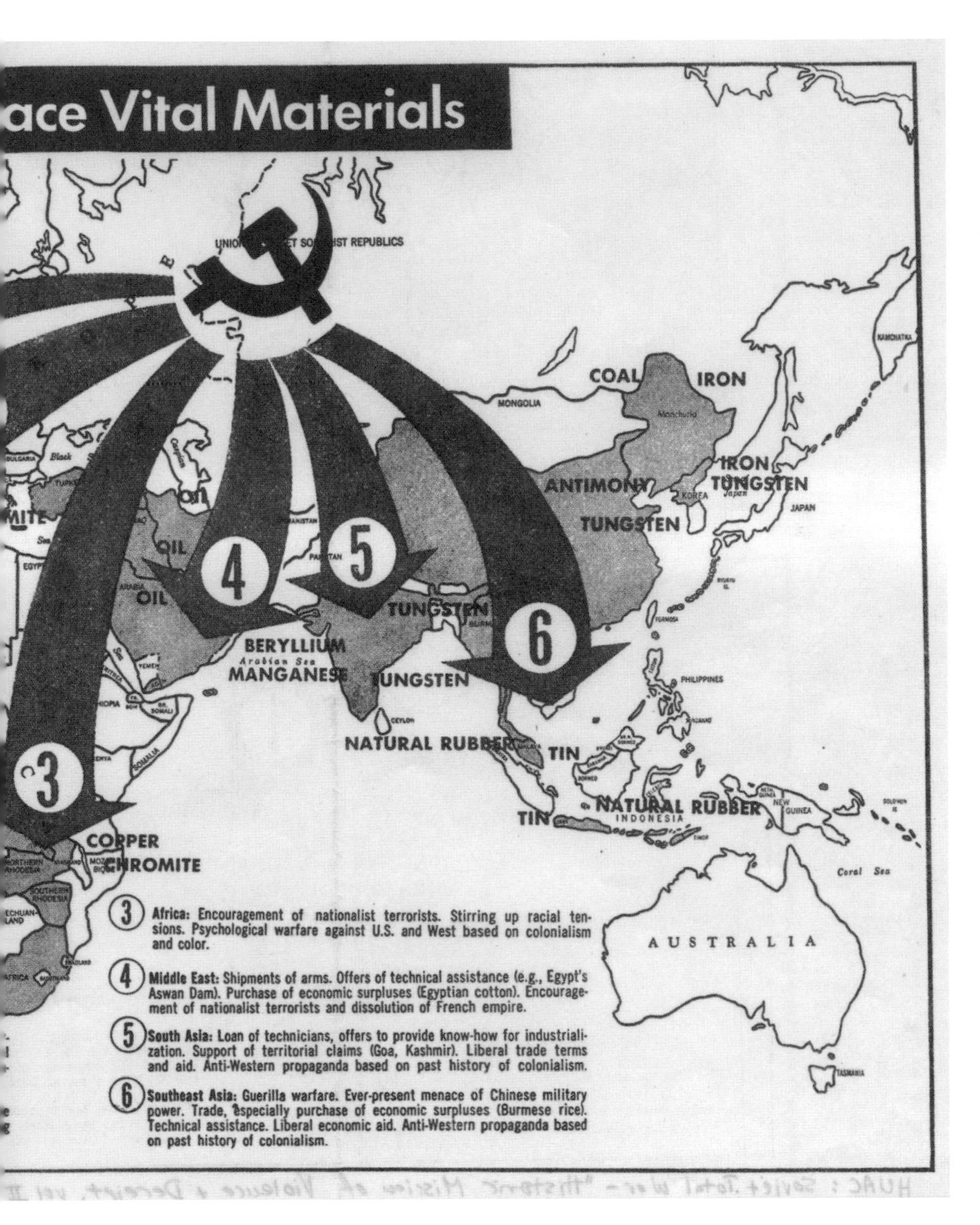
ace Vital Materials
COAL
IRON
MONGOLIA
IRON
TUNGSTEN
ANTIMONY
TUNGSTEN
JAPAN
OIL
OIL
4
5
6
3
TUNGSTEN
BERYLLIUM
Arabian Sea
MANGANESE
TUNGSTEN
NATURAL RUBBER
TIN
NATURAL RUBBER
INDONESIA
TIN
NEW GUINEA
COPPER
CHROMITE
Coral Sea
AUSTRALIA
3 Africa: Encouragement of nationalist terrorists. Stirring up racial tensions. Psychological warfare against U.S. and West based on colonialism and color.
4 Middle East: Shipments of arms. Offers of technical assistance (e.g., Egypt's Aswan Dam). Purchase of economic surpluses (Egyptian cotton). Encouragement of nationalist terrorists and dissolution of French empire.
5 South Asia: Loan of technicians, offers to provide know-how for industrialization. Support of territorial claims (Goa, Kashmir). Liberal trade terms and aid. Anti-Western propaganda based on past history of colonialism.
6 Southeast Asia: Guerilla warfare. Ever-present menace of Chinese military power. Trade, especially purchase of economic surpluses (Burmese rice). Technical assistance. Liberal economic aid. Anti-Western propaganda based on past history of colonialism.

anti-communism. To renew the fight, he compiled a volume of essays purporting to expose the Soviet tendency toward secrecy, indoctrination, and deceit. As he wrote, the communists used "relentless psychological, political, economic, sociological and military strategies." A failure to retaliate would leave Americans condemned "to the Arctic hell of Siberian slave labor camps."

A number of foreign policy leaders contributed to Walter's volume *Soviet Total War*, including Henry Kissinger and the Central Intelligence Agency director, Allen Dulles. But it also included essays by the nation's evangelical Christian leaders and other private citizens. Among the latter was Leo Cherne, an ardent anti-Soviet whose Research Institute of America had long advised companies how to navigate government regulations brought by the New Deal. In his anti-communist essay, Cherne explained that the death of Stalin had ushered in a more aggressive phase of Soviet behavior. Moving forward, the USSR would resort to economic, political, and even psychological warfare against the West, as illustrated by this map. Just as the French Communist Party (page 222) portrayed the United States as reaching across the globe with its military might, here the Soviets are shown penetrating every corner of the world through political and psychological manipulation.

For Cherne, it was not just Soviet military power that threatened the United States but also subterfuge: guerilla warfare, anti-western propaganda, political infiltration—all were part of the communist playbook, especially in the developing world. And, though Cherne's prose was especially virulent, his strategy anticipated the foreign policy posture of presidents Eisenhower, Kennedy, and Johnson. Each of these administrations sought to undermine leftist and communist movements in Africa, Latin America, and Southeast Asia. In fact, this map of 1956 was published at precisely the moment that the United States increased its commitment to South Vietnam in the wake of the French departure from Indochina. Political messages like this, which regarded Soviet influence anywhere as a threat to American security, paved the way for the deployment of armed forces in Vietnam (page 238).

THE MAGIC KINGDOM

Herb Ryman, Bird's-eye view of Disneyland (with inset park plan designed by Marvin Davis), 1953

No individual embodies twentieth-century American culture more than Walt Disney, whose childhood interest in drawing evolved into an early career in animation. Disney created the first animated film with sound with *Steamboat Willie* (1928), and by the mid-1930s he was the creative force behind feature-length animations that dazzled Americans for decades thereafter. World War II brought opportunities to turn that talent toward the Allied cause, most powerfully with *Victory through Air Power*, which also included the first animated map.

In the postwar years, Disney broadened the reach of his entertainment studio by branching out into television. At the same time, he began to imagine a theme park devoted to the same kind of fantasy that had made his films such a success. Amusement parks were not new in America, and had become relatively commonplace by the turn of the twentieth century. But Disney's vision was fundamentally different, as shown in this early conceptual map.

In Disney's view, "Disneyland" was a place where characters from his films would come to life to charm children and their families. His early vision for the park also centered on the re-creation of a small town, along with carnival attractions and a western-themed village. Within a few years, this plan had expanded to include a spaceship and other rides, as well as exhibits based on history and science. His vision attracted little support within the company, however, not even from his brother and collaborator, Roy. Without funding, Disney realized that he would need to convince investors of the concept. Given his recent experience in television, he aimed to convince one of the three major networks to back this grand scheme.

Disney began by forming WED Enterprises and gathering together a team of creative minds to research existing theme parks while brainstorming new ideas. Then, over the last weekend in September 1953, he enlisted one of his talented chief art directors, Herb Ryman, to bring his vision to life in this aerial sketch map of Disneyland.

The map beautifully captures the essence of the park, many of whose elements remain today. The detail at lower right shows the path that guided visitors into the park. In the foreground lies "Main Street, U.S.A."; this crucial point of entry for any visitor encapsulated Disney's interpretation of American values. Greeted by a pleasing town square with a large national flag, visitors strolled down a street of small businesses. Such an experience created a powerful—if nostalgic—sense of community that hearkened back to late-nineteenth-century life, before the advent of the automobile facilitated the suburbanization that would erode these small town centers. Ironically, however, extensive research determined that the best location for this large theme park would be in the heart of Orange County, where explosive suburban development was quickly replacing miles of orange groves that had been cultivated since the late nineteenth century.

At the end of Main Street lay an open plaza that offered several choices: Frontierland, Fantasyland, the World of Tomorrow, and True Life Adventure. Each of these distinct worlds involved an experience far more immersive than the carnivals and amusement parks of the day. Frontierland, for instance, took individuals through a romanticized, compelling view of the American West, framed as a source of national renewal that forged the American character. Steamboats evoked the world of Tom Sawyer, while open expanses of land invited guests to imagine the era of homesteaders and frontiersmen. The World of Tomorrow (eventually Tomorrowland) tapped the contemporary cultural fascination with technological progress and science fiction. In each of these, Disney sought not just to create a fantasy, but to do so in a way that was clean, wholesome, appealing to parents and children alike, and suffused with a vision of values that he considered essential to America's exceptional place in world history.

Ryman's decision to render the park through an oblique perspective is important. Halfway between a traditional map and a picture, this schematic perspective draws the viewer into the experience and translates Disney's unprecedented vision into a tangible enterprise. The park is ringed by a charming railroad that integrates the different worlds into a coherent whole. And it worked. In October 1953 Disney used the map to convince executives at the American Broadcasting Company to help finance this 160-acre dream in Anaheim, California. But perhaps not even Disney could have anticipated the extraordinary and enduring popularity of the park, which remains one of the most visited attractions in the Western United States, a physical cornerstone of a vast entertainment empire that reaches tourists and viewers of all ages around the world.

DISNEYLAND
SCHEMATIC AERIAL VIEW
APPROX. 45 ACRES
WITHIN RAILROAD TRACKS
DESIGNED BY WED ENTERPRISES.
THE HUB
RECREATION PARK
FRONTIER COUNTRY
FANTASY LAND
MICKEY MOUSE CLUB
LILLIPUTIAN LAND
HOLIDAY LAND
WORLD OF TOMORROW
TRUE-LIFE ADVENTURELAND
MAIN STREET
PARKING
W
N
E
S

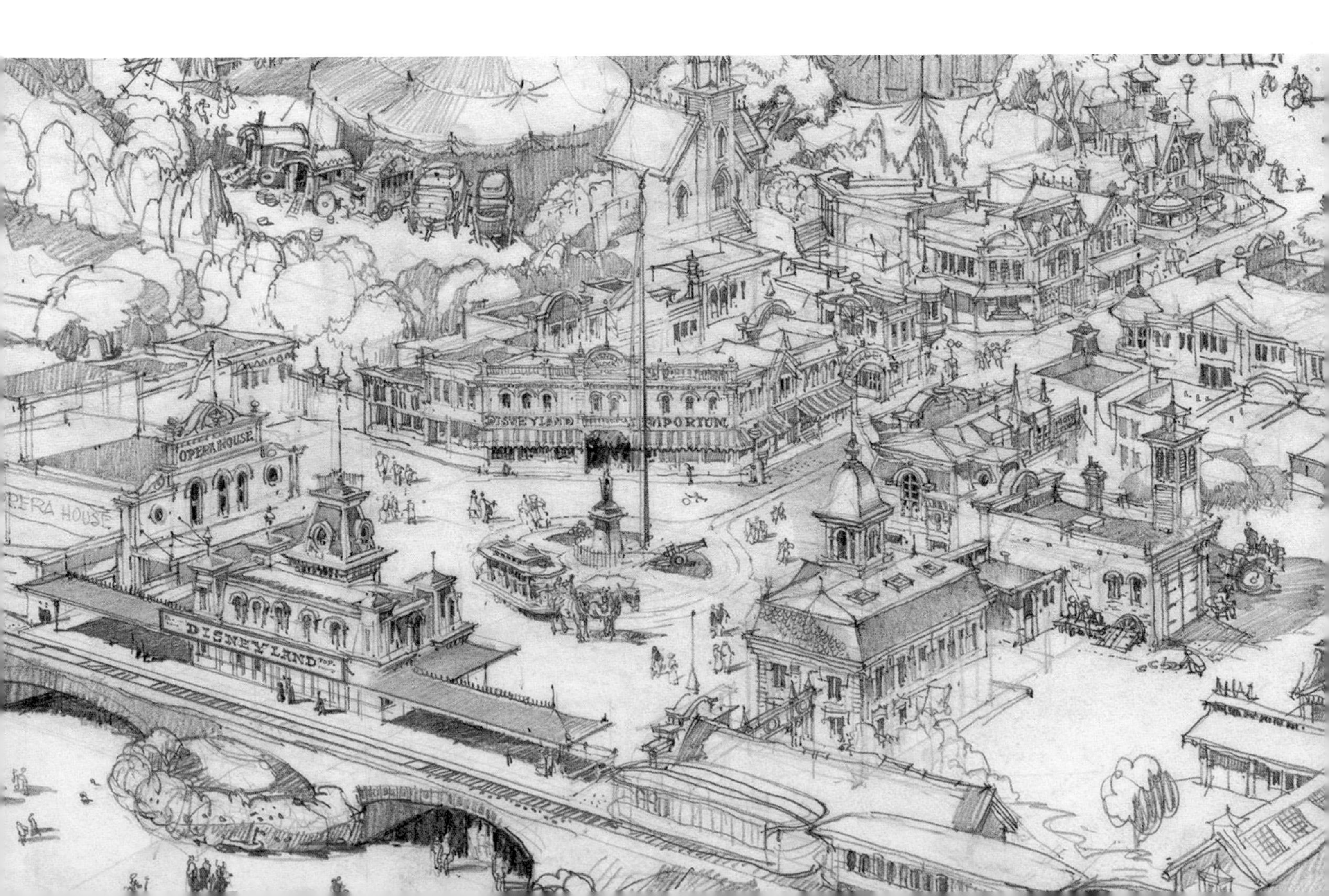
OPERA HOUSE
PERA HOUSE
DISNEYLAND
EMPORIUM
DISNEYLAND

To America's 85 Million Drivers:

A PROGRESS REPORT ON THE INTERSTATE HIGHWAY SYSTEM

- **10,440 miles now open to traffic**
- **Work under way on 14,158 miles**
- **16,019 miles to go!**

Less than five years ago, America launched the greatest construction job in history—the 41,000-mile National System of Interstate and Defense Highways. This vast network of modern freeways will sweep from the Atlantic to the Pacific, Canada to Mexico . . . linking practically every city of more than 50,000.

Progress has been remarkable. As the map shows, the system is taking shape in every part of the nation. However, of the 14,158 miles of work now under way, 10,032 miles are still in the "engineering" and "acquisition of right-of-way" stages. So there is still much construction to be started.

When finished in the early '70s, the system will carry 20% of the nation's traffic, estimated at over 100 million vehicles. Once on the system, you will be able to travel swiftly and safely to any part of the nation without a traffic light or crossroad.

No nation in history has ever completed a project of this magnitude. To stick with this job . . . to see it through to completion as planned . . . is a challenge. It's a job for all of us. After all, if we don't do it . . . who will?

Caterpillar Tractor Co., General Offices, Peoria, Illinois, U. S. A.

CATERPILLAR

REG. U. S. PAT. OFF.

DIESEL ENGINES • TRACTORS • MOTOR GRADERS
EARTHMOVING EQUIPMENT

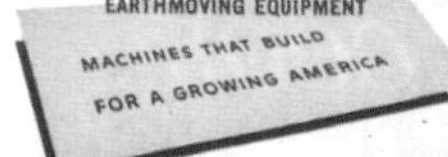

Growing symbol of our explodin
Robert is 18 months old. When h
our nation will need . . . tens of tho
miles of new roads • 25 million
rehabilitation of many metropolita
50% increase in our present supp
• 60% more classroom facilities
double the number of acceptable
double our electric power • 30%
and 50% more pulpwood • over 100
tional farm acres under soil conse
sands of water retention structure
mineral ores • twice our presen

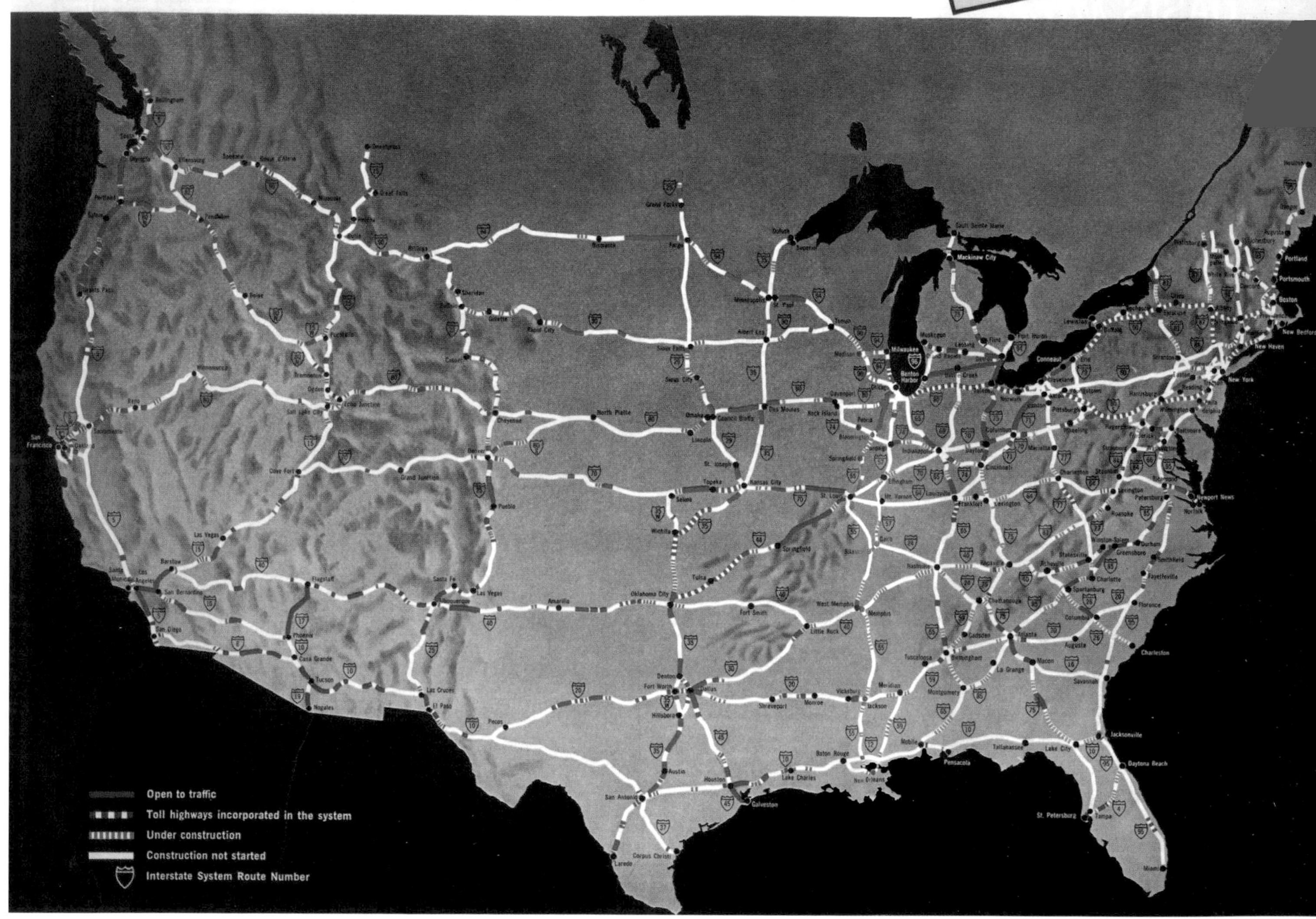

HIGHWAYS AND SUBURBS

See the progress your state is making

(as of Dec. 31, 1960)

STATE	OPEN TO TRAFFIC (miles)	WORK UNDER WAY (miles)	WORK TO BE DONE (miles)
bama	59.5	360.6	453.8
zona	514.3	167.5	479.2
ansas	43.5	458.2	16.1
ifornia	527.8	1,200.9	453.2
orado	228.7	175.2	544.1
nnecticut	138.6	136.8	21.8
aware	3.5	27.9	9.1
rida	86.8	264.3	768.9
orgia	169.4	269.2	670.9
waii	4.6	00	43.5
ho	142.3	232	237.8
nois	492.2	622.9	471.4
iana	262.6	366.9	489.3
a	213.2	258.9	236.6
nsas	390.7	103.5	306.9
ntucky	71	260.2	364.9
uisiana	41.8	324.6	316.2
ine	103.6	42.3	166.1
ryland	120.3	199.9	33.5
ssachusetts	197.2	119.4	145.8
chigan	382.1	370.5	327.1
nnesota	79.8	367.1	451.2
ssissippi	43.2	305.5	329.5
ssouri	367.9	691.4	45.4
ntana	92.5	444.8	641.7
braska	41.1	226.2	222.2
vada	56.2	158.6	319.2
w Hampshire	78.6	28.1	107.1
w Jersey	93.1	115.4	163
w Mexico	291.3	104.5	607.1
w York	668.1	317	242.1
rth Carolina	285.3	145.2	338.4
rth Dakota	197	50.8	320.1
io	565.5	435.9	482.5
lahoma	303.8	258.8	233
egon	425.9	132.5	173.5
nnsylvania	596.4	341.2	603.7
ode Island	20.7	15.6	34.6
uth Carolina	126	241.3	311.9
uth Dakota	115.8	279.4	282.4
nnessee	15.4	518.2	514
xas	879.4	1,426.1	717.9
ah	69.3	183.5	681.5
rmont	23.1	129.2	171.3
rginia	158.1	342.1	552.9
ashington	297.6	287.6	140.1
est Virginia	89.2	106.5	188.9
isconsin	146	306.5	000
yoming	119.4	225.3	572.1
st. of Col.	0.5	11.7	16
	10,439.9	14,157.7	16,019.5

ere are no Interstate Highways in Alaska.

Caterpillar Corporation, "A Progress Report on the Interstate Highway System," *Saturday Evening Post*, April 22, 1961

The interstate highway system remains the most ambitious infrastructure project in American history. It originated just after World War I, when Lieutenant Colonel Dwight Eisenhower was assigned to an army convoy traveling across the country on the Lincoln Highway. The convoy was organized in part to publicize the poor condition of the nation's roads—especially in the West—and the experience made an impression on Eisenhower. Twenty-five years later General Eisenhower observed firsthand the German autobahns, four-lane "superhighways" that ranked among the best in the world and that ultimately facilitated the Allied invasion that ended the war.

As president in 1956, Eisenhower authorized 41,000 miles of interstate roads through the Federal-Aid Highway Act. Yet after five years only 8,000 miles had been completed, and public support for the project began to waver. The main obstacle was fiscal: in 1961 the gasoline tax of four cents per gallon, which had generated billions for the program, was set to drop to three cents. This decline in revenue was compounded by the trend toward smaller and more fuel-efficient cars such as the Volkswagen Beetle. In response, Federal Highway Administrator Rex Whitton launched a public relations campaign in 1961 to stimulate support for the project. He promised that the entire system would be complete by 1972, as originally scheduled. As he told one interviewer in 1962, drivers would soon be able to go from coast to coast without a single stop light. (Decades later, Charles Kuralt quipped that one could drive from coast to coast on the interstates without seeing anything at all.)

At the same time, President John F. Kennedy urged Congress to renew the gas tax and find other sources of revenue to complete the highway system. To make his case, Kennedy echoed Eisenhower, arguing that the interstate system was essential not just to economic growth but also to civil defense. In times of national emergency, such a network would facilitate transportation to all parts of the country. At the height of the Cold War such words were a powerful lever to action. The private sector also lobbied vigorously for the highways, as petroleum and trucking companies, civil engineers, and home builders were all directly affected by the project. The Caterpillar Corporation, which produced heavy construction equipment, issued this advertisement in 1961 to pointedly measure the progress of the highway system. In Caterpillar's rendering, even cities are secondary to the central feature of the landscape, the web of highways across the country. The campaign worked: in June Congress extended the gas tax and expanded funding for the highways, and within three years half of the system was finished.

The Federal-Aid Highway Act was among the century's most consequential legislation. It expanded and realigned settlement, made the automobile the essential form of transportation, and stimulated several new regions and markets. Equally consequential were its effects upon American cities. In Miami, Nashville, St. Paul, and New Orleans, the routes of these new highways went directly through poor and powerless neighborhoods, many of which were predominantly Latino or African American. Kennedy himself estimated in 1962 that each year over 15,000 families and 1,500 businesses were displaced by interstate construction. Some wealthier and more organized communities resisted, such as the residents of San Francisco who successfully halted construction of the Embarcadero Freeway. But for the most part, the interstate dictated subsequent patterns of growth and decline.

The casualty was the nation's urban core. As interstates penetrated the cities, funding and ridership of mass transit fell and many people began to move to the suburbs. These trends further isolated the low-income residents who were left behind. In Chicago, the Dan Ryan Expressway separated a large public housing project that was home to thousands of African Americans from the white neighborhoods to the west. The interstate highway system sparked mobility as well as congestion, optimism as well as blame, prosperity as well as poverty. It shifted people from the Northeast to the Southwest, and from cities to outlying areas. It also superseded the older highway system shown on page 196. Route 66 was surpassed by Route 40, and in countless other areas small towns became ghost towns.

THE BATTLE AGAINST SEGREGATION

Associated Press, background map of the Freedom Rides, 1962

As the map on page 198 shows, African Americans hoping to travel through the South were wise to plan with care. Segregation was enshrined in law, and woven into daily life. But by the 1950s, far fewer were willing to accept a system that excluded them from schools, restaurants, theaters, restrooms, and other public facilities. The 1954 Supreme Court decision in *Brown v. Board of Education* heartened many by ruling that such segregation—particularly in education—violated the Fourteenth Amendment of the Constitution. Separate facilities based on race were inherently unequal.

The *Brown* decision was followed by others, but the failure to enforce this ruling at the local level drove civil rights activists to expose and overturn segregation through direct action. In 1955 the black community's year-long Montgomery bus boycott led to a Supreme Court decision against Alabama's segregated public transit system. In 1960 four black students at the North Carolina Agricultural and Technical College sat down to be served at an all-white lunch counter, and drew violence and humiliation but also national attention to the absurdity and cruelty of segregation.

In 1961 the Congress on Racial Equality took a similar approach of non-violent direct action to test enforcement of the recent court ruling against segregation in interstate travel. The idea was to have whites and blacks sit together on interstate buses, in restaurants, and in terminals. With this ordinary and legal act they risked not just arrest but much worse, for whites routinely responded to the prospect of integration with violence.

On May 4, a century after the outbreak of the Civil War, thirteen passengers—black and white, male and female, Northern and Southern—boarded a bus in Washington, D.C. They first traveled south through Virginia and the Carolinas before turning east through Georgia and Alabama. In Anniston, Klansmen and segregationists attacked and set fire to the bus while the riders were still inside. Days later, another white mob brutally assaulted the Freedom Riders in Birmingham, after which ambulances initially refused to transport injured riders to the hospital. The police occasionally even facilitated these Klan attacks. The *New York Times* and the *Washington Post* reported the Anniston attack on the front page, while television news outlets shocked and shamed viewers with indelible images of the burned bus and its victimized passengers.

A subsequent ride organized by Nashville students encountered violence in Montgomery, prompting Attorney General Robert Kennedy to send in federal marshals. Alabama's governor threatened to arrest the marshals, underscoring the conflict between federal and state law that endured throughout the Civil Rights Movement. Upon arriving in Jackson, Mississippi, over 300 student Freedom Riders were arrested and jailed, an event which drew even more intense national attention. By the end of the year, 400 individuals had participated in the Freedom Rides, but the upshot was unclear.

President Kennedy had been in office only a few months before the first of the Freedom Rides, and initially distanced himself from the action as well as the civil rights bills that had just arrived on his desk. The president was loath to alienate Southern Democrats, and spent most of his energy prosecuting the Cold War. But ironically the nation's moral posture against communism also prodded Kennedy to avoid embarrassing footage of bigoted Southern whites attacking those who were exercising their lawful right to travel.

On November 1, the Interstate Commerce Commission ordered an end to segregation in bus terminals, prompting a second set of rides to test enforcement of the ruling. This map was designed by reporter Sid Moody to accompany a newswire story on the Freedom Rides as a whole, throughout 1961. The diagram traces the routes of the individual rides, and the attacks they met. But in its simplicity it reveals two crucial dynamics of the Civil Rights Movement. First, the map documents the will of the activists themselves, who were prepared to risk humiliation, injury, and death in order to train the nation's attention on these egregious practices. Only by the sustained willingness to venture into these remote and potentially dangerous regions did the Freedom Riders expose the enduring resistance to desegregation.

Second, the map points to the essential role of the media, for the violence that greeted the Freedom Riders was graphically reported across the country. Asterisks on the map mark sites where the passengers met with violence, while annotations list the spots where they were arrested. It was in those spots—where the peaceful riders met obstacles—that the news media found a story to cover. Indeed, over the course of the year civil rights activists began to worry that the country had grown tired of such stories. Yet it was this persistent activism on the ground that forced a nationwide reckoning—however limited—with segregation.

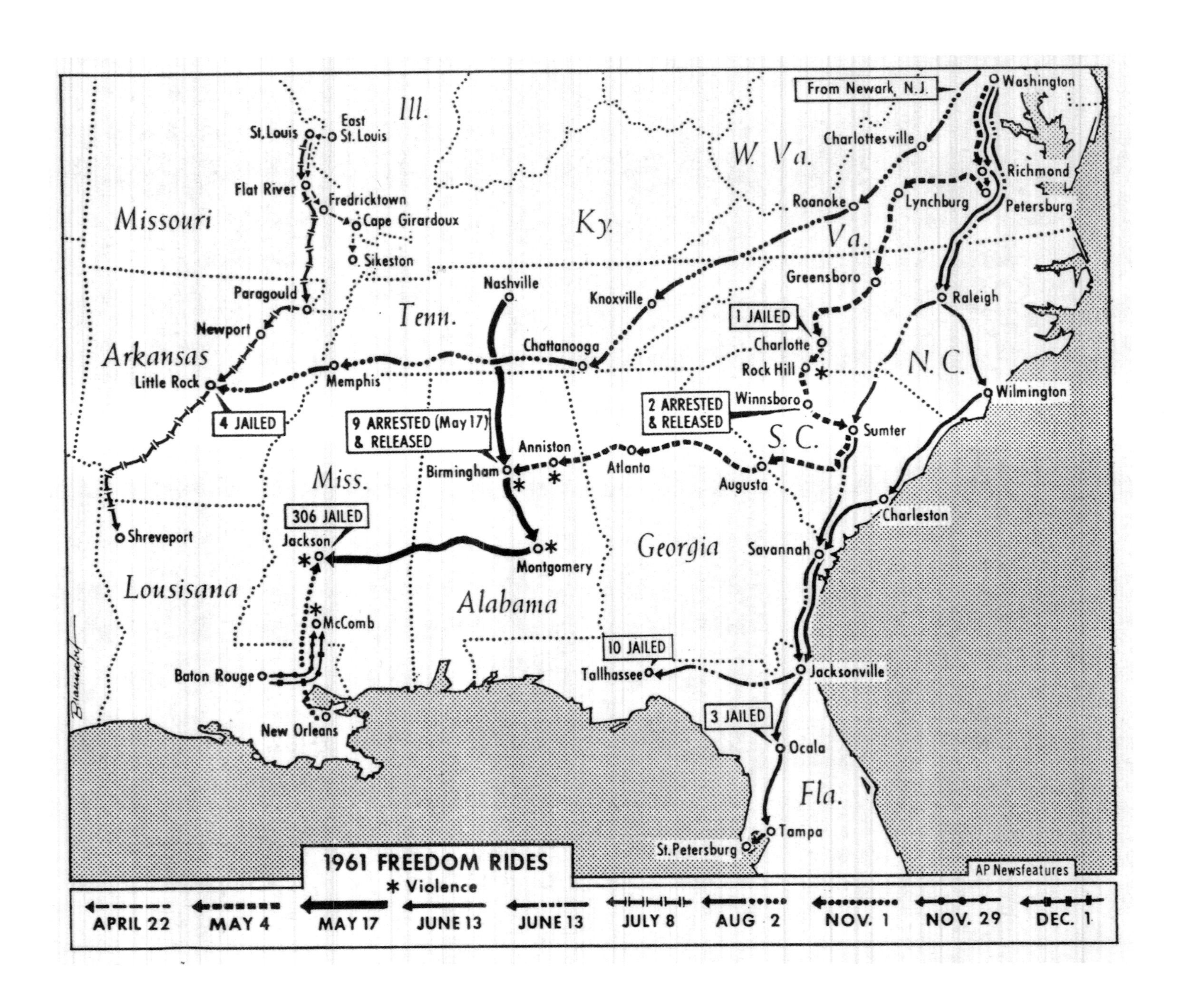
1961 FREEDOM RIDES
* Violence
APRIL 22
MAY 4
MAY 17
JUNE 13
JUNE 13
JULY 8
AUG 2
NOV. 1
NOV. 29
DEC. 1
From Newark, N.J.
Washington
Charlottesville
Richmond
Petersburg
Lynchburg
Roanoke
Greensboro
Raleigh
Knoxville
Nashville
Chattanooga
Memphis
Little Rock
Newport
Paragould
Sikeston
Cape Girardoux
Fredricktown
Flat River
St. Louis
East St. Louis
Shreveport
Jackson
McComb
Baton Rouge
New Orleans
Birmingham
Anniston
Atlanta
Augusta
Montgomery
Charlotte
Rock Hill
Winnsboro
Sumter
Wilmington
Charleston
Savannah
Jacksonville
Tallhassee
Ocala
Tampa
St. Petersburg
1 JAILED
4 JAILED
9 ARRESTED (May 17) & RELEASED
2 ARRESTED & RELEASED
306 JAILED
10 JAILED
3 JAILED
Ill.
Missouri
Ky.
W. Va.
Va.
Tenn.
N.C.
S.C.
Arkansas
Miss.
Alabama
Georgia
Lousisana
Fla.
AP Newsfeatures

9. 1962–2001

An Unsettled Peace

How do we write the history of our own time? Generalizations—and certainly conclusions—are necessarily provisional, for the changes wrought in the late twentieth century are still unfolding, still framing our own experience. But this chapter begins with the 1960s for a reason: in that decade Americans used maps to navigate challenges that persisted through the end of the century. These include the disruptions brought by technology, debates over the nation's role abroad, and the ongoing struggle to realize a more just, equal, and pluralistic society.

At the outset of this period, the Cold War remained the most salient feature of American life. It set foreign priorities, guided domestic politics, and even influenced the economy. Democrats and Republicans alike accepted that the Soviet Union posed the principal threat to the country's security, a consensus that was borne out in the Cuban Missile Crisis. Aerial photographs taken in October 1962 revealed that the Soviets had covertly stationed missiles just ninety miles off the Florida coast. On pages 234 and 236 are examples of the material evidence that President John Kennedy and his advisers used to respond to the terrifying prospect of a nuclear attack.

The Kennedy administration vigorously prosecuted the Cold War in Asia as well, deepening military commitments in Vietnam that would be extended further by President Lyndon B. Johnson. As shown on page 238, the United States primarily framed the conflict in Vietnam in terms of communism, downplaying the equally important strain of nationalism that sought to oust the French—and subsequently the Americans. By 1968 the United States had deployed 500,000 young men to Vietnam in a war that polarized Americans, deeply divided the Democratic Party, and led to the election of Richard Nixon. Though he promised to end the war, Nixon in fact intensified the bombing campaign and launched an invasion of Cambodia. In the spring of 1970, college students across the country coordinated a massive antiwar protest in response, and their tight communication networks and rejection of the Cold War consensus are exhibited on page 240.

The Cold War even extended into outer space. Just a few years after the launch of Sputnik, President Kennedy announced the ambitious goal of landing a man on the moon by the end of the decade. The progress of the National Aeronautics and Space Administration (NASA) riveted Americans in the 1960s, and exposed them to an entirely new extraterrestrial reality. In 1968, the astronauts of Apollo 8 became the first to orbit the moon, yet it was their photographs of earth that captivated the public. The most powerful image of the decade—and one of the most reproduced images of the century—was a picture of our own planet, one that simultaneously evoked fragility and possibility (page 244).

In the same year, the artist Heinrich Berann drew an equally consequential view of the Atlantic Ocean floor that made both an artistic and a scientific contribution. Berann's profile was based on years of research by Marie Tharp and Bruce Heezen, who synthesized information about the oceans in order to advance general theories about plate tectonics. Berann's dazzling artistic technique struck a chord with the American public, and helped to expose a generation to this new idea (page 242). In this regard, the profiles of the ocean floor not only "mapped" some of the last geographical mysteries, but also became evidence for the radical new science of continental drift. The geological research undergirding this theory stretched back to the nineteenth century, but was accelerated significantly by the Cold War.

In guiding investment in science and technology, the Cold War also influenced patterns of migration. NASA centers were built in Texas, Florida, and Alabama, while Southern California boomed with the growth of aerospace and other defense industries. As the maps on page 250 illustrate, these opportunities reconfigured the nation by drawing Americans out of the Northeast and the Midwest to the South and West. But these demographic shifts also sparked unexpected reversals, including a notable migration *out* of California by the 1990s.

Domestic and internal migration only partly explains the nation's reconfiguration in the second half of the twentieth century. From 1965 to 2000, 20 million immigrants entered the United States, chiefly from Latin America, the Caribbean, and Asia. These numbers exceeded the wave at the turn of the nineteenth century, though in that earlier era immigrants constituted a much larger proportion of the total population. This expansion of immigration stemmed from the easing of longstanding entry quotas and restrictions in 1965. Within this more recent phase, it has been the Latino population which has most thoroughly reshaped American society, not just in the Southwest but throughout the country (page 252).

By relaxing restrictions on immigration in 1965, Congress fundamentally transformed the country's racial and ethnic profile. In that same year, Congress helped to realize African American citizenship rights through the landmark Voting Rights Act. The act dramatically expanded black political participation, but it also inadvertently contributed to the thorny problem of redistricting. Gerrymandering has a long history in the United States, but in the 1990s it entered a new phase with the use of computerized technology. Even in the digital era, however, maps remained the lynchpin in those battles over representation (page 254).

The struggle for civil rights in the 1960s inspired several other movements for justice and recognition, notably among gays and lesbians. In the 1970s and 1980s, the gay community began to challenge the treatment of homosexuality as either a sin or a perversion. The emerging recognition of gay rights and sexual identity was partly influenced by the public health crisis around acquired immune deficiency syndrome (AIDS). The maps on pages 246 and 248 represent two distinct yet related efforts to understand the spatial behavior of this deadly disease. In the first instance, the infectious disease specialist Abraham Verghese turned to cartography to investigate the spread of AIDS in rural Tennessee. Maps enabled him to see patterns of patient behavior that would have serious implications for his practice and the treatment of AIDS in rural areas more generally. In the second instance, the geographer Peter Gould experimented with early computer modeling in an urgent quest to track—and thereby predict—the path of the epidemic in the nation's cities.

Gould's "heat maps" underscore the terror of AIDS. They also represent a shift in mapping that points to a more basic—and easily overlooked—legacy of the 1960s. The research required to send a man to the moon also improved computer technology and led to the introduction of the microchip. Similarly, the Internet began as a high-speed communication network for the military in the 1960s. These technological innovations—advanced by the Cold War—also transformed mapmaking, and by extension cartographic and spatial *thinking*. Just as new mapmaking techniques drove the first visualizations of slavery on page 142, the advent of digital mapping opened up new forms of inquiry and investigation. By the turn of the twentieth century, data mapping and modeling had proliferated into areas as diverse as public health, politics, urban planning, and marketing.

We close this chapter with a map that marks the end of one era and the beginning of another. The terrorist attacks of September 11, 2001, point to the unsettled nature of international relations in the aftermath of the Cold War. Laura Kurgan's map of Lower Manhattan represents an effort to guide visitors at Ground Zero as they grappled with the meaning of the attacks (page 256). Kurgan mapped a site that was simultaneously a battlefield and a memorial. The map of Ground Zero—its future uncertain—mirrors the more general questions that bedeviled Americans as they entered a new century.

TO THE BRINK OF NUCLEAR WAR

Aerial photograph of Cuba, annotated map of Cuba, and map of missile range from Cuba, 1962

On October 15, 1962, American intelligence officials discovered that the Soviets were constructing missile sites in Cuba, just ninety miles off the Florida coast. One week later, President Kennedy made these actions public; he then engaged in several days of diplomatic brinksmanship with the Soviet leader Nikita Khrushchev. It was the most dangerous confrontation of the Cold War, and the closest the world has come to nuclear war.

The long and fraught relationship between the United States and Cuba stretched back to the Spanish–American War at the turn of the century. The Americans liberated Cuba from Spanish control but retained the right to intervene for decades thereafter. When a nationalist revolution in Cuba brought the Marxist revolutionary Fidel Castro to power in 1959, the Caribbean became a crucial geopolitical arena of the Cold War. As a candidate for president the next year, Kennedy pointedly criticized the Eisenhower administration for failing to prevent the Cuban revolution. Once inaugurated, Kennedy became even more preoccupied with Castro, authorizing an invasion and coup that failed disastrously at the Bay of Pigs. It was Kennedy's hard line against communism in Cuba that contributed to the missile crisis of 1962.

The Soviet Union began to send military aircraft to Cuba in the summer of 1961, ostensibly to defend the country against the United States. In early October 1962 the Soviets stationed bombers at the far western military base of San Julián, signaling their intent to develop Cuba's offensive capacity. The Central Intelligence Agency responded by increasing its aerial surveillance, and Kennedy warned Khrushchev that any attempt to construct military bases in Cuba would be treated as a direct threat. Soviet shipments continued, partly in response to Kennedy's recent deployment of fifteen Jupiter nuclear missiles in Turkey. No doubt Khrushchev aimed to give the Americans a taste of their own medicine by demonstrating how it felt to live so close to offensive weaponry.

On October 14, an American U-2 plane photographed unusual activity in San Cristóbal, sixty miles west of Havana. The next day the National Photographic Interpretation Center concluded that these images revealed the presence of offensive weapons in Cuba. This photograph at right captured one of several medium-range ballistic missile launch sites around the island; three long-range missiles sites were also under construction. With these photographs and other evidence of military bases and Soviet aircraft, the president and his advisers spent days debating how to proceed.

The map of Cuba shown on the next page was annotated during these tense days of deliberation. It shows the number and location of Soviet MiG fighter jets, helicopters, and—crucially—the squadron of offensive Il-28 bombers at the far western base of San Julián.

The second and even more terrifying map shown on the next page depicts the range of missiles that had been photographed just days earlier. The central ring marks the capacity of the medium-range ballistic missiles, which could reach both Mexico City and Washington, D.C. within twenty minutes. The outer ring—reaching Hudson Bay and Lima, Peru—marked the geographical capability of the long-range missiles, which had arrived in Cuba but were still unassembled. (The missile-range map identifies the major cities in North and South America that could be reached by the Soviet weapons, but why was the small town of Oxford, Mississippi, included? Two weeks earlier, Attorney General Robert Kennedy had dispatched federal marshals to Oxford to quell the riots that had broken out to protest the integration of the University of Mississippi. During the Cuban Missile Crisis, the Attorney General mischievously asked whether the missiles might reach Oxford, prompting its appearance on the map as a kind of macabre joke.)

MRBM LAUNCH SITE 2
SAN CRISTOBAL
1 NOVEMBER 1962
FUEL TRAILERS
MISSILE-READY TENT
FORMER LAUNCH POSITIONS
FORMER LOCATION OF MISSILE-READY TENTS

SUMMARY

San Julian	22 IL-28 (4 partially assembled; 17 in crates)
San Antonio de las Banos	35 MIG 15's, 17's and 19's
Santa Clara	35 MIG 21's plus probably 4 in crates, 13 MIG 15's and 17's.
Camaguey	11 MIG 15's and 17's
Playa Baracda	35 Helicopters, 14 Transports

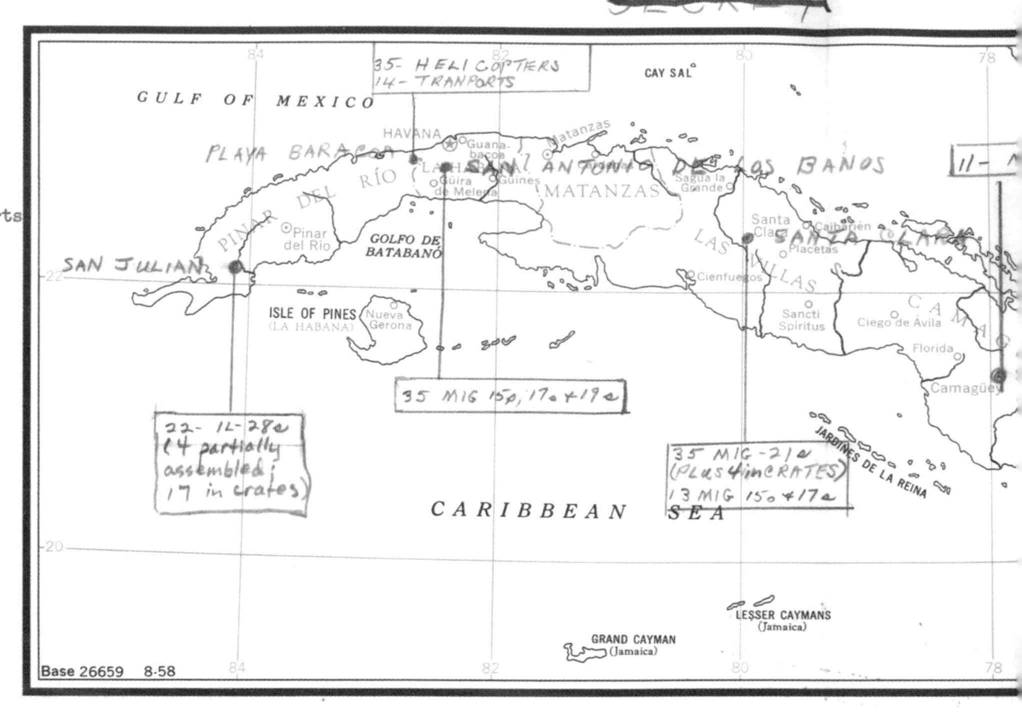

CUBA

Provincia boundary
National capital
Provincia capital

0 20 40 80 Miles
0 20 40 80 Kilometers

The deliberations within the administration ranged widely. The Joint Chiefs of Staff pressed for a preemptive strike to destroy the missiles, followed by an invasion of Cuba. Others counseled restraint, pushing for warnings to Cuba and the Soviet Union. After several days of tense discussion, on October 22 Kennedy announced on radio and television that the Soviets had installed missiles in Cuba. He increased surveillance and declared an immediate blockade of the island to search for any incoming offensive military equipment. The next day, US ambassador Adlai Stevenson presented the photographic evidence shown on the previous page to the United Nations. With support from the Organization of American States, Kennedy publicly demanded that the Soviets withdraw all weapons from Cuba. Privately, he worried that such aggressive posturing might lead to war.

In the following days, the situation worsened. Khrushchev sternly rejected Kennedy's public demands, while American reconnaissance flights confirmed the readiness of the missile sites. Kennedy waited, giving diplomacy more time in a way that ultimately resolved the crisis. Many who were connected to the Kennedy administration have portrayed the president and his brother as cool-headed leaders who overcame the more hawkish advisers. Recently released White House audio recordings suggest, however, that Robert Kennedy was more hawk than dove, and that the president was largely alone in resisting the call to take a hard line with the Soviets. Moreover, the stationing of missiles in Turkey no doubt influenced Soviet actions in Cuba. Ultimately, the United States removed the missiles in Turkey, while the Soviets did the same in Cuba. President Kennedy may have resolved this high-stakes conflict prudently, but it was to some extent a crisis of his own making.

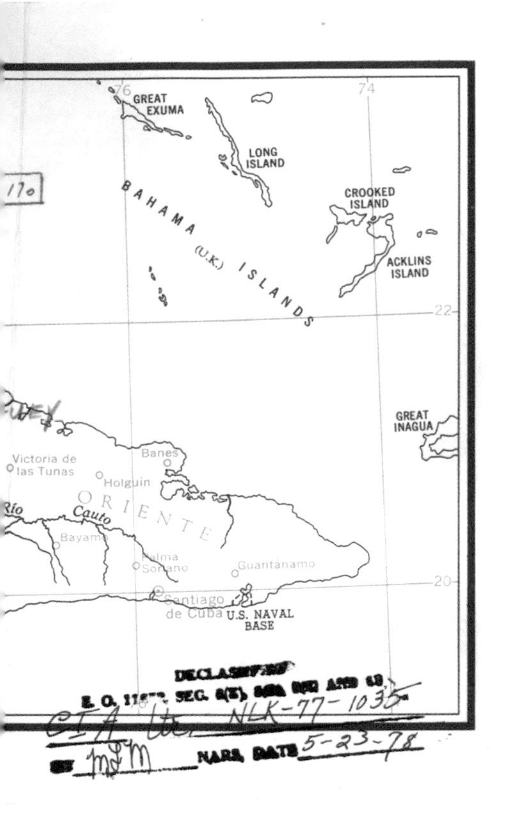

76
74
GREAT EXUMA
LONG ISLAND
BAHAMA (U.K.) ISLANDS
CROOKED ISLAND
ACKLINS ISLAND
22
GREAT INAGUA
Victoria de las Tunas
Banes
Holguin
ORIENTE
Cauto
Palma Soriano
Guantánamo
20
Santiago de Cuba
U.S. NAVAL BASE
CIA ltr. NLK-77-1035
NARS, DATE 5-23-78

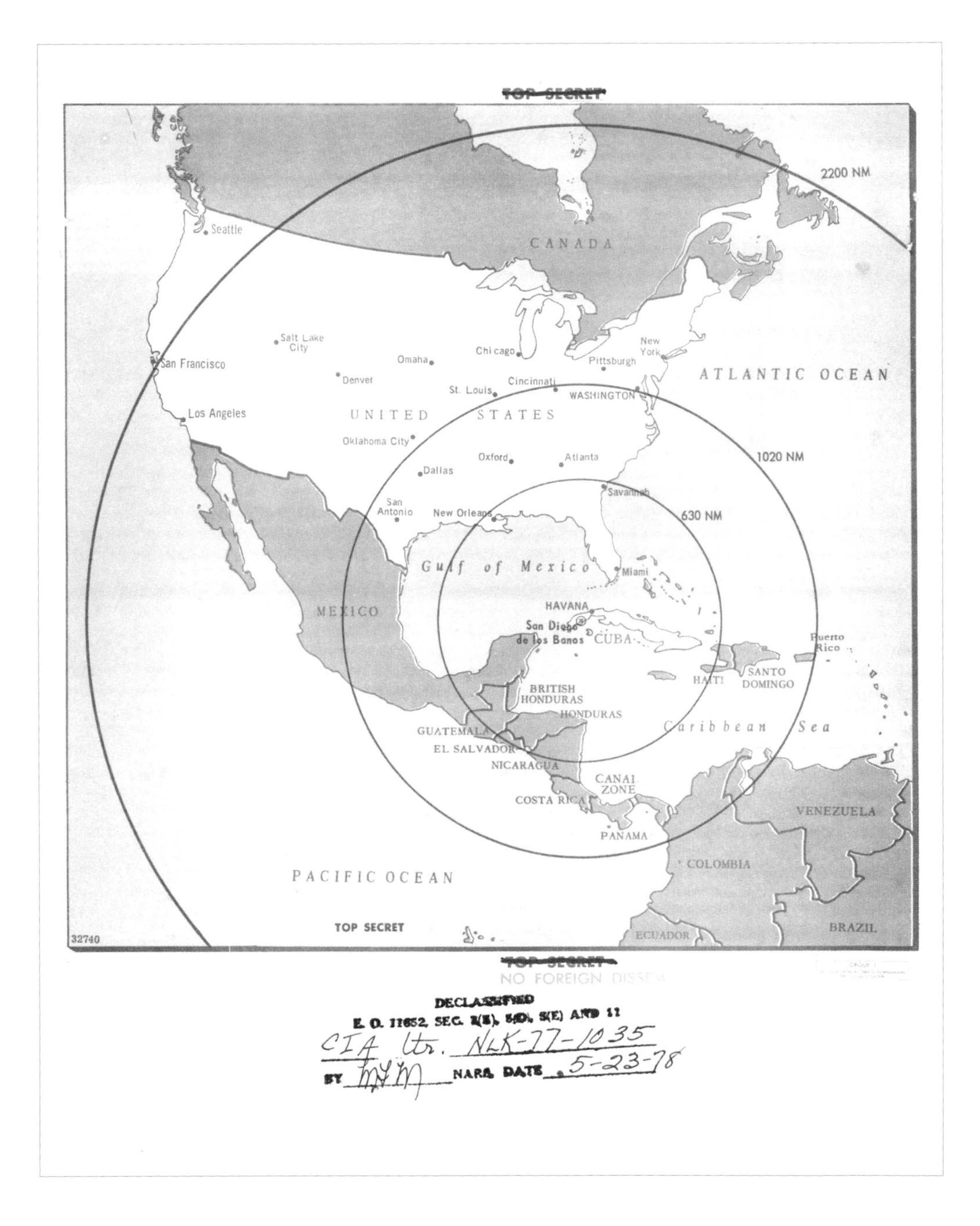

TOP SECRET
2200 NM
CANADA
Seattle
Salt Lake City
Omaha
Chicago
New York
Pittsburgh
San Francisco
Denver
St. Louis
Cincinnati
WASHINGTON
ATLANTIC OCEAN
Los Angeles
UNITED STATES
Oklahoma City
Oxford
Atlanta
1020 NM
Dallas
Savannah
San Antonio
New Orleans
630 NM
Gulf of Mexico
Miami
MEXICO
HAVANA
San Diego de los Banos
CUBA
Puerto Rico
HAITI
SANTO DOMINGO
BRITISH HONDURAS
HONDURAS
GUATEMALA
Caribbean Sea
EL SALVADOR
NICARAGUA
CANAL ZONE
COSTA RICA
VENEZUELA
PANAMA
COLOMBIA
PACIFIC OCEAN
TOP SECRET
32740
ECUADOR
BRAZIL
TOP SECRET
NO FOREIGN DISSEM
DECLASSIFIED
E. O. 11652, SEC. 3(E), 5(D), 5(E) AND 11
CIA ltr. NLK-77-1035
BY MYM NARS, DATE 5-23-78

WHY ARE WE IN VIETNAM?

US Information Agency, "Aggression from the North," 1965

Anti-communism drove US foreign policy after World War II. Every region of the globe was seen through the lens of the Cold War, and this often obscured more complicated dynamics. The 1949 communist revolution in China led Harry Truman to send troops to Korea in 1950. Four years later, President Dwight Eisenhower quietly began to support South Vietnam in its struggle to remain independent from communist North Vietnam. Eisenhower explained that this distant country demanded American aid because its fate might affect that of Thailand, Burma, Malaya, and Indonesia. President Kennedy acted on the same "domino theory" by sending 16,000 American "advisers" to aid the corrupt—but firmly anti-communist—leader of South Vietnam.

The assassination of Kennedy left his successor to reckon with Vietnam. President Lyndon Johnson was keenly aware that it was a Democratic president—Truman—who had "lost" China to the communists, and he was determined to avoid the same with Vietnam. At first Johnson was a reluctant warrior, and in the 1964 presidential campaign he promised not to commit the United States to any land wars in Asia. Then, in August, two American destroyers were attacked off the coast of Vietnam in the Gulf of Tonkin. Johnson responded by authorizing limited bombing of selected North Vietnamese bases and storage facilities, an act which appeared both prudent and restrained. In November he won a landslide victory against Republican Barry Goldwater, who had advocated a much more aggressive reaction to the Gulf of Tonkin incident.

Over the next few months, the North Vietnamese intensified their infiltration of the South by supplying Viet Cong guerillas with weapons and men. An attack on the American air base at Pleiku in February 1965 prompted the United States to launch a sustained bombing operation of North Vietnam, and then deploy 3,500 Marines to defend the American air base at Danang. This bombing campaign had little effect on North Vietnamese resolve, yet it sparked widespread protests on American college campuses.

To buttress the strength of South Vietnam, Johnson's advisers began to press for an increase in American ground troops. The president himself worried that an escalation of the war would undermine his domestic agenda, and spent hours deliberating military strategy with his advisers. A close reading of the transcripts of these deliberations reveals that withdrawal from Vietnam—though advocated by some—was never seriously considered by Johnson or his senior advisers. That, they argued, would be to lose face in the Cold War, which was unacceptable.

While the South Vietnamese government teetered on the brink of collapse, Johnson took his case to the public, arguing that the shadow cast by communist China could not be ignored. To abandon Vietnam, he insisted, would be to compromise American values and invite further aggression worldwide. This poster was issued to generate support for the expanding role of the US in Vietnam at this pivotal moment of decision in 1965. It was made by the United States Information Agency (USIA), and published in its magazine, *Free World*. The USIA was an arm of the executive branch established to promote American ideals and policies abroad during the Cold War. *Free World* was translated into Chinese and Vietnamese, with a circulation of 90,000 squarely aimed at the educated classes of South Vietnam: teachers, students, civil servants, the military, and businessmen.

The poster forcefully characterized the situation in Vietnam as an aggressive attempt by the North to take over the South. Like the State Department memo on which it was based, the broadside presented an increased commitment to South Vietnam as the only viable course of action. The map marks the locations of Viet Cong attacks, including Pleiku and Danang, arguing that the longstanding and "brutal campaign of terror and subversion" by North Vietnam must be stopped. Just as the Committee on Public Information framed American war aims in 1917 (page 180), the USIA propagated intervention in Vietnam in defensive and protective terms.

Just after the broadside was issued in April, Johnson authorized the introduction of ground troops and quietly doubled the draft. In July 1965, 125,000 Americans were sent to fight in Vietnam, and by the end of the year the number had reached 200,000. With this increase came a crucial shift of tactics and purpose, from defense of air bases to the initiation of combat in the field, what many came to call "search and destroy" missions. At precisely this time, Ho Chi Minh stepped up his own funding of insurgency in the South, escalating the conflict further. After 1967, Johnson—and then President Nixon—faced increasing resistance at home. The Cold War consensus that had held since the 1940s was severely tested by the Vietnam War.

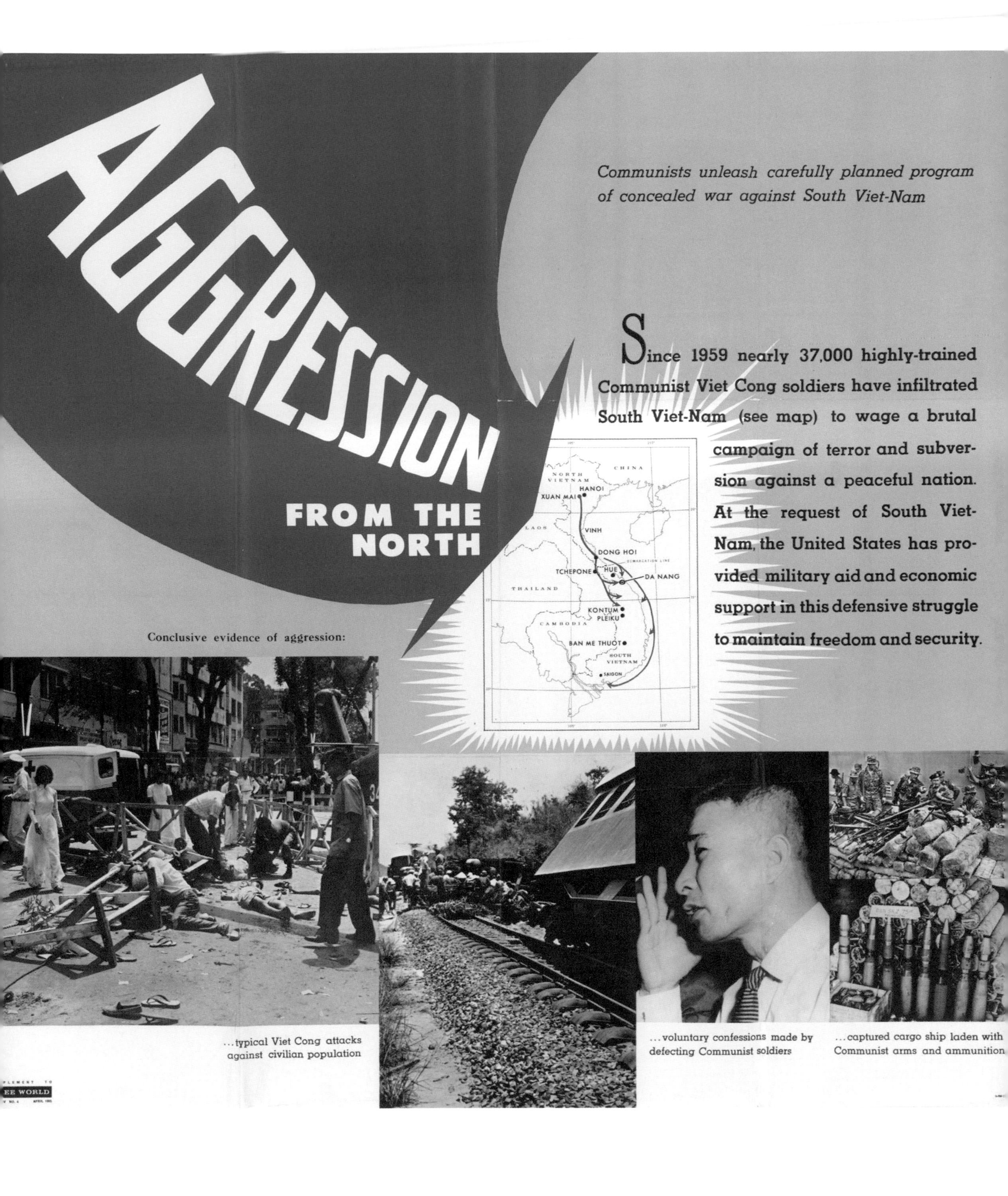
AGGRESSION
FROM THE NORTH
Communists unleash carefully planned program of concealed war against South Viet-Nam
Since 1959 nearly 37,000 highly-trained Communist Viet Cong soldiers have infiltrated South Viet-Nam (see map) to wage a brutal campaign of terror and subversion against a peaceful nation. At the request of South Viet-Nam, the United States has provided military aid and economic support in this defensive struggle to maintain freedom and security.
CHINA
NORTH VIETNAM
HANOI
XUAN MAI
LAOS
VINH
DONG HOI
DEMARCATION LINE
HUE
TCHEPONE
DA NANG
THAILAND
KONTUM
PLEIKU
CAMBODIA
BAN ME THUOT
SOUTH VIETNAM
SAIGON
Conclusive evidence of aggression:
...typical Viet Cong attacks against civilian population
...voluntary confessions made by defecting Communist soldiers
...captured cargo ship laden with Communist arms and ammunition

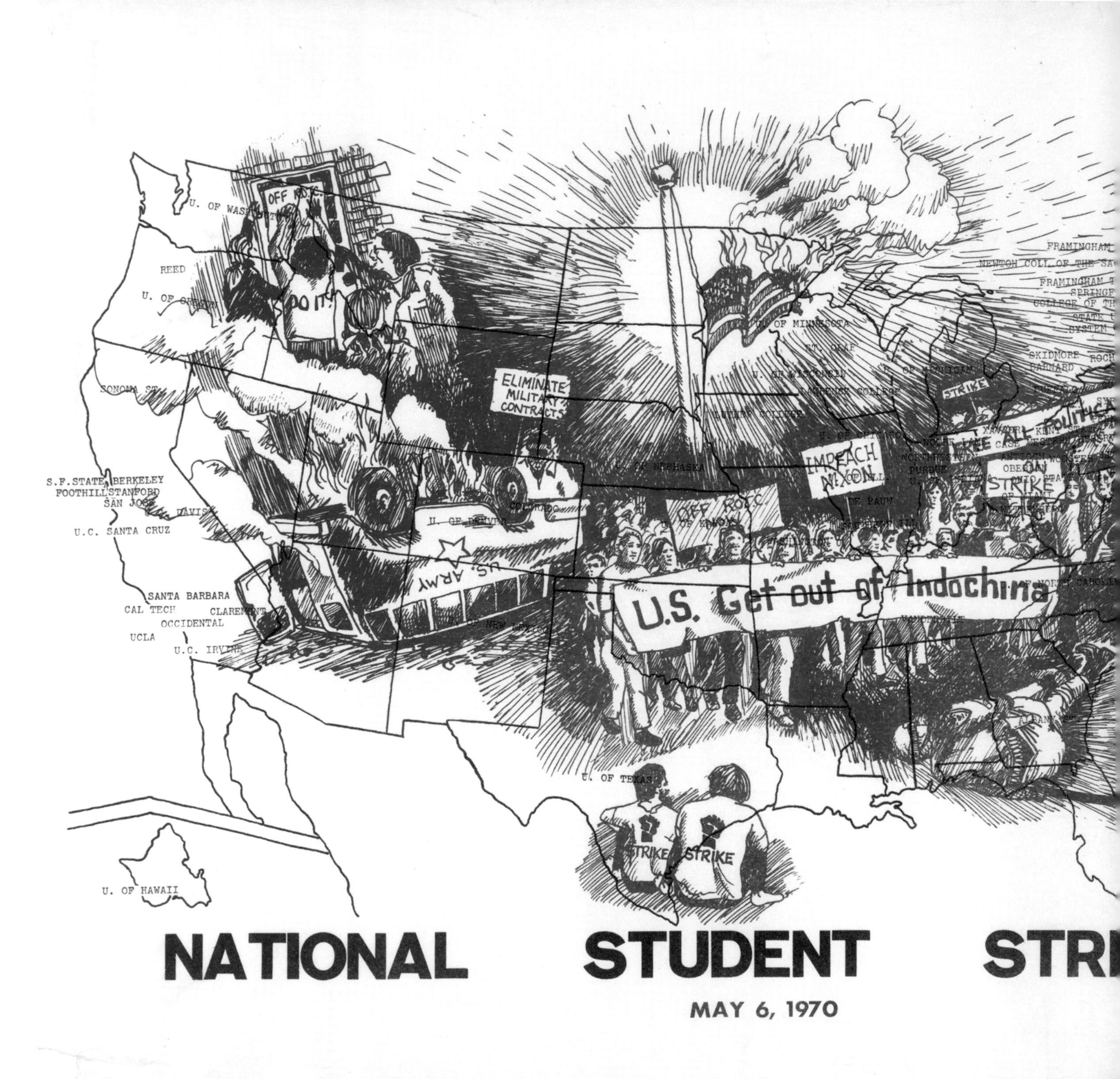
OFF ROTC
DO IT!
U. OF WASHINGTON
REED
U. OF OREGON
SONOMA ST.
S.F. STATE BERKELEY
FOOTHILL STANFORD
SAN JOSE
DAVIS
U.C. SANTA CRUZ
SANTA BARBARA
CAL TECH
CLAREMONT
OCCIDENTAL
UCLA
U.C. IRVINE
U. OF HAWAII
ELIMINATE MILITARY CONTRACTS
U.S. ARMY
U. OF DENVER
U. OF NEBRASKA
U. OF MINNESOTA
ST. OLAF
IMPEACH NIXON
OFF ROTC
DE PAUW
STRIKE
U.S. Get out of Indochina
U. OF TEXAS
STRIKE
STRIKE
FRAMINGHAM
SKIDMORE
BARNARD
NORTHWESTERN
PURDUE
OBERLIN
NATIONAL STUDENT STRI
MAY 6, 1970

REVOLT

National student strike map, 1970

The previous map shows us an official narrative of the Vietnam War, one in which the United States acted as a protector of democracy against a stealthy, subversive, and well-funded enemy. This anti-war map produced five years later squarely rejects that view of the war.

President Johnson's decision in 1965 to bomb North Vietnam and to expand the ground war in South Vietnam sparked massive opposition on the home front. Initially, the anti-war movement was largely limited to teach-ins and draft counseling on college campuses, but by the end of 1967 a majority of Americans considered the war a mistake. The United States had deployed hundreds of thousands of troops, and had dropped a bomb tonnage exceeding that of all the theaters of World War II; yet South Vietnam remained unstable. Even more shocking to Americans at home was the Tet Offensive of January 1968, a coordinated series of attacks throughout South Vietnam that took US forces by surprise. Though the offensive was a military failure for the North Vietnamese, the news footage exposed the frightening gap between the military's optimistic claims of progress and the chaos and confusion on the ground. In February, the administration's call for additional troops confirmed fears of an unwinnable war. More demonstrations followed, and President Johnson stunned the nation in March by announcing he would not run for re-election.

Johnson's decision was just one moment in a year of terrible domestic unrest, division, and violence. In April, Martin Luther King, Jr.—now an opponent of the war—was assassinated in Memphis. In June Robert Kennedy too was assassinated, after his victory in the California Democratic primary. And in August the entire nation watched the Democratic National Convention descend into chaos over the war, both inside the hall and in the streets of Chicago. The fragmentation of the party enabled Richard Nixon to win the White House by campaigning on a platform that included the end of American involvement in Vietnam.

Nixon's promise of "Vietnamization"—a pledge to transfer responsibility for the war to the South Vietnamese army—proved hollow, for in early 1969 the United States began secretly bombing Cambodia. By the end of that year, *Time*, *Newsweek*, and *Life* magazines had covered the terrible massacre at My Lai, where in March 1968, American troops had followed orders to murder hundreds of unarmed Vietnamese civilians. The My Lai revelations outraged the country, and served as a sobering sign that something had gone horribly and intolerably wrong with the American mission.

The opposition to the war remained strongest on college campuses, reaching a fever pitch in the spring of 1970. On April 30, President Nixon announced air strikes to destroy the Cambodian bases from which the communists were operating against South Vietnam. Within two days, a National Strike Committee had been organized in New Haven, and another two days later eleven eastern universities published an open editorial calling for a nationwide student strike. They charged Nixon with violating congressional jurisdiction by acting without a declaration of war, and for perpetrating a "sham" policy of Vietnamization that cynically cloaked an expansion of the war into Cambodia.

In an era before digital communication, the sheer pace of these efforts reflected an impressive level of grassroots coordination and discipline that is captured on this map. Most likely produced by activists at Stanford, it locates more than one hundred campuses where students had pledged to boycott classes on May 6. In a way, the map is an early effort at crowdsourcing, for it depended upon campus activists to report their efforts to the leaders of the movement.

The red imagery across the poster conveys the four ambitious demands of the strike: withdrawal from Southeast Asia, the impeachment of Nixon, the release of imprisoned war protestors, and an end to war-related activities on university campuses, including military contracts and recruitment. In the foreground an officer lies on the ground, while an overturned army bus burns at left. Protestors fan out across the map—as they had across the country—while an American flag stands aflame at half-mast.

The editorial calling for the strike became front-page news in the May 4 *New York Times*. That day was tragically punctuated by the shooting of four students at Kent State University by the Ohio National Guard. The deaths at Kent State galvanized even greater opposition to the war, and over the next week hundreds of campuses erupted in protest. The sheer volume of anti-war demonstrations convinced many across the country that the war was unwinnable. The ground invasion of Cambodia ended the following month, and troop deployments declined rapidly thereafter (even as the bombing of Laos and Cambodia—and the war in Vietnam—continued). As the administration brought the war to a close, Congress passed the War Powers Act, an effort—albeit unsuccessful—to rein in the power of the executive in foreign policy.

CONTINENTAL DRIFT

Heinrich Berann and the National Geographic Society, "Atlantic Ocean Floor," based on studies by Bruce C. Heezen and Marie Tharp, 1968

In 1912 the German geophysicist Alfred Wegener theorized that the world's continents—initially bound together as a single landmass—were slowly drifting apart. Wegener's idea of continental drift was initially met with skepticism, but research into the contours of the ocean floor at mid-century supplied intriguing evidence for this theory. By the late 1960s, earth science had been fundamentally transformed by the theory that the earth was a system of dynamic oceanic and continental plates interacting with one another. This glorious profile of the ocean floor helped to explain this emerging framework of plate tectonics.

The Atlantic seabed was four times the size of the United States, yet until the 1950s it remained more mysterious than the surface of the moon. Research into the ocean floor stretched back to the nineteenth century, but accelerated significantly only after World War II. Some of this was conducted by Bell Labs, which was constructing the first transatlantic telephone cables, from Newfoundland to Scotland. At the same time, the development of nuclear submarines in the early Cold War necessitated a more precise understanding of oceanic depths, which was made possible by advances in sonar technology.

Much of this new research was undertaken at Columbia University's new Lamont Geological Observatory, and sponsored by both the Office of Naval Research and Bell Labs. There, geologists Bruce Heezen and Marie Tharp collaborated to develop a new picture of the ocean floor. In the 1950s, Heezen organized several Atlantic expeditions to generate soundings, and integrated these with existing data from Bell. Despite her advanced degree in geology, Tharp was not permitted to join these expeditions because she was a woman. Yet she played a crucial role in analyzing and synthesizing the data.

Tharp was well equipped for the task. As a child she had traveled extensively with her father, a surveyor for the Department of Agriculture, which exposed her to both mapping and fieldwork. Back at Columbia, she plotted Heezen's soundings into rows, and then organized these into profiles. In the process she began to notice a V-shaped rift *within* the Mid-Atlantic Ridge that ran from north to south along the basin. This led her to posit that the sea floor was acting against itself, with landmasses moving apart in a way that supported Wegener's theory of continental drift. The German oceanographer Günter Dietrich had made similar observations in the 1930s. Armed with this new data and existing theories, Tharp and Heezen began to develop a picture of the rifts and ridges of the ocean floor that advanced theories of continental drift. The shape of the ocean floor *mattered*.

It was Tharp who translated the data into a coherent profile. Her initial map of 1957 was published in Bell Laboratory's technical journal, as it had important implications for communications technology. This early iteration of the map was in fact the first to present the Mid-Atlantic Ridge and the rift within it, which formed a crucial piece of evidence for continental drift. Ten years later, the National Geographic Society hired not a cartographer but an artist to visualize this new research for the public. The Austrian painter Heinrich Berann skillfully presented a comprehensive—if still speculative—picture of the entire Atlantic Ocean floor. Building on decades of research, Berann depicted the seabed in a way that both exposed its complexity and also helped to explain and advance the dynamics of plate tectonics. This was not just an illustration of the ocean floor, but a preliminary explanation of a new geological theory.

The image revealed the full extent of the Mid-Atlantic Ridge. It also identified the transform faults that cut horizontally across the ridge, where two plates slid past one another. Together, these topographic features formed a crucial piece of evidence for plate tectonics. Berann's view of the ocean floor was necessarily general, for the data was itself incomplete: more suggestive than conclusive. Yet much of it would be confirmed within a few years by American and British research. By the late 1960s, Heezen and Tharp had published profiles of all the ocean floors, each of which helped to complete the picture of plate tectonics. Like the "Earthrise" photograph on the next page, these images were a geographical revelation. Equal parts science and art, evidence and speculation, they both visualized a new scientific theory and unveiled the last great unknown reaches of the earth.

ATLANTIC OCEAN FLOOR
-12000 Depth in feet below sea level
9000 Height above sea level
(14000) Height above the 16,000-foot average depth of the abyssal plain
National Geographic Society
MELVIN M. PAYNE, PRESIDENT
for THE NATIONAL GEOGRAPHIC MAGAZINE
MELVILLE BELL GROSVENOR, EDITOR-IN-CHIEF; FREDERICK G. VOSBURGH, EDITOR
WILLIAM N. PALMSTROM, CHIEF, GEOGRAPHIC ART DIVISION
Based on bathymetric studies by Bruce C. Heezen and Marie Tharp of the Lamont Geological Observatory
Painted by Heinrich C. Berann, Compiled by Leo J. Bobertschmidt
HORIZONTAL SCALE 1:30,412,800 OR 480 MILES TO THE INCH AT THE EQUATOR
VERTICAL SCALE EXAGGERATED
Mercator Projection
JUNE 1968
GREENLAND
ICELAND
JAN MAYEN RIDGE
SCANDINAVIA
HUDSON BAY
NORTH SEA
BALTIC SEA
BRITISH ISLES
CANADIAN SHIELD
NORTH AMERICA
EUROPE
REYKJANES RIDGE
MID-OCEAN CANYON
BISCAY ABYSSAL PLAIN
SOHM ABYSSAL PLAIN
OCEANOGRAPHER FRACTURE ZONE
MID-ATLANTIC RIDGE
IBERIAN PENINSULA
ALPS
ATLAS MOUNTAINS
MEDITERRANEAN RIDGE
AEGEAN SEA
GULF OF MEXICO
BERMUDA RISE
HATTERAS ABYSSAL PLAIN
NARES ABYSSAL PLAIN
PUERTO RICO TRENCH
CANARY ISLANDS
MADEIRA ISLANDS
SAHARA
CAPE VERDE ISLANDS
CENTRAL AMERICA
CARIBBEAN SEA
LESSER ANTILLES
GAMBIA ABYSSAL PLAIN
SIERRA LEONE RISE
AFRICA
DEMERARA ABYSSAL PLAIN
AMAZON CONE
GUIANA HIGHLANDS
ROMANCHE FRACTURE ZONE
CEARA ABYSSAL PLAIN
COCOS RIDGE
GALAPAGOS ISLANDS
EQUATOR
AMAZON BASIN
CONGO BASIN
SOUTH AMERICA
PERNAMBUCO ABYSSAL PLAIN
ASCENSION
ANGOLA ABYSSAL PLAIN
BRAZILIAN HIGHLANDS
ST. HELENA
NAZCA RIDGE
PERU-CHILE TRENCH
ANDES
SOUTH PACIFIC OCEAN
RIO GRANDE RISE
WALVIS RIDGE
KALAHARI DESERT
VEMA SEAMOUNT
Cape of Good Hope
TRISTAN DA CUNHA GROUP
AGULHAS PLATEAU
ARGENTINE ABYSSAL PLAIN
FALKLAND PLATEAU
FALKLAND ISLANDS
SCOTIA RIDGE
SOUTH GEORGIA
SOUTH SANDWICH
TIERRA DEL FUEGO

THE PROMISE OF AEROSPACE

NASA, "Earthrise," photo from Apollo 8 mission, 1968, and "Lipton Lunar Space Map," Apollo 11 mission, 1969

In 1957 the Soviet Union launched the world's first artificial satellite. The news of Sputnik landed like a bomb in Cold War America, and triggered fears of Soviet scientific and technological superiority. The Eisenhower administration redoubled its efforts to close this perceived gap, and the following spring established the National Aeronautics and Space Administration (NASA). Across the country, a generation of American students—urged on by their leaders, teachers, and parents—rushed into the sciences. The Cold War now reached into outer space.

Within months of his inauguration, President Kennedy pledged to land a man on the moon by the end of the decade. NASA's Mercury and Gemini missions successfully launched manned spacecraft into orbit, laying the groundwork for the more ambitious goal of reaching the moon. By the mid-1960s, unmanned flights brought back detailed photographs that were used to generate comprehensive maps of the lunar surface, thereby ending a reliance on telescopes that had stretched back for centuries.

The Apollo 8 mission of 1968 was the first to send astronauts to the moon. Initially planned to orbit the earth, it was redesigned at the last minute in response to reports that the Soviets were about to go further. With this change, Apollo 8 became the first mission to leave earth's orbit, capturing the imagination of the entire world. Launched on December 21, it capped an extraordinarily divisive, violent, and confusing year. The January Tet Offensive had forced a reckoning over the United States' chaotic mission in Vietnam, and led President Johnson to end his bid for a second term. Martin Luther King, Jr. had been assassinated in April, just months before presidential candidate Robert Kennedy was killed. The Democrats had torn one another apart at their August convention in Chicago, paving the way for Richard Nixon to win the White House in November.

In this context, the effect of the Apollo 8 mission should not be underestimated. On December 24, the astronauts read from the Book of Genesis, and wished everyone back home a Merry Christmas. But even more moving was what the crew themselves saw on that day. Orbiting the moon, they could not contain their glee as the earth rose over the *lunar* horizon, giving them an unprecedented view of their home planet in the distance. The astronauts were startled: they had been so focused on the lunar mission that they had thought little about the perspective it might bring to the earth itself. They hurried to capture the moment with their cameras.

NASA subsequently released the images to the public. Quickly dubbed "Earthrise," the photograph shown above was the first humanly recorded view of earth from space. The orientation of the photograph shown here is technically correct, with the North Pole at the top and the sunset moving longitudinally from east to west. The photograph is commonly reproduced, however, with earth "rising" above the moon, mimicking a familiar perspective of the sun and moon from earth.

The photograph first appeared in newspapers on December 30, then flooded color magazines in subsequent weeks. It resonated immediately, an evocative picture of a small and distant earth. From this perspective, the conflicts that consumed humanity seemed to pale next to a much larger, even metaphysical reality. A year later, the emerging environmental movement appropriated the photograph to signal the planet's fragility. From the *Whole Earth Catalog* to the first Earth Day, "Earthrise" became not just a symbol but a call to action.

The breathtaking mission of Apollo 8 was soon overshadowed when the astronauts of Apollo 11 landed on the moon the following summer. That exhilarating event fueled even greater enthusiasm for outer space, as shown in this souvenir map issued

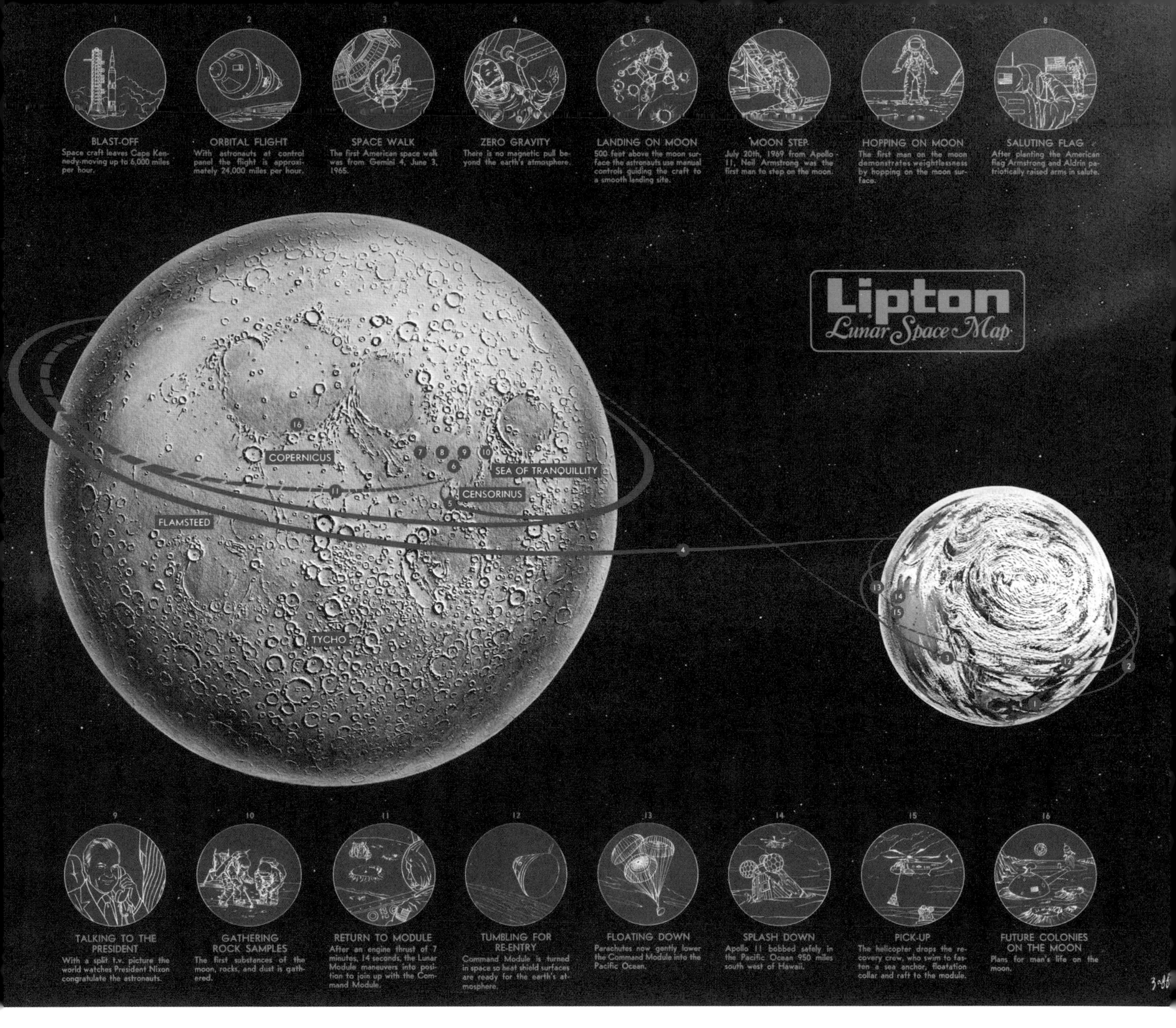
BLAST-OFF
Space craft leaves Cape Kennedy moving up to 6,000 miles per hour.
ORBITAL FLIGHT
With astronauts at control panel the flight is approximately 24,000 miles per hour.
SPACE WALK
The first American space walk was from Gemini 4, June 3, 1965.
ZERO GRAVITY
There is no magnetic pull beyond the earth's atmosphere.
LANDING ON MOON
500 feet above the moon surface the astronauts use manual controls guiding the craft to a smooth landing site.
MOON STEP
July 20th, 1969 from Apollo 11, Neil Armstrong was the first man to step on the moon.
HOPPING ON MOON
The first man on the moon demonstrates weightlessness by hopping on the moon surface.
SALUTING FLAG
After planting the American flag Armstrong and Aldrin patriotically raised arms in salute.
Lipton
Lunar Space Map
COPERNICUS
SEA OF TRANQUILLITY
CENSORINUS
FLAMSTEED
TYCHO
TALKING TO THE PRESIDENT
With a split t.v. picture the world watches President Nixon congratulate the astronauts.
GATHERING ROCK SAMPLES
The first substences of the moon, rocks, and dust is gathered.
RETURN TO MODULE
After an engine thrust of 7 minutes, 14 seconds, the Lunar Module maneuvers into position to join up with the Command Module.
TUMBLING FOR RE-ENTRY
Command Module is turned in space so heat shield surfaces are ready for the earth's atmosphere.
FLOATING DOWN
Parachutes now gently lower the Command Module into the Pacific Ocean.
SPLASH DOWN
Apollo 11 bobbed safely in the Pacific Ocean 950 miles south west of Hawaii.
PICK-UP
The helicopter drops the recovery crew, who swim to fasten a sea anchor, floatation collar and raft to the module.
FUTURE COLONIES ON THE MOON
Plans for man's life on the moon.

by the Lipton Company. In a creative series of images, the noted illustrator George Zaffo visualized each stage of the mission. Notice that Zaffo drew the earth as it had been *photographed* from space, with swirling atmospheric clouds slightly obscuring the terrain and oceans below. In the final image he suggestively pictured "future colonies" on the moon, capturing the contemporary excitement and optimism over this phenomenal mission.

The space program has been romanticized and ridiculed, a source of both hope and scorn. Next to the Panama Canal, it was the largest non-military feat of engineering in American history. To be sure, these missions were driven by the ongoing Cold War and a breakneck effort to fulfill President Kennedy's pledge. But the sense of achievement was felt not just in the US but around the world. Moreover, these photographs and maps spawned a lasting dialogue about the planet. Consider one measure of that moment: in the 1940s and 1950s images of the globe were often used to project power in a world consumed by war. But after 1970 renderings of the earth and the globe more often suggested the interdependence of the world community, the fragility of the earth, and a new level of attention to the environment.

THE TERROR OF AIDS

Abraham Verghese, Steven L. Berk, and Felix Sarubbi, HIV infection across the US, and around Johnson City, Tennessee, 1989

In 1985 Abraham Verghese began to practice medicine in Johnson City, a town of 50,000 near the Great Smoky Mountains in Eastern Tennessee. As an infectious disease specialist, he had learned about acquired immune deficiency syndrome (AIDS) during his medical residency in Boston. But in the mid-1980s Verghese—like most Americans—considered AIDS primarily an urban epidemic, and hardly thought that he would encounter the disease in rural Tennessee. Nonetheless, soon after arriving in town he began to reach out to the gay community to promote voluntary HIV screenings, and was heartened to find that local men tested negative for the virus.

Within three and a half years, however the situation had shifted dramatically. Verghese had treated eighty-one HIV-positive patients, thirteen of whom died from AIDS. He could not understand why so many men were falling ill given that the locals continued to test negative for the virus. In search of an explanation, he began to wonder if there was a geographical pattern at work. At home one evening, he took down a map from the wall of his son's room and began to mark the residence of each of his patients. As shown on the lower map, he found that they clustered around Johnson City, but also extended to Southwest Virginia and Kentucky, the farming towns of Eastern Tennessee, and even the mountains of North Carolina. This was to be expected, since his hospital served a large and geographically extensive area. But the number of cases was still far higher than what the Centers for Disease Control and Prevention would have predicted for a rural region. What accounted for this?

Turning back to the map, Verghese then identified where each of his patients had lived between 1979 and 1985, when they most likely contracted the virus. On the upper map, a pattern began to emerge: by and large, his patients were raised around Eastern Tennessee, but as adults had moved to urban areas around the country in search of opportunity and tolerance. From this he hypothesized that most of them had contracted the virus in the city; those who had never left the area had probably become infected by engaging in high-risk activity locally or by receiving blood transfusions before HIV testing was routine.

Taken together, these two maps showed Verghese a pattern of migration that explained the sharp rise of rural HIV cases: most of his patients had moved *home* to seek care and support once they began to exhibit symptoms of AIDS. Verghese's medical expertise gave him the tools to diagnose the disease, but only by mapping the movement of his patients did he realize the implications it would have for the region and his practice. As patients came home with the illness, families would become crucial networks of care. And if Johnson City's experience was typical, then rural medical centers all over the country would need to prepare. Rural rates of infection remained far lower than urban, but AIDS devastated families everywhere.

Verghese's maps and research were published in the *Journal of Infectious Diseases*, which itself is revealing. Previously, the journal had typically focused on the large centers of the epidemic in urban areas. Yet as Verghese stressed, HIV infection presented unique problems for rural communities. For one, the stigma of the virus remained higher in rural areas. And because incidence of the disease was lower, rural communities might be less prepared for its inevitable rise. Verghese found that families of his patients in Johnson City almost invariably rallied around their loved ones, challenging assumptions about rural attitudes toward homosexuality. Yet at the same time, he stressed the need for rural communities to mobilize in order to anticipate the inevitable rise of HIV infection.

While Verghese was treating patients in Eastern Tennessee, the geographer Peter Gould was in Pennsylvania researching the national scope of the epidemic. Gould was particularly frustrated by the failure of medical professionals to understand—and thereby address—the geographical dimension of this public health crisis. Even the epidemiologists who specialized in the virus made little effort to understand its spatial distribution. To address this, Gould painstakingly gathered data from different agencies around the country to trace the evolution of the disease across the nation. The maps on the next page showcase Gould's adoption of digital techniques to capture the geographical dynamics of the epidemic.

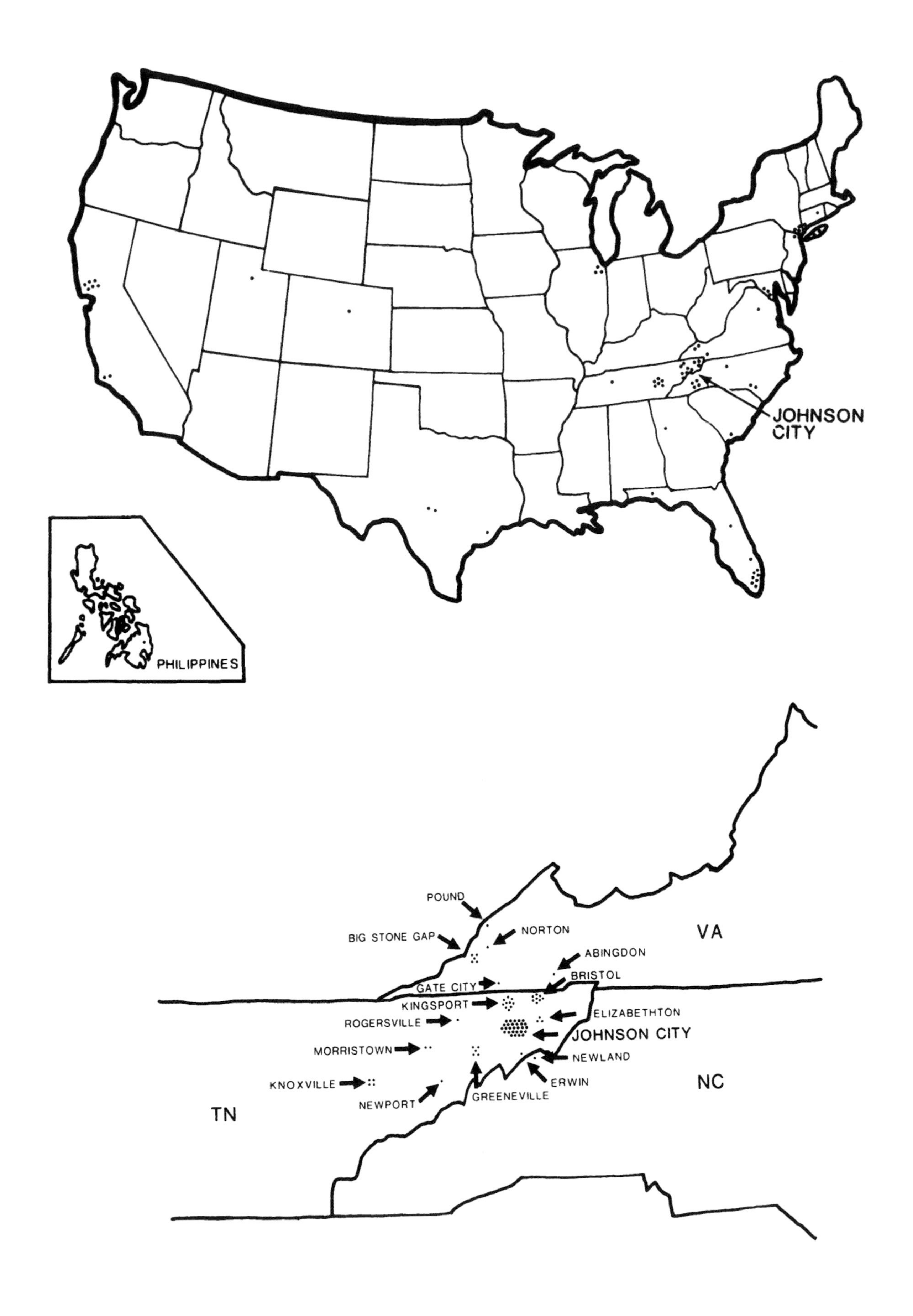

JOHNSON CITY
PHILIPPINES
POUND
BIG STONE GAP
NORTON
VA
ABINGDON
BRISTOL
GATE CITY
KINGSPORT
ELIZABETHTON
ROGERSVILLE
JOHNSON CITY
MORRISTOWN
NEWLAND
KNOXVILLE
ERWIN
NC
NEWPORT
GREENEVILLE
TN

Peter Gould, number of AIDS cases in 1982, 1984, 1986, 1988, 1990 (published 1999)

As shown on the previous page, the physician Abraham Verghese used maps to understand the outbreak of AIDS in his rural community of Eastern Tennessee. At the same time, Pensylvania State University geographer Peter Gould used early digital techniques to map from 1982 to 1988 the geographical distribution of HIV infection at two-year intervals. Gould organized the data into what geographers refer to as "heat maps," showing the spread of the epidemic from cities to the entire nation.

To capture the soaring rates of the epidemic, Gould measured the incidence of the disease not incrementally, but in qualitative shifts. Areas with the lowest rates of the outbreak are marked in blue, with each successive color marking a level of infection that is *seven times* greater than the last. In this regard, Gould felt restrained by his flat, two-dimensional representations. He imagined a more powerful visualization in three dimensions, one that would accurately capture the magnitude of the epidemic in urban centers by showing a vertical measure that dwarfed its incidence elsewhere. This type of geographical modeling was well established within a few years, as shown in the 2001 map of labor in the meat processing industry on page 252.

Seen in sequence, the maps demonstrated the pervasive and national threat of HIV by the late 1980s. Gould noticed that the virus initially exploded in urban areas, but when it inevitably spread beyond the cities it did so not just by seeping outward as epidemics had in the past, but also by leaping across the country via air routes and along other transportation corridors. In a context where every infected patient is also a carrier, and where carriers might live years before manifesting symptoms of the disease, HIV was particularly difficult to contain. But without a proactive and coordinated effort to share information, Gould argued, there was little hope of limiting the disease.

To bring home the seriousness of the situation, Gould designed a final map projecting the extent of the epidemic by 1990. Though his worst predictions were not realized, the death toll of AIDS continued. By 2000, it had killed 400,000 Americans, and nearly nineteen million people worldwide. Like Verghese, Gould turned to the map to uncover paths and patterns of AIDS that were otherwise invisible. And both men used maps not just to document its history, but to guide future decisions and policies. However flawed, these maps powerfully facilitated spatial *thinking*.

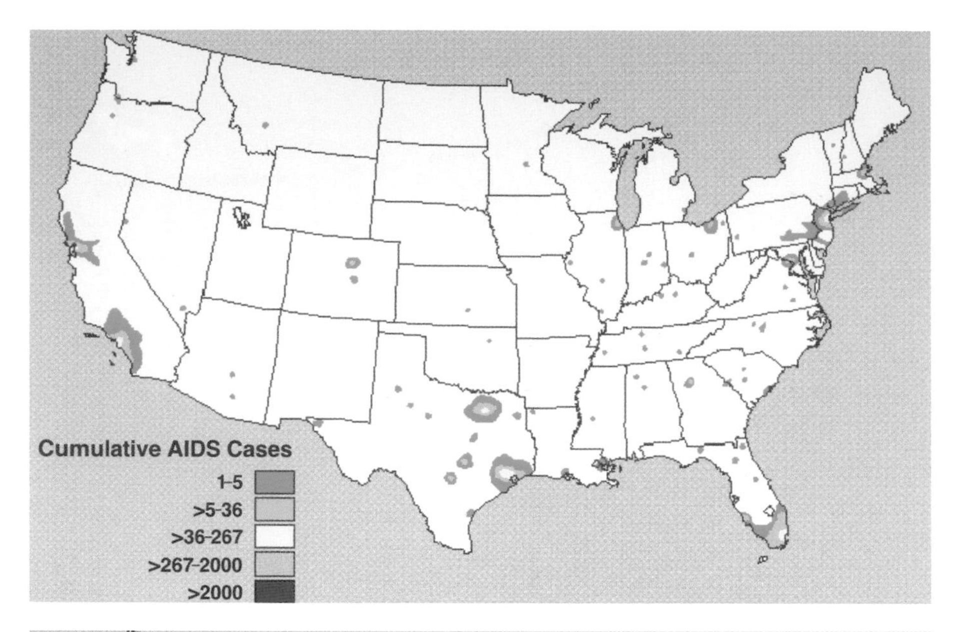

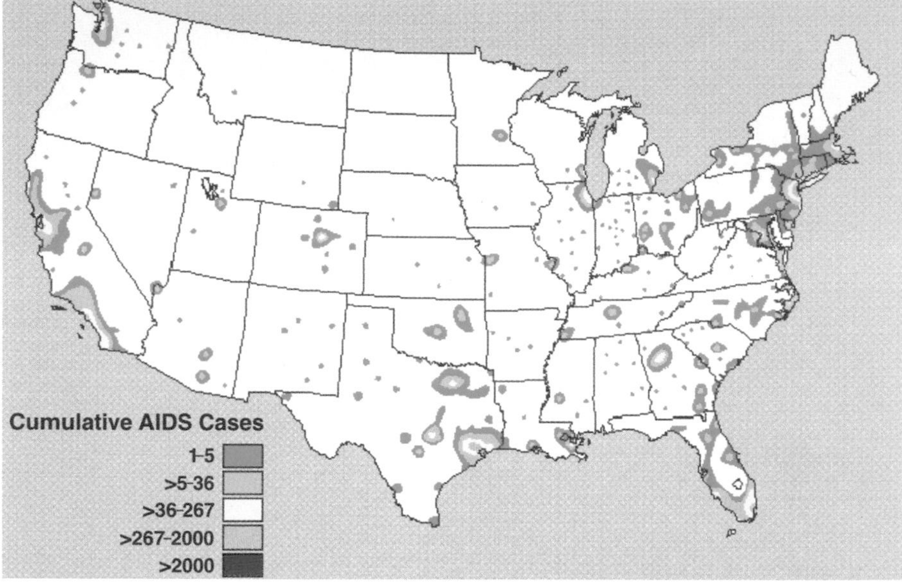

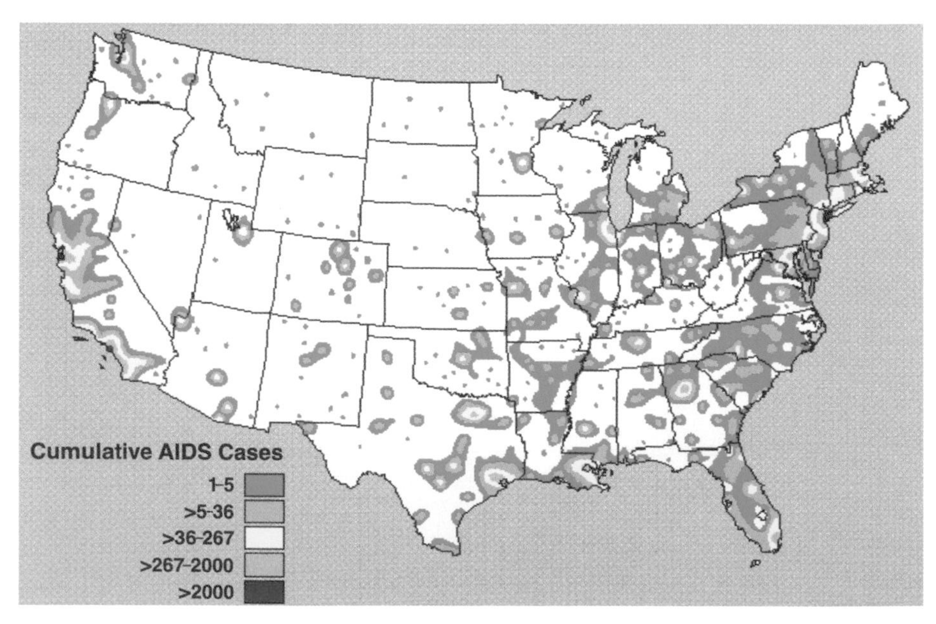

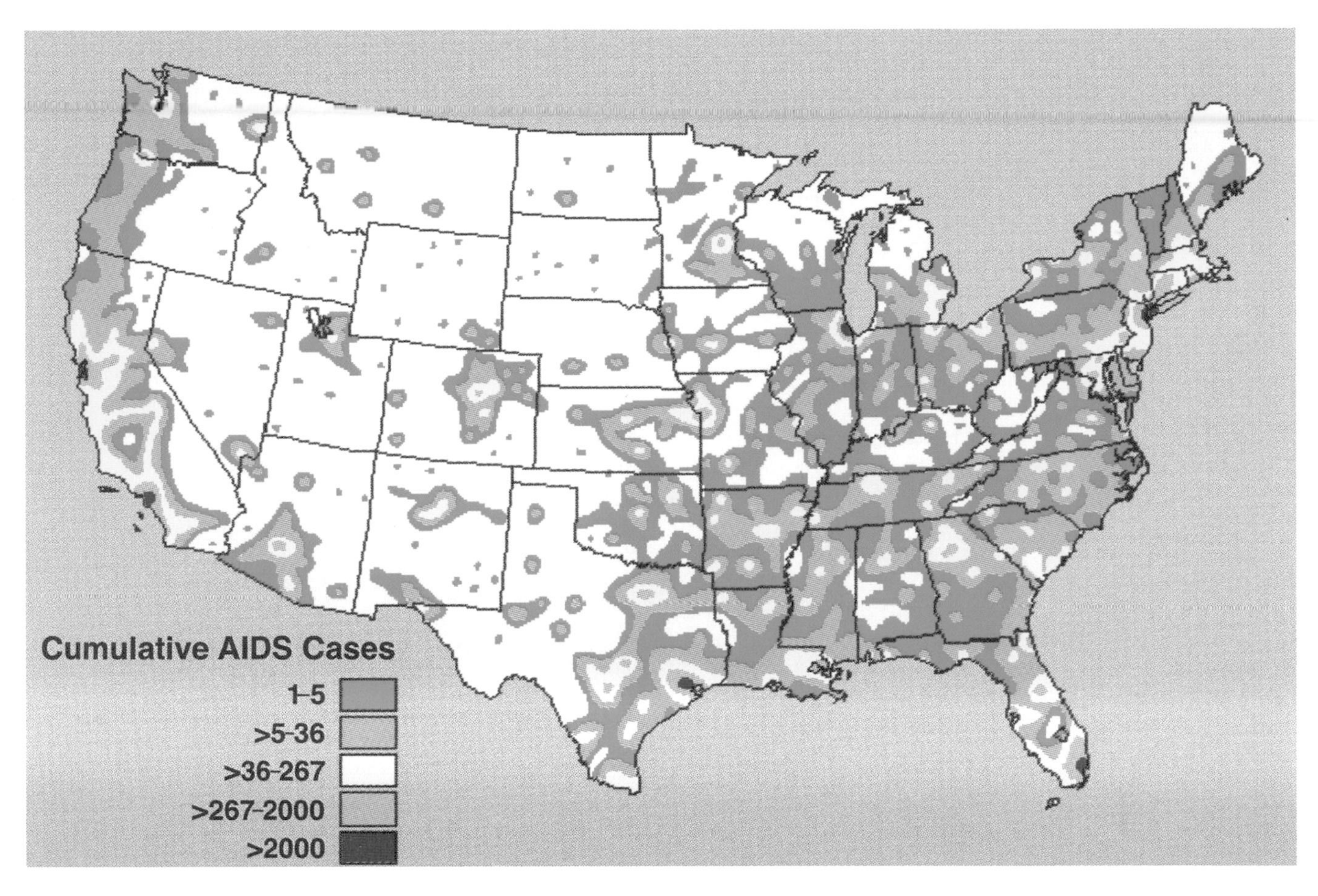
Cumulative AIDS Cases
1-5
>5-36
>36-267
>267-2000
>2000

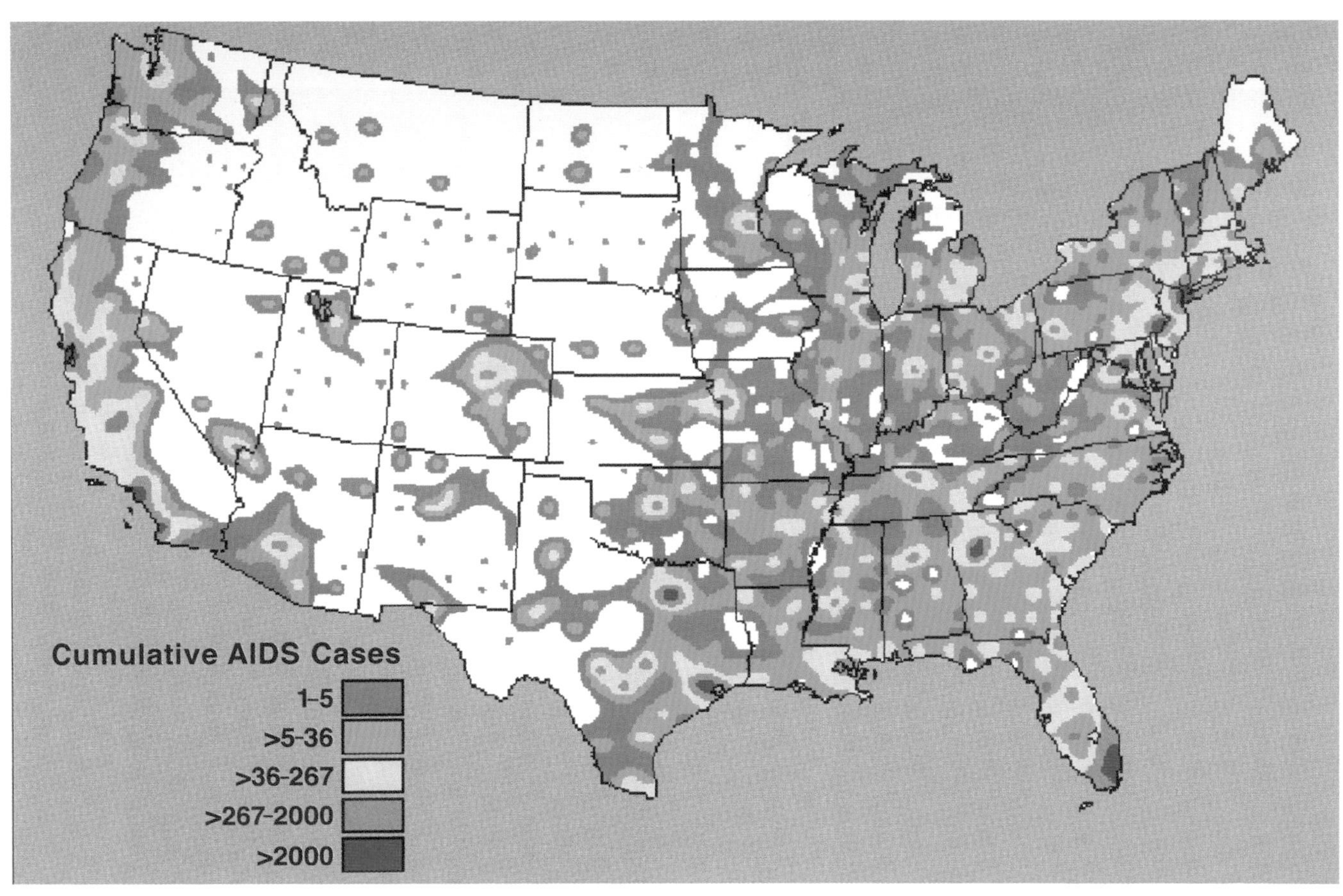
Cumulative AIDS Cases
1-5
>5-36
>36-267
>267-2000
>2000

A NATION ON THE MOVE

US Census, "Regional Migration, 1955 to 1960" and "Regional Migration, 1995 to 2000," 2007

Americans remain highly mobile. On average, nearly one in seven moves each year—to a new block, city, region, or country. And it has ever been thus: migration, whether forced or voluntary, driven by catastrophe or by opportunity, has been a hallmark of American history. John Smith's 1612 map of Virginia, Malachy Postlethwayt's 1757 map of the slave trade, Charles Gratiot's 1837 map of Indian territory, and Charles Preuss' 1846 map of the Oregon Trail are just a few examples of the continuous settlement—and resettlement—of North America.

In the twentieth century, domestic migration profoundly realigned not just the population, but the American political and economic landscape as well. As late as 1900, the nation's people and industrial capacity were still concentrated in the northeast quadrant of the country, bounded to the east by Illinois and to the south by Pennsylvania. As that region continued to grow through World War II, however, so too did California and Florida. Los Angeles could never have expanded beyond a city of a few hundred thousand without an engineered water supply; once that resource was secured, California's capacity seemed limitless.

In the 1930s the devastating dust storms on the Great Plains created a virtual exodus to Southern California that deeply shaped its political culture. The extraordinary labor demands and federal investment of World War II brought another flood of migrants, which continued with the robust federal investment during the Cold War. By 1990 California received more than 20 percent of the contracts awarded by the Department of Defense. The first flow map here hints at some of that movement from the manufacturing belt of the Midwest and the Northeast toward the Far West at mid-century. Hidden by the map is the degree to which California was the overwhelming destination within the West, gaining 1.1 million residents between 1955 and 1960. By the early 1960s, it was the most populous state in the country.

While California's rise was certainly the most dramatic, other regions also shifted significantly in the twentieth century. The most remarkable of these was the South. For the first half of the century, that region consistently lost population, particularly its rural areas. African Americans led the migration out of the South to the Northeast and the Midwest, and to a lesser extent the Far West. Within the South, farmers and sharecroppers migrated to small towns, particularly as agricultural prices remained low in the 1920s and 1930s.

But after World War II, improvements in infrastructure and a rebounding economy began to slow that trend. Just as the Northeast and Midwest were experiencing industrial decline, the South picked up momentum, buoyed by federal contracts and lower wages that attracted employers from other regions. By the 1990s, the South recorded the highest rate of domestic migration of any region in the country, especially in Atlanta, Charlotte, and Austin. That demographic shift from the Northeast to the South shown on the second map was also augmented by the migration of retirees to Florida. All of this, of course, was made possible by innovations in air conditioning.

The brisk growth of the South and West in the second half of the century also transformed the landscape of political power across the country. Between 1950 and 1990, California gained twenty-two seats in the House of Representatives while New York *lost* twelve. The Southwest, Florida, and the Pacific Northwest gained as well, while states of the Northeast, Midwest, and Great Plains all declined in political representation. This nationwide reapportionment occurred alongside the shift of the South from a Democratic to a Republican stronghold. A return to the 1880 electoral map on page 156 shows what an astonishing reversal this was. Though the Civil War still shaped some of the region's social and cultural sensibilities, it no longer determined its politics. White Southerners now embraced the Republican Party in a way that would have been unimaginable in the first half of the twentieth century.

Political strategist Kevin Phillips observed the early stirrings of this change in the late 1960s, and coined the term "Sunbelt" to denote the South and Southwest. To be sure, there are limits to the Sunbelt as a coherent region given sharp distinctions among states and between urban and rural cultures. Moreover, trends can change quickly: in the 1980s and 1990s the most noticeable trend was the *outmigration* from California. While the second flow map identifies interregional movement, it hides the significant migration from California to neighboring states, principally Colorado, Nevada, and Washington. Few at mid-century would have predicted that 750,000 people would leave California between 1995 and 2000, a loss exceeded only by the 900,000 who left New York.

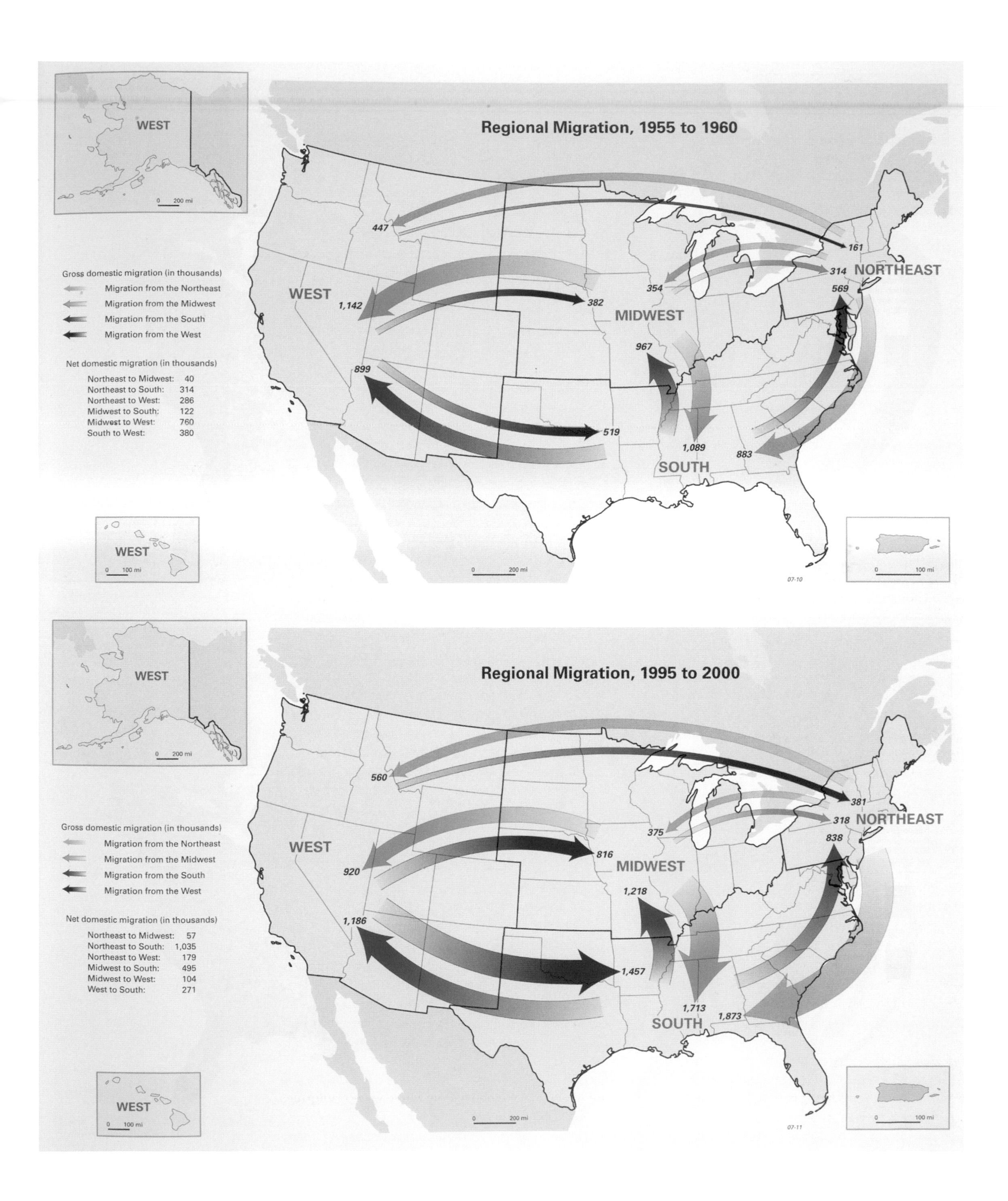
Regional Migration, 1955 to 1960
WEST
Gross domestic migration (in thousands)
Migration from the Northeast
Migration from the Midwest
Migration from the South
Migration from the West
Net domestic migration (in thousands)
Northeast to Midwest: 40
Northeast to South: 314
Northeast to West: 286
Midwest to South: 122
Midwest to West: 760
South to West: 380
447
161
314
NORTHEAST
569
354
WEST
1,142
382
MIDWEST
967
899
519
1,089
883
SOUTH
WEST
07-10
Regional Migration, 1995 to 2000
WEST
Gross domestic migration (in thousands)
Migration from the Northeast
Migration from the Midwest
Migration from the South
Migration from the West
Net domestic migration (in thousands)
Northeast to Midwest: 57
Northeast to South: 1,035
Northeast to West: 179
Midwest to South: 495
Midwest to West: 104
West to South: 271
560
381
318
NORTHEAST
838
375
WEST
920
816
MIDWEST
1,218
1,186
1,457
1,713
1,873
SOUTH
WEST
07-11

IMMIGRATION AND WORK IN RURAL AMERICA

Applied Population Laboratory, "Meatpacking and Hispanic Population Percent Change 1990–2000: Midwest Counties," 2001

From 1945 to 2000, the nation's population doubled from 140 to 281 million. Much of that increase was driven by immigration, the largest single source of which was from Mexico. According to the 1990 census, natives of Mexico constituted over 20 percent of the country's foreign-born population. For much of the twentieth century, that population centered in the Southwest. Indeed, the presence of Latinos in the Southwest long predates the westward expansion of the United States. As Mexican Americans often quip, it was the border that moved, not them. But the 2000 census returns revealed two remarkable changes. First, while the Latino population continued to grow, it was now soaring *outside* the Southwest. Second, through the 1990s that growth was primarily rural rather than urban. Indeed, Latinos were moving to rural areas of the United States at greater rates than any other racial or ethnic group.

In part this was facilitated by legislation passed in 1986, which gave over 2 million undocumented workers a more stable legal status, and thereby more geographical mobility. At the same time, tighter enforcement at the traditional spots along the US–Mexico border meant that migrants were crossing further east. Together, these changes exposed Mexican immigrants—as well as Latinos already in the country—to areas outside the Southwest and West. Labor demands in these regions were quickly filled by Latinos, primarily in meat processing, carpet production, oil refining, and other light industries.

As a result of these factors, the Latino population increased by 13 percent in the Midwest during the 1990s, and nearly 19 percent in the Southeast. Between 1992 and 1997, the number of Latinos in ten states of the Midwest—Ohio, Indiana, Illinois, Michigan, Wisconsin, Minnesota, Iowa, Missouri, Kansas, and Nebraska—rose from 1.8 million to 2.3 million.

These shifting dynamics are highlighted in this three-dimensional digital map, which was created by demographers using new techniques of geographic information systems (GIS). As the map indicates, this shift was particularly apparent in the meatpacking industry. In the early twentieth century, companies such as Armour and Swift sought to make Chicago the hub of meatpacking by modernizing techniques of butchering and transportation. These streamlined techniques produced economies of scale that put beef and pork at the center of the American diet, and made Chicago the most important city of the Midwest.

By the 1980s, however, several forces conspired to disrupt the meatpacking industry yet again. Americans had begun to adopt a diet that was lower in fat, choosing chicken over beef. The cost of poultry also dropped after producers expanded and streamlined their operations, and from 1970 to 2000 per capita consumption doubled. All of this translated into more jobs in poultry processing throughout the Southeast. This growth of poultry consumption in turn put pressure on the beef industry, which was further compounded by the growth of imported—and cheaper—meats. American meat-processing firms responded by lowering wages, eliminating unions, and hiring contract workers.

This competition also led to industry-wide consolidation, just as it had in the early decades of the twentieth century. By the late 1990s, four companies accounted for half of American poultry production, and 80 percent of beef. Tyson Foods, for instance, now slaughters 5 million chickens and a quarter of a million head of cattle every week. Consolidation also led companies to move closer to the livestock feedlots. From 1980 to 2000 several of the largest American meat and poultry processors relocated to rural Colorado, Kansas, Nebraska, Oklahoma, and Texas. But these regions had been declining in population for decades, which made it difficult to find low-skilled and low-wage labor to staff the plants. Meat-processing work is highly unpleasant and dangerous—wet and noisy, with high rates of injury. All of these factors led to the aggressive recruitment of immigrant and migrant labor.

The result has been a remarkable migration of Latinos into rural regions and towns that had historically been struggling to survive. In Nebraska, almost every one of the state's 93 counties lost population in the 1980s, yet the number of Latinos grew. That increase became an explosion in the 1990s. The IBP meatpacking plant in Dawson County, Nebraska—shown on the map—drew more Latinos than have been recorded in the census. By 1998, Latinos constituted half of the student population for the school district, and drove economic growth in the county. Latinos now constitute the largest minority group in Nebraska.

That rural revival was seen in Kansas as well. Throughout the twentieth century, the town of Liberal

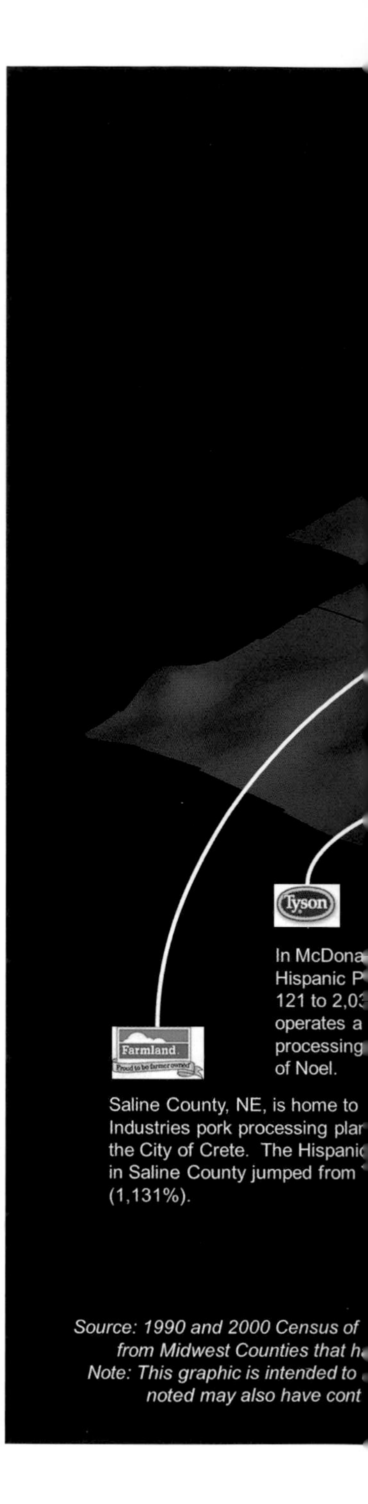
Tyson
In McDona
Hispanic P
121 to 2,0
operates a
processing
of Noel.
Farmland
Saline County, NE, is home to
Industries pork processing plar
the City of Crete. The Hispani
in Saline County jumped from
(1,131%).
Source: 1990 and 2000 Census of
from Midwest Counties that h
Note: This graphic is intended to
noted may also have cont

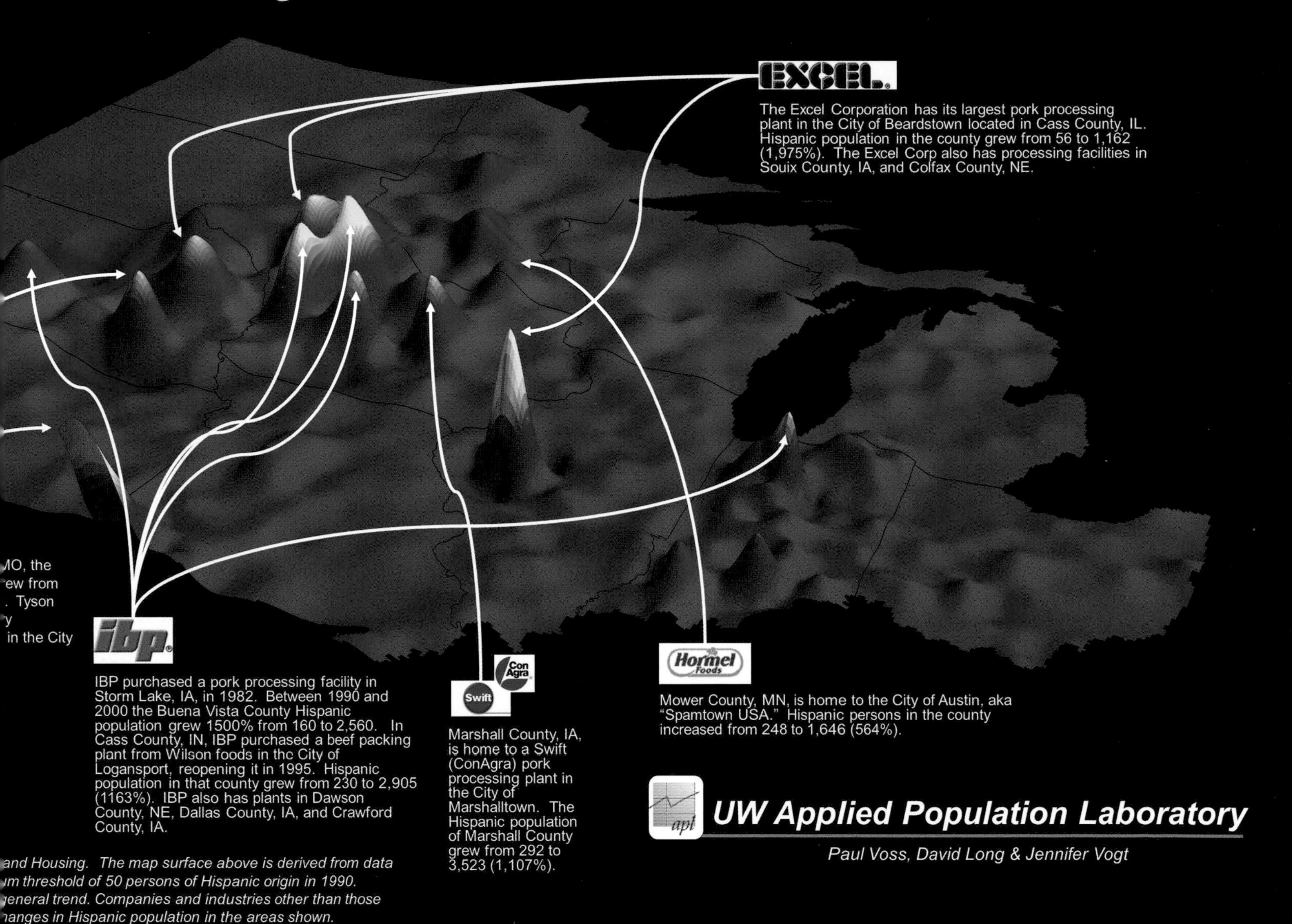

lost residents. From 1980 to 2000, however, its population quadrupled, as did the Latino share of that population. Nationwide, the non-Latino white workforce in meat processing dropped from 74 to under 50 percent in those same two decades, while the Latino proportion grew from under 10 to nearly 30 percent. Today, Latinos constitute more than 80 percent of the nation's meatpacking workforce.

atpacking and Hispanic Population
Percent Change 1990-2000 Midwest Counties
EXCEL
The Excel Corporation has its largest pork processing plant in the City of Beardstown located in Cass County, IL. Hispanic population in the county grew from 56 to 1,162 (1,975%). The Excel Corp also has processing facilities in Souix County, IA, and Colfax County, NE.
IO, the
ew from
. Tyson
y
in the City
ibp
IBP purchased a pork processing facility in Storm Lake, IA, in 1982. Between 1990 and 2000 the Buena Vista County Hispanic population grew 1500% from 160 to 2,560. In Cass County, IN, IBP purchased a beef packing plant from Wilson foods in the City of Logansport, reopening it in 1995. Hispanic population in that county grew from 230 to 2,905 (1163%). IBP also has plants in Dawson County, NE, Dallas County, IA, and Crawford County, IA.
Con Agra
Swift
Marshall County, IA, is home to a Swift (ConAgra) pork processing plant in the City of Marshalltown. The Hispanic population of Marshall County grew from 292 to 3,523 (1,107%).
Hormel Foods
Mower County, MN, is home to the City of Austin, aka "Spamtown USA." Hispanic persons in the county increased from 248 to 1,646 (564%).
apl
UW Applied Population Laboratory
Paul Voss, David Long & Jennifer Vogt
and Housing. The map surface above is derived from data
m threshold of 50 persons of Hispanic origin in 1990.
eneral trend. Companies and industries other than those
anges in Hispanic population in the areas shown.

The map captures all of these trends, showcasing not absolute populations but rather the change over time. Admittedly, this leaves us with a slightly exaggerated picture given that the Latino population of these regions was so small before 1990. But it maps several dynamics that have transformed the region: global competition and migration, a shifting labor market, consumer demand, and the commodification of food.

[Note: while "Hispanic" and "Latino" are terms used interchangeably by the Census Bureau, here we use the term "Latino."]

GERRYMANDERING IN THE DIGITAL AGE

Maps of North Carolina's 12th congressional district, 1992–2016

The Voting Rights Act (VRA) of 1965 protected the African American right to vote after a century of disfranchisement through poll taxes, literacy clauses, and intimidation. In response, many Southern states redrew political districts in order to limit the power of black voters, updating the longstanding practice of gerrymandering. In 1973, for example, the Hind County Board of Supervisors in Mississippi "cracked" the black population of Jackson into *five* separate districts to dilute the power of that constituency. When the VRA came up for renewal in 1982, Congress responded by reaffirming the right of minorities "to elect the representatives of their choice." These words profoundly—if unintentionally—altered American electoral politics.

To fulfill the mandate of the VRA, Democrats began to craft districts where minorities constituted a majority. Yet these "majority-minority" districts also required gerrymandering. In 1992 the North Carolina state legislature created a new district shown at right from Charlotte through Winston-Salem and Greensboro to Durham. The geography of district 12 was connected in some places by little beyond Interstate 85, yet it worked: for the first time in the twentieth century, North Carolina elected an African American to the House of Representatives.

Even though this new district elected a Democrat, fellow Democrats argued that by forcing the state legislature to draw boundaries based upon race, the federal government had violated the Equal Protection Clause of the Fourteenth Amendment. The Supreme Court tentatively agreed, finding that majority-minority districts *may* violate the Constitution. Supreme Court Justice Sandra Day O'Connor wrote that the bizarrely shaped district 12 echoed the ugly racial gerrymanders of the Jim Crow era, and came close to political apartheid by segregating African Americans. But the court also acknowledged that race must necessarily be taken into consideration to advance minority representation.

This dilemma is striking. "Concentrating" racial minorities advanced their representation, but also weakened their party *elsewhere*. Because African Americans tended to vote Democrat, the effort to create secure majority-minority districts unintentionally made the surrounding districts more Republican. In part the problem lies with representative democracy, which hinges on geography: what constitutes a coherent constituency, and how are lines to be fairly drawn?

As the map on page 116 shows, gerrymandering is endemic to American politics. But technology—coupled with an increasingly partisan culture—has sent it into overdrive. Census returns provide demographic information, while redistricting software adds highly refined political data. Expert mapmakers integrate all of these variables, then test the likely outcomes of various redistricting scenarios. As the maps here show, small geographical changes produce important results.

In 2010 Republicans won both houses of the North Carolina legislature, which gave them exclusive control over redistricting based on the new census returns. They redrew boundaries to "pack" Democratic votes into no more than three districts, one of which was district 12. In doing this, the party claimed to be fulfilling the spirit of the VRA by moving African Americans into the district. But they also diluted the strength of blacks, who mostly vote Democrat, in neighboring districts. With sophisticated precinct data, they also shaved off Republican voters, strengthening that party elsewhere. The map designed for the 113th Congress at lower left may not appear much different from its predecessor, but small tweaks had enormous consequences: Republicans received less than 49 percent of the statewide vote in 2012, yet won nine of the thirteen seats in the House of Representatives. This trend extended across the South: in 1991 Democrats held 81 out of 133 Southern seats in the House, but by 2013 that number had fallen to 18 out of 145.

Gerrymandering makes *all* races less competitive. Candidates facing homogeneous constituencies have little incentive to moderate their positions, which exacerbates polarization. Extreme gerrymandering has also created a situation where politicians increasingly choose their voters rather than the reverse. Recently, however, the courts have begun to use the same digital tools and maps that facilitate gerrymandering to assess its constitutionality. In 2017 the Supreme Court struck down North Carolina's map at lower right for excessively packing African Americans into district 12. Then in 2018 a panel of federal judges ruled that "partisan advantage" had improperly been used as the primary criterion in modifying that same district.

Incumbents also use gerrymandering to protect the "safety" of their seats. The advent of ever more precise marketing and online data promises to complicate gerrymandering even further. And, at the center of the problem—but perhaps also its solution—is the map.

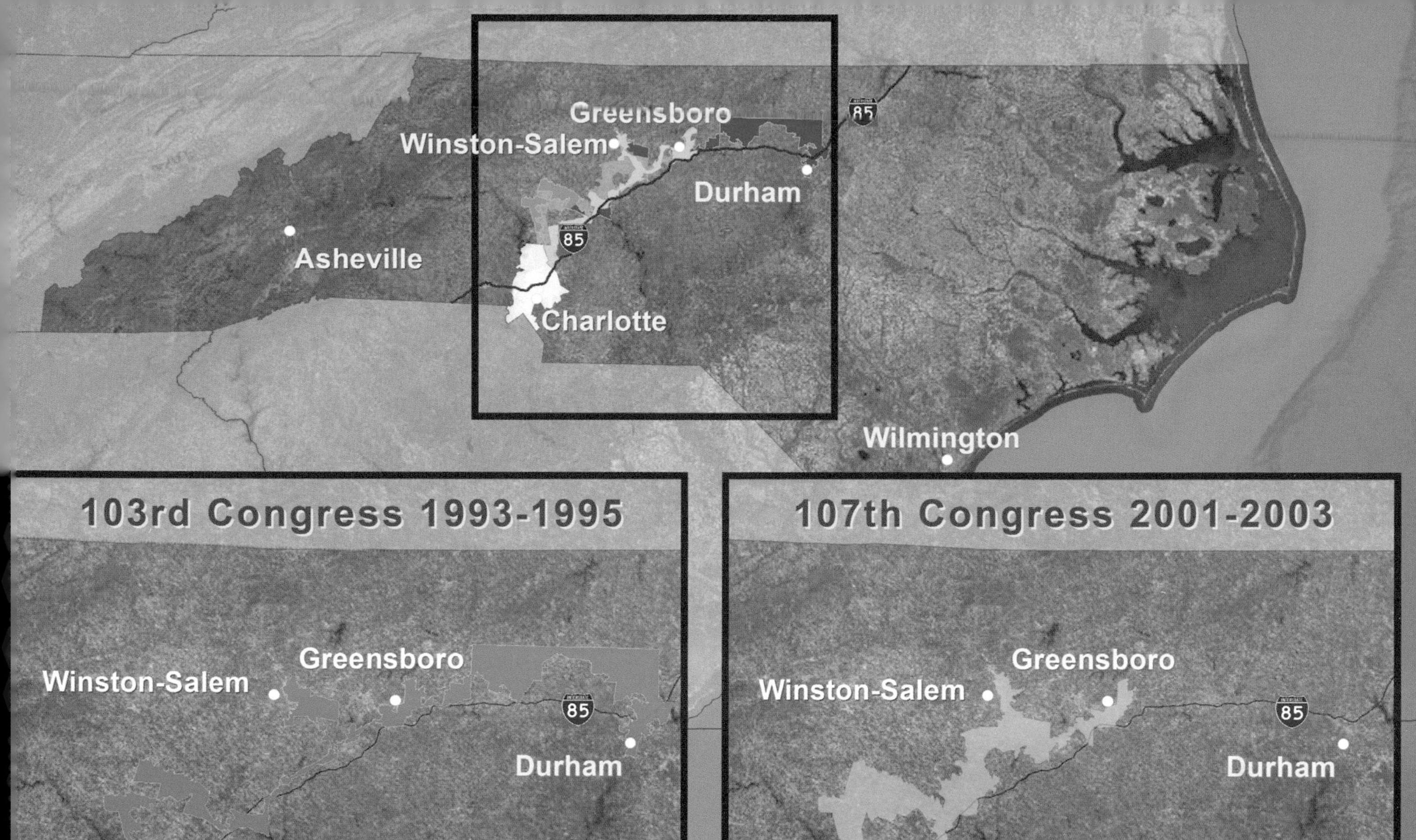

Greensboro
Winston-Salem
85
Durham
Asheville
Charlotte
Wilmington

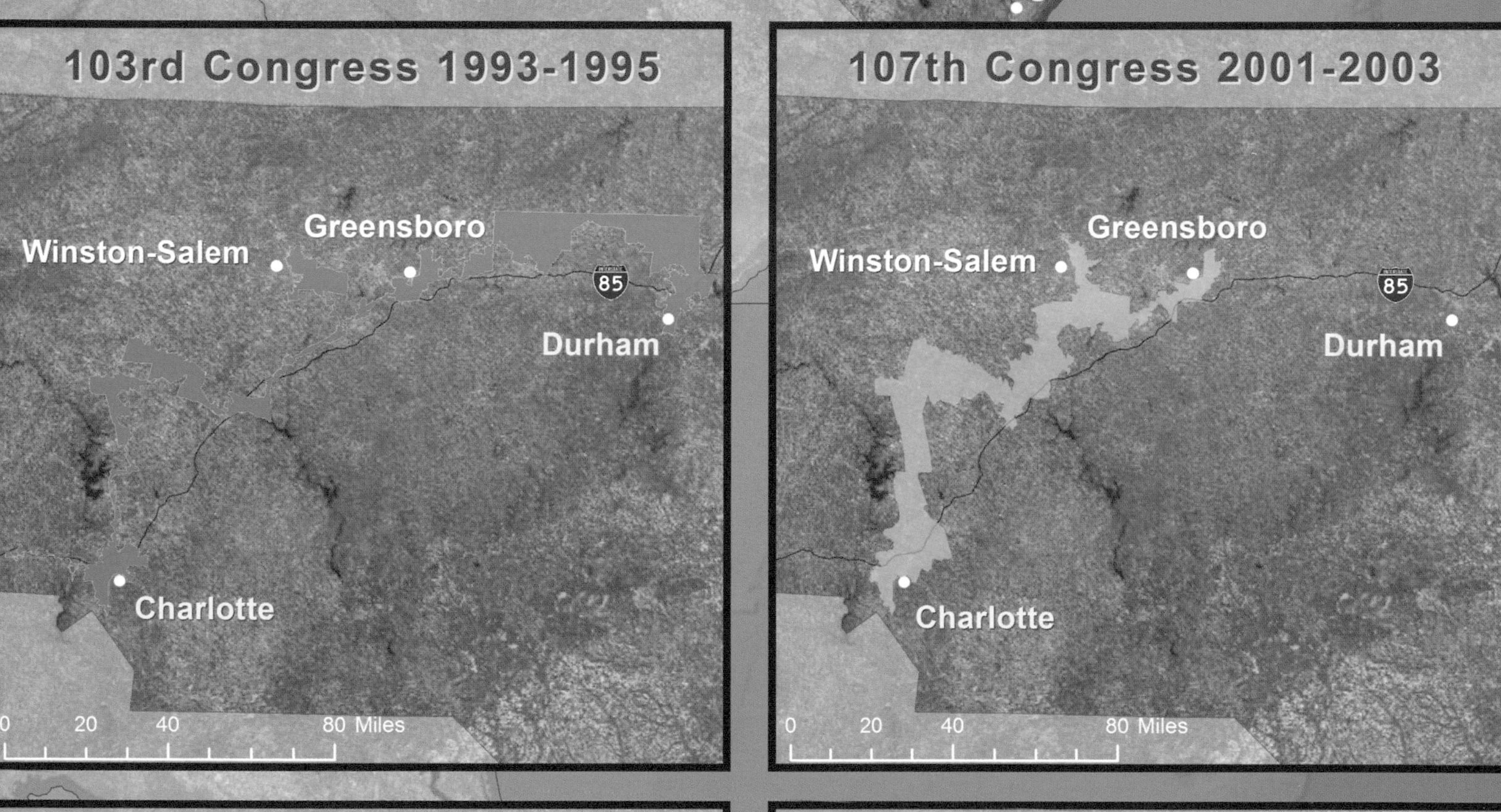

103rd Congress 1993-1995
Winston-Salem
Greensboro
85
Durham
Charlotte
0
20
40
80 Miles
107th Congress 2001-2003
Winston-Salem
Greensboro
85
Durham
Charlotte
0
20
40
80 Miles

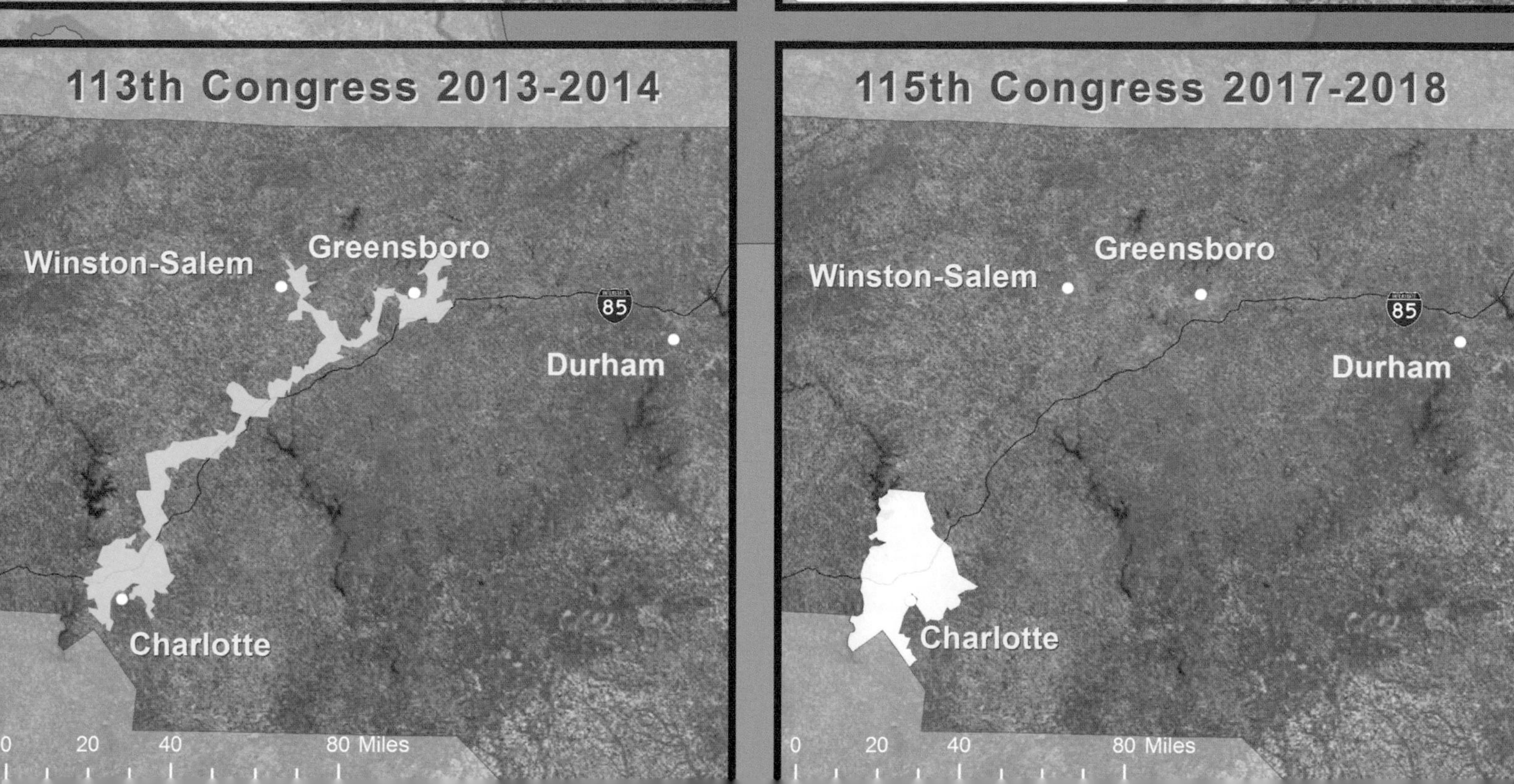

113th Congress 2013-2014
Winston-Salem
Greensboro
85
Durham
Charlotte
0
20
40
80 Miles
115th Congress 2017-2018
Winston-Salem
Greensboro
85
Durham
Charlotte
0
20
40
80 Miles

BETWEEN PAST AND FUTURE

Laura Kurgan, "Around Ground Zero," 2001

On the morning of September 11, 2001, nineteen men hijacked four separate airplanes to commit the worst terrorist attack in US history. Two of those planes were flown into the twin towers at the World Trade Center in New York City, killing nearly 3,000 civilians, firefighters, and police officers.

The attacks brought most businesses in the area to a halt, leaving 80,000 people without work. Over the next few months, rescue and recovery efforts continued around the clock to clear the sixteen-story-high pile of debris created by the collapse of the towers. Laura Kurgan, an architect from South Africa, had been living in New York since 1985, and in the weeks following the attack she watched thousands converge on Ground Zero every day. First came those posting flyers of missing loved ones, followed by those who had traveled to see the site for themselves. All of these visitors, Kurgan realized, were trying to both process and commemorate this senseless act.

In the immediate aftermath of the attack, Ground Zero was entirely disorienting. The scope of the wreckage had left much of the neighborhood off-limits. Fences erected around the perimeter of the site made it difficult to see what was going on inside. Even those who could catch a glimpse had difficulty understanding the operations within the site. Moreover, the situation changed almost daily: barricades went up, viewing platforms moved, and makeshift memorials appeared everywhere. In such a shifting, disrupted landscape, existing street maps were of little use. The neighborhood was in limbo.

Kurgan responded to this logistical confusion by making a map. The idea of mapping Ground Zero in the immediate aftermath of the attack may strike some as macabre, even voyeuristic. But it was the general sense of confusion that drove Kurgan's efforts; she hoped that a map might bring a small measure of order to this chaos. More importantly, she sought to guide visitors, helping them make sense of what had happened so that Ground Zero became a site of reflection rather than a spectacle.

Kurgan was not a mapmaker, but as she walked around Lower Manhattan that fall she was reminded of a map she had used to navigate war-torn Sarajevo in the 1990s. The map was designed to record what had been and what had happened, so that the city might rebuild. Kurgan began to enlist a cadre of volunteers in the fall to compile something similar for Lower Manhattan. A portion of that map is reproduced here, the result of a collective effort among architects, designers, and researchers.

This paper map was designed for those visiting the site. The legend is especially revealing: personal memorials are mapped and marked even though they were temporary and regularly removed. The locations of unobstructed views are identified, as is a walking path around the entire perimeter. In other words, this is a map designed to be *used* by those navigating the site. But even making this fairly straightforward map proved challenging: the constantly shifting barricades and viewing platforms forced one of the volunteers to redraw the map three times before it went to print. Such a situation might have suggested the need for a digital map that could be quickly and easily revised to reflect changing circumstances. But Kurgan insisted on a paper map, deliberately producing a permanent map of an ephemeral site.

On a cold and windy Saturday in late December, Kurgan distributed thousands of these maps to visitors around Ground Zero. A few months later, changes in the site necessitated a new edition. In this sense Kurgan and her team anticipated the crowdsourcing efforts of a few years later, where fluid data contributes to an ever-changing—and hopefully accurate—map. Her paper map was, of course, destined to become a historical artifact, even within a few months. But that was part of the goal. She had mapped a site somewhere between a battlefield and a memorial in a moment of time, creating a record of the past for the future.

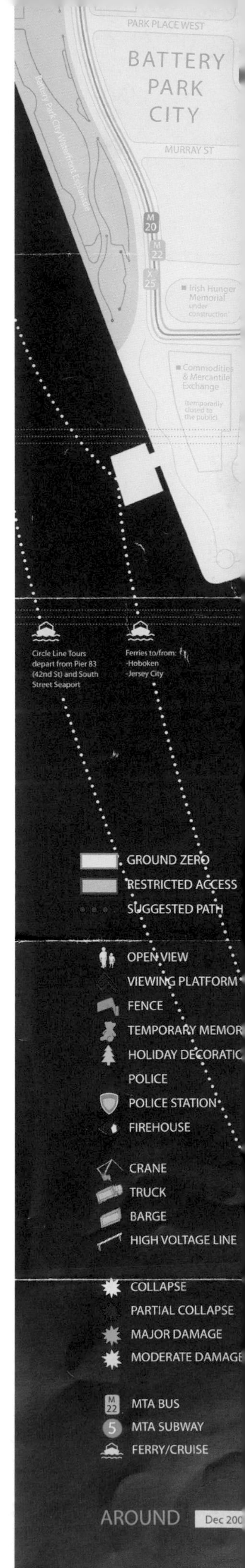

City Hall
City Hall Park
MURRAY STREET
College of Insurance
US Dept. of Internal Revenue
BROADWAY
PARK PLACE
Park Place
Bank of New York
US Immigration & Naturalization Bldg
Red Cross staging area
WEST STREET
BARCLAY STREET
Woolworth Building
PARK ROW
7 WTC
struck by portion of north tower, and collapsed, 6:00pm. Debris collapsed into 1/9 subway tunnel below. former site of Mayoral Emergency Command Center
140 West St Verizon Bldg
Federal Office Building (Post Office)
St. Peter's Church
THEATER AL
VESEY ST
ANN ST
St. Paul's Chapel
World Trade Center Station (not in service)
WORLD FINANCIAL CENTER
4 WFC
3 WFC
6 WTC: Custom House
struck by facade of north tower, debris partially removed due to government offices in building
5 WTC
struck by collapse of towers, has since been demolished former site of Port Authority Police Station
Public Viewing Platform
Fulton St
FULTON STREET
North Bridge
1 WTC north tower
8:46 a.m. impact 10:28 a.m. collapse debris partially removed
Cortlandt St (not in service)
Millennium Hotel
Winter Garden
DEY ST
NASSAU ST
JOHN ST
PATH Trains to New Jersey (not in service)
World Trade Center Plaza
2 WFC
Cortlandt St (not in service)
4
5
6
J
M
Z
MAIDEN LN
Cove Harbor
3 WTC Marriott Hotel
struck by portion of south tower, collapsed
2 WTC south tower
9:03 a.m. impact 9:50 a.m. collapse debris partially removed
4 WTC
struck by collapse of towers, has since been demolished
CORTLANDT ST
One Liberty Plaza
Federal Reserve Bank
NY City Police Memorial
Port Authority Viewing Platform (restricted access)
Gateway Plaza
South Bridge
St. Nicholas Greek Orthodox Church
Banker's Trust
Struck by portion of south tower, 09.11.01, 9:59am
LIBERTY ST
Liberty Plaza
CEDAR STREET
CEDAR ST
US Realty Building
THAMES ST
SOUTH END AVENUE
1 WFC
Green Exchange Bldg
THAMES STREET
ALBANY STREET
PINE ST
Marriott Hotel
American Stock Exchange
CARLISLE STREET
ALBANY STREET
TRINITY STREET
BROADWAY
Trinity Church
South Esplanade
RECTOR PLACE
Wall St
WALL ST
Federal Hall
NYPD
Rector St (not in service)
Rector St
New York Stock Exchange
RECTOR STREET
WASHINGTON ST
GREENWICH ST
BATTERY PARK CITY
WEST STREET
FINANCIAL DISTRICT
EXCHANGE PL
Broad St
WEST THAMES ST
BROOKLYN BATTERY TUNNEL
NEW ST
BROAD ST
3RD PL
MORRIS ST
BATTERY PLACE
BEAVER ST
South Cove
MARKETFIELD ST
2ND PL
Police Museum
200 feet
400 feet
0
.05 mile
.10 mile
GROUND ZERO

AFTERWORD: THE ROAD AHEAD

DeepMap, data visualization for autonomous vehicles, 2018

If this volume seems eclectic, it is so by design. To showcase the wide influence of maps in American history, I have selected examples that run the gamut. Across five centuries, maps drove statecraft and diplomacy, exploration and imperialism. They shaped settlement patterns, political strategy, and moral reform. They galvanized social movements and stimulated patriotism, asserted territory and investigated disease. They advertised destinations and products, explained scientific theories, and recorded personal histories. Occasionally they made people laugh.

Yet however diverse their motives—not to mention their circulation and appearance—maps have almost always been physical artifacts that could be touched, used, updated, and filed away. That generalization no longer holds true. Since the turn of the century, maps have been more likely to be generated by software and to live online; some are never printed at all, much less archived. Geographic information systems (GIS) and other platforms have also democratized mapmaking, and we live with those results on a daily basis. The news and mass culture are littered with graphic information, and even animated and interactive maps now seem par for the course. We have fluid maps of Internet traffic and Twitter feeds, wind patterns, demographic shifts, and even consumer preferences.

This avalanche of digital maps can make it difficult to separate the signal from the noise. But one innovation on the horizon promises to change not just the way we live, but the very nature of mapping: self-driving cars. The implications of this technology are startling. In the short term, autonomous vehicles will transform commuting and bring mobility to the elderly, the disabled, and the young. Even more consequential are the long-term structural consequences for an economy that has been inextricably linked to the automobile for nearly a century. Auto repair, trucking, and insurance will all be fundamentally affected.

Equally radical are the implications for maps, since self-driving cars cannot operate without them. More precisely, autonomous vehicles require elaborate, detailed, three-dimensional systems of measurement to shepherd them through space. This gargantuan challenge prompted the recent founding of DeepMap in Silicon Valley, where engineers are now working to support autonomous vehicles on a wide scale. The foundation of this platform is the Global Positioning System (GPS), a technology now commonly used on cellular phones and dashboard navigation systems. DeepMap integrates this GPS base data with dynamic information incorporated in real time through vehicles outfitted with high-definition cameras and a range of sensors. This equipment reads and monitors both the fixed and the fluid environment on the road, enabling the vehicle to assemble multiple data points and to form a "memory" of its surroundings. By comprehensively recording the immediate environment, the vehicle is safely guided through space.

But that is just the beginning, for every piece of information read by those instruments on that individual vehicle is then shared with a larger platform. This makes the car itself a mapmaking tool, one that constantly gathers information to improve the larger system. In this way, we are taken back to the era of discovery and reconnaissance: just as field surveyors took measurements that were then translated into cartographic form, self-driving cars will generate situational data that constantly advance the general geographical system of any given locality. In a rather ingenious loop, the vehicles gather information, use that information to move through space, and in so doing generate new data. As map scholar David Rumsey observed, the result is nearly an organic entity, a constant exchange of information in real time that is designed to achieve maximum accuracy.

Of course what guides the vehicle safely and successfully is not a visual "map," but rather a digital system of instructions, the data themselves. Which prompts the question: Are "maps" for self-driving cars really maps at all? DeepMap's chief operating officer Wei Luo explains that even for autonomous vehicles, there is still a need to visualize the data. Some of these visualizations are generated to show passengers what the car "sees" and "knows" at any given moment; others are designed for engineers developing the technology; and still others are needed for quality control. In other words, the visual maps are designed for human consumption. Luo rendered this visual picture of the data to convey how the system "teaches" the rules of the road to the vehicle, with green lines guiding it through an intersection.

In this regard, autonomous vehicles are not so different from humans: they need to sense their surroundings in

the same way as we see and hear. They then process this information in real time in order to move through space. In fact, in the course of generating this new technology the engineers at DeepMap were reminded of the extraordinary power of the human brain. From speed limits to navigating rules of intersections, merging, and taking a car through curves, humans rely on their senses and their prior experience to process an enormous amount of information while driving. Automated vehicles will require similarly sophisticated systems to replicate this operation. And, for this reason, the measurements involved in this mapping database must be far more precise than any road map in history, down to a centimeter.

All of this presents a rather colossal challenge for DeepMap. Yet it is also an oddly old-fashioned task of creating wayfinding aids to get from one point to another. This same quest drove so many mapmaking efforts over the past five centuries, in the service of discovery, conquest, settlement, or general mobility. To teach cars to drive themselves, we are in a way replicating a very basic cartographic operation. But the creation of platforms for autonomous vehicles also reminds us that maps are far more than instruments of navigation. Maps make this technological innovation possible. In turn, the technology demands an entirely different kind of map. This reciprocal dynamic has been at work throughout American history: across five centuries, maps have both reflected and mediated change.

pp. 2–3: Image courtesy of the F. D. R. Presidential Library & Museum.
pp. 6–7: Edward Savage, "The Washington Family, 1789–1796." Courtesy, Andrew W. Mellon Collection, National Gallery of Art.

1. 1490–1600: Contact and Discovery

THE WORLD THAT COLUMBUS KNEW
Additional sources: Kenneth T. Nebenzahl, *Atlas of Columbus and the Great Discoveries* (Chicago: Rand McNally, 1990); Chet Van Duzer, "Multispectral Imaging for the Study of Maps: The Example of Henricus Martellus's World Map at Yale," *Imago Mundi* 68 (2016), 62–6; Genevieve Carlton, *Worldly Consumers: The Demand for Maps in Renaissance Italy* (Chicago: University of Chicago Press, 2015); Scot McKendrick, Kathleen Doyle, and John Lowden, *Royal Manuscripts: The Genius of Illumination* (London: British Library Publications, 2011).

A GENERATION OF CONFUSION
Additional sources: Roberto Almagia, "On the Cartographic Work of Francesco Rosselli," *Imago Mundi* 8 (1951), 27–35; Edward Heawood, "A Hitherto Unknown World Map of A.D. 1506," *Geographical Journal* 62: 4 (1923), 279–93.

HOW AMERICA (INADVERTENTLY) GOT ITS NAME
Additional sources: David Buisseret, *From Sea Charts to Satellite Images: Interpreting North American History through Maps* (Chicago: University of Chicago Press, 1990); Seymour Schwartz and Ralph Ehrenberg, *The Mapping of America* (New York: Harry N. Abrams, 1980); John Hébert, "The Map That Named America," *Library of Congress Information Bulletin* 62: 9 (2003), https://www.loc.gov/loc/lcib/0309/maps.html (accessed February 20, 2018).

AMERICA LOOMS INTO VIEW
p.21 *"a human bridge between Portuguese and Spanish"*: Arthur Davies, "The Egerton MS.2803 Map and the Padron Real," *Imago Mundi* 11 (1954), p. 54.
Additional sources: Henry Harrisse, *The Discovery of North America: A Critical, Documentary, and Historic Investigation* (Amsterdam N. Israel, 1961 [reprint of the 1892 edition]); G. Caraci, "A Little Known Atlas by Vesconte Maggiolo, 1518," *Imago Mundi* 2 (1937), 37–54; Edward L. Stevenson, *Atlas of Portolan Charts* (New York: Hispanic Society, 1911).

THE INVASION AND DESTRUCTION OF MEXICO
Additional sources: Barbara E. Mundy, "Mapping the Aztec Capital: The 1524 Nuremberg Map of Tenochtitlan, Its Sources and Meanings," *Imago Mundi* 50 (1998), 11–33; Hernán Cortés, *Letters from Mexico*, trans. and ed. Anthony Pagden (New Haven, CT: Yale University Press, 2001); Barbara E. Mundy, "Indigenous Civilization," in Jordana Dym and Karl Offen (eds.), *Mapping Latin America: A Cartographic Reader* (Chicago: University of Chicago Press, 2011); Robert S. Weddle, *Spanish Sea: The Gulf of Mexico in North American Discovery, 1500–1685* (College Station, TX: Texas A&M University Press, 1985); Philip D. Burden, *The Mapping of North America: A List of Printed Maps, 1511–1670* (Rickmansworth, Herts.: Clive A. Burden 1996).

THE HEMISPHERE TAKES SHAPE
Additional sources: Philip D. Burden, *The Mapping of North America: A List of Printed Maps, 1511–1670* (Rickmansworth, Herts.: Clive A. Burden 1996).

THE SPANISH REACH NORTHWARD
Additional sources: Rodney Shirley, *The Mapping of the World: Early Printed World Maps, 1472–1700* (London: Holland Press, 1983); David Woodward, "The Italian Map Trade," in David Woodward (ed.), *Cartography in the European Renaissance*, vol. 3 of *The History of Cartography* (Chicago: University of Chicago Press, 2007); Michael Wyatt, *The Cambridge Companion to the Italian Renaissance* (Cambridge: Cambridge University Press, 2014).

THE SPANISH ASSERTION OF AMERICA
Additional sources: John Hébert, "The 1562 Map of America by Diego Gutiérrez", http://www.loc.gov/rr/hispanic/frontiers/gutierrz.html? (accessed August 20, 2017); Clara Egli LeGear, "Sixteenth-Century Maps Presented by Lessing J. Rosenwald," *Quarterly Journal of Current Acquisitions* 6: 3 (May 1949), 18–22.

THE FATHER OF THE BRITISH EMPIRE
Additional sources: Lesley Cormack, *Charting an Empire: Geography at the English Universities, 1580–1620* (Chicago: University of Chicago Press, 1997); Ken MacMillan, "Brytanici Imperii Limites," *Huntington Library Quarterly* 64: 1–2 (2001), 151–9; Ken MacMillan, "Discourse on History, Geography, and Law: John Dee and the Limits of the British Empire 1576–80," *Canadian Journal of History* 36: 1 (2001), 1–25; David Livingstone, *The Geographical Tradition: Episodes in the History of a Contested Enterprise* (Oxford: Blackwell, 1992).

2. 1600–1700: Early Settlement and the Northwest Passage

THE ORIGINS OF THE VIRGINIA COLONY
p.36 *"where never Christian before hathe been"*: Letter of Robarte Tindall to Prince Henry, June 22, 1607, British Library, Harley MS 7007, folio 139, reproduced in Alexander Brown (ed.), *The Genesis of the United States* (London, 1890), p.108–109.
p.38 *"There were never Englishmen left"*: George Percy, journal, in Lyon Gardiner Tyler (ed.), *Narratives of Early Virginia: 1606–1625* (New York: Scribner's, 1907), p.9.
Additional sources: Maurice A. Mook, "The Ethnological Significance of Tindall's Map of Virginia," *William and Mary Quarterly* 23: 4 (1943), 371–408; "Tyndall's Map of Virginia," *Proceedings of the Massachusetts Historical Society*, 3rd series, 58 (1924–5), 244–7.

THE SURVIVAL OF VIRGINIA
p.41 *"So great was our famine"*: John Smith, *The Generall Historie of Virginia, New England & the Summer Isles* (Glasgow, Scotland: James MacLehose & Sons, 1907), v.1, p. 204.
Additional sources: William Cumming, *The Southeast in Early Maps* (Princeton, NJ: Princeton University Press, 1958); Susan Schulten, "Mapping History," in James Akerman and Robert Karrow (eds.), *Maps: Finding Our Place in the World* (Chicago: University of Chicago Press, 2007); Donald Meinig, *Atlantic America, 1492–1800*, vol. 1 of *The Shaping of America* (New Haven, CT: Yale University Press, 1986); Peter Mancall, *Envisioning America: English Plans for the Colonization of North America, 1580–1640* (Boston: St. Martin's Press, 1995).

THE INVENTION OF NEW ENGLAND
Additional sources: John Smith, *A Description of New England; or, The Observations, and Discoveries, of Captain John Smith* (London: Printed by Humfrey Lownes for Robert Clerke, 1616); Walter W. Woodward, "Captain John Smith and the Campaign for New England: A Study in Early Modern Identity and Promotion," *New England Quarterly* 18: 1 (2008), 91–125; John Smith, *The Generall Historie of Virginia, New England & the Summer Isles*, vol. 1 (Glasgow, 1907).

THE LURE OF A NORTHWEST PASSAGE
p.45 *"very near as far toward the west"*: *A Treatise of the Northwest Passage to the South Sea, through the Continent of Virginia and by Fretum Hudson*, in Edward Waterhouse, *A Declaration of the State of the Colony and Affaires in Virginia* (London, 1622), p. 48.
p.45 *"all those rich countries"*: Ibid., p. 50.
p.45 *"through the continent of Virginia"*: Ibid., p. 49.
Additional sources: Edward Waterhouse, *A Declaration of the State of the Colony and Affaires in Virginia* (London, 1622); Samuel Purchas, *Purchas His Pilgrimes: In Five Books* (London: Printed by William Stansby for Henrie Fetherstone, 1626); Ken MacMillan, "Sovereignty 'More Plainly Described': Early English Maps of North America, 1580–1625," *Journal of British Studies* 42 (2003), 413–47.

THE FRENCH EXPLORE AMERICA
Additional sources: David Hackett Fischer, *Champlain's Dream* (New York: Simon & Schuster, 2009); Lawrence C. Wroth, "An Unknown Champlain Map of 1616," *Imago Mundi* 11 (1954), 85–94; David Buisseret, *Mapping the French Empire in North America* (Chicago: University of Chicago Press, 1991).

THE ORIGINS OF THE ATLANTIC SLAVE TRADE
Additional sources: William Waller Hening (ed.), *The Statutes at Large: Being a Collection of All the Laws of Virginia*, vols. 2 and 3 (New York: R. & W. & B. Bartow, 1823); Robin Law, "The Slave Trade in Seventeenth-Century Allada: A Revision," *African Economic History* 22 (1994), 59–92; New-York Historical Society, "Slavery in New York" (exhibition), http://www.slaveryinnewyork.org (accessed January 19, 2018).

I'LL TAKE MANHATTAN
Additional sources: Joyce D. Goodfriend, *Before the Melting Pot: Society and Culture in Colonial New York City, 1664–1730* (Princeton, NJ: Princeton University Press, 1994); Paul E. Cohen and Robert T. Augustyn, *Manhattan in Maps, 1527–1995* (New York: Rizzoli, 2006).

VIOLENCE AND DEVASTATION IN EARLY NEW ENGLAND
Additional sources: J. B. Harley, "New England Cartography and the Native Americans," in *The New Nature of Maps* (Baltimore, MD: Johns Hopkins University Press, 2001); Jill Lepore, *The Name of War: King Philip's War and the Origins of American Identity* (New York: Knopf, 1998); Matthew Edney and Susan Cimburek, "Telling the Traumatic Truth: William Hubbard's 'Narrative' of King Philip's War and His 'Map of New-England'," *William and Mary Quarterly* 61: 2 (2004), 317–48.

PENN'S HOLY EXPERIMENT
p.56 *"such a Scituation is scarce to be parallel'd"*: *A Portraiture of the City of Philadelphia*, as part of *A Letter from William Penn Proprietor and Governour of Pennsylvania ...* (London: Printed and sold by Andrew Sowle, 1683).
p.56 *"two Miles in Length and one in Breadth"*: Ibid.
p.58 *"we want a map to the degree that I am ashamed"*: quoted in Walter Klinefelter, "Surveyor General Thomas Holme's 'Map of the Improved Part of the Province of Pennsilvania'," *Winterthur Portfolio* 6 (1970), p. 42.
Additional sources: Albert Cook Myers, *Narratives of Early Pennsylvania* (New York, Charles Scribner's Sons ,1912); Irma Corcoran, "Thomas Holme, 1624–1695: Surveyor General of Pennsylvania," *Memoirs of the American Philosophical Society* 200 (1992); James T. Lemon, *The Best Poor Man's Country: A Geographical Study of Early Southeastern Pennsylvania* (Baltimore, MD, 1972).

FRENCH EXPANSION IN AMERICA
Additional sources: Philip D. Burden, *The Mapping of North America: A List of Printed Maps*, vol. 2, 1671–1700 (Rickmansworth, Herts.: Clive A. Burden 2007).

3. 1700–1783: Imperialism and Independence

THE WAR OF THE MAPS I
Additional sources: Anne Godlewska, *Geography Unbound: French Geographic Science from Cassini to Humboldt* (Chicago: University of Chicago Press, 1999); James Pritchard, *In Search of Empire: The French in the Americas, 1670–1730* (Cambridge: Cambridge University Press, 2004); Christine Petto, *When France Was King of Cartography: The Patronage and Production of Maps in Early Modern France* (Lanham, MD: Lexington Books, 2007).

THE WAR OF THE MAPS II
Additional sources: William P. Cumming (ed.), *The Southeast in Early Maps* (Chapel Hill, NC: University of North Carolina Press, 1998).

NATIVE AMERICANS NAVIGATE THE DEERSKIN TRADE
Additional sources: Alan Taylor, "Squaring the Circles," in Eric Foner and Lisa McGirr (eds.), *American History Now* (Philadelphia: Temple University Press, 2011); Gregory Waselkov, "Indian Maps for the Colonial Southeast," in Gregory A. Waselkov, Peter H. Wood, and Tom Hatley (eds.), *Powhatan's Mantle: Indians in the Colonial Southeast*, rev. ed. (Lincoln, NE: University of Nebraska Press, 2006); Ian Chambers, "A Cherokee Origin for the 'Catawba' Deerskin Map," *Imago Mundi* 65 (2013), 207–16.

ENGLAND AND THE SLAVE TRADE
p.74 *"our Planters a constant Supply of Negroe-Servants"*: Malachy Posthlethwayt, *The National and Private Advantages of the African Trade Considered* (London, repr. 1772 [1746]), p. 1.
Additional sources: Malachy Posthlethwayt, *The African Trade, the Great Pillar and Support of the British Plantation Trade in America* (London:

J. Robinson, 1745; Christopher Leslie Brown, *Moral Capital: Foundations of British Abolitionism* (Chapel Hill, NC: University of North Carolina Press, 2006).

WAR AND DISCORD IN THE SOUTHEAST
Additional sources: Donald Meinig, *Atlantic America, 1492–1800*, vol. 1 of *The Shaping of America* (New Haven, CT: Yale University Press, 1986); Mark E. Smith (ed.), *Stono: Documenting and Interpreting a Southern Slave Revolt* (Chapel Hill, NC: University of North Carolina Press, 2005).

IROQUOIS DIPLOMACY
p.79 *"extreamly Revengeful the Indians naturally are"*: Cadwallader Colden, *The History of the Five Indian Nations depending on the Province of New-York in America* (New York: Printed and sold by William Bradford in New-York, 1727), p. 16.
Additional sources: John Dixon, *The Enlightenment of Cadwallader Colden* (Ithaca, NY: Cornell University Press, 2016).

TOBACCO AND VIRGINIA
p.80 *"a river at his door" from* "A Compendious Account of the British Colonies in North-America," in *The Theatre of War in North America*, quoted in Margaret Beck Pritchard and Henry G. Taliaferro, *Degrees of Latitude: Mapping Colonial America* (Colonial Williamsburg Foundation and Harry N. Abrams, 2002), p. 157.
Additional sources: Donald Meinig, *Atlantic America, 1492–1800*, vol. 1 of *The Shaping of America* (New Haven, CT: Yale University Press, 1986); Peter H. Wood, "Slave Labor Camps in Early America: Overcoming Denial and Discovering the Gulag," in Carla Gardina Pestana and Sharon V. Salinger (eds.), *Reencounters with Colonialism: New Perspectives on the Americas* (Lebanon, NH: University Press of New England, 1999).

A YOUNG GEORGE WASHINGTON MAPS THE CLASH OF EMPIRES
p.84 *"The Lands upon the River Ohio"*: Letter of Dinwiddie to French Commander reproduced in *Journal of Major George Washington, Sent by the Hon. R. Dinwiddie* (Williamsburg printed, London reprinted, 1754), p. 25.
p.84 *"no Englishman had a Right"*: Ibid.,p. 22.
p.84 *"absolute Command of both Rivers"*: Ibid., p. 6.
Additional sources: Jared Sparks, *The Writings of George Washington*, vol. 1 (Boston: American Stationers' Company, 1837).

THE PRESENT CONJUNCTURE OF AFFAIRS IN AMERICA
p.86 *"The English have several Ways to Ohio"*: Lewis Evans, *Geographical, Historical, Political . . . Essays: The First, containing an Analysis of a General Map of the Middle British Colonies in America. . .*, 2nd ed. (Philadelphia: Printed by B. Franklin and D. Hall, 1755), p. iii.
p.87 *"whatever is theirs, is expressly acceded to the English"*: Ibid., p. iv.
p.87 "*French power*": Ibid., p.32.
Additional sources: Henry Stevens, *Lewis Evans His Map of the Middle British Colonies in America*, 2nd ed. (London: Henry Stevens, Son & Stiles, 1905).

SPANISH GEOGRAPHICAL INTELLIGENCE IN THE SOUTHWEST
Additional sources: John L. Kessell, *Miera y Pacheco: A Renaissance Spaniard in Eighteenth-Century New Mexico* (Norman, OK: University of Oklahoma Press, 2015); Herbert Bolton, *Pageant in the Wilderness: The Story of the Escalante Expedition to the Interior Basin, 1776* (Salt Lake City, UT: Utah State Historical Society, 1950).

THE SHOT HEARD ROUND THE WORLD
Additional sources: Richard Brown and Paul E. Cohen, *Revolution: Mapping the Road to American Independence, 1755–1783* (New York: W. W. Norton, 2015) Elizabeth Fenn, *Pox Americana: The Great Smallpox Epidemic of 1775–82* (New York: Hill & Wang, 2002).

WHERE THE BRITISH LAID DOWN THEIR ARMS
Additional sources: Robert Middlekauff, *Washington's Revolution: The Making of America's First Leader* (New York: Vintage Books, 2015); Charles E. Hatch, Jr., "Gloucester Point in the Siege of Yorktown, 1781," *William and Mary Quarterly* 20: 2 (1940), 265–84; Richard Brown and Paul E. Cohen, *Revolution: Mapping the Road to American Independence, 1755–1783* (New York: W. W. Norton, 2015).

INDEPENDENCE
p.97 *"the most important map in American history"*: in Lawrence Martin, "John Mitchell," *Dictionary of American Biography*, Dumas Malone, (ed.), vol. 13 (New York: C. Scribner's Sons, 1934), p. 50.
Additional sources: Susan Schulten, "Mapping American History," in James Akerman and Robert Karrow (eds.), *Maps: Finding Our Place in the World* (Chicago: University of Chicago Press, 2007); Matthew H. Edney, "The Mitchell Map, 1755–1782: An Irony of Empire," http://oshermaps.org/special-map-exhibits/mitchell-map (accessed February 20, 2018); Lawrence Martin, "John Mitchell's Map of the British and French Dominions in North America," in Walter W. Ristow and Richard W. Stephenson (eds.), *A la Carte: Selected Papers on Maps and Atlases* (Washington, DC: Library of Congress, 1972).

4. 1783–1835: A Nation Realized

NATIONAL ASPIRATIONS
p.100 *"tyranny of Britain"*: Jedidiah Morse, *Geography Made Easy* (New-Haven: Printed by Meigs, Bowen & Dana, 1784), p. 24.
Additional sources: Library of Congress, "Mapping a New Nation: Abel Buell's Map of the United States, 1784" (exhibition, 2013–16), https://www.loc.gov/exhibits/mapping-a-new-nation (accessed February 20, 2018); Paul E. Cohen, "Abel Buell, of Connecticut, Prints America's First Map of the United States, 1784," *New England Quarterly* 86: 3 (2013), 357–97; Lawrence C. Wroth, *Abel Buell of Connecticut: Silversmith, Type Founder and Engraver* (Middletown, CT: Wesleyan University Press, 1958).

AN INVITATION TO SETTLEMENT
p.102 *"the most extraordinary country"*: John Filson, *The Discovery, Settlement, and Present State of Kentucke* (Wilmington, DE: Printed by James Adams, 1784), p. 21.
Additional sources: John Walton, *John Filson of Kentucke* (Lexington, KY: University of Kentucky Press, 1956); P. Lee Phillips, *The First Map of Kentucky by John Filson: A Bibliographical Account* (Washington, DC: W. H. Lowdermilk & Co., 1908).

THE CURRENTS OF THE ATLANTIC WORLD
Additional sources: Benjamin Franklin to Anthony Todd, October 29, 1768, in *The Papers of Benjamin Franklin*, vol. 15 (New Haven, CT: Yale University Press, 1972), pp. 246–8; W. G. de Brahm, "Hydrographical Map of the Atlantic Ocean, extending from the Southernmost Part of North America to Europe," in *The Atlantic Pilot* (London, 1772); "A Letter from Dr. Benjamin Franklin, to Mr. Alphonsus le Roy ...," *Transactions of the American Philosophical Society* 2 (1786), 294–329; Louis De Vorsey, "Pioneer Charting of the Gulf Stream: The Contributions of Benjamin Franklin and William Gerard de Brahm," *Imago Mundi* 28 (1976), 105–20.

ENGINEERING THE NATION'S CAPITAL
Additional sources: Richard W. Stephenson, *"A Plan Whol[l]y New": Pierre Charles L'Enfant's Plan of the City of Washington* (Washington, DC: Library of Congress, 1993); Iris Miller, *Washington in Maps, 1606–2000* (New York: Rizzoli, 2002).

FORGING A NATIONAL NETWORK
Additional sources: John Calhoun, Speech of February 4, 1817; Richard John, *Spreading the News: The American Postal System from Franklin to Morse* (Cambridge, MA: Harvard University Press, 1995); Brian Balogh, *A Government out of Sight: The Mystery of Authority in Nineteenth-Century America* (Cambridge: Cambridge University Press, 2009).

BEFORE LEWIS AND CLARK
p.113 *"explore the Missouri river"*: President Thomas Jefferson, letter to Meriwether Lewis, June 20, 1803, Thomas Jefferson Papers, Series 1: General Correspondence, 1651–1827, Microfilm Reel 028; Manuscript Division, Library of Congress.
p.113 *"the most direct & practicable water"*: Ibid.
Additional sources: Carl I. Wheat, *Mapping the Transmississippi West, 1540–1861*, vol. 1 (San Francisco: Institute of Historical Cartography, 1957–63); W. Raymond Wood, *An Atlas of Early Maps of the American Midwest* (Springfield, IL: Illinois State Museum, 1983).

AFTER LEWIS AND CLARK
Additional sources: Meriwether Lewis, *History of the Expedition under the Command of Captains Lewis and Clark* ... (Philadelphia: Bradford & Inskeep, 1814); Ralph E. Ehrenberg and Herman J. Viola, *Mapping the West with Lewis and Clark* (Washington, DC: Levenger Press and the Library of Congress, 2015); John Logan Allen, *Passage through the Garden: Lewis and Clark and the Image of the American Northwest* (Urbana, IL: University of Illinois Press, 1975).

DEMOCRACY SUBVERTED
Additional sources: Elmer C. Griffith, *The Rise and Development of the Gerrymander* (Chicago: Scott, Foresman & Company, 1907); Kenneth C. Martis, "The Original Gerrymander," *Political Geography* 27 (2008), 833–9; Mark Monmonier, *Bushmanders and Bullwinkles: How Politicians Manipulate Electronic Maps and Census Data to Win Elections* (Chicago: University of Chicago Press, 2001).

A SCHOOLGIRL MAPS HER COUNTRY
p.118 *"principles of association"*: Sarah Pierce, Address at the Close of School (1818), quoted in Emily Noyes Vanderpoel, *Chronicles of a Pioneer School from 1792 to 1833*, Elizabeth C. Barney Buel (ed.) (Cambridge, MA: University Press, 1903), p. 177.
p.118 *"Catherine Beecher ... recalled"*: Beecher quoted in Emily Noyes Vanderpoel, *More Chronicles of a Pioneer School from 1792 to 1833* (New York: Cadmus Book Shop, 1927), p. 179.
Additional sources: Susan Schulten, "Map Drawing, Graphic Literacy, and Pedagogy in the Early Republic," *History of Education Quarterly* 57: 2 (2017), 187–220; John Pinkerton, *A Modern Atlas from the First and Best Authorities* (Philadelphia: Thomas Dobson & Son, 1818).

A LITTLE SHORT OF MADNESS
p.120 *"a little short of madness"*: David Hosack, *Memoir of DeWitt Clinton* (New York: J. Seymour, 1829), p. 346.
Additional sources: Cadwallader Colden, *Memoir ... at the Celebration of the Completion of the New York Canals* (New York: Printed by the order of the Corporation of New York, 1825); David Hosack, *Memoir of De Witt [sic] Clinton* (New York, 1829); Carter Goodrich, *Government Promotion of American Canals and Railroads, 1800–1890* (New York: Columbia University Press, 1960); Donald Meinig, *Continental America, 1800–1967*, vol. 2 of T he *Shaping of America* (New Haven, CT: Yale University Press, 1993).

A CONTINENTAL FUTURE
p.122 *"Melish explained that his map was a 'picture'"*: John Melish, *A Geographical Description of the United States: With the Contiguous British and Spanish Possessions, Intended as an Accompaniment to Melish's Map of These Countries* (1816), p. 4.
Additional sources: Walter W. Ristow, *A la Carte: Selected Papers on Maps and Atlases* (Washington, DC: Library of Congress, 1972); Martin Bruckner, *The Social Life of Maps in America, 1750–1860* (Chapel Hill, NC: Omohundro Institute and the University of North Carolina, 2017).

5. 1835–1874: Expansion, Fragmentation, and Reunification

THE GEOGRAPHY OF SIN
Additional sources: W. J. Rorabaugh, *The Alcoholic Republic: An American Tradition* (Oxford: Oxford University Press, 1981).

AN ATLAS FOR THE BLIND
Additional sources: Samuel Gridley Howe, *Atlas of the United States Printed for the Use of the Blind* (Boston, New England Institution for the Education of the Blind, 1837); *Letters and Journals of Samuel Gridley Howe*, vol. 2, Laura Elizabeth (ed.) Howe (Boston: Dana Estes, 1906); Edward J. Waterhouse, *History of the Howe Press of Perkins School for the Blind* (Watertown, MA: Howe Press, 1975).

THE ENEMY WITHIN
p.130 *"happiness in their own way"*: Andrew Jackson, "President's Message," 21st Congress, 2nd Session, in Appendix to *Gales & Seaton's Register*, Library of Congress, pp. ix–x.
Additional sources: Report of the Secretary

of War, "Defence of the Western Frontier," December 30, 1837, in *American State Papers*, vol. 7 (Washington, DC, 1861); Donald Meinig, *Continental America, 1800–1967*, vol. 2 of *The Shaping of America* (New Haven, CT: Yale University Press, 1993); Steve Inskeep, *Jacksonland: President Andrew Jackson, Cherokee Chief John Ross, and a Great American Land Grab* (New York: Penguin, 2015).

OPENING THE OREGON TRAIL
p.133 *"If 1846 was indeed the 'year of decision'"*: Bernard DeVoto, *The Year of Decision: 1846* (New York: Little, Brown & Co., 1943), pp. 3–5.
Additional sources: John Fremont, *The Life of Col. John Charles Fremont, and His Narrative of Explorations and Adventures in Kansas, Nebraska, Oregon and California* (New York: Miller, Orton & Mulligan, 1856); William Goetzmann, *Exploration and Empire: The Explorer and the Scientist in the Winning of the American West* (New York: Knopf, 1966); Carl I. Wheat, *Mapping the Transmississippi West, 1540–1861* (San Francisco: Institute of Historical Cartography, 1957–63).

A CONTINENTAL NATION
Additional sources: Donald Meinig, *Continental America, 1800–1967*, vol. 2 of *The Shaping of America* (New Haven, CT: Yale University Press, 1993); Richard Francaviglia, *The Mapmakers of New Zion: A Cartographic History of Mormonism* (Salt Lake City, UT: University of Utah Press, 2015).

GOLD IN CALIFORNIA
Additional sources: Col. Richard B. Mason to Brigadier-General R. Jones, August 17, 1848, in Rodman Paul, *The California Gold Discovery* (Georgetown, CA: Talisman Press, 1966); Carl I. Wheat, *Mapping the Transmississippi West, 1540–1861* (San Francisco: Institute of Historical Cartography, 1957–63); Andrew Scott Johnston, *Mercury and the Making of California: Mining, Landscape, and Race, 1840–1890* (Boulder, CO: University Press of Colorado, 2014).

THE GEOGRAPHY OF IMMIGRATION
Additional sources: Carl Schurz, *The Reminiscences of Carl Schurz*, vol. 2 (New York: J. Murray, 1909); F. W. Bogen, *The German in America; or, Advice and Instruction for German Emigrants in the United States of America* (Boston: B. H. Greene, 1851); Donald Meinig, *Continental America, 1800–1967*, vol. 2 of *The Shaping of America* (New Haven, CT: Yale University Press, 1993).

THE GEOPOLITICS OF SLAVERY
Additional sources: John Jay, *America Free, or America Slave* (New York, 1956); Susan Schulten, "The Cartography of Slavery and the Authority of Statistics," *Civil War History* 56: 1 (2010), 5–32.

SLAVERY, SECESSION, AND WAR
Additional sources: Susan Schulten, *Mapping the Nation: History and Cartography in Nineteenth-Century America* (Chicago: University of Chicago Press, 2012); Francis Bicknell Carpenter, *Six Months at the White House with Abraham Lincoln: The Story of a Picture* (New York: Hurd & Houghton, 1866).

LINCOLN'S MAP
p.144 *"showing the slave population"*: Francis Bicknell Carpenter, *Six Months at the White House with Abraham Lincoln: The Story of a Picture* (New York: Hurd & Houghton, 1866), pp. 215–16.
p.144 *"you have appropriated my map"*: Ibid., pp.215–16.
Additional sources: Susan Schulten, "The Cartography of Slavery and the Authority of Statistics," *Civil War History* 56: 1(2010), 5 – 32.

GENERAL SHERMAN AND THE LOGIC OF DESTRUCTION
p.47 *"which otherwise would have been subjected to"*: from letter of William T. Sherman to Census Superintendent Joseph Kennedy, August 15, 1865, reprinted in "Value of the United States Census," *New-York Daily Tribune*, August 22, 1865,p. 4.
p.147 *"I knew exactly where to look for food"*: Ibid., p. 4.
Additional sources: *Memoirs of General William T. Sherman* (New York: D. Appleton, 1875); Margo Anderson, *The American Census: A Social History*, 2nd ed. (New Haven, CT: Yale University Press, 2015 [1988]); Joseph C. G. Kennedy, *Agriculture of the United States in 1860; Compiled from the Original Returns of the Eighth Census* (Washington: Government Printing Office, 1864); Susan Schulten, "Sherman's Maps," *New York Times* (November 20, 2014); Scott Nesbit, "The Irony of Emancipation in the Civil War South," unpublished PhD dissertation, University of Virginia (2013).

THE DEFEAT OF RECONSTRUCTION
Additional sources: James K. Hogue, *Uncivil War: Five New Orleans Street Battles and the Rise and Fall of Radical Reconstruction* (Baton Rouge, LA: Louisiana State University Press, 2006); Justin Nystrom, *New Orleans after the Civil War: Race, Politics, and a New Birth of Freedom* (Baltimore, MD: Johns Hopkins University Press, 2010).

6. 1874–1914: Industrialization and Its Discontents

UNEARTHING COAL
Additional sources: *Virginia: A Geographical and Political Summary* (Richmond, VA: R. F. Walker, 1876); Jedediah Hotchkiss, *The Virginias: A Mining, Industrial, and Scientific Journal Devoted to the Development of Virginia and West Virginia* (Staunton, VA: Printed by S.M. Yost & Son, 1880–85).

MAPPING THE MOTHERLODE
Additional sources: Joseph T. Lambie, *From Mine to Market: The History of Coal Transportation on the Norfolk and Western Railway* (New York: New York University Press, 1954); Ronald D. Eller, *Miners, Millhands, and Mountaineers: Industrialization of the Appalachian South, 1880–1930* (Knoxville, TN: University of Tennessee Press, 1982).

RED AND BLUE AMERICA
Additional sources: *Scribner's Statistical Atlas of the United States, Showing by Graphic Methods Their Present Condition and Their Political, Social and Industrial Development* (New York: C. Scribner's Sons, circa. 1883); Glen C. Altschuler and Stuart M. Blumin, *Rude Republic: Americans and Their Politics in the Nineteenth Century* (Princeton, NJ: Princeton University Press, 2001).

STRANGLED BY THE RAILROADS
p.159 *"improvidently granted to railroad corporations"*: *The Political Reformation of 1884. A Democratic Campaign Book. By Authority of the National Democratic Committee* (New York: 1884), p. 6.
Additional sources: Sean M. Kammer, "Land and Law in the Age of Enterprise: A Legal History of Railroad Land Grants in the Pacific Northwest, 1864–1916," unpublished PhD dissertation, University of Nebraska at Lincoln (2015); Richard White, *Railroaded: The Transcontinentals and the Making of Modern America* (New York: W.W. Norton & Company, 2011).

TO THE BRINK OF EXTINCTION
Additional sources: Joel A. Allen, *The American Bisons, Living and Extinct* (Cambridge: Cambridge University Press, 1876); William T. Hornaday, *The Extermination of the American Bison with a Sketch of Its Discovery and Life History* (Washington, DC: Smithsonian Institution, 1889); Andrew Isenberg, *The Destruction of the Bison: An Environmental History, 1750–1920* (Cambridge: Cambridge University Press, 2001).

AN ALTERNATIVE VISION FOR THE AMERICAN WEST
Additional sources: John Wesley Powell, *Report on the Lands of the Arid Region* (Washington, DC: Government Printing Office, 1879); *Eleventh Annual Report of the Director of the United States Geological Survey*, part 2, *Irrigation: 1889–1890*, in *Report of the Secretary of the Interior* v. IV, part 2 (Washington: Government Printing Office, 1890); Donald Worster, *A River Running West: The Life of John Wesley Powell* (New York: Oxford University Press, 2001).

MAPPING VICE IN SAN FRANCISCO
p.64 the *"unvarnished truth"*: Willard B. Farwell, *The Chinese at Home and Abroad* (San Francisco: A. L. Bancroft & Co., 1885), p. 5.
p.164 *"immorality, vice, and disease"*: Ibid., p. 4.
Additional sources: Nayan Shah, *Contagious Divides: Epidemics and Race in San Francisco's Chinatown* (Berkeley, CA: University of California Press, 2001).

THE HUMAN LANDSCAPE OF CHICAGO
Additional sources: *Hull-House Maps and Papers*, intro. Rima Lunin Schultz (Urbana: University of Illinois Press, 2005); Mary Jo Deegan, *Jane Addams and the Men of the Chicago School* (New Brunswick, NJ: Transaction Books, 1988).

RACE AND THE LIMITS OF MOBILITY
p.168 *"we must study, we must investigate"*: W. E. B. DuBois, *The Philadelphia Negro: A Social Study* (Philadelphia: Published for the University, 1899), p. 3.
p.168 *"a menace to a civilized people"*: Ibid., p. 390.
Additional sources: Michael B. Katz and Thomas J. Sugrue (eds.), *W. E. B. DuBois, Race, and the City: "The Philadelphia Negro" and Its Legacy* (Philadelphia: University of Pennsylvania Press, 1998).

AN AMERICAN EMPIRE?
Additional sources: Woodrow Wilson, "The Ideals of America," *Atlantic Monthly* 90 (December 1902), 721–34; Walter La Feber, *The New Empire: An Interpretation of American Expansion, 1860–1898* (Ithaca, NY: Cornell University Press, 1963); Susan Schulten, *The Geographical Imagination in America, 1880–1950* (Chicago: University of Chicago Press, 2001).

BALBOA'S DREAM REALIZED
Additional sources: "Report of the Board of Consulting Engineers and of the Isthmian Canal Commission on the Panama Canal" (Washington, DC: Government Printing Office, 1906); 59th Congress, 1st Session, Senate Document No. 231; J. Saxon Mills, *The Panama Canal: A History and Description of the Enterprise* (London: Thomas Nelson & Sons, 1913); David McCullough, *The Path between the Seas: The Creation of the Panama Canal, 1870–1914* (New York: Simon & Schuster, 1977).

BEFORE THE NINETEENTH AMENDMENT
Additional sources: Christine E. Dando, "'The Map Proves It': Map Use by the American Woman Suffrage Movement," *Cartographica* 45: 4 (2010), 221–40.

7. 1914–1940: Prosperity, Depression, and Reform

OVER THE TOP
p.179 *"two years ahead of his country"*: "'Over the Top' by an American Soldier Who Went," Morgan County (Colorado), *Republican* 18: 12 (March 22, 1918), p. 2 [unpaginated newspaper].
Additional sources: Arthur Guy Empey, *"Over the Top" by an American Soldier Who Went* (New York: G. P. Putnam's Sons, 1917); "Don'ts from the Author of 'Over the Top'," *New York Times* (October 14, 1917), p. 68.

THE CREATION OF A GERMAN ENEMY
Additional sources: George Creel, *How We Advertised America* (New York: Harper & Brothers, 1920); Wallace Notestein and Elmer E. Stoll (eds.), *Conquest and Kultur: Aims of the Germans in Their Own Words* (Washington, DC: Committee on Public Information, 1917, 1918); "What Every Soldier Should Know about the World War," Trench & Camp: edition for Fort Beauregard, Alexandria (February 11, 1918); "Kaiser Aimed at South America," *New York Times* (September 2, 1917).

DEADLIER THAN WAR
Additional sources: Alfred Crosby, *America's Forgotten Pandemic: The Influenza of 1918* (Cambridge: Cambridge University Press, 2003 [1989]); Carol R. Byerly, "The U.S. Military and the Influenza Pandemic of 1918–1919," *Public Health Report* 125: supplement 3 (2010), 82–9; Laura Spinney, *Pale Rider: The Spanish Flu of 1918 and How It Changed the World* (New York: PublicAffairs, 2017).

THE MASS PRODUCTION OF FOOD
p.185 *"I aimed at the public's heart"*: Upton Sinclair, "What Life Means to Me," *Cosmopolitan Magazine*, vol. 41 (October 1906), p. 594.
Additional sources: Armour & Company, "Nation's Ever-Growing Food Problem Has Been Met by Packing Industry" (David Rumsey Map Collection, Stanford University Libraries); Emma Tolman East, "Dreams That Came True," Romance of Big Business Series,

Number One (Chicago: Armour's Bureau of Agricultural Research and Economics, 1918); Donald Meinig, *Transcontinental America, 1850–1915*, vol. 3 of *The Shaping of America* (New Haven, CT: Yale University Press, 1998); Robert M. Aduddell and Louis P. Cain, "Public Policy Toward 'The Greatest Trust in the World'," *Business History Review*, 55: 2(Summer 1981), 217–42.

WATER FOR LOS ANGELES
Additional sources: Marc Reisner, *Cadillac Desert: The American West and Its Disappearing Water* (New York: Penguin, 1986); Kevin Starr, *Material Dreams: Southern California through the 1920s* (Oxford: Oxford University Press, 1990).

UNDERSTANDING THE UNDERWORLD
Additional sources: Frederic M. Thrasher, *The Gang: A Study of 1,313 Gangs in Chicago* (Chicago: University of Chicago Press, 1927).

MUSIC AND MAYHEM IN NEW YORK
Additional sources: Robert C. Harvey, *Insider Histories of Cartooning: Rediscovering Forgotten Famous Comics and Their Creators* (Jackson: University Press of Mississippi, 2014); Lisa McGirr, *The War on Alcohol: Prohibition and the Rise of the American State* (New York: W. W. Norton & Company, 2016).

THE MAPS IN OUR HEADS
Additional sources: Peter Gould and Rodney White, *Mental Maps* (Harmondsworth: Penguin Books, 1974).

THE GEOGRAPHY OF HOLLYWOOD
Additional sources: *The Motion Picture Industry as a Basis for Bond Financing* (Chicago: Halsey, Stuart, Inc., 1927); Chris Lukinbeal, "Teaching Historical Geographies of American Film Production," *Journal of Geography* 101: 6 (2007), 250–60.

IN SEARCH OF FREEDOM ON THE OPEN ROAD
p.196 *"As Donald Meinig once observed"*: Donald Meinig, *Atlantic America, 1492–1800*, vol. 4 of *The Shaping of America* (New Haven, CT: Yale University Press, 1986), p. 33.
Additional sources: "The Great American Roadside," *Fortune* (September 1934); *The Negro Motorist Green Book* (New York, published 1936–1967); Mark S. Foster, "In the Face of 'Jim Crow': Prosperous Blacks and Vacations, Travel, and Outdoor Leisure, 1890–1945," *Journal of Negro History* 84: 2 (1999), 130–49; Marguerite S. Shaffer, *See America First: Tourism and National Identity 1880–1940* (Washington, DC: Smithsonian Books, 2001).

REDLINING, HOME OWNERSHIP, AND CIVIL RIGHTS
Additional sources: "Residential Security Maps" and narrative descriptions accessed through the American Geographical Society and the Ohio State University; Amy E. Hillier, "Redlining and the Homeowners' Loan Corporation," *Journal of Urban History* 29: 4 (2003), 394–420; Kenneth T. Jackson, *Crabgrass Frontier: The Suburbanization of the United States* (New York: Oxford University Press, 1985).

LET THERE BE LIGHT
Additional sources: James T. Patterson, *Congressional Conservatism and the New Deal: The Growth of the Conservative Coalition in Congress, 1933–1939* (Lexington: University of Kentucky Press, 1966); Roger Biles, *A New Deal for the American People* (DeKalb: Northern Illinois University Press, 1991).

8. 1940–1962: Between War and Abundance

WORLD WAR II AND THE REINVENTION OF CARTOGRAPHY
p.206 *"a city of maps"*: *Newsweek* (January 26, 1942), p. 30.
Additional sources: Susan Schulten, T *he Geographical Imagination in America, 1880–1950* (Chicago: University of Chicago Press, 2001); Susan Schulten, "Richard Edes Harrison and the Challenge to American Cartography," *Imago Mundi* 50 (1998), 174–88.

THE WARTIME ROOTS OF THE AMERICAN CENTURY
Additional sources: Donald Meinig, *Global America, 1915–2000*, vol. 4 of *The Shaping of America* (New Haven, CT: Yale University Press, 2004).

THE DEFEAT OF GERMANY
Additional sources: "XIX Corps Cracks Siegfried Line," *Le Tomahawk* (XIX Corps newsletter) 2: 6, David Rumsey Map Collection, Stanford University; "The 83rd Infantry Division," United States Holocaust Memorial Museum, https://www.ushmm.org/wlc/en/article.php?ModuleId=10006145 (accessed February 14, 2018).

HOLOCAUST
p.216 *"he recalled how 'gorgeous' it was"*: Michael Kraus, *Drawing the Holocaust: A Teenager's Memory of Terezin, Birkenau, and Mauthausen* (Cincinnati, OH: Hebrew Union College Press; and Pittsburgh, PA: University of Pittsburgh Press, 2016), p. 56.

MAPPING THE MIGHTY MISSISSIPPI
Additional sources: Rufus J. LeBlanc, Sr., "Harold Norman Fisk as a Consultant to the Mississippi River Commission, 1948–1964—An Eye-Witness Account," *Engineering Geology* 45 (1996), 15–36; Michael C. Robinson, "Harold N. Fisk: A Luminescent Man," *Engineering Geology* 45 (1996), 37–44; Ellis L. Krinitzsky, "The Contributions of H. N. Fisk to Engineering Geology in the Lower Mississippi Valley, *Engineering Geology* 45 (1996), 45–58; Richard J. Russell, "Memorial to Harold Norman Fisk," *Bulletin of the Geological Society of America* vol. 76 no. 4 (April 1965), 53–8.

THE ADVENT OF AIR TRAVEL
p.221 *"the mountains jump up"*: "Picture Maps Put the U.S. in Focus," *New York Times* (August 8, 1954), p. 76.
Additional sources: Tom Patterson and Nathanial Vaughn Kelso, "Hal Shelton Revisited: Designing and Producing Natural-Color Maps with Satellite Land Cover Data," *Cartographic Perspectives* 47 (2004), 28–55.

THE ORIGINS OF THE COLD WAR
Additional sources: George F. Kennan ("X"), "The Sources of Soviet Conduct," *Foreign Affairs* 25: 4 (July 1947), 566–82; Timothy Barney, *Mapping the Cold War: Cartography and the Framing of America's International Power* (Durham, NC: University of North Carolina Press, 2015).

THE VIGILANT FIGHT AGAINST COMMUNISM
p.224 *"peaceful coexistence"*: Francis E. Walter, "Foreword" to "Soviet Total War: 'Historic Mission' of Violence and Deceit," September 23, 1956, prepared by the Committee on Un-American Activities, US House of Representatives, p.ix.
Additional sources: Andrew F. Smith, *Rescuing the World: The Life and Times of Leo Cherne* (Albany, NY: State University of New York Press, 2002); "The Great Pretense: A Symposium on Anti-Stalinism and the 20th Century Congress of the Soviet Communist Party," May 19, 1956, prepared by the Committee on Un-American Activities, US House of Representatives.

THE MAGIC KINGDOM
Additional sources: Richard Francaviglia, "Walt Disney's Frontierland as an Allegorical Map of the American West," *Western Historical Quarterly* 30: 2 (Summer 1999): 155–82; Steven Watts, *The Magic Kingdom: Walt Disney and the American Way of Life* (Columbia, MO: University of Missouri Press, 2001).

HIGHWAYS AND SUBURBS
Additional sources: Tom Lewis, *Divided Highways: Building the Interstate Highways, Transforming American Life* (Ithaca, NY: Cornell University Press, 2013); Richard F. Weingroff, "The Greatest Decade," Highway History Federal Highway Administration: https://www.fhwa.dot.gov/infrastructure/50interstate2.cfm (accessed October 25, 2017).

THE BATTLE AGAINST SEGREGATION
Additional sources: Raymond Arsenault, *Freedom Riders: 1961 and the Struggle for Racial Justice* (New York: Oxford University Press, 2011); Sid Moody, "Freedom Rides Brought More than Violence," AP Newsfeature, Title Collection. U.S. Subjects. U.S.-South-Social problems, 1961, Geography and Map Division, Library of Congress.

9. 1962–2001: An Unsettled Peace

TO THE BRINK OF NUCLEAR WAR
Additional sources: Sheldon M. Stern, *The Cuban Missile Crisis in American Memory: Myths versus Reality* (Stanford, CA: Stanford University Press, 2012).

WHY ARE WE IN VIETNAM?
Additional sources: Larry Berman, *Planning a Tragedy: The Americanization of the War in Vietnam* (New York: W. W. Norton, 1982); *Aggression from the North: The Record of North Viet-Nam's Campaign to Conquer South Viet-Nam* (Washington, DC: Department of State, 1965).

REVOLT
Additional sources: "Historical Archive," April Third Movement, http://www.a3mreunion.org/archive/archive.html (accessed February 15, 2018); *Harvard Crimson* (May 4, 1970); *New York Times* (May 4, 1970), p. 1.

CONTINENTAL DRIFT
Additional sources: Albert E. Theberge, "Seeking a Rift," *Hydro International*, https://www.hydro-international.com/content/article/seeking-a-rift (accessed February 15, 2018); Ronald E. Doel, Tanya J. Levin, and Mason K. Marker, "Extending Modern Cartography to the Ocean Depths: Military Patronage, Cold War Priorities, and the Heezen–Tharp Mapping Project, 1962–1969," *Journal of Historical Geography* 32 (2006), 605–26.

THE PROMISE OF AEROSPACE
Additional sources: Denis Cosgrove, *Apollo's Eye: A Cartographic Genealogy of the Earth in the Western Imagination* (Baltimore, MD: Johns Hopkins University Press, 2003); Robert Poole, *Earthrise: How Man First Saw the Earth* (New Haven, CT: Yale University Press, 2008).

THE TERROR OF AIDS
Additional sources: Abraham Verghese, Steven L. Berk, and Felix Sarubbi, "Urbs in Rure: Human Immunodeficiency Virus Infection in Rural Tennessee," *Journal of Infectious Diseases* 160: 6 (1989), 1051–5; Peter Gould, *The Slow Plague: A Geography of the AIDS Pandemic* (Oxford: Blackwell, 1993); Tom Koch, *Cartographies of Disease: Maps, Mapping, and Medicine* (Redlands, CA: ESRI Press, 2005).

A NATION ON THE MOVE
Additional sources: Donald Meinig, *Global America, 1915–2000*, vol. 4 of *The Shaping of America* (New Haven, CT: Yale University Press, 2004); *Census Atlas of the United States* (Washington, DC: US Census Bureau, 2007); James N. Gregory, "Internal Migration: Twentieth Century and Beyond," *Oxford Encyclopedia of American Social History* (New York: Oxford University Press, 2012), pp. 540-55; Kevin Phillips, *The Emerging Republican Majority* (Princeton, NJ: Princeton University Press, 2014 [1969]).

IMMIGRATION AND WORK IN RURAL AMERICA
Additional sources: William Kandel and Emilio A. Parrado, "Restructuring of the US Meat Processing Industry and New Hispanic Migrant Destinations," *Population and Development Review* 31: 3 (2005), 447–71; Lourdes Gouveia and Rogelio Saenz, "Global Forces and Latino Population Growth in the Midwest: A Regional and Subregional Analysis," *Great Plains Research* 10 (Fall 2000), 305–28; Upton Sinclair, *The Jungle*, intro. Eric Schlosser (New York: Penguin, 2006).

GERRYMANDERING IN THE DIGITAL AGE
Additional sources: Mark Monmonier, *Bushmanders & Bullwinkles: How Politicians Manipulate Electronic Maps and Census Data to Win Elections* (Chicago: University of Chicago Press, 2001); *Common Cause et al. v. Rucho* (2018).

BETWEEN PAST AND FUTURE
Additional sources: Laura Kurgan, "Around Ground Zero," *Grey Room* 7 (Spring 2002), 96–101; David Handelman, "History's Rough Draft in a Map of Ground Zero," *New York Times* (January 3, 2002).

AFTERWORD: THE ROAD AHEAD
Additional sources: Wei Luo and David Rumsey, "a16z Podcast: Exploding the Map," interview of September 17, 2017, https://a16z.com/2017/09/16/exploding-map-evolution-cartography-deep-map (accessed February 20, 2018).

LIST OF MAPS

Henricus Martellus, world map, in *Insularium Illistratum*, Florence, c.1490. Add. MS 15670, fols. 68v-69r. 30 x 47 cm.

Giovanni Contarini and Francesco Rosselli, "Mundu spericum ... cognosces diligentia joani mathei Contareni, arte et ingenio francisci Roselli florentini 1506 notu" [Florence?], c. 1506. Maps C.2.cc.4. 42 x 63 cm.

Martin Waldseemüller, "Universalis cosmographia secundum Ptholomaei traditionem et Americi Vespucii alioru[m] que lustrationes." [Strasbourg, France?] 1507. Library of Congress. 128 x 233 cm.

Vesconte de Maggiolo, world map in portolan atlas, c. 1508. Egerton MS 2803. 21 x 28 cm.

Hernán Cortés, map of the Gulf of Mexico and plan of Tenochtitlan, in *Praeclara Ferdin di Cortesii de Noua Maris Oceani Hyspania narratio sacratissimo*, 1524 (Nuremberg, Germany: Friedrich Peypus, 1524). Photo courtesy of the Newberry Library, Chicago. Ayer 655.51.C8 1524d. 30 x 47 cm.

Sebastian Münster, Map of the Americas, in *Novae insulae, XVII nova tabula* (Basel: H. Petri, 1540). Maps C.1.c.2. 28 x 35 cm.

Giacomo Gastaldi, "Cosmographia Universaliset Exactissima iuxta postremam neotericorum tradition[n]em. [Venice?] c.1561. Maps C.18.n.1. 90 x 225 cm.

Diego Gutiérrez, "Americae sive qvartae orbis partis nova et exactissima descriptio." (Antwerp? 1562) Maps * 69810.(18.) 92 x 93 cm.

John Dee, chart of part of the northern hemisphere, 1580. Cotton MS. Augustus I.i.1. 66 x 99 cm.

Robarte Tindall, a colored chart of the entrance to Chesapeake Bay (1608). Cotton MS Augustus I.ii.46. 46 x 81 cm.

John Smith, "Virginia" (Discovered and Described by Captayn John Smith. Graven by William Hole, 1612). G.7121. 31.5 x 40 cm.

John Smith, "New England," in *A Description of New England: or The Observations, and Discoveries, of Captain John Smith* (London: Printed for Humfrey Lownes, for Robert Clerke, 1616) C.33.c.12. 29.5 x 39.5 cm.

Henry Briggs, "The North Part of America: conteyning Newfoundland, New England, Virginia, Florida, New Spaine, and Nova Francia ... /R. Elstracke sculpsit." (London: Printed by William Stansby, for Henrie Fetherstone, 1625). 679.h.11-14. 27.5 x 34.5 cm.

Hugo Allardt, "Effigies ampli Regni auriferi Guineae in Africa siti ..." (Amsterdam, c.1650). Based on the manuscript map of Luis Teixeira of 1602. Courtesy of Stanford University Library. 44 x 61 cm.

Samuel de Champlain, "Le Canada, faict par le Sr. de Champlain, ou sont la Nouvelle France, la Nouvelle Angleterre, la Nouvelle Holande, la Nouvelle Svede, la Virginie, & Canada Avec les Nations voisines, et autres Terres nouuellement decouurtes suiuant les Memoires e P. Du Val, Geographe du Roy" (Paris, 1653). Maps 70615.(8). 35 x 65 cm.

Robert Holmes, "A Description of the Towne of Mannados or New Amsterdam as it was in September 1661 ..." [London? 1664]. Maps K.Top.121.35. 52 x 66 cm.

William Hubbard and John Foster, "A Map of New England, Being the first that ever was here cut ... ," in *A Narrative of the Troubles with the Indians in New-England ... by W. Hubbard* (Boston: Printed by John Foster, 1677). G.7146. 30 x 39.5 cm.

"A Portraiture of the City of Philadelphia in the Province of Pennsylvania in America" in: *A Letter from William Penn, Proprietary and Governour of Pennsylvania in America to the Committee of the Free Society of Traders of that Province* ... By Thomas Holme, Surveyor General. (London: Andrew Sowle, 1683). C.32.l.2.(4.). 30.5 x 45.5 cm.

"A map of the province of Pennsilvania, containing the three countyes of Chester, Philadelphia, and Bucks, as far as yet surveyed and laid out, &c., with the names of the owners, by Thomas Holme, surveyor-general" (c.1687). Add. MS. 5414 (23.). 81 x 137 cm.

"America Settentrionale colle nuoue scoperte fin al-anno 1688," in *Atlante Veneto ... Opera e studio del Padre Maestro Coronelli, etc.* (Venice, 1688). Maps C.44.f.6. volume 1, following page 56. Two maps, each measuring 62 x 46 cm.

Guillaume de L'Isle, "Carte de la Louisiane et du Course du Missisipi," in untitled atlas (Paris, 1718). Maps C.36.f.4 item 17. 47 x 63.5 cm.

"A new map of the north parts of America claimed by France under the names of Louisiana, Mississipi, Canada and New France, with the adjoyning Territories of England and Spain." By H. Moll. (London, 1720). Maps *69917.(29.). 60 x 100 cm.

"This Map describing the Scituation of the Several Nations of Indians to the NW of South Carolina was copyed from a Draught drawn & painted on a Deer Skin, by an Indian Cacique and presented to Francis Nicholson, Esq., Governour of South Carolina, by whom it is most humbly Dedicated to his Royal Highness George Prince of Wales." Add. MS. 4723. 81 x 112 cm.

Mark Catesby, "A Map of Carolina, Florida and the Bahama Islands," in *The Natural History of Carolina, Florida and the Bahama Islands*, 1731–43 (London: the Author, 1748). 44.k.7–9. 44.5 x 60 cm.

Cadwallader Colden, "A Map of the Country of the Five Nations, belonging to the Prince of New York; and of the Lakes near which the Nations of Far Indians live, with part of Canada." [in *The History of the Five Indian Nations depending on the Province of New-York in America*] (London, Lockyer Davis, 1755). 9555.a.1. 19 x 23 cm.

Malachy Postlethwayt, "A New and Correct Map of the Coast of Africa ... " (London: Printed for J. & P. Knapton, [1757]. Maps K.Top.117.90. 47 x 37 cm.

"A Map of the Most inhabited Part of Virginia containing the whole Province of Maryland with part of Pensilvania, New Jersey and North Carolina. Drawn by Joshua Fry & Peter Jefferson in 1751. Engraved ... by Thos. Jefferys." Maps 188.l.3.(2.). One map on four sheets, together measuring 78 x 124 cm.

"A map of the country between Will's Creek and Lake Erie, shewing the designs of the French for erecting forts southward of the lake; drawn ... before the erection of Fort Duquesne," circa 1754. Add. MS. 15563 B. 43 x 36 cm.

Lewis Evans, "A General Map of the Middle British Colonies, in America" in *Geographical, Historical, Political, Philosophical and Mechanical Essays, the First, Containing an Analysis of a General Map of the Middle British Colonies* (Philadelphia, Printed by B. Franklin, and D. Hall, 1755). 145.d.3. 49 x 65 cm.

"Plano geographico de la tierra descubierta nuevamente, a los rumbos norte, moroeste y oeste, del Nuevo Mexico, demarcado por mi, Don Bernardo de Miera y Pacheco ... ". Add. MS. 17661 D. 69 x 81 cm.

"A Plan of the Town and Harbor of Boston. And the Country adjacent with the Road from Boston to Concord. Shewing the Place of the late Engagement between the King's Troops & the Provincials, together with the several Encampments of both Armies and & about Boston." (London: J. Hand, 1775). Courtesy of the Newberry Library, Chicago, Call #Map 2F3701. S3.113. 37 x 49 cm.

"To His Excellency Genl Washington Commander in Chief of the Armies of the United States of America. This plan of the investment of York and Gloucester has been surveyed and laid down, and is Most humbly dedicated by his Excellencys Obedient and very humble servant. Sebast.n Bauman Major of the New York or 2nd Reg.t of Artillery. Add. MS. 57715 (13.). 49 x 45 cm.

"A Map of the British Colonies in North America with the Roads, Distances, Limits and Extent of the Settlements, Humbly Inscribed to the Right Honourable The Earl of Halifax, and the other Right Honourable The Lords Commisioners for Trade & Plantations, / By their Lordships Most Obliged and very humble Servant Jn.o Mitchell," (London: the author, 1755, Printed for Jefferys & Faden Geographers to the King ... [about 1775, with manuscript additions 1782]). Maps K. Top.118.49.b. 136 x 193 cm.

"This Map of Kentucke, Drawn from actual Observations, is inscribed with the most perfect respect to the Honorable the Congress of the United States of America, and to his Excell.cy George Washington, late Commander of the Chief of their Army. By their Humble Servant, John Filson," in Filson, *The Discovery, Settlement, and Present State of Kentucke* (Wilmington: Printed by James Adams, 1784). C.55.c.38. 48 x 44 cm.

Abel Buell, "A New and Correct Map of the United States of North America," (New Haven, CT: 1784). Maps *71490.(150.) 127 x 160 cm.

James Poupard and Benjamin Franklin, "A Chart of the Gulf Stream," in *Transactions of the American Philosophical Society* (Philadelphia, 1786). Courtesy of the Library of Congress. 21 x 26 cm.

Andrew Ellicott [and Pierre Charles L'Enfant], "Plan of the City of Washington in the Territory of Columbia, ceded by the States of Virginia and Maryland to the United States of America and by them established as the seat of their Government after the year 1800." (Philadelphia: Thackara & Vallance, 1792). Maps *72310.(1.). 57 x 70 cm.

"A Map of the United States, exhibiting the Post-Roads, the situations, connections & distances of the Post-Offices, Stage Roads, Counties, Ports of Entry and Delivery for foreign vessels, and the principal Rivers. By Abraham Bradley, junr. W. Barker Sculp. (Philadelphia, 1796). Maps *71490.(10.). 93 x 86.5 cm.

Aaron Arrowsmith and Samuel Lewis, "Louisiana," in *A New and Elegant General Atlas, Comprising All the New Discoveries, to the Present Time* (Philadelphia: J. Conrad, 1804). 25 x 20 cm. Courtesy of David Rumsey Map Collection, David Rumsey Map Center, Stanford Libraries.

"A Map of Lewis and Clark's Track, across the Western Portion of North America. From the Mississippi to the Pacific Ocean; By Order of the Executive of the United States in 1804, 5 & 6.," Copied by Samuel Lewis from the Original Drawing of Wm. Clark., in *History of the Expedition under the Command of Captains Lewis and Clark to the Sources of the Missouri* ... (Philadelphia: Bradford & Inskeep, 1814). 1431.h.2-3. 30 x 70 cm.

"The Gerry-Mander, or Essex South District Formed Into a Monster!," in *Salem Gazette* (April 2, 1813). Courtesy of Cornell University—PJ Mode Collection of Persuasive Maps. Map 19 x 16 cm.

Catharine Cook, "A Map of the United States" from "Catharine M. Cook's Book of Penmanship at Mr. Dunham's School, Windsor Vermont June 15, 1818." Courtesy of Osher Map Library, University of Southern Maine. Map on page measuring 24 x 19 cm.

"New York," Drawn by F. Lucas Jr., in *Memoir, Prepared at the Request of a Committee of the Common Council of the City of New York ... at the Celebration of the Completion of the New York Canals. By Cadwallader D. Colden* (New York, 1825). 714.g.25. 30.5 x 47 cm.

"Map of the United States, with the contiguous British & Spanish Possessions." Compiled ... by J. Melish (Philadelphia, 1823). Courtesy of David Rumsey Map Collection, David Rumsey Map Center, Stanford Libraries. 109 x 143 cm.

John Christian Wiltberger Jr., "Temperance Map." Courtesy of Cornell University—PJ Mode Collection of Persuasive Maps. 31 x 36 cm.

Map of Vermont, in *Atlas of the United States Printed for the Use of the Blind ...Under the direction of S. G. Howe, etc.* (Boston: Institution for the Education of the Blind, 1837). Maps 1.c.2. 27 x 22 cm.

"Map Illustrating the plan of the defences of the Western & North-Western Frontier, as proposed by Charles Gratiot, in his report of Oct. 31, 1837." Compiled in the U.S. Topographical Bureau under the direction of Col. J. J. Abert, U.S.T.E. by [Washington]. Hood." (Washington, DC: U.S. War Department, 1837). 54 x 38 cm.

S. Augustus Mitchell, "A New Map of Texas Oregon and California, with the regions adjoining" (Philadelphia: S. A. Mitchell, 1846). Courtesy of David Rumsey Map Collection, David Rumsey Map Center, Stanford Libraries. 57 x 52 cm.

Section IV of "Topographical Map of the Road from Missouri to Oregon Commencing at the Mouth of the Kansas in the Missouri River and Ending at the Mouth of the Wallah-Wallah in the Columbia. In VII Sections. From the field notes and journal of Capt. J. C. Fremont, and from sketches and notes made on the ground by his assistant, C. Preuss. Compiled by Charles Preuss, 1846, by order of the Senate of the United States." (Washington, DC: E. Weber & Co., 1846). Maps 32.d.9. 41 x 67 cm.

"Topographical Sketch of the Gold & Quicksilver District of California. July 25, 1848." E.O.C. Ord., Lt. U.S.A. Courtesy of Library of Congress. 54 x 42 cm.

Gotthelf Zimmerman, "Auswanderer-karte und Wegweiser nach Nordamerika." (Stuttgart: J. B. Metzler'schen Buchh., 1853). Courtesy of Library of Congress. 49 x 65 cm.

"Freedom and Slavery, and the Coveted Territories," in John Jay, *America Free, or America Slave. An Address on the State of the Country Delivered at Bedford, Westchester County, New York.* (New York, 1856). 8177.e.32.(6.). Sheet 15 x 22 cm.

Edwin Hergesheimer and the U.S. Coast Survey, "Map Showing the Distribution of the Slave Population of the Southern States of the United States. Compiled from the Census of 1860." (Washington: Engraved by Th. Leonhardt, 1861). Courtesy of Library of Congress. 59 x 86 cm.

Francis Bicknell Carpenter, "First Reading of the Emancipation Proclamation by President Lincoln," Oil on canvas. Courtesy of United States Senate. 274 x 457 cm.

"Map of Georgia & Alabama. Representing Rail-ways, Post-roads, Population, and Agricultural Productions. Prepared at the Census Office. Under direction of Jos. C.G. Kennedy Superintendent." Adapted circa 1864. Original map: "Map of Georgia & Alabama exhibiting the Post Offices, Post Roads, Canals, Railroads, &c. by David H. Burr, late topographer to the Post Office," 1839. Courtesy of National Archives and Records Administration RG77. 94 x 62 cm.

"Battle of New Orleans for Freedom. September 14, 1874. Compiled by T.S. Hardee, C.E." Courtesy of Boston Rare Maps and Barry Ruderman Antique Maps. 46 x 58.5 cm.

"Map of Virginia, by J. Hotchkiss" (Richmond, VA: Lith. By A. Hoen & Co., 1874), in *Virginia: a Geographical and Political Summary* (Richmond, VA: R. F. Walker, 1876). 10410. eee.7. 25 x 56 cm.

Map of Part of the Great Flat-top Coal-field of Va. & W. Va. Showing Location of Pocahontas & Bluestone Collieries May, 1886." Jed. Hotchkiss, Staunton, Va. Courtesy of Library of Congress. 22 x 22 cm.

"Popular Vote. Ratio of Predominant to Total Vote, by Counties. 1880." In Henry Gannett and Fletcher W. Hewes, *Scribner's Statistical Atlas of the United States* (New York: Scribner, 1883). 42 x 63 cm. Courtesy of David Rumsey Map Collection, David Rumsey Map Center, Stanford Libraries.

"Under a Black Cloud!" (1883). Courtesy of Library of Congress. 43 x 66 cm.

"Map Illustrating the Extermination of the American Bison. Prepared by W. T. Hornaday." Compiled under the direction of Henry Gannett. In *Annual Report of the Board of Regents of the Smithsonian Institution* (Washington, DC: Government Printing Office, 1889). A.S.910. 61 x 58 cm.

"Arid Region of the United States Showing Drainage Districts," in John Wesley Powell, *Report of the Secretary of the Interior, United States. Department of the Interior.* 51st Congress, Second Session, v.IV, part II, 1890. 35.5 x 26.5 cm.

"Official Map of 'Chinatown' in San Francisco. Prepared under the supervision of the Special Committee of the Board of Supervisors, July 1885. Willard B. Farwell, John E. Kunkler, E. B. Pond," in *The Chinese at Home and Abroad* (San Francisco: A. L. Bancroft, 1885). Courtesy of David Rumsey Map Collection, David Rumsey Map Center, Stanford Libraries. 62 x 140 cm.

Agnes Sinclair Holbrook and Florence Kelley, "Map of Nationalities," in *Hull-House Maps and Papers* (New York: T. Y. Crowell, 1895). 08225.e.1/5. 36 x 25.4 cm.

"The Seventh Ward of Philadelphia: The Distribution of Negro Inhabitants throughout the Ward, and Their Social Condition," in W. E. B. DuBois, *The Philadelphia Negro: A Social Study*, University of Pennsylvania Publications. Ac.2692.p. 15 x 100 cm.

"United States of America, 1904," reproduced in *Atlas of the Mexican Conflict Containing Detailed Maps Showing the Territory Involved ...* (New York: Rand McNally & Company, 1914) [map copyright 1900 and 1904]. Courtesy of David Rumsey Map Collection, David Rumsey Map Center, Stanford Libraries. 25 x 32 cm.

C.P. Gray, "Aeronautical View of the Panama Canal" (New York: Central Novelty Company, 1911). Maps 29.c.34. Map only 11 x 33 cm.

"Votes for women a success: the map proves it." (New York: National American Woman Suffrage Publishing Company, Inc., circa 1914). Maps CC.5.a.551. 15 x 11 cm.

"Diagram illustrating typical Fire Trench, Second Line and Communication Trenches, First Aid Stations, &c &c.." in Arthur Guy Empey, *"Over the Top," by an American Soldier Who Went* (New York, London: G. P. Putnam's Sons, 1917). Courtesy of University of Denver Penrose Library. 18 x 22.5 cm.

"Why Germany Wants Peace Now," in Frederic L. Paxson, Edward S. Corwin & Samuel B. Harding, *War Cyclopedia: A Handbook for Ready Reference on the Great War.* Issued by the Committee on Public Information (Washington: Government Printing Office, 1917). Courtesy of Cornell University—PJ Mode Collection of Persuasive Maps. 22 x 25 cm.

"Chronological map of the influenza epidemic of 1918," *Annual Report of the Surgeon General of the Public Health Service* (Washington, DC: Government Printing Office, 1919). A.S.516. 15 x 10 cm.

"Armour's Food Source Map: The Greatness of the United States is Founded on Agriculture" (Chicago: Armour & Company, 1922). Lithograph by Joseph Pennell. Courtesy of David Rumsey Map Collection, David Rumsey Map Center, Stanford Libraries. 64 x 99 cm.

"Why Not Free Water?" [utility bill], Department of Public Service, City of Los Angeles. Courtesy of Cornell University—PJ Mode Collection of Persuasive Maps. 9 x 12 cm.

"Chicago's Gangland," in Frederic Milton Thrasher, *The Gang: A Study of 1,313 Gangs in Chicago* (Chicago: University of Chicago Press, 1927). Ac.2691.d/34.(2.). 65 x 45 cm.

E. Simms Campbell, "A night-club map of Harlem" (New York: Dell Publishing Company, circa 1933). Courtesy of Library of Congress. 35 x 56 cm.

Daniel K. Wallingford, "A New Yorker's Idea of the United States of America" (New York: Daniel K. Wallingford, 1939). Courtesy of David Rumsey Map Collection, David Rumsey Map Center, Stanford Libraries. 30 x 41 cm.

"Around the World in California in 4 Days," *Los Angeles Times* (March 4, 1934). © 1934 *Los Angeles Times*. Used with permission. Image courtesy of Margaret Herrick Library, Academy of Motion Pictures and Sciences. 56 x 81 cm.

"The Great American Roadside," *Fortune* (September 1934). Author's collection. 27 x 46 cm.

"Afro American Travel Map," in *Travel Guide of Negro Hotels and Guest Houses. Published by Afro-American Newspapers* (1942). Courtesy of David Rumsey Map Collection, David Rumsey Map Center, Stanford Libraries. 44 x 44 cm.

"Metropolitan Cleveland Security Map No 1 ... Supervised and Submitted by: C. C. Boyd, Field Agent, Mortgage Rehabilitation Division, Home Owner's Loan Corporation" (Cleveland: Mountcastle Map Co., 1936). National Archives and Records Administration. 77 x 99 cm.

Stephen Vorhies and Rand McNally, "Raw Materials for a U.S. Ruhr," *Fortune* (October 1933). Author's collection. 34 x 53 cm.

Richard Edes Harrison, "Three Approaches to the U.S.," *Fortune* (September 1940). Author's collection. Reproduced courtesy of the estate of Richard Edes Harrison. 33 x 27 cm.

Richard Edes Harrison, "The World Divided," *Fortune* (August 1941). Image courtesy of Cornell University—PJ Mode Collection of Persuasive Maps. Reproduced courtesy of the estate of Richard Edes Harrison. 33 x 43 cm.

MacDonald Gill, "The Time & Tide Map of the Atlantic Charter" (London: George Philip & Son, 1942). Maps 950.(211.). 76 x 109 cm.

Henry J. MacMillan, "XIX Corps in Action" (Fort Lewis, Washington: U.S. Army 62nd Engineer Topographic Company, 1945). Courtesy of David Rumsey Map Collection, David Rumsey Map Center, Stanford Libraries. 59 x 73 cm.

Henry J. MacMillan, "XIX Corps in Action: From Siegfried Line to Victory" (Fort Lewis, Washington: U.S. Army 62nd Engineer Topographic Company, 1945). Courtesy of David Rumsey Map Collection, David Rumsey Map Center, Stanford Libraries. 59 x 73 cm.

1945 III. Div. Mauthausen and Z Lince do Nachoda 1945, from the diary of Michael J. Kraus. Courtesy of United States Holocaust Memorial Museum, Michael J. Kraus Papers. Pages measuring 21 x 15 cm.

"Ancient Courses Mississippi River Meander Belt," Plate 9, in Harold N. Fisk, *Geological Investigation of the Alluvial Valley of the Lower Mississippi River* (U.S. Department of the Army and the Mississippi River Commission, 1944). Document Supply Wf1/7959. 107 x 76 cm

Hal Shelton and Jeppesen Map Company, Denver – Chicago Chart 6, in *Air Maps of United Air Lines. Main Line Airway* (1949). Author's collection. Reproduced with permission of Jeppesen Sanderson, Inc. © Jeppesen Sanderson, Inc. 2018. 21 x 57 cm.

"Voici les bases Americaines dans le monde," Parti Communiste Francais [Communist Party of France] (1951). Courtesy of Cornell University—PJ Mode Collection of Persuasive Maps. 82 x 122 cm.

"How Communists Menace Vital Materials," in *Soviet Total War: 'Historic Mission of Violence and Deceit* (Washington, DC, 1956). Courtesy of Cornell University—PJ Mode Collection of Persuasive Maps. 22 x 39 cm.

Herb Ryman and Walt Disney, Schematic Aerial View of Disneyland, 1953. Image courtesy of Walt Disney Imagineering © Disney. 99 x 171 cm on page 128 x 199 cm.

"A Progress Report on the Interstate Highway System," Caterpillar Corporation, *Saturday Evening Post* (April 22, 1961). Author's collection. 27 x 53 cm.

Background map of the Freedom Rides, Associated Press Features, 1962. Courtesy of Associated Press/Shutterstock. 12 x 14 cm.

Photograph of missile sites. U.S. Department of Defense Cuban Missile Crisis Briefing Materials. Courtesy of John F. Kennedy Presidential Library, National Archives and Records Administration. 12.7 x 10 cm.

Map of aircraft locations in Cuba, 1962. Courtesy of Theodore Sorensen Papers, John F. Kennedy Presidential Library, National Archives and Records Administration. 15 x 43 cm.

Map of missile range in Cuba, 1962. Courtesy of Theodore Sorensen Papers, John F. Kennedy Presidential Library, National Archives and Records Administration. 26.6 x 20 cm.

United States Information Agency, "Aggression from the North" (1965). National Archives and Records Administration. 85 x 85 cm.

Map of the National student strike, April 3 Movement (1970). Courtesy of Cornell University—PJ Mode Collection of Persuasive Maps. 43 x 56 cm.

"Atlantic Ocean Floor" (1968), Heirich Berann, *National Geographic Magazine* (June 1968). Maps CC.5.b.42.(1.). © National Geographic. 68 x 48 cm.

"Earthrise" photograph (1968), National Aeronautics and Space Administration.

George Zaffo and Thomas J. Lipton Inc., "Lipton Lunar Space Map," (1969). Courtesy of David Rumsey Map Collection, David Rumsey Map Center, Stanford Libraries. 50 x 60 cm.

Abraham Verghese, Steven L. Berk and Felix Sarubbi, maps of HIV infection across the United States, and around Johnson City, Tennessee. *Journal of Infections Diseases* vol. 160 no. 6 (December 1989), pp. 1052 and 1053 © 1989, Oxford University Press. 9 x 11 cm.

Peter Gould, serial maps of HIV infection from 1982–1990, in *Becoming a Geographer* (1999). Courtesy of Syracuse University Press. Maps 7.5 x 11.5 cm. except 1990, measuring 11.5 x 18 cm.

"Regional Migration, 1955 to 1960" and "Regional Migration, 1995 to 2000," from *Census Atlas of the United States: Census 2000 Special Reports.* (Washington, DC: U.S. Census Bureau, 2007). Maps 238.b.70. Each map 15 x 26.5 cm.

"Meatpacking and Hispanic Population Change 1990–2000 Midwest Counties" (2001). Map made by Paul Voss, David Long and Jennifer Vogt. Courtesy of Applied Population Laboratory, University of Wisconsin.

Map of North Carolina's 12th Congressional District from 1992–2016. Compiled by Kelly Measom at the University of Denver. U.S. Census Bureau MAF/TIGER; basemap: Esri, DigitalGlobe. Projection: WGS 1984 Web Mercator Auxiliary Sphere.

Mary Kurgan, "Around Ground Zero: Lower Manhattan after September 11" (2001). Maps X.6952. 61 x 46 cm.

DeepMap, data visualization for autonomous vehicles (2018). Courtesy Wei Luo and DeepMap.

INDEX

ACKNOWLEDGMENTS

Many people contributed to this project. Rob Davies at British Library Publications proposed the idea for the book, and then worked with me to shape both its arc and its many details. Sally Nicholls helped me unearth some of the British Library's great treasures, while Jacqueline Harvey, Elizabeth Woabank, and Karin Fremer each brought patience and skill to the manuscript. Early encouragement by Christie Henry at the University of Chicago Press was continued by Mary Laur, Rachel Kelly, Christine Schwab, and Kristen Raddatz, all of whom saw this project through to completion.

The University of Denver supported this book from the outset, and I thank Dean Daniel McIntosh and Provost Gregg Kvistad in particular for their steadfast encouragement. The university—as well as a J. B. Harley-Delmas Fellowship—funded early stages of research in the British Library. A Public Scholar Fellowship from the National Endowment for the Humanities provided me with time to complete the manuscript.

While at the British Library, I was aided regularly by Peter Barber, Ashley Baynton-Williams, and Tom Harper. Others who kindly advanced my research include Kristine Krueger at the Fairbanks Center for Motion Picture Study; John Powell at the Newberry Library; Ed Redmond at the Library of Congress; Ian Fowler (formerly) at the Osher Map Library; Maryrose Grossman at the John F. Kennedy Library; David Long at the University of Wisconsin; Maxine Raley and Margaret Adamic at Disney Enterprises; and the able staff of Penrose Library at the University of Denver.

The community of map scholars, collectors, archivists, and dealers is an exceedingly gracious one. For their expertise and guidance, I thank David Rumsey, Abby Smith Rumsey, Jim Akerman, Barry Ruderman, Michael Buehler, Henry Taliaferro, Mary Pedley, Wes Brown, P. J. Mode, Matthew Edney, Mark Monmonier, and Ralph Ehrenberg. For their help with particular maps I thank Chet Van Duzer, Rob Nelson, John Cloud, Albert Theberge, Ronald Eller, David Bosse, Paul Colomy, John Lindemann, Tom Touchton, Ed Redmond, Wei Luo, Kelly Measom, and those individuals whose work is cited in the endnotes. My sincere thanks to Samuel Edes Harrison, who kindly shared the extraordinary maps drawn by his father, Richard Edes Harrison.

For their ongoing interest in this project I thank Robert Anderson, Judy Schulten, Jennifer Karas, Ingrid Tague, Elizabeth Escobedo, Carol Helstosky, Karen Iker, Megan Bertron, Gregg Kvistad, Andrei Kutateladze, Bin Ramke, Denise O'Leary, Stalker Henderson, Seth Masket, Rebecca Chopp, Yasmaine Ford and my colleagues in the history department at the University of Denver. Ed Ayers shares my love of old maps, but has taught me just as much about new frontiers in historical cartography. I owe a special debt to Elliott Gorn, Tim Spears, and Tim Gilfoyle, longstanding colleagues whose nuanced understanding of the past continues to inform my own.

Robert, Sam, and David Anderson have—as always—been a source of strength, joy, and good humor. I am equally grateful to the Schulten family—especially Vonnie and Warren—and dedicate this book to them with love.